AP Precalculus:
The Teacher's Compendium

Looking for an AP Precalculus Exam Preparation Book for your students?

Get David Hornbeck's
Acing the AP Precalculus Exam
2nd edition

This book includes 8 review lessons:

Rates of Change
Graphs of Polynomial and Rational Functions
Equivalent Forms of Expressions
Using the Calculator
Periodic Phenomena
Polar Functions and Modeling
Solving Equations and Inequalities
FRQ "Point"-ers

The first 7 lessons each conclude with 15 original multiple-choice problems and 2 original free-response problems to reinforce the concepts. The eighth lesson comes with 15 partial free-response questions.

Also included are 2 complete, full-length AP Precalculus practice exams!

Available at Lulu.com (no solutions) for $14.49 and Amazon.com (with solutions) for $28.99.

Also available on Lulu, Amazon, or both:
AP Precalculus: Student Edition by David Hornbeck and Chuck Garner
Five Weeks to a Five: Preparation for the AP Calculus AB Exam by Chuck Garner
Five Weeks to a Five: Preparation for the AP Calculus BC Exam by Chuck Garner
The AP Statistics Study Companion by David Hornbeck

AP Precalculus:
The Teacher's Compendium

David Hornbeck and Chuck Garner, Ph.D.

Rockdale Magnet School for Science and Technology

Lulu Publishing

AP Precalculus: The Teacher's Compendium

by David Hornbeck and Chuck Garner

Published by Lulu.com.

Set in *Palatino* and Optima, designed by Hermann Zapf.

This book was produced directly from the authors' LaTeX files.
Figures were drawn by the authors using the Ti*k*Z package.
Cover image is a photo of the Kauffman Center in Kansas City, Missouri, taken by Andreas Staver. Photo courtesy of www.pexels.com.

ISBN: 978-1-304-35915-5

10 9 8 7 6 5 4

© 2024, David Hornbeck and Charles Garner.

AP and the Advanced Placement Program are registered trademarks of the College Board, which was not involved in the production of, and does not endorse, this book.

Contents

Preface	ix
Introduction	xi

1 Polynomial and Rational Functions — 1

- Topic 1.1 ~ Change in Tandem (Day 1) 3
  - Topic 1.1 (Day 1) Homework 6
  - Topic 1.1 (Day 1) Solutions/Notes 7
- Topic 1.1 ~ Change in Tandem (Day 2) 10
  - Topic 1.1 (Day 2) Homework 14
  - Topic 1.1 (Day 2) Solutions/Notes 16
- Topics 1.2-1.3 ~ Rates of Change in Linear and Quadratic Functions (🖩) 22
  - Topics 1.2-1.3 Homework 26
  - Topics 1.2-1.3 Solutions/Notes 27
- Topic 1.4 ~ Polynomial Functions 32
  - Topic 1.4 Homework 36
  - Topic 1.4 Solutions/Notes 37
- Topics 1.1-1.4 ~ Review 42
- Topics 1.5-1.6 ~ Zeros and End Behavior of Polynomials (🖩) 45
- Calculator Skills (🖩) 57
- Topic 1.7 ~ End Behaviors of Rational Functions (🖩) 60
  - Topic 1.7 Homework 63
  - Topic 1.7 Solutions/Notes 65
- Topics 1.8-1.10 ~ Zeros, Vertical Asymptotes, and Holes (Day 1) (🖩) 70
  - Topics 1.8-1.10 (Day 1) Homework 73
  - Topics 1.8-1.10 (Day 1) Solutions/Notes 74
- Topics 1.8-1.10 ~ Zeros, Vertical Asymptotes, and Holes (Day 2) 78
  - Topics 1.8-1.10 (Day 2) Homework 80
  - Topics 1.8-1.10 (Day 2) Solutions/Notes 81
- Topic 1.11 ~ Equivalent Representations 86
  - Topic 1.11 Homework 90
  - Topic 1.11 Solutions/Notes 91
- Topics 1.7-1.11 ~ Review 97
- Topic 1.12 ~ Transformations of Functions 108
  - Topic 1.12 Homework 112
  - Topic 1.12 Solutions/Notes 113
- Topic 1.13 ~ Selecting a Function Model 118
  - Topic 1.13 Homework 121

Topic 1.13 Solutions/Notes 123
Topic 1.14 ~ Constructing a Function Model 128
Topic 1.14 Homework 131
Topic 1.14 Solutions/Notes 133
Unit 1 Test ~ Polynomial and Rational Functions 137
Unit 1 Test Solutions and Scoring 143

2 Exponential and Logarithmic Functions — 147

Topic 2.1 ~ Change in Arithmetic and Geometric Sequences 149
Topic 2.1 Homework 151
Topic 2.1 Solutions/Notes 152
Topics 2.2-2.3 ~ Change in Linear and Exponential Functions 157
Topics 2.2-2.3 Homework 160
Topics 2.2-2.3 Solutions/Notes 161
Topic 2.4 ~ Equivalent Exponential Forms 166
Topic 2.4 Homework 169
Topic 2.4 Solutions/Notes 170
Topics 2.5-2.6 ~ Exponential Function Modeling (▣) 174
Topics 2.5-2.6 Homework 178
Topics 2.5-2.6 Solutions/Notes 180
Topic 2.7 ~ Compositions of Functions 185
Topic 2.7 Homework 188
Topic 2.7 Solutions/Notes 190
Topic 2.8 ~ Inverse Functions 195
Topic 2.8 Homework 199
Topic 2.8 Solutions/Notes 201
Topics 2.9-2.10 ~ Logarithms, the Inverse of Exponentials 207
Topics 2.9-2.10 Homework 210
Topics 2.9-2.10 Solutions/Notes 211
Topic 2.11 ~ Logarithmic Functions 216
Topic 2.11 Homework 219
Topic 2.11 Solutions/Notes 220
Topic 2.12 ~ Equivalent Logarithmic Forms 224
Topic 2.12 Homework 226
Topic 2.12 Solutions/Notes 227
Topic 2.13 ~ Exponential and Logarithmic Equations and Inequalities 231
Topic 2.13 Homework 233
Topic 2.13 Solutions/Notes 234
Topic 2.14 ~ Constructing a Logarithmic Model 239
Topic 2.14 Homework 241
Topic 2.14 Solutions/Notes 242
Topic 2.15 ~ Semi-log Plots 246
Topic 2.15 Homework 249
Topic 2.15 Solutions/Notes 249
Unit 2 Test ~ Exponential and Logarithmic Functions 253
Unit 2 Test Solutions and Scoring 259

3 Trigonometric and Polar Functions — 263

- Topics 3.1-3.2 ~ Periodic Phenomena and Radian Measure 265
  - Topics 3.1-3.2 Homework 270
  - Topics 3.1-3.2 Solutions/Notes 273
- Topics 3.2-3.3 ~ Sine, Cosine, and Tangent 279
  - Topics 3.2-3.3 Homework 284
  - Topics 3.2-3.3 Solutions/Notes 286
- Topic 3.3 ~ Sine and Cosine Function Values 292
  - Topic 3.3 Homework 295
  - Topic 3.3 Solutions/Notes 296
- Topics 3.4-3.5 ~ Graphs of Sine and Cosine (🖩) 301
  - Topics 3.4-3.5 Homework 305
  - Topics 3.4-3.5 Solutions/Notes 306
- Topic 3.6 ~ Sinusoidal Transformations 312
  - Topic 3.6 Homework 317
  - Topic 3.6 Solutions/Notes 318
- Topic 3.7 ~ Modeling with Sinusoidal Functions 323
  - Topic 3.7 Homework 326
  - Topic 3.7 Solutions/Notes 328
- Topic 3.8 ~ The Tangent Function 333
  - Topic 3.8 Homework 337
  - Topic 3.8 Solutions/Notes 338
- Topic 3.9 ~ Inverse Trigonometric Functions 343
  - Topic 3.9 Homework 346
  - Topic 3.9 Solutions/Notes 347
- Topic 3.10 ~ Trigonometric Equations and Inequalities (Day 1) 353
  - Topic 3.10 (Day 1) Homework 355
  - Topic 3.10 (Day 1) Solutions/Notes 356
- Topic 3.10 ~ Trigonometric Equations and Inequalities (Day 2) 360
  - Topic 3.10 (Day 2) Homework 363
  - Topic 3.10 (Day 2) Solutions/Notes 364
- Topic 3.11 ~ The Secant, Cosecant, and Cotangent Functions 367
  - Topic 3.11 Homework 370
  - Topic 3.11 Solutions/Notes 371
- Topic 3.12 ~ Equivalent Representations of Trigonometric Functions (Day 1) 376
  - Topic 3.12 (Day 1) Homework 379
  - Topic 3.12 (Day 1) Solutions/Notes 380
- Topic 3.12 ~ Equivalent Representations of Trigonometric Functions (Day 2) 384
  - Topic 3.12 (Day 2) Homework 389
  - Topic 3.12 (Day 2) Solutions/Notes 390
- Topics 3.10-3.12 Circuit 396
  - Topics 3.10-3.12 Circuit Solutions 398
- Topic 3.13 ~ Trigonometry and Polar Curves 401
  - Topic 3.13 Homework 404
  - Topic 3.13 Solutions/Notes 405
- Topic 3.14 ~ Polar Function Graphs 410
  - Topic 3.14 Homework 414
  - Topic 3.14 Solutions/Notes 416

Topic 3.15 ~ Rates of Change in Polar Functions 424
    Topic 3.15 Homework 427
    Topic 3.15 Solutions/Notes 428
Unit 3 Test ~ Trigonometric and Polar Functions 434
    Unit 3 Test Solutions and Scoring 441

Pacing and Scheduling   445

80 days, 90-minute periods 446
70 days, 90-minute periods 447
160 days, 45-minute periods 448
140 days, 45-minute periods 449

So You Want to Write a Test?   451

Preface

I recall being present for the College Board Forum at the AP Calculus Reading in the summer of 2017 in Kansas City. In that forum, Trevor Packer was asked by an audience member if there were plans to introduce an AP Multivariable Calculus. Trevor said that maybe if there was enough demand. He then asked those assembled, "Raise your hand if your school would be interested in such a course, or if your institution would likely give credit for such a course." I seem to remember that slightly more than half the hands raised. The person who asked the initial question said "It seems you have your demand." "It sure does," Trevor replied, and then he quipped, with a smile on his face, "too bad there are no plans to develop such a course." The crowd laughed, realizing that he was just leading us on. Then he said "But we are working on an AP Precalculus course which we plan to roll out in a few more years." The crowd was a little shocked, and more questions were lobbed at Mr. Packer. This was the first I had ever heard of AP Precalculus!

I've taught AP Calculus since 2002. I have created many materials for AP Calculus (a series of textbooks, an AP Exam preparation book, and an AP Calculus Problem Book) and used them with success in my classes. Others who have used my materials also report them as being useful and beneficial for students. When David Hornbeck applied for a position at my school in 2021, one of the things that stood out for me was that David also created his own materials for teaching AP Statistics at his previous school. Of course, we hired David immediately to teach Precalculus, AP Statistics, and a section of AP Calculus AB for which he used my materials. We also jointly sponsored our school's competitive Math Team. We spent many hours together during our common planning period, during Math Team meetings after school, at math conferences, and district professional development discussing ways to teach mathematics. I had never had a colleague who is as interested as I am in finding the best to approach every topic we teach, and in creating the materials for that approach. Our discussions resulted in making each of us better teachers.[1] Indeed, his creation of a Precalculus textbook (something I had never done in my seven years teaching that course) motivated me to create a textbook for my Discrete Mathematics course.

So when AP Precalculus was announced, and our school decided we should offer it, we knew David had to teach it, and he didn't think twice about creating his own materials. He started in the Summer of 2023. As we discussed his plan for the book sitting in a Kansas City restaurant (I was there for the AP Calculus reading and David was there for the AP Statistics reading), we realized that this could be a rich resource for any teacher of AP Precalculus. We could include tests, activities, lessons, and extra problems. We were determined to make this a useful book for teachers, and we determined that in order to do that, the lessons, activities, and problems should be used in class first.

Therefore, throughout the 2023-24 school year, David wrote this book by creating the lesson or activity for his AP Precalculus class each day and then reflecting upon the activity the next day. Throughout, we had regular discussions concerning almost every aspect of the approach to the material. Once a unit's worth of lessons was completed, I edited it, rewrote a few things, fixed typos, and wrote around 60% of

---

[1] I feel it is appropriate to mention another colleague, Julie Matthews, who was the best supporter of our work, the best sounding board, and the best person to question us when we needed it. Julie listened to many of the discussions David and I had, and she brought up ideas she had for teaching Geometry and Algebra II and how they could fit with how we are teaching. She even volunteered to change the way she teaches certain things so they align with what her students see in later courses. The urge to create materials manifested in Julie as well, and she has put together an Algebra II Problem Book. The three of us have been an unstoppable trio of mathematics teachers.

the multiple-choice problems (and solutions for those and some of the ones David wrote). Doing this gave me great insight into how clearly and cleverly David built up each topic so the student could grasp it more easily. I was so blown away by the quality of his materials, I petitioned our school administrator to allow us to teach all precalculus (AP and non-AP) using this book as the basis for, and the approach to, the curriculum. I found out recently that my petition was granted.

As you read the book, you may wonder why things look a little better than you are used to seeing in many kinds of teacher-generated materials. That is because instead of using Word or GoogleDocs, this book was written in LaTeX, the free document typesetting language. The figures were all drawn in TikZ, a LaTeX graphics package. LaTeX can be viewed as a markup language for documents with lots of mathematics, and we highly recommend using it for your mathematical documents.

I truly hope you find this book useful. I would encourage you to use the lessons straight out of the book with your students *at first*, particularly if you have never taught AP Precalculus before. I would then encourage you to tweak the materials in this book to align with your teaching practices and your students' needs, and begin to create your own materials. Only you know how your students learn, how quickly or slowly they grasp concepts, and what you have taught them. You should use materials that reflect these situations. Please use the ones in this book, and then modify them to suit your students' needs! This including writing your own tests and quizzes. While we provide sample tests and many good multiple-choice problems for both quizzes and tests, a test that you create in the style and substance of the AP exam is more appropriate than using someone else's test (including ours). To that end, we have included a short guide at the end of the book concerning how to write good problems.

I wish your students success in AP Precalculus!

CHUCK GARNER
CONYERS, GA, APRIL 2024

Introduction

In this book, you will find my lessons, activities, homework, reflections, and sample test items for an entire year of AP Precalculus. There are a few things to know about how the book is laid out and how these lessons were, or could be, implemented.

This book is ordered identically to (a) how College Board lays out the course in the Course and Exam Description and (b) how I taught the lessons. Each lesson corresponded to one 90-minute period, but most of those presented could be split into two 45- or 50-minute lessons. The vast majority of lessons have one of two structures: student-driven activities or teacher-led lessons. For the student-driven activities, I had students in groups of 3 or 4, and they would work through the activity for anywhere from 45 to 80 minutes. For teacher-led lessons, I would simply be at the board working through various examples. Any of the student-driven activities could easily be converted into a whole-class lesson, and teacher-led lessons could also be utilized as student-driven activities or practice days.

Many teachers decided to deviate from College Board's ordering of the material: placing Unit 3 first, rearranging Unit 3, or moving around individual topics. I chose to follow the course as designed by College Board for the first implementation of the course for the sole purpose of being able to utilize Progress Checks and to see how it went. Now, having taught the course, I completely stand by the original ordering. Though the lessons in this book could be taught in a different order, I would be very hesitant to change much, as I would fear students not making certain connections.

The bulk of the lessons are student-driven activities, which was a worthwhile departure from my own teaching style for 10 years. The activities try to thread the needle between completely open-ended tasks that students may give up on quickly and those insidious "activities" that are just lectures on paper. Throughout the course of the year, I saw students really engage with these activities, and the level of scaffolding seemed generally appropriate without sacrificing rigor. The activity structure does come at the expense of "practice" time, though, and this is where a teacher's discretion and knowledge of their own students' skills come into play.

In quite a few of the activities, there are links to Geogebra sketches. I am a firm believer in using animations to help mathematics come alive for students, so that they may truly *see* the mathematics, and I found that students responded well to them. These sketches are public, and I would encourage any interested teacher to try downloading them to look under the hood and see how they were made. Even something as simple as animating of the graphing of a function can transform how a student understands a concept, and such things aren't nearly as complicated in Geogebra as they may seem on first glance.

You may find that the book is short on "review" and "practice." This was not borne out of some philosophical reason, but rather just a shortage of time. My classes are on a modified block schedule, meaning I only saw them every other day. With disruptions – field trips, assemblies, etc. – and the timing of the AP exam, I really only had about 75 days with my students. When I saw a need for more skill practice, I would frequently assign Delta Math or write worksheets. For review, I would have my students work through the Progress Checks in AP Classroom (which I would print out, as I'm a firm believer in doing math on paper!). The sample pacing provided details when I would utilize these Progress Checks. If I had more class days, though, I would most certainly include more days for skill practice and review.

The homework assignments in this book were intended to give students 30 minutes to an hour of homework for each class period. The problems in these assignments are intended to mimic the style and rigor of AP exam questions, only without the multiple choice or free response format. By and large, the problems eschew any rote practice, but when this was needed, supplemental online assignments were created. The beginning of each class period was usually dedicated to going over any questions students had from these assignments. For my students, I actually posted the solutions to the homework online in their learning management system.

This book makes use of the ▦ symbol at various times. When the ▦ symbol is featured in the title of a lesson, it's meant to indicate that the lesson itself will rely heavily on a graphing calculator (or may even be a calculator-based investigation). The absence of this symbol does not, of course, mean that the calculator is prohibited or will not be used at all. In lessons in which the ▦ symbol is missing, the symbol may again appear on individual questions within the lesson. The reflections at the end of each lesson should indicate the extent to which calculators should be or were used.

The book makes the assumption that students are working with a TI-84 (or TI-83) graphing calculator. This is partly selfish, as my own students had access to a class set of TI-84 calculators, and it certainly is the industry standard. That said, I also stuck with the TI-84 because including instructions for other common calculators – the NSpire, Numworks, various Casio models – would have ballooned the lessons to an unacceptable length. For students in my own class with other calculator models, I would simply provide assistance where I could and direct them to online manuals or videos.

You will find in this book a reflection at the end of each lesson. These reflections were written within a day of teaching or facilitating each lesson, and they're meant to provide insight into what went well, what didn't go so well, and what instructional techniques I may have used that aren't explicitly featured in the lesson. Each reflection also contains 5 sample multiple choice items. These items could be used on formative assessments like exit tickets or quizzes or on summative tests. Each and every item should be very similar to those that students might see on the AP exam.

Solutions to the activities, lessons, and problems are all included, and it is worth noting that, in most cases, multiple solutions may be appropriate. The solutions written here were meant to reflect either the most efficient or mathematically sound approach.

I would like to thank my co-author, Chuck, for his work in proofreading and editing this book and writing tremendous problems and solutions. Beyond his tireless and thorough work, I credit many of the conversations I had with him for the inspiration for many of these lessons. Had I never worked with Chuck, I never would have retaught myself LaTeX and definitely never would've thought to self-publish. For years now, he has been incredibly patient with me, answering my near constant questions about everything from precalculus and calculus to publishing and LaTeX.

I would additionally like to thank all of my students this year. They were the guinea pigs for these lessons, and their insights were both fascinating and invaluable. Without them, none of these lessons would exist!

I hope this book serves you and your students well.

David Hornbeck
Conyers, GA, April 2024

Unit 1
Polynomial and Rational Functions

Topic 1.1 ~ Change in Tandem (Day 1)

Learning Objectives

1. (1.1.A) Describe how the input and output values of a function vary together by comparing function values.

2. (1.1.B) Construct a graph representing two quantities that vary with respect to each other in a contextual scenario.

Success Criteria

1. I can identify and express the domain, range, and zeros of a function from a graph.

2. I can determine whether a function is increasing or decreasing.

3. I can describe whether a function is increasing or decreasing at a constant, increasing, or decreasing rate.

A new local insurance company is looking to bolster its client base by advertising with television and online ads, billboards, and social media. Over time, it tracks how many clients it has based on how much money it has spent on advertising. The function $C(x)$ shows the number of clients C, in hundreds, based on x, the amount spent in advertising, in tens of thousands of dollars.

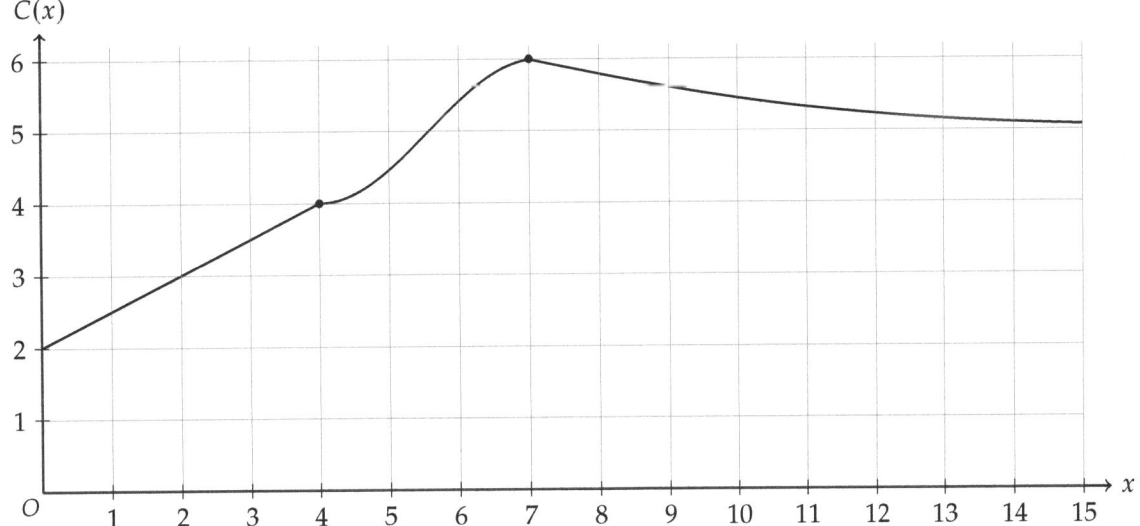

1. Based on the graph, what is $C(4)$? What does this mean in context?

2. Based on the graph, what is the domain of $C(x)$?

3. Based on the graph, what is the range of $C(x)$?

4. What is the maximum number of clients the insurance company had? How much money were they spending on advertising at this point?

The company is obviously looking to see what happens if it spends more on advertising. The following questions are about how the graph of $C(x)$ changes as the input values (advertising money) increase.

5. Describe what happens to the client base as the amount the company spent increased from 0 to $40,000 (so $0 \leq x \leq 4$).

6. Describe what happens to the client base for $4 \leq x \leq 5.5$.

7. How were the ways in which the client base increased *different* from 0 to 4 and from 4 to 5.5?

8. Below is a table of values for $C(x)$ from $x = 5.5$ to $x = 7$. In the table, record how much $C(x)$ changed for each change of 0.3 in x. (The first one has been filled out already!)

x	$C(x)$	Change
5.5	4.93	N/A
5.8	5.23	$5.23 - 4.93 = 0.3$
6.1	5.5	
6.4	5.74	
6.7	5.9	
7	5.99	

While the values of C are increasing as x increases by 0.3, what is happening to the *amount* by which C is increasing?

9. For $5.5 < x < 7$, we say that $C(x)$ is *increasing at a decreasing rate*. Try to fill in the blanks with similar vocabulary for the following.

    (a) For $4 < x < 5.5$, $C(x)$ is _____.

    (b) For $0 < x < 4$, $C(x)$ is _____.

10. One of the managers of the insurance company says, "Forget all this data. We should put $200,000 into our advertising budget. Then the clients will really come!" How do you assess this claim? What would you tell the manager? What, if any, assumptions are you making?

11. Now, try to put these concepts to use for a different function. Determine which of the following are true for the function $h(x)$ graphed below. More than one are true.

    (A) h has three zeros.

    (B) $h(x) \leq 4$ for all x

    (C) h is decreasing for $x > 0$

    (D) h is decreasing at an increasing rate for $0 < x < 2$

    (E) h is increasing at a decreasing rate for $-2 < x < 0$

    (F) h is constant for $x < -2$

    (G) The rate of change of h is constant for $x < -2$

    (H) h is decreasing at an increasing rate for $x > 2$

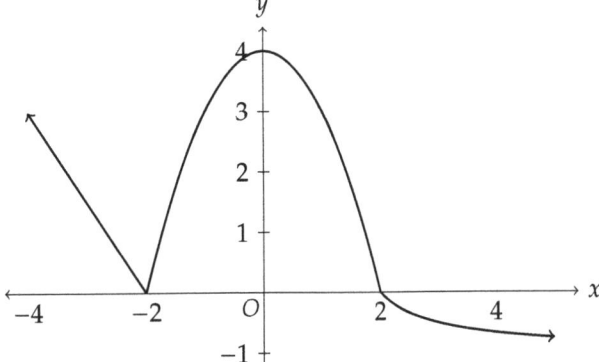

Notes

Topic 1.1 (Day 1) Homework

1. Shown at the right is the graph of a function $f(x)$. Answer the following.

    (a) What appears to be the domain of f?

    (b) What appears to be the range of f?

    (c) Identify the zeros of f.

    (d) Describe how the function f is behaving on each of the following intervals. Use rate of change vocabulary.

    (i) $-4 < x < -3.49$

    (ii) $-3.49 < x < -2.93$

    (iii) $x > -2.93$

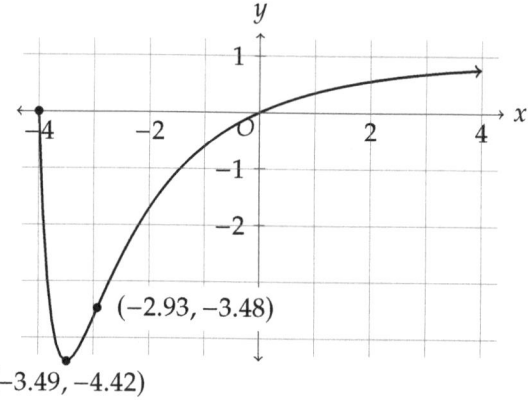

2. A table of values for $g(x)$ is shown below.

x	-1	3	7
$g(x)$	14	?	2

    If $g(x)$ decreases at a constant rate for $-1 \leq x \leq 7$, what is $g(3)$?

3. Draw the graph of a function that meets the following criteria.

    - f is increasing for all $x < 2$.
    - f is increasing at a decreasing rate only for $-3 < x < 2$.
    - f is decreasing at a decreasing rate for $x > 2$.
    - $f(x) \leq 5$ for all x

4. Below is a table of values for a function $f(x)$.

x	-4	-2	0	2	4	6
$f(x)$	4	0	-3	1	5	3

    The maximum of f is $f(4) = 5$ and the minimum is $f(0) = -3$. Determine if the following are definitely true, definitely false, or possibly true. Explain.

    (a) f is increasing for $-4 < x < 0$.

    (b) f is increasing at an increasing rate on the interval $[0, 4]$.

    (c) f has one zero.

Topic 1.1 (Day 1) Solutions/Notes

Activity

1. $C(4) = 4$. When the company spent $40,000, they had 400 clients.

2. The domain is $0 \leq x \leq 15$.

3. The range is $2 \leq C(x) \leq 6$.

4. The company had a maximum of 600 clients when they spent $70,000.

5. The client base increased at a constant rate of 50 clients per $10,000 spent.

6. The client base increases at an increasing rate.

7. For $0 \leq x \leq 4$, the client base increased at a constant rate, but for $4 \leq x \leq 5.5$, the client base was increasing at an increasing rate (the growth rate was speeding up).

8. For $5.5 \leq x \leq 7$, the amounts by which $C(x)$ increases after a change of 0.3 in x are decreasing.

x	$C(x)$	Change
5.5	4.93	N/A
5.8	5.23	$5.23 - 4.93 = 0.3$
6.1	5.5	0.27
6.4	5.74	0.24
6.7	5.9	0.16
7	5.99	0.09

9. (a) For $4 < x < 5.5$, $C(x)$ is *increasing at an increasing rate*.

   (b) For $0 < x < 4$, $C(x)$ is *increasing at a constant rate*.

10. It appears that this claim is foolish. Based on the graph, the numbers of clients appears to be approaching a constant of about 500 as the amount spent on advertising increases from $100,000. However, this is assuming that this trend will continue, which is unknown. Perhaps another infusion of $50,000 would cause a spike in the client base.

11. C, E, G, H

> **Notes**
>
> A function maps *input values* of an *independent variable* to *output values* of a *dependent variable*. Common function notation includes $f(x), g(x), h(t), P(x)$, etc.
>
> The set of *all* input values for a function is the DOMAIN of a function and the set of all output values is the RANGE. Domain and range can be written as *intervals* or *inequalities*, with examples below.
>
> | Example | Interval | Inequality |
> |---|---|---|
> | The domain of f is all values between 0 and 4 inclusive. | $[0, 4]$ | $0 \leq x \leq 4$ |
> | The range of f is all values less than -2. | $(-\infty, -2)$ | $f(x) < -2$ |
>
> The set of inputs at which the graph of a function hits the x-axis is called the ZEROS of the function.
>
> A function is INCREASING if, as input values increase, the output values always increase. Analytically, this means that, if $a < b$, then $f(a) < f(b)$. Similarly, a function is DECREASING if, as input values increase, the output values always decrease. Analytically, this means that, if $a < b$, then $f(a) > f(b)$. Graphically, f is increasing if it goes up as you scan left to right and decreasing if it goes down as you scan left to right. (Draw pictures!)
>
> Functions can increase at an *increasing rate* or at a *decreasing rate*. In essence, they can be going up and slowing down, or going up and speeding up. Similarly, functions may decrease at an increasing rate (go down, but slow down) or a decreasing rate (go down and speed up). (Draw pictures!)

Homework

1. (a) $[-4, \infty)$ or $x \geq -4$
   (b) $[-4.42, 1)$ potentially, or $[-4.42, \infty)$! It cannot be determined.
   (c) $x = -4$ and $x = 0$
   (d) (i) f is decreasing at an increasing rate.
       (ii) f is increasing at an increasing rate.
       (iii) f is increasing at a decreasing rate.

2. The average rate of change on $[-1, 7]$ is $\frac{2-14}{7-(-1)} = -\frac{3}{2}$. The change in x from -1 to 3 is 4, so the net change will be $4\left(-\frac{3}{2}\right) = -6$. Therefore $g(3) = 14 - 6 = 8$.

3. A possible graph is shown below.

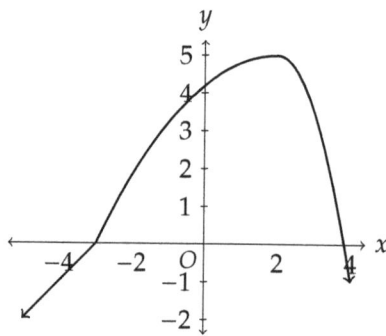

4. (a) is definitely false. (b) is definitely false. (c) is possibly true.

Reflection

Suggested Class Time. 60 minutes

Prerequisites. Students should be familiar with domain, range, and function notation.

Instructional Strategies. Numerous students wrote domain and/or range using "to" vocabulary, as in "0 to 4." On both an individual basis and when going over the notes with the whole class, we discussed more formal ways of writing this as both intervals and inequalities.

Many students only wrote things like "The client base increased," without addressing the rate being constant, increasing, or decreasing. For a few student groupings, it was helpful to show them three superimposed graphs that each started at the origin and ended in a generic place in Quadrant I but had differing rates of rates of change. This got them to understand that not all "increasing" is alike.

Students inevitably may feel that "decreasing rate" should always mean slowing down (as this makes some intuitive sense). To help students understand, I drew a diagram using arrows: for instance, a function increasing at a decreasing rate is moving to the right but being "pulled back" by its decreasing rate...

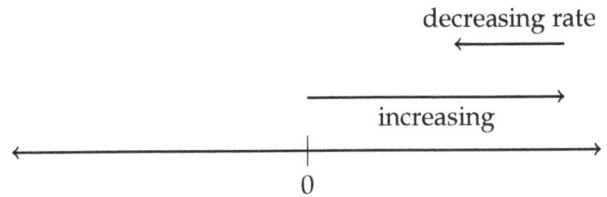

Then, for decreasing at a decreasing rate, the arrows helped students understand that this is actually speeding up, but in a direction away from the origin.

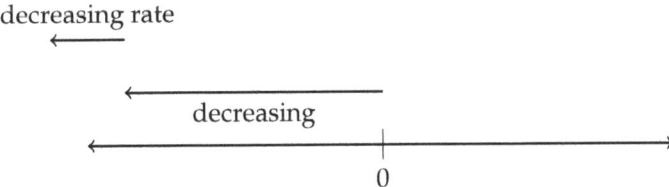

We then discussed as a class how a function is "speeding up" only if both the rate of change and the rate of rate of change are the same sign or "agree" (which is a nice preview of velocity and acceleration in calculus).

Interestingly enough, I actually went ahead and informally introduced the students to the notion of a tangent line as a way of determining whether a function is increasing or decreasing at an increasing or decreasing rate. The example I gave was related to inertia: when in a car that is making a sharp left, one's bodies moves "to the right," or function in the direction of a tangent curve. Students took to this very quickly, and it helped one student in particular quickly determine whether the function was increasing at an increasing or decreasing rate on one question of the task.

Technology. Students didn't use any technology except for a few who used a calculator to compute some differences.

Topic 1.1 ~ Change in Tandem (Day 2)

Learning Objectives

1. (1.1.A) Describe how the input and output values of a function vary together by comparing function values.

2. (1.1.B) Construct a graph representing two quantities that vary with respect to each other in a contextual scenario.

Success Criteria

1. Determine whether a graph is concave up or concave down given an analytical, graphical, or verbal description of a function.

2. Express concavity in terms of the rate of the rate of change of a function.

In the previous activity, you investigated functions and their rates of change. Functions can be constant, increasing, or decreasing, and they can increase or decrease at increasing or decreasing rates. This latter part can be seen as what is called *concavity*.

Suppose the height of a roller coaster for the first 20 seconds is graphed by the function $h(t)$ shown below.

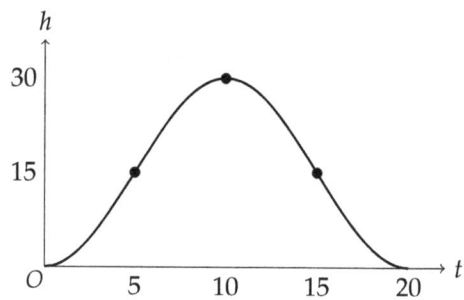

1. Complete the table for how the rollercoaster's height is changing as a function of time.

For t in ____,	$h(t)$ is	increasing/decreasing at a/an	increasing/decreasing rate.	
$0 < t < 5$	$h(t)$ is			
$5 < t < 10$	$h(t)$ is			
$10 < t < 15$	$h(t)$ is			
$15 < t < 20$	$h(t)$ is			

2. We say that the graph of a function is CONCAVE UP when its rate of change is *increasing*. On the other hand, the graph of a function is CONCAVE DOWN when its rate of change is *decreasing*. In the table above, write "Concavity" in the top of the last column. Then, based on what you wrote in the table, write either "up" or "down" in each row.

3. On the graph, label each section of the graph as either "C. up" or "C. down" based on your last column.

4. In your own words, what do the concave up pieces of the graph have in common? How do they look?

5. How about the concave down pieces? How do they look?

6. Consider the function $f(x)$ graphed below.

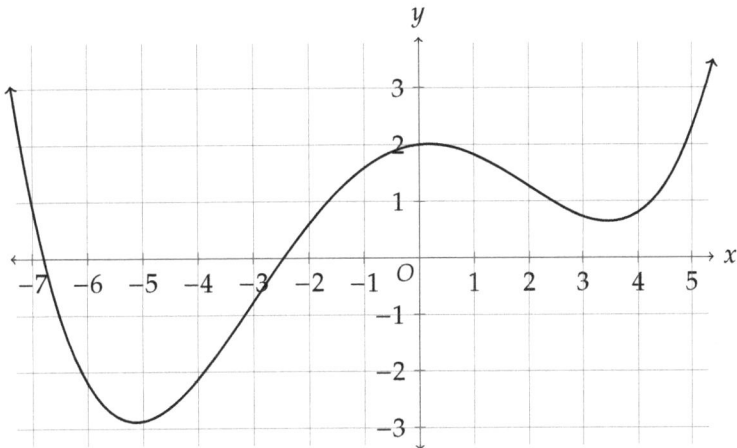

   (a) On what interval/s is $f(x)$ concave up?

   (b) On what interval/s is $f(x)$ concave down?

7. Try a tabular example now: the values of a function $g(x)$ are given below. Assume that g does not change concavity on the interval $[1,3]$.

x	1	1.5	2	2.5	3
$g(x)$	5	2	0	-1	-1.5

Is $g(x)$ concave up or concave down on $[1,3]$? Explain.

8. This question will have each person in your group draw a graph. Don't tell your groupmates what you're drawing!

   (a) On the grid below, draw a function $f(x)$ that meets the following criteria.
   - f is increasing for $x < 3$.
   - f is concave down only for $-1 < x < 3$ and concave up for all other x.
   - f is decreasing for $x > 3$.

   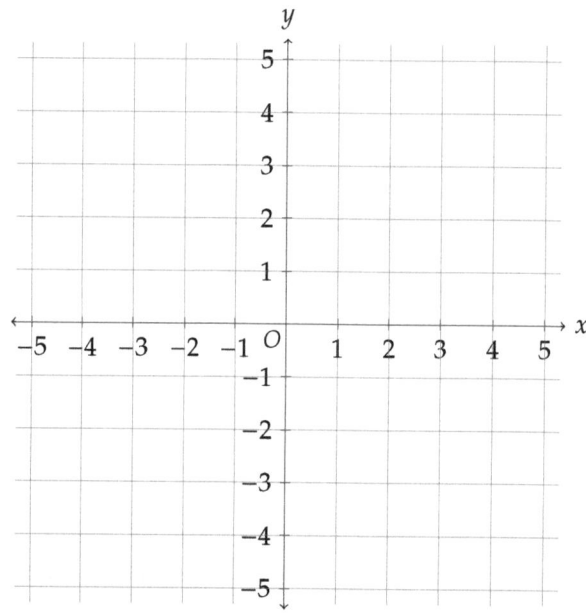

   (b) Pass your graph to the left and have that groupmate check your graph. If there are any issues, discuss as a group.

9. A function $f(x)$ is graphed at right.

   Use the x-values of A through G to determine the following.

   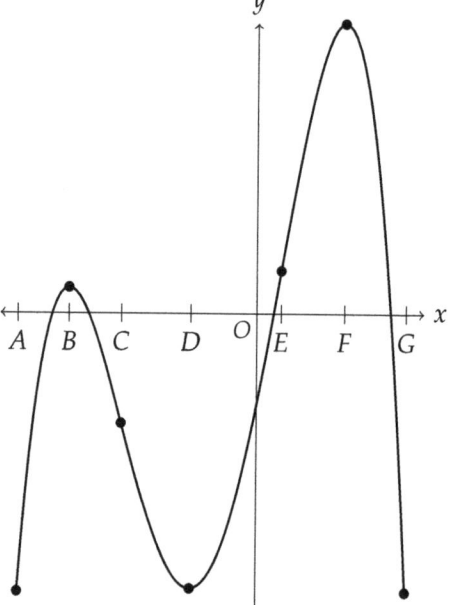

   (a) On what interval/s is $f(x)$ increasing?

   (b) On what interval/s is the rate of change of $f(x)$ increasing?

   (c) On what interval/s is $f(x)$ concave down?

Notes

Topic 1.1 (Day 2) Homework

1. A teenager who doesn't have their driver's license yet is told by their parents that they're allowed to drive the parents' car up the street and back. For the first minute, the teenager leaves and drives very slowly, but they quickly gain confidence and ramp up their speed for the next minute. They then have to gradually slow down to stop at the end of the street, which takes 30 seconds. On the way back, the teenager again ramps up their speed for a full 1.5 minutes before slowing down in the 30 seconds before getting home.

    (a) On the graph below, where the x-axis is time and the y-axis is distance from home, draw a possible graph representing the teenager's distance d based on time t in minutes. Assume the teenager leaves at time $t = 0$.

    (b) On what interval/s is the function you graphed concave up?

    (c) On what interval/s is the function you graphed concave down?

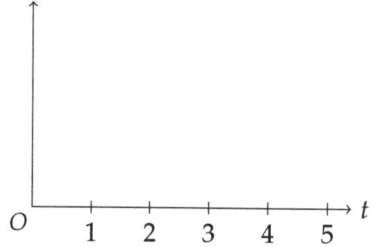

2. Match each of the following descriptions on the left with all applicable terms on the right.

    (1) The rate of change of f is positive.

    (2) The rate of change of f is increasing.

    (3) The rate of change of f is decreasing.

    (4) The rate of change of f is negative.

    (A) f is concave up.

    (B) f is concave down.

    (C) f is increasing.

    (D) f is decreasing.

3. Consider the function graphed below.

    (a) Complete the table below with the correct interval.

Description	Interval/s
f is increasing.	
f is decreasing.	
f is concave up.	
f is concave down.	
The rate of change of f is increasing.	

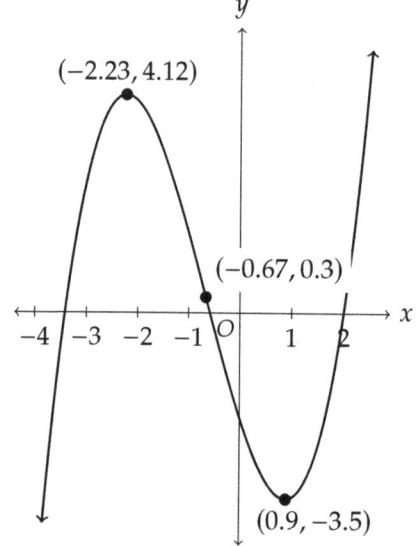

    (b) At what x-value does the function have an inflection point?

4. The rates of change of the function graphed in #3 for some select x-values are $-3.7, -2.1, 0, 1.4,$ and 5.9. Match each rate of change with the corresponding x-value.

x	−3.2	−2.23	−0.7	0.37	1.17
Rate of change of f					

5. A table of values of the function $f(x)$ is given below. Suppose that $f(x)$ is decreasing and concave down on the interval $(1, 3)$.

x	1	2	3
$f(x)$	-4	?	-10

If $f(2)$ is an integer, how many possible values of $f(2)$ are there?

Topic 1.1 (Day 2) Solutions/Notes

Activity

1.

For t in ___,	$h(t)$ is	increasing/decreasing at a/an	increasing/decreasing ROC.	Concavity
$0 < t < 5$	$h(t)$ is	increasing at an	increasing rate.	Up
$5 < t < 10$	$h(t)$ is	increasing at a	decreasing rate.	Down
$10 < t < 15$	$h(t)$ is	decreasing at a	decreasing rate.	Down
$15 < t < 20$	$h(t)$ is	decreasing at an	increasing rate.	Up

4. The concave up pieces are curved upwards.

5. The concave down pieces are curved downwards.

6. (a) $x < -3$ and $x > 2$ (b) $-3 < x < 2$

7. The function is concave up on $[1, 3]$ as the rate of change is increasing (getting less negative).

x	1	1.5	2	2.5	3
$g(x)$	5	2	0	-1	-1.5
Change		-3	-2	-1	-0.5

8. A sample graph is provided.

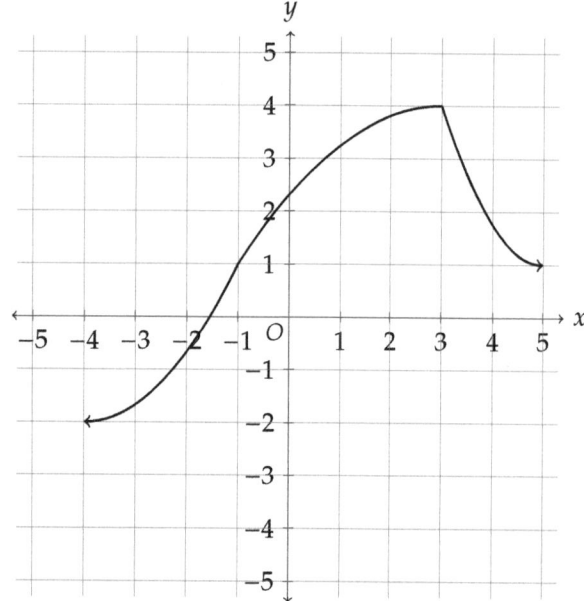

9. (a) (A, B) and (D, F)
   (b) (C, E)
   (c) (A, C) and (E, G)

Notes

When the rate of change of a function is increasing, the graph of a function is called CONCAVE UP. When the rate of change of a function is decreasing, the graph is called CONCAVE DOWN.

Graphically, a function is concave up when it is "curved up" and concave down when it is "curved down."

If a function's rate of change is increasing (the graph of the function is concave up), then the RATE OF RATE OF CHANGE is positive; if a function's rate of change is decreasing (the graph of the function is concave down), then the RATE OF RATE OF CHANGE is negative.
When a function changes from concave up to concave down (or vice-versa) at a given input, the point on the graph is called an INFLECTION POINT.

With rates of change, particularly negative ones, it's important to keep in mind that decreasing means *getting lower* - this includes getting *more negative*. So, if a line with a negative slope gets an even more negative slope, we say that the rate of change has *decreased* despite the line getting steeper.

Homework

1. (a) A sample graph is provided.

    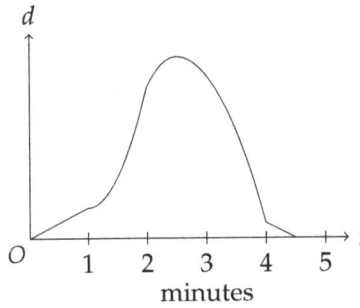

    (b) Answers will vary based on whether students drew a linear function or not for $0 \leq t \leq 1$ and $4 \leq t \leq 4.5$.

    (c) Students should at least have $2.5 < t < 4$.

2. (1, C), (2, A), (3, B), (4, D)

3. (a)

Description	Interval/s
f is increasing.	$x < -2.23, x > 0.9$
f is decreasing.	$-2.23 < x < 0.9$
f is concave up.	$x > -0.67$
f is concave down.	$x < -0.67$
The rate of change of f is increasing.	$x > -0.67$

    (b) $x = -0.67$

4.

x	−3.2	−2.23	−0.7	0.37	1.17
Rate of change of f	5.9	0	−3.7	−2.1	1.4

5. The rate of change must be decreasing. If the rate of change were constant, then $f(2) = -7$. Therefore, $f(2)$ must be less than -7. Since it must be an integer, we can have $f(2) = -5$ or $f(2) = -6$. There are 2 possible values.

Reflection

Suggested Class Time. 30-45 minutes for activity

Prerequisites. Students should be comfortable with discussing increasing/decreasing rates of change.

Instructional Strategies. Students largely cruised through this activity. There were some debates on #6 about whether concavity was changing at *extrema*, though: this made sense, as students were thinking about the *function* changing from increasing to decreasing or vice-versa, rather than *the rate of change* of the function. Referring students back to the inflection point in #1 really helped.

Multiple representations are key throughout this lesson. Students will find the graphical understanding incredibly easy, as it's literally just looking for positive or negative curvature; finding it in a tabular, verbal, or analytical representation is more difficult. For instance, many students solved #7 by graphing the data. While I told them this was completely fine, I also encouraged them to just compute the net changes on each interval to see if they could further support their answer.

When I first introduce ROROC, I tie it to unit analysis, something my students are somewhat comfortable with from science classes. In particular, I discussed dropping an object and saying that its changing position might be measured in meters m. Its velocity or speed, or its *rate of* change would then be measured in meters per second, or m/s. Students felt good about this, and I then asked about acceleration; at least a few students said they had seen m/s^2 before. I mentioned gravity and said that m/s^2 could be thought of as $m/s/s$, or *meters per second per second*, which is the *rate of* change of the rate of change, or the ROROC. At first, students were overwhelmed or displeased with the term ROROC, but we then tied it back to #7.

In fact, some students actually found the ROROC argument the easiest for #7! Students may struggle to see an increasing sequence formed out of negative values, but being able to see the positive changes of changes (negative ROROC) makes it much easier to see that the function's graph in #7 is concave up. I underestimated how helpful the ROROC would be.

Lastly, an applet for practice! Check the link at: https://www.geogebra.org/m/wptcuvug

Technology. No technology is required.

Problems.

1. The function $K(a)$ models the amount of koi K in a pond that has a pounds of algae in it. The table below provides values of K for certain inputs a.

a	2	5	10	20	50
$K(a)$	50	70	75	60	25

    Which of the following could be true?

    (A) As more algae is added to the pond, the koi population grows.

    (B) More algae increases koi population up to a point, but too much algae reduces the koi population.

    (C) Because $K(20) < K(10)$, K is a decreasing function.

    (D) As the amount of algae increases from 2 pounds to 20 pounds, the amount of koi increases.

2. A function $V(t)$ gives the volume of air in a compressed tank after t minutes. The tank will alternatively let in or release air to maintain a certain amount of pressure. The graph of $V(t)$ is decreasing and concave up for $0 < t < 5$, increasing and concave up for $5 < t < 7$, increasing and concave down for $7 < t < 10$, and decreasing and concave down for $10 < t < 12$. Which of the following is true?

   (A) From $t = 0$ to $t = 5$ minutes, the tank loses air more and more rapidly.

   (B) From $t = 5$ to $t = 7$, the tank gains air more and more rapidly.

   (C) From $t = 7$ to $t = 10$, the tank gains air more and more rapidly.

   (D) From $t = 10$ to $t = 12$, the tank loses air less and less rapidly.

3. A table of values of the function $y = g(x)$ is provided below.

x	1	2	3	4
$g(x)$	28	20	?	0

   If $g(x)$ is concave down for $1 < x < 4$ when graphed, which of the following is a possible value of $g(3)$?

   (A) 10 (B) 11 (C) 12 (D) 13

4. Audrey is going to fill up a kiddie pool with water. She gradually increases the water flow for a minute, after which she quickly turns it up to full power. She then leaves the water running for 3 minutes until the pool is completely full. Which of the following graphs could represent the volume of water in the pool based on time?

   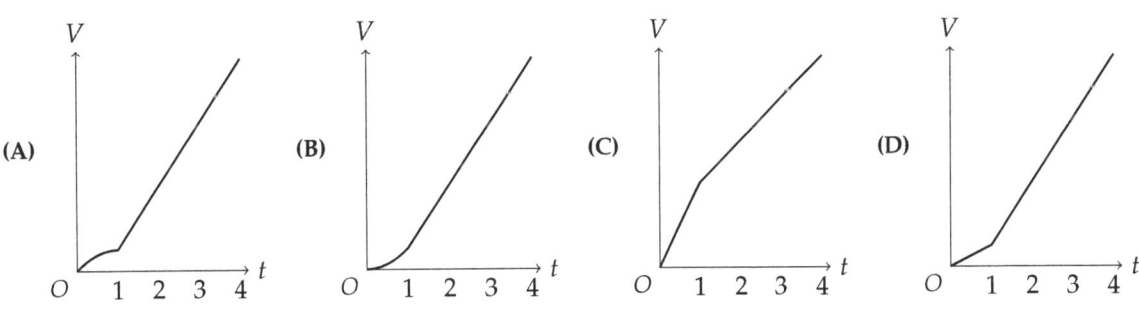

5. The function $y = f(x)$ is graphed below. The points of inflection of f are labeled A and B.

   The rate of change of f is increasing on which interval/s?

   (A) $x < -1$ only

   (B) $x > 2$ only

   (C) $-1 < x < 2$

   (D) $x < -1$ and $x > 2$

   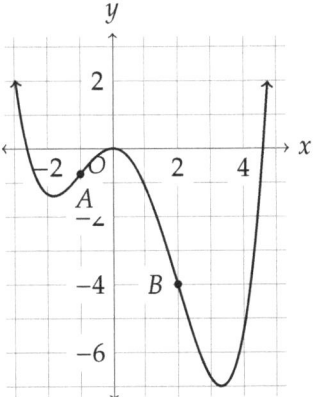

6. The function f given by $y = f(x)$ is graphed below.

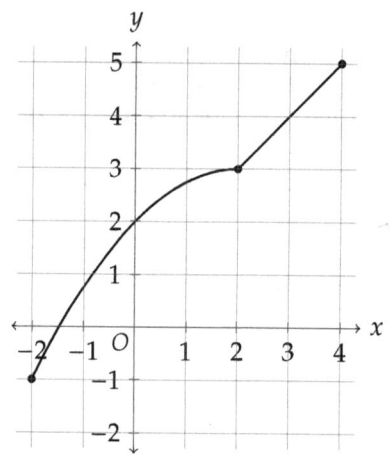

Which of the following is true about the function f?

(A) f is increasing at a decreasing rate for $-2 < x < 4$.

(B) f is increasing at a decreasing rate for $-2 < x < 5$.

(C) f is increasing at a decreasing rate for $-2 < x < 2$.

(D) f is increasing at a decreasing rate for $-2 < x < 3$.

Solutions.

1. The output values appear to increase to 75, and then decrease. This indicates that the amount of algae increases the koi population, but too much algea decreases the population. The correct answer is therefore **(B)**.

2. From $t = 0$ to $t = 5$, the tank loses air less rapidly; from $t = 5$ to $t = 7$, the tank gains air more rapidly; from $t = 7$ to $t = 10$, the tank gains air less rapidly; and from $t = 10$ to $t = 12$, the tank loses air more rapidly. The correct answer is therefore **(B)**.

3. The average rate of change from $x = 2$ to $x = 4$ is
$$\frac{4-20}{4-2} = \frac{-16}{2} = -8.$$
Then, if g is linear, $g(3) = 20 - 8 = 12$. However, g is not linear, it is concave down, so g must decrease more rapidly. This will happen if $g(3) = 11$, since the rate of change from $x = 2$ to $x = 3$ is
$$\frac{11-20}{3-2} = -9$$
and the rate of change from $x = 3$ to $x = 4$ is
$$\frac{0-11}{4-3} = -11.$$
This indicates that the rate of change is decreasing. The correct answer is therefore **(B)**.

4. Audrey increases the flow of water and also increases the rate at which it flows, so for the first minute, the function increases at an increasing rate. Then Audrey leaves the water running at a constant rate, so for the next three minutes, the flow of water increases at a constant rate. The correct answer is therefore **(B)**.

5. The rate of change is increasing where the graph of the function is concave up. This happens for $x < -1$ and $x > 2$. The correct answer is therefore **(D)**.

6. On the interval $-2 < x < 2$, the function is increasing but the rate of change is decreasing. On the interval $2 < x < 4$, the function has a constant rate of increase. The correct answer is therefore **(C)**.

Topics 1.2-1.3 ~ Rates of Change in Linear and Quadratic Functions

Learning Objectives

1. (1.2.A) Compare the rates of change at two points using average rates of change near the points.
2. (1.2.B) Describe how two quantities vary together at different points and over different intervals of a function.
3. (1.3.A) Determine the average rates of change for sequences and functions, including linear, quadratic, and other function types.
4. (1.3.B) Determine the change in the average rates of change for linear, quadratic, and other function types.

Success Criteria

1. I can compute average rates of change and compute net changes on intervals.
2. I can use average rates of change to approximate instantaneous rates of change.
3. I can use first and second differences to determine whether a linear or quadratic function fits the relationship between two variables.

Mr. Hornbeck was driving on a lonely country road, pleasantly going 55 miles per hour, when he passed a police officer. The speed limit was 50 mph, so Mr. Hornbeck thought he'd be fine. Exactly 8 minutes later, Mr. Hornbeck was driving 55 miles per hour again and passed by a different police officer. The officer pulled him over and cited him for speeding. When Mr. Hornbeck asked, "Do I really get a citation for going only 5 miles over the limit?," the officer responded, "Sir, another officer spotted you 8 miles ago. You must've been going at least 60 at some point."

1. What is the justification for the officer's claim that Mr. Hornbeck was speeding?

One important concept related to rates of change is the *average* rate of change, or *ARC*.

The AVERAGE RATE OF CHANGE of a function f on an interval $[a, b]$ is given by

$$\text{ARC} = \frac{f(b) - f(a)}{b - a}$$

2. What does this equation resemble?

The graph below shows Mr. Hornbeck's distance driven $f(x)$, in miles, x minutes after passing the first police officer.

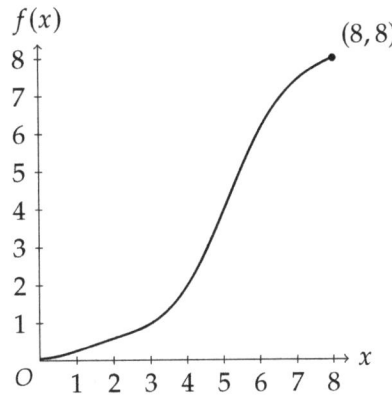

3. Compute the ARC of $f(x)$ on the interval $[0, 8]$. Then, draw the line from $(0, 0)$ to $(8, 8)$.

4. The line you just drew is called a *secant line*. How is the ARC related to this line?

5. Two additional points on the graph of f are $(3, 1)$ and $(5, 4)$. Find the ARC of $f(x)$ on the interval $3 \leq x \leq 5$ and draw the corresponding secant line.

6. We can use average rates of change to compute distances or, in general, *net changes*. For instance, suppose the function $g(x)$ models how far Mr. Hornbeck's friend, Mr. Cook, drove for time x in hours. If Mr. Cook's average rate of change for $3 \leq x \leq 4.5$ was 50 miles per hour, how far did he drive?

When it comes to the speedometer on a car, Mr. Hornbeck was looking at his speed in "real time." When he passed the second officer going 55 mph, he was traveling 55 miles per hour at that *instant*. This leads to the concept of *instantaneous* rate of change, or *IRC*.

7. How do you think a car computes your speed at any one instant?

The IRC of a function can actually be approached using ARCs. For instance, to determine how fast Mr. Hornbeck was traveling $x = 4$ minutes after passing the police officer, we could use the ARC of $f(x)$ from $x = 3$ to $x = 5$.

8. To get an even better estimate, though, how should we choose the interval for the ARC?

Below is a table of values for $f(x)$ near $x = 4$.

x	3.9	3.99	3.999	4.001	4.01	4.1
$f(x)$	1.822	1.952	1.966	1.969	1.983	2.124

9. Compute the ARC of $f(x)$ on each of the following intervals.

   (a) $3.9 \leq x \leq 4.1$

   (b) $3.99 \leq x \leq 4.01$

   (c) $3.999 \leq x \leq 4.001$

10. What would you estimate the IRC of $f(x)$ is at $x = 4$? What does this convert to in miles per hour?

Rates of change can also be used to determine what *kind* of function best models the relationship between two variables. Today, we'll first consider linear and quadratic models. For a refresher:

> A LINEAR FUNCTION has the general form $f(x) = a + bx$, where $(0, a)$ is the y-intercept and b is the slope, which is the constant rate of change of the function.
>
> A QUADRATIC FUNCTION has the general form $g(x) = ax^2 + bx + c$, where $(0, c)$ is the y-intercept.

11. Let's first consider the function $f(x) = 3x + 4$. Complete the second column of the table below with the output value of f for each input given.

x	$f(x)$	First Differences
0		N/A
1		3
2		
3		
4		
5		

12. The average rates of change over equal length intervals are called FIRST DIFFERENCES. For instance, the first difference between $x = 0$ and $x = 1$ is $f(1) - f(0) = 7 - 4 = 3$. Complete the last column in the table with the remaining first differences. Note that each interval has the same length of 1.

13. What do you observe about the first differences of $f(x)$?

14. Fill in the conjecture:

    For a linear function, the first differences are _____ and equal to _____.

15. Now, consider the quadratic function $g(x) = x^2 + 2x + 3$. Repeat the process from #11-14 in the second and third column of the table above.

x	$g(x)$	First Differences	
0		N/A	N/A
1			N/A
2			
3			
4			
5			

16. What do you observe about the first differences of $g(x)$?

17. Fill in the conjecture:

    For a quadratic function, the first differences are a _____.

18. In the final column of the table, compute the differences between the first differences. These are called the SECOND DIFFERENCES. For instance, the second difference at $x = 2$ is $5 - 3 = 2$.

19. What do you observe about the second differences of $g(x)$?

20. Fill in the conjecture:

> For a quadratic function, the second differences are _____.

Notes

Topics 1.2-1.3 Homework

1. Consider the graph of $f(x)$ below.

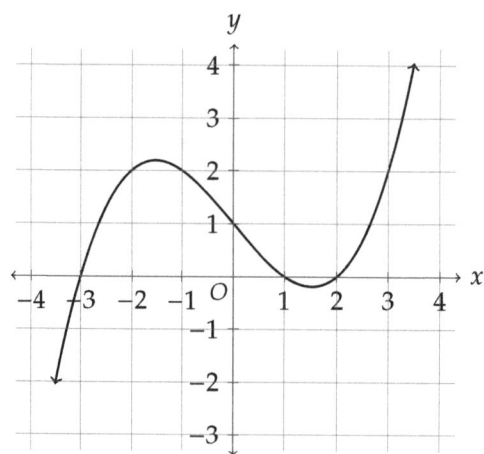

   (a) What is the average rate of change of f on the interval $-2 \leq x \leq 1$?

   (b) On which of the following interval is the average rate of change the greatest? The least?
      i. $-3 \leq x \leq -2$
      ii. $-2 \leq x \leq 0$
      iii. $0 \leq x \leq 3$
      iv. $1 \leq x \leq 3$

   (c) The quantity $\dfrac{f(3.1) - f(2.9)}{0.2}$ approximates what?

   (d) Find two intervals $a \leq x \leq b$ and $c \leq x \leq d$, where a, b, c, d are integers, such that the average rate of change of f on $[a,b]$ and $[c,d]$ are both 0.

2. A startup company has been tracking its profit every 2 months. A table of profit for the first 10 months is provided below, as well as a scatterplot of the data.

Month	2	4	6	8	10
Profits (tens of thousands)	3	5.7	8.7	12	15.6

   Determine whether a linear function or a quadratic function is better suited to model the relationship between profit and month for this company.

3. The table below provides average rates of change of f on various intervals within $-3 \leq x \leq 9$.

Interval of x	$-3 \leq x \leq 0$	$0 \leq x \leq 2$	$2 \leq x \leq 7$	$7 \leq x \leq 9$
Average rate of change of f on interval of x	3	-2	2	0

   (a) On which, if any, of the intervals could $f(x)$ be decreasing?

   (b) On which of the four given intervals does f increase the most?

   (c) Suppose $f(0) = 5$. Compute $f(2)$, $f(7)$, and $f(9)$.

4. For the function $g(x)$, let $h(x) = g(x+1) - g(x)$. If $h(x) = -6x + 1$, what kind of function is $g(x)$? Justify.

Topics 1.2-1.3 Solutions/Notes

Activity

1. Mr. Hornbeck drove 8 miles in 8 minutes for an average of 60 miles per hour. If he was driving at 55 mph at at least two different times, then he must have driven above 60 at some point to have an average speed of 60 mph.

2. Slope

3. The ARC is $\dfrac{8-0}{8-0} = 1$.

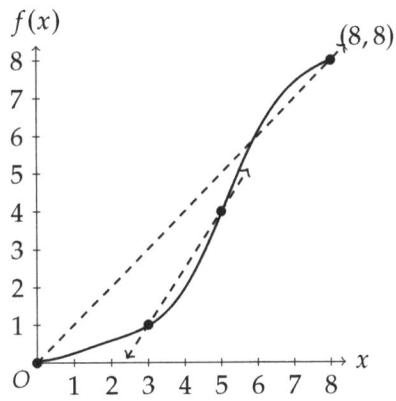

4. The ARC is the slope of the secant line.

5. ARC $= \dfrac{4-1}{5-3} = \dfrac{3}{2}$. Drawn above.

6. He drove 50 miles per hour for 1.5 hours, so he drove 50(1.5) = 75 miles.

7. Student answers will vary.

8. Choose an interval consisting of two points very close to $x = 4$.

9. (a) 1.51 (b) 1.55 (c) 1.5

10. About 1.5. This is 1.5 miles per minute, or 1.5(60) = 90 miles per hour (yikes!).

11.

x	$f(x)$	First Differences
0	4	N/A
1	7	3
2	10	3
3	13	3
4	16	3
5	19	3

13. They are constant and equal the slope of the line.

14. For a linear function, the first differences are *constant* and equal to *the slope of the line*.

15.

x	f(x)	First Differences	
0	3	N/A	N/A
1	6	3	N/A
2	11	5	2
3	18	7	2
4	27	9	2
5	38	11	2

16. The first differences are linear.

17. For a quadratic function, the first differences are a *linear function*.

19. The second differences are constant.

20. For a quadratic function, the second differences are *constant*.

> **Notes**
>
> The AVERAGE RATE OF CHANGE, or ARC, of a function $f(x)$ on an interval $[a,b]$ is given by
>
> $$\text{ARC} = \frac{f(b) - f(a)}{b - a}.$$
>
> This is the slope of the secant line between $(a, f(a))$ and $(b, f(b))$.
>
> The net change of a function over an interval $[a,b]$ is precisely the average rate of change on $[a,b]$ times the change in x, or
>
> $$\Delta f = \frac{f(b) - f(a)}{b - a}(b - a)$$
>
> To estimate the instantaneous rate of change, or IRC, of a function $f(x)$ at $x = a$, one can compute the average rate of change of $f(x)$ for two x-values getting closer and closer to $x = a$.
>
> Looking at average rates of change over equal-length intervals can reveal which type of function would best model a relationship between two variables. The average rates of change over successive equal-length intervals of the inputs are called FIRST DIFFERENCES. The average rates of change *of the average rates of change* over successive equal-length intervals of the inputs are called SECOND DIFFERENCES.
>
> For a linear function $f(x) = a + bx$, the first differences are *constant* and equal to *the slope of the line*.
>
> For a quadratic function $g(x) = ax^2 + bx + c$, the first differences are a *linear function* and the second differences are *constant*.

Homework

1. (a) $\dfrac{f(1)-f(-2)}{1-(-2)} = \dfrac{0-2}{3} = -\dfrac{2}{3}$

    (b) It is the least on $-2 \leq x \leq 0$ (ARC = $-\frac{1}{2}$) and the greatest on $-3 \leq x \leq -2$ (ARC = 2).

    (c) The instantaneous rate of change of f at $x = 3$ (students could argue for the IRC at any value around 2.9 to 3.1)

    (d) There are five intervals on which the ARC is zero. The response must include two of them. The five intervals are $[-3, 2]$, $[-3, 1]$, $[-2, -1]$, $[-1, 3]$, and $[1, 2]$.

2. We examine the first and second differences.

Month	2	4	6	8	10
Profits (tens of thousands)	3	5.7	8.7	12	15.6
First differences		2.7	3	3.3	3.6
Second differences			0.3	0.3	0.3

    As the second differences are constant, the profits are best modeled by a quadratic function (despite the apparent linearity of the scatterplot).

3. (a) f could be decreasing on $0 \leq x \leq 2$, as this is the only interval with a negative ARC.

    (b) The net changes are $3(3) = 9, 2(-2) = -4, 5(2) = 10$, and $2(0) = 0$, respectively. The largest is 10, which occurs on $2 \leq x \leq 7$.

    (c) $f(2) = f(0) + 2(-2) = 5 - 4 = 1; f(7) = f(2) + 5(2) = 1 + 10 = 11; f(9) = f(7) + 2(0) = 11$

4. The function $g(x + 1) - g(x)$ gives the first differences of g on successive intervals of length 1 of the inputs. Because these differences are linear, g is a quadratic function.

Reflection

Suggested Class Time. 60 minutes

Prerequisites. Students should have familiarity with the formula for the slope of a line.

Instructional Strategies. Many students surprisingly struggled with #6, computing the distance traveled given the interval of time and the speed. What helped some students, particularly those who are more algebraically-minded, compute this was using the equation ARC = $\frac{\text{distance}}{\text{time}}$ and substituting in the known quantities.

When going through the notes, some students noted they had a hard time visualizing "what was going on" with the ARC calculations in #8-10. To aid these students, a Geogebra sketch was created: we went over this during the Notes portion of the lesson. The sketch can be found here: https://www.geogebra.org/m/yunqnmwk.

Technology. Students used their graphing calculators to compute the average rates of change. It helps to show students how to use alph (X,t,θ,n) to create traditional fractions in the calculator.

Problems.

1. The function $f(x)$ is a quadratic function with a positive leading coefficient. If $f(2) = 6, f(5) = 12$, and $f(11) = 30$, which of the following is the value of $f(8)$?

    (A) 18 (B) 20 (C) 21 (D) 24

2. (▤) Values of the function $g(x)$ for certain x-values are provided in the table below.

x	1	3	6	10
$g(x)$	-2	7	19	31

   On which of the following intervals is the average rate of change of g the greatest?

   (A) $[1, 3]$ **(B)** $[1, 6]$ **(C)** $[3, 6]$ **(D)** $[6, 10]$

3. (▤) The table below provides values of the function $f(x)$ for x-values near $x = -2$.

x	-2.001	-2.0001	-1.9999	-1.999
$f(x)$	6.0401	6.004	5.996	5.9601

   Which of the following is true?

   (A) The rate of change of f at $x = -2$ is approximately 6.
   (B) The rate of change of f at $x = -2$ is approximately -40.
   (C) The rate of change of f at $x = -2$ is increasing at a rate of approximately 6.
   (D) The rate of change of f at $x = -2$ is increasing at a rate of approximately -40.

4. The function f satisfies $f(x + 3) - f(x + 1) = 4$ for all x. Which of the following is true about $f(x)$?

   (A) The graph of f is a line with a positive slope because the average rate of change of f is a positive constant over successive equal-length intervals of the inputs.
   (B) The graph of f is a line with a positive slope because the average rate of change of f is increasing at a constant rate over successive equal-length intervals of the inputs.
   (C) The graph of f is a parabola that opens up because the average rate of change of f is a positive constant over successive equal-length intervals of the inputs.
   (D) The graph of f is a parabola that opens up because the average rate of change of f is increasing at a constant rate over successive equal-length intervals of the inputs.

5. The table below displays the average rate of change for a function $f(x)$ over certain intervals of x.

Interval of x	Average rate of change of f
$-3 \leq x \leq -1$	9
$-1 \leq x \leq 1$	5
$1 \leq x \leq 3$	1
$3 \leq x \leq 5$	-3

   Which of the following could describe the graph of f?

   (A) The graph of f is a line with a positive slope.
   (B) The graph of f is a line with a negative slope.
   (C) The graph of f is a parabola that is concave up.
   (D) The graph of f is a parabola that is concave down.

6. A technology startup company keeps track of the number of daily active users of its new app. At a pitch to an investor, the CEO of the startup reports, "Every month, we see our daily active users grow by 10,000 more than it grew the previous month." Which of the following does this statement best describe?

   (A) The relationship between month and daily active users could be modeled by a linear function because the rate of change is positive.

   (B) The relationship between month and daily active users could be modeled by a linear function because the rate of change is increasing at a constant rate.

   (C) The relationship between month and daily active users could be modeled by a quadratic function because the rate of change is positive.

   (D) The relationship between month and daily active users could be modeled by a quadratic function because the rate of change is increasing at a constant rate.

Solutions.

1. The first differences must be linear, say with constant rate of change d. The first difference from $x = 2$ to $x = 5$ is 6, so the difference from $x = 5$ to $x = 8$ is $6 + d$ and the difference from $x = 8$ to $x = 11$ is $6 + 2d$. This makes the difference from $x = 2$ to $x = 11$ a total of $(6 + d) + (6 + 2d) = 12 + 3d$. Given that this difference is $30 - 12 = 18$, we get $18 = 12 + 3d$, or $d = 2$. Therefore, $f(8) = 12 + (6 + 2) = 20$. The correct answer is therefore **(B)**.

2. The ARC on $[1, 3]$ is $(7 + 2)/(3 - 1) = 9/2 = 4.5$. The ARC on $[1, 6]$ is $(19 + 2)/(6 - 1) = 21/5 = 4.2$. The ARC on $[3, 6]$ is $(19 - 7)/(6 - 3) = 12/3 = 4$. The ARC on $[6, 10]$ is $(31 - 19)/(10 - 6) = 12/4 = 3$. The greatest is 4.5 on $[1, 3]$. The correct answer is therefore **(A)**.

3. The ARC on the interval $[-2.0001, -1.9999]$ is

$$\frac{5.996 - 6.004}{-1.9999 - (-2.0001)} = -\frac{0.008}{0.0002} = -40.$$

   This is the approximate rate of change at $x = -2$. The correct answer is therefore **(B)**.

4. The ARC on the interval $[x + 1, x + 3]$ is

$$\frac{f(x + 3) - f(x + 1)}{x + 3 - (x + 1)} = \frac{4}{2} = 2.$$

   Hence, this is a positive constant over successive equal-length intervals. The correct answer is therefore **(A)**.

5. The average rate of change is decreasing at a constant rate on successive equal-length intervals of the inputs, so this describes a concave down parabola. The correct answer is therefore **(D)**.

6. That users grow by 10,000 more than it grew the previous indicates that the rate of change is increasing at a constant rate (10,000 per month). This implies that the relationship can be modeled by a quadratic function. The correct answer is therefore **(D)**.

Topic 1.4 ~ Polynomial Functions

Learning Objectives

1. (1.4.A) Identify key characteristics of polynomial functions related to rates of change.

Success Criteria

1. I can determine the degree and leading coefficient of a polynomial.
2. I can determine characteristics of a polynomial based on its degree and leading coefficient.
3. I can answer questions about extrema and inflection points based on rates of change and zeros.

Today, you're going to investigate various characteristics of polynomial functions.

A POLYNOMIAL FUNCTION is a function of the form

$$p(x) = a_n x^n + a_{n-1} x^{n-1} + a_{n-2} x^{n-2} + \cdots + a_2 x^2 + a_1 x + a_0$$

where each of the values $a_n, a_{n-1}, \ldots, a_0$ are the COEFFICIENTS. The highest exponent n is called the DEGREE of the polynomial and the coefficient of the x^n term is called the LEADING COEFFICIENT. The term a_0 with no variable is called the CONSTANT TERM.

1. Go to the Geogebra sketch at www.geogebra.org/m/phagwe8y.

2. There is a polynomial $g(x) = ax^4 + bx^3 + cx^2 + dx + f$ shown, where you will be able to adjust $a, b, c, d,$ and f (don't do it yet!). You should see $a = 1, b = 2, c = -3, d = -5$ and $f = 3$.

3. The beginning polynomial is $g(x) = x^4 + 2x^3 - 3x^2 - 5x + 3$.

    (a) What is the degree of g?
    (b) What is the leading coefficient of g?

4. The red points are called EXTREMA (singular: extremum), as they are "extremes" of the function. All extrema are either *maxima* or *minima*, and they can be *local* or *global*. A LOCAL MINIMUM or LOCAL MAXIMUM is simply the lowest or highest value in a *neighborhood* of an x value, while a GLOBAL MINIMUM or GLOBAL MAXIMUM is the lowest or highest value on the entire domain of the function.

    Classify each of the extrema of g.

$g(x)$	Maximum or Minimum?	Local or Global?
0.971		
4.618		
−2.027		

5. Now, look at the rates of change of g between the extrema. Fill in the table below with either "positive" or "negative" for the rate of change of g on each interval.

Interval	$x < -1.941$	$-1.941 < x < -0.612$	$-0.612 < x < 1.053$	$x > 1.053$
Rate of change of g				

6. Complete the conjectures below.
   - At a local minimum, the rate of change of a function _____.
   - At a local maximum, the rate of change of a function _____.

7. There is another way that a minimum or maximum can occur. Change the "domain lower bound" to 0 and the "domain upper bound" to 1.5.

8. Explain why both $g(0)$ and $g(1.5)$ can be considered extrema. Are they maxima or minima, and are they local or global?

9. In general, then, we can say that extrema occur at one of two kinds of places:
   - where the rate of change of a function changes signs, or
   - at an endpoint of the domain.

   It's also worth looking at minima and maxima with regards to the *zeros* of a function. Return the domain lower and upper bounds to -10 and 10, respectively. Then, investigate: where do the maxima and minima occur with regards to the zeros of g?

10. You may have noticed that g had a global minimum, but no global maximum.

    (a) Play around with the values of a, b, c, d, and f to try to create a graph of g that has *only* a global maximum. How'd you do it?

    (b) Now, try to create a graph of g that has neither a global maximum *nor* a global minimum. How'd you do it?

    (c) Try to create a graph that has *both* a global maximum *and* a global minimum. How'd you do it?

    Now, reset the sliders to $a = 1, b = 2, c = -3, d = -5, f = 3$ (or just refresh the page!).

11. We looked at rates of change of g: how about the *rates of change* of the rates of change? Press the "Show Extrema" checkbox to hide the extrema and press the "Show Inflection Points" checkbox. Additionally, reset the lower and upper domain bounds to -5 and 5.

12. Complete the table below with "positive" or "negative" for the rates of rates of change of g on each given interval.

Interval	$x < -1.366$	$-1.366 < x < 0.366$	$x > 0.366$
Rate of rate of change of g			

13. As you can see in the graph, the points $(-1.366, 2.616)$ and $(0.366, 0.884)$ are *inflection points*. Based on how you filled out the table above, what occurs at inflection points?

Now for a couple of practice questions!

14. Deselect both the "show extrema" and "show inflection points" checkboxes so that neither sets of points appears. Now, change the coefficients as follows:

    $a = -1, b = 2, c = 0, d = -2, f = 4,$ domain lower bound $= -1,$ domain upper bound $= 2$

    Work individually on the following.

    (a) What is the degree?
    (b) What is the leading coefficient?
    (c) How many extrema?
    (d) How many local minima? (Note: Global minima count as local minima!)[2]
    (e) How many local maxima? (See note above.)
    (f) How many global minima?
    (g) How many global maxima?
    (h) How many inflection points?

15. Select both the extrema and inflection point boxes and discuss your answers with your group.

16. Now, change the coefficients as follows and repeat what you did for the previous question (alone, then group). Don't forget to de-select the extrema and inflection point boxes!

    $a = 0, b = 1, c = -2, d = 2, f = 2,$ domain lower bound $= -1,$ domain upper bound $= 1.5$

    (a) What is the degree?
    (b) What is the leading coefficient?
    (c) How many extrema?
    (d) How many local minima?
    (e) How many local maxima?
    (f) How many global minima?
    (g) How many global maxima?
    (h) How many inflection points?

17. Finally, how about one without a graph? Deselect the "show graph" button and change the coefficients as follows:

    $a = -2, b = 3, c = 1, d = 1, f = -2,$ domain lower bound $= -5,$ domain upper bound $= 5$

    As a group, determine the more limited set of characteristics below.

    (a) What is the degree?
    (b) What is the leading coefficient?
    (c) How many extrema *at most*?
    (d) How many global minima?
    (e) How many global maxima?

    Then, check your work.

---

[2]This may seem strange, but think of it this way: if the Burj Khalifa in Dubai is the tallest building in the world, then it's also the tallest building in its own city, for instance. In other words, if a value is a maximum *globally*, then it must also be a maximum *locally*. This also applies to minima.

Notes

Topic 1.4 Homework

1. Consider the polynomial function $f(x) = -2x^6 + 8x^3 + 5x + 1$ defined for all real x. Does x have a global maximum, global minimum, both, or neither?

2. The table below describe the rate of change for a polynomial $g(x)$.

Interval of x	$x < 0$	$0 < x < \frac{3}{2}$	$\frac{3}{2} < x < \frac{5}{2}$	$x > \frac{5}{2}$
Rate of change of g	Negative	Positive	Positive	Negative
Rate of change of g	Increasing	Increasing	Decreasing	Decreasing

   (a) At what x-value does a local minimum occur?
   (b) At what x-value does a local maximum occur?
   (c) At what x-value does an inflection point occur?
   (d) What could the degree of g be?

3. The function $p(x)$ is a polynomial defined for $-3 \leq x \leq 4$. The table below gives values of p for select values of x.

x	-3	0	1	2	3	4
$p(x)$	2	-1	0	5	0	-7

   (a) What is the least possible number of zeros that p could have?
   (b) What is the least possible number of local extrema that p could have?

4. The polynomial $f(x)$ is graphed below.

   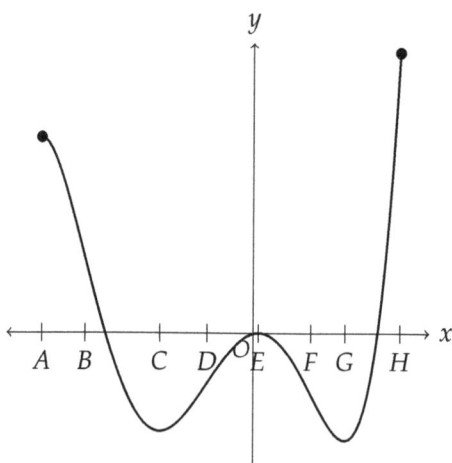

   (a) On what interval/s is the rate of change of f positive and decreasing?
   (b) At what input value/s does f reach a local minimum?
   (c) At what input value/s does f reach a local maximum?
   (d) On what interval/s is the rate of change of f negative?
   (e) At what input value does f reach a global maximum?
   (f) At what input value does f reach a global minimum?
   (g) On what interval/s is the graph of f concave up?

5. Let $f(x) = x(2x - 3)^2(1 - x)$. Determine the degree and leading coefficient of x.

Topic 1.4 Solutions/Notes

Activity

3. (a) 4   (b) 1

4.

$g(x)$	Maximum or Minimum?	Local or Global?
0.971	Minimum	Local
4.618	Maximum	Local
−2.027	Minimum	Global

5.

Interval	$x < -1.941$	$-1.941 < x < -0.612$	$-0.612 < x < 1.053$	$x > 1.053$
Rate of change of g	Negative	Positive	Negative	Positive

6.
   - At a local minimum, the rate of change of a function <u>changes from negative to positive</u>.
   - At a local maximum, the rate of change of a function <u>changes from positive to negative</u>.

8. The global maximum is $g(0)$. The local maximum is $g(1.5)$ because it is greater than the values of $g(x)$ for x near 1.5.

9. The maxima and minima occur between the zeros of g.

10. (a) $a < 0$ will work, or $a = b = 0$ and $c < 0$
    (b) $a = 0, b \neq 0$ or $a = b = c = 0$ and $d \neq 0$
    (c) This requires a domain restriction.

12.

Interval	$x < -1.366$	$-1.366 < x < 0.366$	$x > 0.366$
Rate of rate of change of g	Positive	Negative	Positive

13. The rate of rate of change of the function changes signs.

14. (a) 4   (b) −1   (c) 3   (d) 2   (e) 1   (f) 1   (g) 1   (h) 2

16. (a) 3   (b) 1   (c) 2   (d) 1   (e) 1   (f) 1   (g) 1   (h) 1

17. (a) 4   (b) −2   (c) 2   (d) 0   (e) 1

TOPIC 1.4 ~ POLYNOMIAL FUNCTIONS

> **Notes**
>
> A polynomial $p(x) = a_n x^n + a_{n-1} x^{n-1} + a_{n-2} x^{n-2} + \cdots a_2 x^2 + a_1 x + a_0$ has degree n, leading coefficient a_n, and constant term a_0.
>
> When a point $p(x_0) > p(x)$ for x near x_0, we say $p(x_0)$ is a LOCAL MAXIMUM. Similarly, if $p(x_0) < p(x)$ for x near x_0, we say $p(x_0)$ is a LOCAL MINIMUM. If either of these are true for *all* x in the domain of p, then we say $p(x_0)$ is a GLOBAL maximum or minimum. **Note that all global extrema are also local extrema, but not all local extrema are global extrema.** (If a building is the tallest in the U.S.A, then it would certainly be the tallest in Georgia; but if a building is the tallest in Georgia, it doesn't necessarily mean it would be the tallest in the U.S.A.)
>
> A minimum or maximum can occur at the endpoints of the domain of a polynomial. If they are not at the endpoints, they will satisfy the following.
>
> - At a minimum, the rate of change of the polynomial will change from negative to positive.
> - At a maximum, the rate of change of the polynomial will change from positive to negative.
> - The maxima or minima will occur at or between the zeros.
> - Between any two zeros, there must be a minimum or maximum.
>
> Functions with an *even* degree will only have a global maximum or a global minimum - never both, unless there is a domain restriction. Functions with an *odd* degree will have neither a global maximum nor a global minimum unless there is a domain restriction.
>
> INFLECTION POINTS occur when the graph of a polynomial changes from concave up to concave down (or vice-versa). This is equivalent to when the rate of rate of change changes from positive to negative (or vice-versa).
>
> As an AP tip: when given a function in its analytical form, it is often helpful to sketch a rudimentary graph to get a feel for what the function looks like!

Homework

1. The degree is even and the leading coefficient is negative, so f has only a global maximum.

2. (a) $x = 0$ (b) $x = \dfrac{5}{2}$ (c) $x = \dfrac{3}{2}$ (d) Any odd number greater than 3

3. (a) There are two known zeros, and one more must occur in the intervals $(-3, 0)$. Therefore there must be at least three zeros.

   (b) The endpoints must be local extrema, and there must be two other local extrema (between the three zeros). There must be at least four extrema.

4. (a) $D \leq x \leq E$
   (b) $x = C, G$
   (c) $x = A, E, H$
   (d) $A \leq x \leq C, E \leq x \leq G$
   (e) $x = H$
   (f) $x = G$
   (g) $B \leq x \leq D, F \leq x \leq H$

5. The leading term is $x(2x)^2(-x) = -4x^4$, so f has degree 4 and leading coefficient -4.

Reflection

Suggested Class Time. 45-50 minutes

Prerequisites. Students need to be comfortable with Topics 1.1-1.3 to have the most success.

Instructional Strategies. When teaching local and global extrema, it helped to refer students to a simpler example, like height. For instance, ask a student "Who's the tallest person in class?" Then, ask "Are they the tallest in the whole school?" Assuming you don't have *the* tallest person in the school, you can quickly point out that this person is a local maximum (with respect to height) but not a global one. This can also be used to reinforce the fact that global extrema are also local extrema: "Well, if Student X is the tallest in the school, would they be the tallest in just this class?"

Some students felt uncomfortable with considering endpoints of closed intervals as global extrema (they were oddly okay with the idea of them as local extrema, though), to which the best I could do was to emphasize that *many* functions realistically have restricted domains.

A very common issue through the first four topics was students not realizing the difference between "the rate of change is positive" and "the rate of change is increasing." To help students understand, it is very useful to train them to be able to replace "increasing" with "positive rate of change." For instance, in a multiple choice question that reads

The rate of change of f is increasing...

students would rewrite this as

The rate of change of f *has a positive rate of change*...

This really helped students understand that this is referring to the *rate of rate of change*, or concavity, of f. Similarly, students could rewrite

The function g is increasing at a decreasing rate...

as

The function g has a positive rate of change changing at a negative rate...

which again emphasizes that this is referring to the concavity or rate of rate of change.

Throughout Topic 1.4 and the ensuing homework and topic questions, really encourage students to *draw pictures*. When presented with tabular information, they should almost always be sketching a rudimentary graph to get an idea of how the function may be behaving. Additionally, though, encourage students to not draw the most "predictable" graphs. Many assume that, for instance, a graph with zeros at $x = 2$ and $x = 4$ must have a local max or min *precisely* at the midway point of $x = 3$.

Technology. Students were directed to the Geogebra sketch at www.geogebra.org/m/phagwe8y.

Problems.

1. (▦) Let f be the function given by $f(x) = \frac{1}{4}x^4 - \frac{2}{3}x^3 - \frac{5}{2}x^2 + 6x$. Which of the following is true?

    (A) f has a global maximum of approximately 3.083.

    (B) f has a global minimum of approximately -12.667.

    (C) f has global minima at both approximately -12.667 and -2.25.

    (D) f has neither a global minimum nor global maximum.

2. The polynomial function f is defined for $0 \leq x \leq 5$. Information about the rate of change of f is given for select intervals of x below.

   Which of the following *must* be true?

   (A) f has a global minimum at $x = 0$.

   (B) f has a local maximum at $x = 3$.

   (C) f has a local minimum at $x = 4$.

   (D) $f(x) = 0$ for some x with $2 < x < 3$.

Interval	Rate of Change
$0 < x < 2$	Positive and decreasing
$2 < x < 3$	Negative and decreasing
$3 < x < 4$	Negative and increasing
$4 < x < 5$	Positive and increasing

3. The polynomial function $g(x)$ is graphed at right.

   How many points of inflection does the graph of $g(x)$ have?

   (A) 2

   (B) 3

   (C) 4

   (D) 5

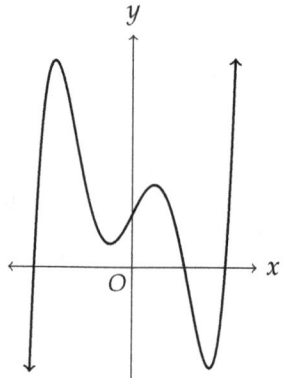

4. The function $y = f(x)$ is graphed at right.

   Which of the following could be an expression for $f(x)$?

   (A) $f(x) = \frac{1}{2}x^2 - \frac{1}{2}x^2 - 2x + 2$

   (B) $f(x) = -2x^3 + 6x^2 + 2x - 6$

   (C) $f(x) = x^4 - 4x^3 + 3x^2$

   (D) $f(x) = -2x^4 + 22x^2 + 36x + 16$

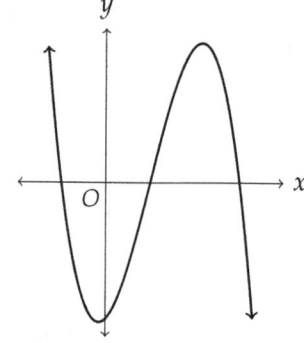

5. The rate of change of the function f is negative and increasing. This means that f is

   (A) increasing and concave up.

   (B) increasing and concave down.

   (C) decreasing and concave up.

   (D) decreasing and concave down.

Solutions.

1. Using a graphing calculator to graph the function, we find it has a global minimum of approximately -12.667. The correct answer is therefore **(B)**.

2. A minimum occurs when the rate of change changes from negative to positive. This happens at $x = 4$. The correct answer is therefore **(C)**.

3. The concavity changes three times so there are three inflection points. The correct answer is therefore **(B)**.

4. The function has three zeros and an even number of extrema, so it is of degree 3. It also has the form of a cubic with a negative leading coefficient. The correct answer is therefore **(B)**.

5. If the rate of change of f is negative, then f is decreasing. If the rate of change of f is increasing, then f is concave up. The correct answer is therefore **(C)**.

Topics 1.1-1.4 ~ Review

The first four topics included multiple behaviors of functions: increasing, decreasing, concave up, and concave down, to name a few. These can be be expressed or represented multiple ways.

Behavior	Graphical	Verbal, in terms of ROC
Increasing		
Decreasing		
Concave up		
Concave down		

Rates of change behave specific ways for linear and quadratic functions.

- For a linear function, the average rates of change over successive equal-length intervals are _____.

- For a linear function, the rate of the average rates of change over successive equal-length intervals is _____.

- For a quadratic function, the average rates of change over successive equal-length intervals are _____.

- For a quadratic function, the average rates of change over successive equal-length intervals are changing at _____.

- For a quadratic function, the rate of the average rate of changes over successive equal-length intervals is _____.

Polynomials of any degree, including linear and quadratic functions, also have characteristics that can be found by looking at rates of change.

- When the rate of change changes from positive to negative, a function has a _____.

- When the rate of change changes from negative to positive, a function has a _____.

Maxima and minima also abide by a few rules.

- Local extrema occur where rates of change change sign or _____.

- There will always be a local maximum or local minimum between any two _____ of a polynomial.

- Polynomials with _____ degree will always have either a global maximum or global minimum.

Solutions

The first four topics included multiple behaviors of functions: increasing, decreasing, concave up, and concave down, to name a few. These can be be expressed or represented multiple ways.

Behavior	Graphical	Verbal, in terms of ROC
Increasing		$f(b) > f(a)$ for $b > a$ OR positive rate of change
Decreasing		$f(b) < f(a)$ for $b > a$ OR negative rate of change
Concave up	increasing at incr. ROC / decreasing at incr. ROC	Rate of rate of change positive OR rate of change is increasing OR increasing/decreasing at an increasing rate
Concave down	decreasing at decr. ROC / increasing at decr. ROC	Rate of rate of change negative OR rate of change is decreasing OR increasing/decreasing at an decreasing rate

Rates of change behave specific ways for linear and quadratic functions.

- For a linear function, the average rates of change over successive equal-length intervals are *constant*.
- For a linear function, the rate of the average rates of change over successive equal-length intervals is *zero*.
- For a quadratic function, the average rates of change over successive equal-length intervals are *linear*.
- For a quadratic function, the average rates of change over successive equal-length intervals are changing at *a constant rate*.
- For a quadratic function, the rate of the average rate of changes over successive equal-length intervals is *constant*.

Polynomials of any degree, including linear and quadratic functions, also have characteristics that can be found by looking at rates of change.

- When the rate of change changes from positive to negative, a function has a *local maximum*.
- When the rate of change changes from negative to positive, a function has a *local minimum*.

Maxima and minima also abide by a few rules.

- Local extrema occur where rates of change change sign or *at endpoints of closed intervals.*.
- There will always be a local maximum or local minimum between any two *zeros* of a polynomial.
- Polynomials with *an even* degree will always have either a global maximum or global minimum.

Topics 1.5-1.6 ~ Zeros and End Behavior of Polynomials

Learning Objectives

1. (1.5.A) Identify key characteristics of a polynomial function related to its zeros when suitable factorizations are available or with technology.
2. (1.5.B) Determine if a polynomial function is even or odd.
3. (1.6.A) Describe end behaviors of polynomial functions.

Success Criteria

1. I can identify end behavior and express it analytically based on the degree and leading coefficient.
2. I can determine the degree of a polynomial using input-output pairs and successive differences.
3. I can determine whether a function is even or odd and use characteristics of such functions to answer questions.
4. I can explain why zeros correspond to particular factors and use this to sketch graphs of factored polynomial functions.

Today, you're going to investigate more characteristics of polynomials.

1. In your calculator, first go to `apps`, go to the bottom, and select `:Transfrm`.
2. Next, press `mode`, scroll down a few rows, and select `GRAPH-TABLE`.
3. To finish setup, press `y=` and input `AX`^{D} for `Y1`.
4. Press `2nd window` (`tblset`) and change `TblStart` to `-4`. Now, hit `graph`.
5. The polynomial graphed is $y = ax^d$, where a is the leading coefficient and d is the degree. As you'll see, there are a few characteristics of the graph that can be predicted from just a and d.

Note: to change the value of `D`, press the down arrow to highlight the `D=1` equation. Then, use the left/right arrows to decrease/increase `D`. The same process works for `A`.

6. To begin, start changing the value of `D` from 1 to 2 to 3 and so on.

    (a) What do you notice in the graph as `D` changes? Be as specific as possible.

    (b) What do you notice in the table as `D` changes? Be as specific as possible.

7. Go set `D = 3`. Now, change the value of `A`: try increasing it up to 5, then decreasing it down to −5. Describe what you see.

8. Make a little sketch of what you *think* the graph of $y = -2x^5$ would look like. **Do not input this into the calculator yet.**

9. Some output values of $y = -2x^5$ have been filled out below. Try filling in the table. Again, **do not input this into or use the calculator yet**.

x	−3	−2	−1	0	1	2	3
$y = -2x^5$	486	64			−2		

10. Now, check your graph and table. How'd you do?

11. Make a little sketch of what you *think* the graph of $y = 5x^4 - 3$ will look like. **Do not input this into the calculator yet.**

12. Some output values of $y = 5x^4 - 3$ have been filled out below. Try filling in the table. Again, **do not input this into or use the calculator yet**.

x	−3	−2	−1	0	1	2	3
$y = 5x^4 - 3$	402		2			77	

13. Now, go to Y1 and enter 5X^4-3 and check your graph and table. How'd you do?

14. Make a little sketch of what *think* the graph of $y = x^3 - 3x + 2$ will look like.

15. Again, a table is provided. Try filling it out (you know the drill).

x	−3	−2	−1	0	1	2	3
$y = x^3 - 3x + 2$	−16				0	4	

16. Now, go to Y1 and enter your function and check the graph and table. How'd you do?

There is something interesting about that last function: it hits the *x*-axis *multiple* times.

17. For what *x*-values does $f(x) = x^3 - 3x + 2$ hit the *x*-axis? What output does this correspond to?

18. It can be shown that $f(x) = x^3 - 3x + 2$ can be factored into three expressions. One of these expressions is $x - 1$, as shown below:
$$x^3 - 3x + 2 = (x-1)(\quad)(\quad)$$

   (a) Why does $x - 1$ *have* to be a factor?
   (b) What do you think the other factors are?
   (c) Input what you think the factored form is into Y2 and graph it. Repeat this process until you've got the correct factored form.
   (d) Can you explain how the factored form agrees with the graph? (Hint: Zoom in using zoom and 2:Zoom In around the *x*-intercepts.

19. Try sketching the graph of $y = \frac{1}{20}x(x-2)^3(2x+5)^2$ on the grid below. It will help to ignore the scale of the *y*-values, as they will get very large very quickly.

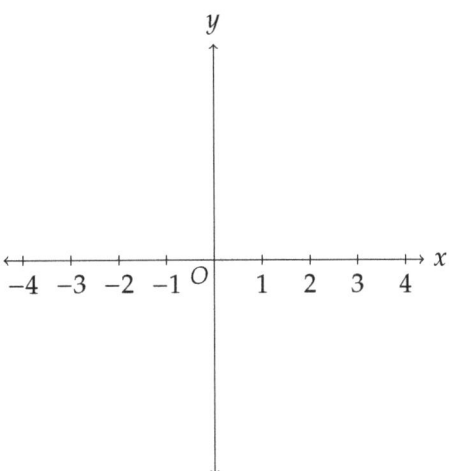

20. Check your graph. How'd you do?

We have one final property of polynomials that is rather fascinating. Recall the process of finding *first* and *second differences*.

21. For a linear function, what can you say about the first differences?

22. For a quadratic function, what can you say about the first differences? How about the second differences?

23. Shown below are some inputs and outputs for the function $f(x) = x^3 - 3x + 2$. Compute the remaining first, second, and *third* differences.

x	$f(x)$	1st Diff.	2nd. Diff	3rd Diff.
1	0	$4 - 0 = 4$	$16 - 4 = 12$	$18 - 12 = 6$
2	4	$20 - 4 = 16$	$34 - 16 = 18$	
3	20	$54 - 20 = 34$		
4	54			
5	112			N/A
6	200		N/A	N/A
7	324	N/A	N/A	N/A

24. Describe what you see. Does this make sense?

25. Complete the following statements.

    The degree of a linear function is ____, and the _____ differences are constant.

    The degree of a quadratic function is ____, and the _____ differences are constant.

    The degree of a cubic function is ____, and the _____ differences are constant.

26. Complete the following conjecture.

    For a polynomial of degree n, the _____.

Notes

Degree	Sign of Leading Coefficient	$\lim_{x \to \infty}$	$\lim_{x \to -\infty}$	Sketch
Even (= $2k$)	Positive			
Even (= $2k$)	Negative			
Odd (= $2k+1$)	Positive			
Odd (= $2k+1$)	Negative			

Topics 1.5-1.6 Homework

Make sure to turn the `Transfrm` app off (go to app, then `Transfrm`, then `2: Quit Transfrm Graphing`.

You are to log into AP Classroom and take **Unit 1 Progress Check: MCQ Part A** and **Unit 1 Progress Check: FRQ Part A**. You may use the space below for scratch work, and there is a box for you to record any particular questions you had.

Questions

Solutions/Notes

Activity

6. (a) As d oscillates between even and odd, the function oscillates from being symmetric over the y-axis and increasing without bound both as x increases and decrease without bound to being symmetric about $(0,0)$ and increasing/decreasing without bound as x increases/decreases without bound.

    (b) The output values are the same for opposite input values of x when d is even; the output values are opposites for opposite input values of x when d is odd. Additionally, the values get much larger much more quickly for larger d.

7. As a increases to 5, the polynomial gets thinner and output values increase more quickly. As a approaches 0, the outputs grow more slowly until the graph collapses into a line. Then, as a decreases towards -5, the graph reflects over the x-axis and starts to decrease more quickly.

8. Students should get the correct end behavior (increasing/decreasing without bound as x decreases/increases without bound) and the fact that the graph passes through the origin.

9. Because $y = -2x^5$ is odd, we can simply change output signs for inputs with opposite signs.

x	-3	-2	-1	0	1	2	3
$y = -2x^5$	486	64	2	0	-2	-64	-486

11. Students should get the correct end behavior (increasing/decreasing without bound as x decreases/increases without bound) and y-intercept (the origin).

12. We can use the fact that opposite inputs will have the same outputs.

x	-3	-2	-1	0	1	2	3
$y = 5x^4 - 3$	402	77	2	-3	2	77	402

14. Students may only get the end behavior (increases without bound both as x increases and decreases without bound) and y-intercept (the point $(0, -3)$): this is to be expected!

15. Students will have to make a decision as to whether they believe the function is even, odd, or neither. Some will anticipate it is odd due to the degree (it is neither). Correct values are below.

x	-3	-2	-1	0	1	2	3
$y = x^3 - 3x + 2$	-16	0	4	2	0	4	20

17. For $x = 1, -2$. Outputs are $f(1) = f(-2) = 0$.

18. (a) Since $f(1) = 0$, one of the factors must have a solution of 1.

    (b) The correct factorization is $(x-1)(x-1)(x+2)$.

    (d) Around $x = 1$, the graph looks like the graph of a quadratic! Around $x = -2$, it looks like the graph of a line.

19.

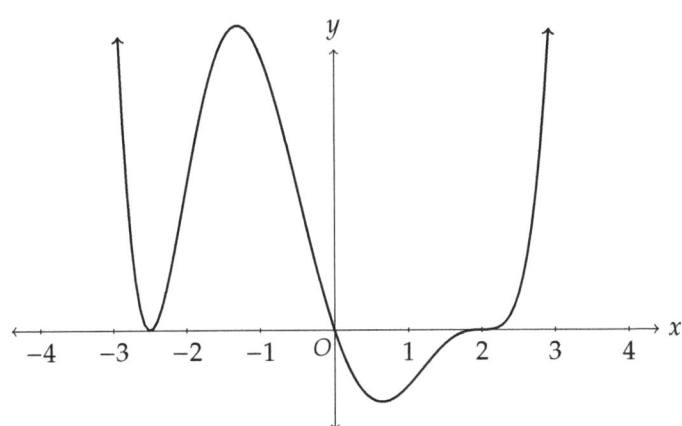

21. The first differences are constant.

22. The first differences are linear and the second differences are constant.

23.

x	$f(x)$	1st Diff.	2nd. Diff	3rd Diff.
1	0	$4 - 0 = 4$	$16 - 4 = 12$	$18 - 12 = 6$
2	4	$20 - 4 = 16$	$34 - 16 = 18$	6
3	20	$54 - 20 = 34$	24	6
4	54	58	30	6
5	112	88	36	N/A
6	200	124	N/A	N/A
7	324	N/A	N/A	N/A

24. This is a third-degree function, and its third differences are constant.

25. The degree of a linear function is 1, and the *first* differences are constant. The degree of a quadratic function is 2, and the *second* differences are constant. The degree of a cubic function is 3, and the *third* differences are constant.

26. For a polynomial of degree n, the *nth differences are constant*.

Notes

The END BEHAVIOR of a function is what happens to the output values as the input values *increase without bound* or *decrease without bound*. We can write end behavior using what are known as *limits*. For instance, suppose the following is true: "As x increases without bound, the output values of $f(x)$ get more and more negative, decreasing without bound." We would write this as $\lim_{x\to\infty} f(x) = -\infty$, which reads "the limit of $f(x)$ as x approaches is infinity is negative infinity."

End behavior of a polynomial is completely determined by the degree and leading coefficient.

Degree	Sign of Leading Coefficient	$\lim_{x\to\infty}$	$\lim_{x\to-\infty}$	Sketch
Even (= $2k$)	Positive	∞	∞	
Even (= $2k$)	Negative	$-\infty$	$-\infty$	
Odd (= $2k+1$)	Positive	∞	$-\infty$	
Odd (= $2k+1$)	Negative	$-\infty$	∞	

Certain functions have special symmetry. An EVEN FUNCTION, like $f(x) = x^2$ or $f(x) = x^4$, is graphically symmetric about the line $x = 0$ (the *y-axis*) and analytically satisfies $f(x) = f(-x)$ for all x. On the other hand, an ODD FUNCTION, like $f(x) = x^3$ or $f(x) = x$, is graphically symmetric about the *point* $(0,0)$ and analytically satisfies $f(-x) = -f(x)$.

Functions can have more than one term and be even or odd. For polynomials, it can be clearly stated:

- Even polynomials must have *only* terms with *even* exponents (including x^0, a constant).
- Odd polynomials must have *only* terms with *odd* exponents, and must therefore have only a constant term of 0.
- Because of the previous constraints, many polynomials are neither even nor odd.

The zeros of a polynomial are uniquely linked to *factors* of that polynomial. Therefore, if $x = a$ is a zero of $f(x)$, then $(x - a)$ is a factor of $f(x)$. If $(x - a)^k$ is a factor of f, then k is called the MULTIPLICITY of $x = a$. For even multiplicity, the sign of $f(x)$ will be the same for x-values near a; for odd multiplicity, the sign of $f(x)$ will be different for x-values less than/greater than $x = a$.

For any polynomial with degree n, the successive nth differences will be constant and the $(n-1)$th differences will be linear. On the flip side, the degree of a polynomial is the *least* value n for which the successive nth differences are constant.

Finally, though not formally addressed in the activity (for length), a function can have *complex* zeros of the form $a + bi$. When a polynomial has a root $a + bi$, it must also have the CONJUGATE $a - bi$ as a root. A useful way of remembering this is that *the number of complex zeros of a polynomial must be* even.

Reflection

Suggested Class Time. 80 minutes

Prerequisites. Students should have knowledge of $i = \sqrt{-1}$ and the fact that $i^2 = -1$. Additionally, it is helpful if students are already familiar with the conjugate $a - bi$ of a complex number $a + bi$.

Instructional Strategies.
When this lesson came up, I actually assigned it as homework and instead decided to provide direct instruction, starting with the Notes of the activity. Note that the homework requires use of the Transfrm app on the TI-84 - I had to provide two of my students with one to take home for the evening.

In my direct instruction, I started with a discussion of the characteristics of polynomials we had learned already: concavity, inflection points, maxima and minima, etc. We then transitioned immediately into end behavior, which students generally found very easy and largely remembered from the previous school year. The only really new thing was limit notation, which students found straightforward. What was also slightly new as well was the unique phrasing College Board uses: "increases/decreases *without bound*." Students didn't have trouble with this, but I tried to phrase as many of my questions like this as possible. At one point, students were saying something along the lines of "as x goes up, y goes up," and I challenged them to phrase it in terms of inputs and outputs as well as in terms of x and "the function," both times using "increases/decreases without bound."

After end behavior, I used Geogebra to graph two functions: $f(x) = x^4 - 3x^2 - 2$ and $g(x) = 2x^3 + x^2 - x + 4$. I asked students to guess which of these functions was *even*. Nearly all of the students chose the correct function f. We discussed what was "even" about it, and most students focused on the end behavior and the degree. I then graphed another function, $h(x) = x^4 - 3x^2 - x + 2$ that was not an even function. Pretty quickly, a student said "symmetrical." From there, we discussed the analytical expression of an even function, $f(x) = f(-x)$. What helped students was using colors to highlight "opposite inputs have the same output" for the aforementioned f. I asked students to point out any differences between the analytical representations of the even function f and the non-even function h, and students were quick to pick up on the fact that one had a term with an odd exponent. We then played around with Geogebra, showing how a single odd-termed exponent would make an even function no longer even.

Our discussion then translated into odd functions, with me asking students to try to, as a class, create a function that would be odd. Students were successful, giving me $y = x^7 + x^5 + 2x^3$. We then discussed what made the graph of this function special. Students were surprisingly more likely to identify "opposite inputs have opposite outputs" than the rotational symmetry. We discussed this representation of an odd function and tied it all back to *why* even and odd polynomials behave the way they do: as a feature of $(-1)^n$, or, in essence, how a negative to an even power is positive, while a negative to an odd power is negative.

As we started to talk about zeros, I drew circles at three x-intercepts: $x = -4, x = \frac{1}{2}$, and $x = 2$. I asked if we could *create* a polynomial that would have these zeros. Students found this concept surprisingly difficult: we had to discuss what it *means* to be a zero in terms of inputs and outputs. Soon enough, students were able to get a factor of $x + 4$ for the zero $x = -4$. I then asked students to generate a factor for $x = 2$ and students quickly got $x - 2$: for $x = \frac{1}{2}$, there was a hesitation, but one student finally said $x - \frac{1}{2}$. In discussing this last one, we discussed whether this was correct and how we could modify it if we desired for all coefficients of our polynomial to be integers. This led to doubling the factor into $2x - 1$.

For determining factors that would produce certain zeros, I taught the idea of "unsolving" an equation. Typically, students take linear factors and solve for x: something like

$$2x - 1 = 0$$
$$2x = 1$$
$$x = \frac{1}{2}$$

I then showed how, to identify a linear factor that would *have* a certain zero, we could simply reverse this process.

$$x = \frac{1}{2}$$
$$2x = 1$$
$$2x - 1 = 0$$

Students weren't 100% sold, but we looked at a couple more, and I told them we'd come back momentarily. We then wanted to check our function, so we graphed $p(x) = (x+4)(x-2)(2x-1)$ and it passed perfectly through each of our drawn x-intercepts.

Next up was *complex* zeros. I asked students what the degree of our polynomial p was. Students said 3, and then I asked how many zeros our polynomial had; again, they said 3. I asked how many zeros a 4th degree polynomial should have, and students quickly said 4. When I asked again, "How many *x-intercepts* should a 4th degree polynomial have?," students again said 4. I then graphed a modified version of $p(x)$: $y = (x+4)^2(x-2)(2x-1)$. We discussed that the degree was 4, but that the graph had only three zeros: where had the fourth one gone? This led to a discussion of multiplicity, and we played around with the graph of $y = (x+4)^a(x-2)^b(2x-1)^c$ by using sliders in Geogebra and changing the values of $a, b,$ and c. While doing so, we discussed Essential Knowledge 1.5.A.5 and, more generally, how a polynomial at a given zero x behaves like $x^{\pm m}$, where m is the multiplicity of that zero.

We finally discussed complex zeros. I used Geogebra to graph $y = x^2 + 4$, which obviously had no x-intercepts. We analytically solved $x^2 + 4 = 0$ and got $x = 2i, -2i$. My students in particular were quite rusty on the definition and properties of i and had never heard of complex numbers before, so we had to take a few minutes to go back through real numbers, imaginary numbers, and complex numbers. From here, we then used the "unsolving" process to try and create a polynomial with a given complex zero. We set $x = 2 + 3I$ and "unsolved":

$$x = 2 + 3i$$
$$x - 2 = 3i$$

For the next step, we discussed how simply subtracting the $3i$ over would not do, as this would result in a polynomial with a nonreal coefficient. This led to the idea of where the i "came from," which was taking the square root of a negative. To undo this, we squared both sides.

$$(x-2)^2 = (3i)^2$$
$$x^2 - 4x + 4 = -9$$
$$x^2 - 4x + 13 = 0$$

"But wait!," I said. Shouldn't a quadratic have *two* zeros? In my faux surprise, we discussed the fact that somehow, we "lost" a zero in the solving process. We looked back through the equation, trying to determine where there could have been an omission, and one student eventually pointed out that square roots can be positive or negative. Finally, we arrived at the other solution, $2 - 3i$. I then defined the conjugate $a - bi$ of a complex number $a + bi$ and gave the fact that the number of complex zeros must always be even.

We ended class by going over the Topic Questions together. Time was running out to teach Essential Knowledge 1.5.A.6, so we worked the first Topic Question. Students were very quick to understand that EK.

Technology. I used Geogebra for the entirety of the lesson.

Problems.

1. The polynomial function f has zeros including $x = 1, x = -2, x = 3i$, and $x = 4 - 2i$. Which of the following could be the degree of f?

    (A) 3   **(B)** 4   **(C)** 5   **(D)** 6

2. The function f is given by $f(x) = (x - 1)^2(3x + 4)$. For what x is $f(x) > 0$?

    (A) $-\frac{4}{3} < x < 1$ only   **(B)** $x > 1$ only   **(C)** $x < -\frac{4}{3}$ and $x > 1$ **(D)** $-\frac{4}{3} < x < 1$ and $x > 1$

3. The function $y = h(x)$ has a local minimum at $(3, -4)$. If $h(x)$ is an even function, which of the following is true?

    (A) h has a local minimum at $(-3, -4)$.

    (B) h has a local maximum at $(-3, -4)$.

    (C) h has a local minimum at $(-3, 4)$.

    (D) h has a local maximum at $(-3, 4)$.

4. If the function $f(x) = x^p(x^2 + k)(3 - x)(4 - 5x)$ has a zero of $x = -2i$ and a degree of 6, which of the following is true?

    (A) $p = 2$ and $k = -4$ **(B)** $p = 4$ and $k = -4$ **(C)** $p = 2$ and $k = 4$ **(D)** $p = 4$ and $k = 4$

5. The polynomial function $f(x)$ has only two zeros, both of which are real: $x = 2$ with multiplicity 4, and $x = 1$. If the leading coefficient of f is negative, which of the following is true?

    (A) As x increases without bound, the output values of $f(x)$ increase without bound.

    (B) As x increases without bound, the output values of $f(x)$ decrease without bound.

    (C) f is an even function.

    (D) f is an odd function.

6. Which of the following correctly give the end behavior for $f(x) = (3x + 1)(x - 2)(3 - x)^2$?

    (A) $\lim\limits_{x \to \infty} f(x) = \infty$ and $\lim\limits_{x \to -\infty} f(x) = \infty$

    (B) $\lim\limits_{x \to \infty} f(x) = \infty$ and $\lim\limits_{x \to -\infty} f(x) = -\infty$

    (C) $\lim\limits_{x \to \infty} f(x) = -\infty$ and $\lim\limits_{x \to -\infty} f(x) = \infty$

    (D) $\lim\limits_{x \to \infty} f(x) = -\infty$ and $\lim\limits_{x \to -\infty} f(x) = -\infty$

Solutions.

1. Since complex zeros appear in conjugate pairs, there are also roots of $x = -3i$ and $x = 4 + 2i$. So there are six zeros which implies the degree is at least 6. The correct answer is therefore **(D)**.

2. The zeros are $x = -4/3$ and $x = 1$. The multiplicity of the zero at $x = 1$ is 2, and the function is a positive cubic. Hence, the function is positive for $-4/3 < x < 1$ and $x > 1$. The correct answer is therefore **(D)**.

3. Since it is even, the graph has symmetry around the y-axis. Hence, there is also a minimum at $(-3, -4)$. The correct answer is therefore **(A)**.

4. We "unsolve" $x = -2i$:

$$x = -2i$$
$$x^2 = (-2)^2 i^2$$
$$x^2 = -4$$
$$x^2 + 4 = 0.$$

We find that $k = 4$. Since the degree is 6, we have $p + 2 + 1 + 1 = 6$ so that $p = 2$. The correct answer is therefore **(C)**.

5. The degree of the polynomial is $4 + 1 = 5$, which is odd. The leading coefficient being negative implies that the end behavior is that the output values of $f(x)$ increases/decreases without bound as x decreases/increases without bound. The correct answer is therefore **(B)**.

6. The degree is $2 + 1 + 1 = 4$ and the leading coefficient is positive. Hence, the function increases without bound both as x increases and decreases without bound. The correct answer is therefore **(A)**.

Calculator Skills (🧮)

The AP Precalculus exam will ask you to use a graphing calculator in a number of ways. We'll address how to do some of these things here. Note that, on the AP exam, all rounding should be done *only* at the final step and should be to at least *three* decimal places.

Skills

1. Perform calculations (e.g., exponents, roots, trigonometric values, logarithms)

    (a) *Typing fractions.* Press alpha then X,t,θ,n

    (b) *Typing a function.* To type a function like Y1 or Y2 on the home screen, press alpha then trace

    (c) *Evaluating a function.* On home screen, you can simply use Y1, Y2, etc.

2. Graph functions and analyze graphs

    (a) *Changing the window.* Press window and adjust Xmin, Xmax, Ymin, and Ymax.

    (b) *Return to standard window.* The standard window is with $-10 \leq x \leq 10$ and $-10 \leq y \leq 10$. To get that, press zoom then 6:ZStandard

    (c) *Hide a graph without deleting it.* Press y=, then go to the equal sign of the function you want to hide the graph of. Press enter so that the equal sign is not highlighted.

    (d) *Evaluating the function on a graph.* Press 2nd, then trace (calc), then 1:value. Enter the x-value and hit enter.

    (e) *Finding real zeros.* Graph the function. Then, press 2nd, then trace, then 2:zero. Use the arrows to go to the *left* of the zero and press enter. Now, use the arrows to go to the *right* of the zero and press enter. Press enter a final time for Guess?.

    (f) *Finding minima/maxima.* Graph the function. Then, press 2nd, then trace, then 3:maximum or 4:minimum. Use the arrows again to trace to the left, press enter, trace to the right, press enter, and press enter a final time.

    (g) *Finding points of intersection of graphs of functions.* Graph the two functions. Press 2nd, then trace, then 5:intersect. A cursor will appear on one function's graph. Trace near the intersection point of interest and press enter. The cursor will appear on the other function's graph. Trace near the intersection point of interest again and press enter. Finally, press enter for Guess?.

    (h) *Storing value computed from graph.* Immediately after computing a value/intersection/minimum/maximum, press 2nd mode (quit) to go to the home screen. Press 2nd then (-) (Ans). The value stored in Ans is the x-value of whatever was just computed in the graph screen. To store this as a new letter, press sto-> and then select any letter of your choosing (and press enter).

    (i) *Find numerical solutions to equations in one variable.* To solve an equation of the form $f(x) = k$ for some constant k, you can input $f(x)$ into Y1 and the constant k into Y2. Then, simply find the intersection/s of the graph of $f(x)$ and the line $y = k$.

3. Generate a table of values for a function

    (a) *Generate the table.* Press 2nd, then graph (table).

    (b) *Change table settings.* Press 2nd, then window (tblset). You can change the starting value (TblStart) or the increment of change of input values (ΔTbl).

Example. Let $f(x) = \frac{1}{3}x^2 + 2x - 5\ln(x)$ and $g(x) = \dfrac{x^3 + x^2 + 30}{x^2 + 2x + 6}$.

(a) Find the global minimum of $f(x)$.

(b) Find the x-value where the graph of g goes from increasing to decreasing.

(c) Find all extrema of $g(x)$ for $1 \leq x \leq 6$.

(d) For what value/s of x does $f(x) = g(x)$?

(e) Solve $x^3 - 4x^2 - 8x + 4 = 35$.

(f) Find the positive x-intercepts of $h(x) = x^3 - 4x^2 - 8x + 4$.

(g) If $h(x) = 0$ for $x < 0$, compute $h(x + 10)$.

Solutions

(a) $y = 1.703$

(b) $x = -0.937$

(c) Local max of $y = \frac{32}{9}$ at $x = 1$; local/global min of $y = 2.985$ at $x = 2.220$; local/global max of $y = \frac{47}{9}$ at $x = 6$

(d) $x = 0.564$ and $x = 2.806$

(e) $x = 6.13$

(f) $x = 0.421$ and $x = 5.355$

(g) $x = -1.775$, store it, then $h(x + 10) = 223.981$

Topic 1.7 ~ End Behaviors of Rational Functions

Learning Objectives

1. (1.7.A) Describe end behaviors of rational functions.

Success Criteria

1. I can explain how the end behavior of a rational function can be understood by examining the corresponding quotient of the leading terms.
2. I can determine whether a rational function has a horizontal or slant asymptote and compute such asymptotes.
3. I can express the end behavior of rational functions using limits.

In the last activity, you investigated the end behavior of polynomials. Now, you're going to investigate the end behavior of *rational functions*. Recall that a RATIONAL FUNCTION is analytically represented as a quotient of two polynomial functions. More formally, a rational function r has the form $r(x) = \frac{p(x)}{q(x)}$, where p and q are polynomial functions.

Now, suppose a cattle rancher is going to install a fence around a square, x feet by x feet field. The fence costs $50 per foot, and the company installing the fence requires a $2,000 deposit for labor.

1. If p is the function for the total cost of fence for an x feet by x feet field, write an analytical expression for $p(x)$.

2. If q is the function for the area of the field, write an analytical expression for $q(x)$.

3. Let $r(x) = \dfrac{p(x)}{q(x)}$. Use your functions from #1 and #2 to write an analytical expression for $r(x)$.

4. In your own words, what exactly does $r(x)$ measure?

5. As the size of the fence grows, what do you think will happen to the size of $r(x)$? Explain.

6. In your graphing calculator, enter $r(x)$ into Y1. (Note: You can input a fraction by pressing alph then X,T,θ,n.) Then, press zoom 6:ZStandard.

7. Now, you shouldn't even be able to see $r(x)$ for any $x > 0$. To zoom out, press zoom, then 3:Zoom Out. Your cursor will be at the origin – press enter.

8. What appears to happen to $r(x)$ as the linear dimension x of the field increases?

9. Predict the value of $\lim\limits_{x \to \infty} r(x)$.

10. Press mode, go down a few rows, and change to GRAPH-TABLE. Hit graph.

11. Press trace and then hold down the right arrow key. Is what you see consistent with what you wrote in #9?

To see why this is happening, it's helpful to discuss the idea of *dominance*. Let's evaluate the numerator and denominator polynomials separately.

12. The polynomial $p(x) = 200x + 2000$ has two terms, $200x$ and 2000. Fill in the table below with the value of each term for the various values of x.

x	1	5	10	25	50
Value of term $200x$					
Value of term 2000					

13. As x increases, we say that one of these terms *dominates*. Which term do you think it is?

14. When one term of a polynomial *dominates*, we can basically *ignore* all other terms for large enough x and only consider the one that dominates. If we looked only for large x and considered only the dominant term, what would be a new expression for $p(x)$?

15. Now, let's compare the new $p(x)$ with the function $q(x)$. Fill in the table below with the values of $p(x)$ and $q(x)$. Notice how we're looking at relatively large values of x.

x	100	300	500	700	900
$p(x)$					
$q(x)$					

16. What happens to the sizes of $p(x)$ and $q(x)$ as x increases? What is your best explanation for why this is happening?

17. We stated that the function $p(x)$ could be rewritten as just $p(x) = 200x$ for large x. Using this, write a *simplified* form of $r(x)$.

18. Input this new simplified $r(x)$ into Y2 and hit graph. What do you see?

In this case, the line $y = 0$, which is precisely $\lim\limits_{x \to \infty} r(x) = 0$, is called a *horizontal asymptote*.

Now, let's see if you can apply this without a contextual situation.

19. Consider the function $f(x) = \dfrac{14x^2 + 25x + 4}{2x^2 - 15x - 3}$.

    (a) Which term will dominate in the numerator?
    (b) Which term will dominate in the denominator?
    (c) Use (a) and (b) to create a simplified expression for $f(x)$ for large x.
    (d) What will happen to $f(x)$ as x increases or decreases without bound?
    (e) Evaluate $\lim\limits_{x \to \infty} f(x)$.

20. Consider the function $g(x) = \dfrac{3x^3 + 9x^2 - 4x - 2}{x^2 + x - 1}$.

   (a) Which term will dominate in the numerator?
   (b) Which term will dominate in the denominator?
   (c) Use (a) and (b) to create a simplified expression for $f(x)$ for large x.
   (d) What will happen to $f(x)$ as x increases without bound?
   (e) What will happen to $f(x)$ as x decreases without bound?
   (f) Evaluate $\lim\limits_{x \to \infty} f(x)$.
   (g) If you were to graph g and zoom out, what would you see?
   (h) We have a name for the type of asymptote identified in (c). What name would you give it?

Notes

Topic 1.7 Homework

1. Determine whether each function below has a horizontal asymptote. If it does, identify the equation of this asymptote.

   (a) $f(x) = \dfrac{9x^3 + 2x - 1}{4x^3 + 8x - 49}$

   (b) $g(x) = \dfrac{3x^2(x-1)^2}{5x^3 + 2x^2 - 4x + 8}$

   (c) $h(x) = \dfrac{(4x-1)(2x+3)(x+4)}{x(8x-1)^2}$

   (d) $f(x) = \dfrac{x^3 - 4x^2 + 5x - 2}{(x^2+3)^2}$

   (e) $g(x) = \dfrac{25 - 9x^2}{x(5-x)}$

2. The function $f(x)$ is a fifth degree polynomial with a leading coefficient of 4, and the function $g(x)$ is a third degree polynomial with a leading coefficient of -2. Complete the following sentences about the function $h(x) = \dfrac{f(x)}{g(x)}$.

   (a) As x increases without bound, $h(x)$ _____.

   (b) As x decreases without bound, $h(x)$ _____.

   (c) Rewrite each of the sentences in (a) and (b) as limits.

3. The function $f(x)$ has a slant asymptote parallel to the line passing through $(2, 5)$ and $(-7, 23)$. If
$$f(x) = \dfrac{10x^2(x-2)(x-1)}{ax^b + cx + d}$$
   for nonzero integer values of $a, b, c,$ and d, what are a and b?

4. Construct a function that has a horizontal asymptote of $y = -4$, a vertical asymptote of $x = 2$, and a zero at $x = 5$.

5. The function
$$f(x) = \dfrac{ax^2 + b}{(x-c)(x-d)},$$
   where $c < d$, has been graphed on the following page. Identify the values of $a, b, c,$ and d.

TOPIC 1.7 ~ END BEHAVIORS OF RATIONAL FUNCTIONS

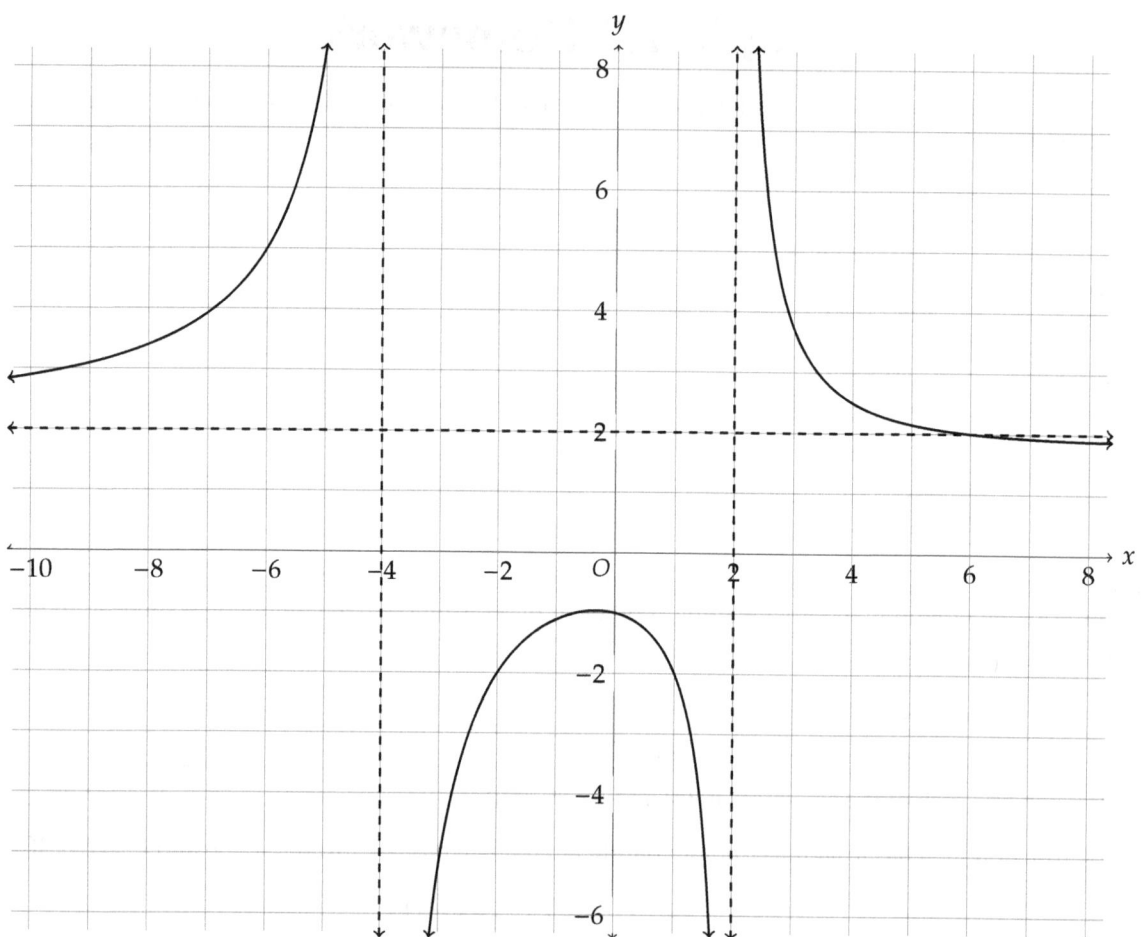

Topic 1.7 Solutions/Notes

Activity

1. $p(x) = 4(50x) + 2000 = 200x + 2000$

2. $q(x) = x^2$

3. $r(x) = \dfrac{200x + 2000}{x^2}$

4. $r(x)$ measures the cost of fence per square foot of the field.

5. Student answers will vary.

8. $r(x)$ decreases.

11. $r(x)$ approaches 0.

12.
x	1	5	10	25	50
Value of term $200x$	200	1000	2000	5000	10,000
Value of term 2000	2000	2000	2000	2000	2000

13. The $200x$ term

14. $p(x) = 200x$

15.
x	100	300	500	700	900
$p(x)$	20000	60000	100000	140000	180000
$q(x)$	10000	90000	250000	490000	810000

16. The size of $q(x)$ starts to drastically outweigh the size of $p(x)$.

17. $r(x) = \dfrac{200x}{x^2} = \dfrac{200}{x}$

18. The graphs look identical for large enough x.

19. (a) $14x^2$

    (b) $2x^2$

    (c) $f(x) = \dfrac{14x^2}{2x^2} = 7$

    (d) The value of $f(x)$ will approach 7.

    (e) 7

20. (a) $3x^3$

    (b) x^2

    (c) $g(x) = \dfrac{3x^3}{x^2} = 3x$

    (d) The function will resemble $3x$ and approach ∞.

    (e) The function will resemble $3x$ and approach $-\infty$.

    (f) ∞

    (g) The graph would resemble the line $3x$.

Notes

In a polynomial with degree n, the Ax^n term is said to *dominate*.

In a rational function $r(x) = \frac{p(x)}{q(x)}$, the end behavior can be determined by looking at the quotient of the *leading terms*, assuming the polynomials are written in standard form.

Certain combinations of degrees of numerators and denominators have special asymptotes.

- If the degree of p and q are equal (say n), then for x with large magnitude, the rational function $r(x) = \frac{p(x)}{q(x)}$ can be approximated by

$$r(x) = \frac{p_a x^n}{q_a x^n} = \frac{p_a}{q_a},$$

a *constant function*. This is called a HORIZONTAL ASYMPTOTE.

- If the degree of q (call it n) is greater than the degree of p (call it m), then for x with large magnitude, the rational function $r(x) = \frac{p(x)}{q(x)}$ will behave like

$$r(x) = \frac{p_a x^m}{q_a x^n} = \frac{p_a}{q_a x^{n-m}},$$

which will approach 0. This is also a horizontal asymptote.

- If the degree of p is precisely 1 *greater than* the degree of q, then for x with large magnitude, the rational function $r(x) = \frac{p(x)}{q(x)}$ will behave like a *linear function*. This is called a SLANT or OBLIQUE ASYMPTOTE. The slant asymptote will be *parallel* to the ratio of the leading terms.

Recall that two or more rational functions can be added/subtracted by finding a *common denominator*. It is often helpful to do this addition/subtraction *before* looking at degrees and leading terms.

Homework

1. (a) There is a horizontal asymptote of $y = \frac{9}{4}$.
   (b) The degree of the numerator (4) is greater than the degree of the denominator, so there is not a horizontal asymptote.
   (c) The numerator is a third degree polynomial with leading coefficient 8 and the denominator is a third degree polynomial with leading coefficient 64, so there is a horizontal asymptote of $y = \frac{8}{64} = \frac{1}{8}$.
   (d) The degree of the denominator (4) is greater than the degree of the numerator, so there is a horizontal asymptote of $y = 0$.
   (e) The degrees of both the numerator and denominator are 2, so g has a horizontal asymptote of $y = \frac{-9}{-1} = 9$.

2. The end behavior of $h(x)$ will resemble that of a quadratic with leading coefficient -2.

   (a) As x increases without bound, $h(x)$ decreases without bound.
   (b) As x decreases without bound, $h(x)$ decreases without bound.
   (c) $\lim\limits_{x \to \infty} h(x) = -\infty$, $\lim\limits_{x \to -\infty} h(x) = -\infty$

3. The function will only have a slant asymptote if the degree of the numerator (4) is one greater than the degree of the denominator, so $b = 3$. The slope of the line is $(23-5)/(-7-2) = -2$. Therefore, the ratio of the leading terms of the numerator and denominator of f must be $-2x$. This leads to $\frac{10x^4}{ax^3} = -2x$, or $a = -5$.

4. Many options are possible, but the simplest is to first generate a vertical asymptote with a $x - 2$ in the denominator. Then, to generate the horizontal asymptote, the numerator can be a linear polynomial with a leading coefficient of -4. Finally, to get a zero of $x = 5$, the numerator must equal 0 at the input of 5: the linear function $-4(x-5)$ achieves this. The result is $f(x) = \frac{-4(x-5)}{x-2}$.

5. The vertical asymptotes occur at $x = -4$ and $x = 2$, which must be zeros of the denominator: this leads to the factors $(x+4)$ and $(x-2)$, so $c = -4$ and $d = 2$ (since $c < d$). The horizontal asymptote occurs at $y = 2$, so $\frac{ax^2}{x^2} = 2$ gives $a = 2$. Finally, the y-intercept of $(0, -1)$ can be used to determine b:

$$\frac{2(0)^2 + b}{(0+4)(0-2)} = -1$$

$$\frac{b}{-8} = -1$$

$$b = 8.$$

Reflection

Suggested Class Time. 55 minutes

Prerequisites. Students need to be comfortable writing linear and quadratic expressions related to perimeter and area.

Instructional Strategies. One thing that became apparent immediately: students struggled to write accurate expressions in #1 and #2. In #1, many students only wrote $p(x) = 2000 + 50x$, not considering the four sides of the square. Some also thought it was $50 per square foot, writing $p(x) = 50x^2$. In #2, many students did not correctly compute area, but instead computed perimeter, giving $q(x) = 4x$. As I observed these mistakes repeatedly occurring as I walked around the room, I decided to uncharacteristically stop everyone after a few minutes and make sure they all had the correct function. This was necessary, as the activity relies on these functions being correct.

Students did not struggle with the vast majority of the rest of the activity. I did notice some students were writing limits as full sentences. For instance, when asked $\lim_{x \to \infty} r(x)$ in #9, many students responded with something like "as $x \to \infty$, $r(x) \to 0$." We had to discuss the fact that this was an *interpretation* of the limit, not its actual value. In essence, students were "overthinking it." It may also have simply been a misunderstanding of what the notation means.

When it came to the Notes section, I used the traditional Algebra I/Algebra II concept of "top heavy" and "bottom heavy." Many students remembered this from a previous course, so it accessed their prior knowledge in a useful way.

To really highlight what an asymptote is, I graphed the function from #19 in Geogebra. I told them that an asymptote is not simply "a line that the function can't touch" - on the contrary, many functions do indeed intersect their asymptotes. Rather, I taught the students that an asymptote is what the function resembles if you only consider extreme (very large or very negative) inputs: this is tantamount to *zooming out*. Placing my cursor at the origin, I zoomed out a lot from the graph, and after a few seconds, the graph completely collapsed and became just a single line. This was precisely the line $y = 7$, and there were audible "ooh's" and "aah's"!

One tip in explaining the concept of dominance to students: talk about money! I told students that it would be a *big* deal if I got $2,000 (the number from the activity), but then I asked them if Jeff Bezos would care about getting $2,000. The students unanimously said no because he already has so much. In this way, students instantly got the idea of one seemingly large value being negligible in the presence of *much* larger values.

Some students did mistakenly associate dominance with the *coefficient*, though. One student in particular thought that $25x$ would dominate over $14x^2$ in #19. We turned this into a teachable moment and showed the entire class the graph of both functions. For (very) small x, $25x$ was greater than $14x^2$, but the $14x^2$ quickly outpaced the $25x$. To hammer this point home, I then modified $25x$ into $100,000x$. In Geogebra, this appeared to be essentially a vertical line, so I changed the scale of the y-axis (as a ratio to the x-axis) to be $1 : 20$. Then, students got to see how, again, the $14x^2$ would eventually grow at such a fast rate that it would dominate the $100,000x$. This drove home that domination, when it comes to power terms, is all about the exponent.

Technology. Students will use the TI-84 for the activity. For teaching, it helps to use a dynamic geometry software/website like Geogebra or Desmos to best highlight what exactly an asymptote means and to analyze graphs.

Problems.

1. Let the function f be given by $f(x) = \dfrac{2x(x-1)(x-2)^2}{3x^3 + 3x^2 - 18x}$. This function has

    (A) a horizontal asymptote of $y = 0$.

    (B) a horizontal asymptote of $y = \frac{2}{3}$.

    (C) a slant asymptote with a slope of $\frac{2}{3}$.

    (D) neither a horizontal nor a slant asymptote.

2. The function f is given by $f(x) = \dfrac{9x^2 - 4x + 10}{3x^2 + 8x + 2}$. What happens as x increases without bound?

   (A) The value of $f(x)$ approaches 3.
   (B) The value of $f(x)$ approaches 9.
   (C) The value of $f(x)$ increases without bound.
   (D) The value of $f(x)$ decreases without bound.

3. The polynomial function f has zeros of $x = 4$ with multiplicity 2 and $x = -3$. The function g has a degree of 4 and a negative leading coefficient. Which of the following is true about the function $h(x) = \dfrac{f(x)}{g(x)}$?

   (A) As x increases without bound, $h(x)$ approaches 0.
   (B) As x increases without bound, $h(x)$ decreases without bound.
   (C) As x increases without bound, $h(x)$ increases without bound.
   (D) End behavior cannot be determined without knowing the sign of the leading coefficient of $f(x)$.

4. Which of the following functions has the same end behavior as the function f given by $f(x) = \dfrac{2x^3 + 17}{x^3 - 3x^2 + 1}$?

   (A) $g(x) = \dfrac{2}{x}$ (B) $h(x) = \dfrac{x}{2}$ (C) $j(x) = 2$ (D) $k(x) = \dfrac{1}{2}$

5. The rational function f is given by $f(x) = \dfrac{g(x)}{h(x)}$ where g and h are polynomial functions. The degree of g is less than the degree of h. Which of the following could be true?

   (A) f has only a slant asymptote.
   (B) f has only a horizontal asymptote.
   (C) f has both a slant asymptote and a horizontal asymptote.
   (D) f has neither a slant asymptote nor a horizontal asymptote.

Solutions.

1. The numerator is a fourth degree polynomial with a leading coefficient of 2, and the denominator is a third degree polynomial with a leading coefficient of 3. The function will behave like $\dfrac{2x^4}{3x^3} = \dfrac{2}{3}x$. Thus, $f(x)$ has a slant asymptote with slope 2/3. The correct answer is therefore **(C)**.

2. The numerator and denominator are each polynomials of degree 2, so the function will behave like $\dfrac{9x^2}{3x^2} = \dfrac{9}{3} = 3$. Thus the values of $f(x)$ approach 3. The correct answer is therefore **(A)**.

3. The numerator of h is a polynomial of degree $2 + 1 = 3$ and the denominator is a polynomial of degree 4. Thus, $h(x)$ approaches 0 as x increases without bound. The correct answer is therefore **(A)**.

4. The numerator and denominator are each polynomials of degree 3, so the function will behave like $\dfrac{2x^3}{x^3} = \dfrac{2}{1} = 2$. Thus the values of $f(x)$ approach 2, and the function $j(x) = 2$ has the same end behavior. The correct answer is therefore **(C)**.

5. The polynomial in the numerator has degree smaller than the polynomial in the denominator. Thus, $f(x)$ approaches 0 as x increases without bound. Hence, f has a horizontal asymptote. The correct answer is therefore **(B)**.

Topics 1.8-1.10 ~ Zeros, Vertical Asymptotes, and Holes (Day 1)

Learning Objectives
1. (1.8.A) Determine the zeros of rational functions.
2. (1.9.A) Determine vertical asymptotes of graphs of rational functions.
3. (1.10.A) Determine holes in graphs of rational functions.

Success Criteria
1. I can find zeros, vertical asymptotes, and holes of a rational function.
2. I can explain the difference between a zero, vertical asymptote, and hole.
3. I can express the behavior of a function near a hole or vertical asymptote using limit notation.

This activity is all about the number *zero*.

For a rational function $r(x) = p(x)/q(x)$, because of the separate numerator and denominator polynomials, there are three different cases in which one or more of the polynomials could output 0. They are:

- $p(x) = 0$, but $q(x) \neq 0$
- $q(x) = 0$, but $p(x) \neq 0$
- $p(x) = 0$ and $q(x) = 0$

You're going to investigate each of these, all starting with the function $f(x) = \dfrac{p(x)}{q(x)}$ given by

$$f(x) = \frac{(x+4)(x-1)(x+3)}{2(x-3)(x+1)(x-1)}$$

For starters, do the following.

1. Enter the *numerator* function $p(x)$ into Y1 in your calculator.
2. Enter the *denominator* function $q(x)$ into Y2 in your calculator.
3. In Y3, enter Y1/Y2 (you can access these by pressing `alpha trace`).

Zeros of the Numerator

4. Find a value of x, call it x_1, for which $p(x_1) = 0$ but $q(x_1) \neq 0$.
5. Press `2nd window` (tblset) and make sure TblStart=0 and ΔTbl=0.05.
6. Press `2nd graph` (table) to open the table.

The second through fourth columns represent $p(x)$, $q(x)$, and $\dfrac{p(x)}{q(x)}$, respectively.

7. Scroll slowly to make x approach the value of x_1 you got in #4.

    (a) What does $p(x)$ approach as $x \to x_1$?

UNIT 1   TOPICS 1.8-1.10 ~ ZEROS, VERTICAL ASYMPTOTES, AND HOLES (DAY 1)

    (b) What does $q(x)$ approach as $x \to x_1$?

    (c) What does $\dfrac{p(x)}{q(x)}$ approach as $x \to x_1$?

    (d) Explain why (c) makes sense in light of (a) and (b).

8. Complete the statement:

    When the numerator of a rational function has a zero at a given input and the denominator does not, the rational function has a _____ at that input.

Zeros of the Denominator

9. Find a value of x, call it x_2, for which $q(x_2) = 0$ but $p(x_2) \neq 0$.

10. Go back and change `TblStart=1` and Δ`Tbl=0.01`. Press `2nd graph` (table).

11. Scroll slowly to make x approach the value of x_2 you got in #9.

    (a) What does $p(x)$ approach as $x \to x_2$ from each side of x_2?

    (b) What does $q(x)$ approach as $x \to x_2$ from each side of x_2?

    (c) What does $\dfrac{p(x)}{q(x)}$ approach as $x \to x_2$ from each side?

    (d) Can you explain why $\dfrac{p(x)}{q(x)}$ behaves this way near x_2?

    (e) What does the calculator give for $\dfrac{p(x)}{q(x)}$ precisely at $x = x_2$?

12. Press `y=`, then go to the `=` on `Y1`. Press `enter` to un-highlight it - this will turn the graph of `Y1` off *without* having to delete the function. Repeat this for `Y2`.

13. Now, press `zoom`, then `6:ZStandard` to graph *just* `Y3`, which is $\dfrac{p(x)}{q(x)}$. What do you see occurring at the value of x_2?

14. Try to complete the following statement:

    When the denominator of a rational function has a zero at a given input and the numerator does not, the rational function has a _____ at that input.

Zeros of Both the Numerator and the Denominator

15. Find a value of x, call it x_3, for which *both* $p(x_3) = 0$ but $q(x_3) = 0$.

16. Go back and re-highlight each of the equal signs in `Y1` and `Y2` (otherwise, they won't show up in the table).

17. Go back and change `TblStart=0`. Press `2nd graph` (table).

18. Scroll slowly to make x approach the value of x_3 you got in #15.

    (a) What does $p(x)$ approach as $x \to x_3$ from each side of x_3?

    (b) What does $q(x)$ approach as $x \to x_3$ from each side of x_3?

    (c) What does $\dfrac{p(x)}{q(x)}$ approach as $x \to x_3$ from each side?

    (d) Can you explain why $\dfrac{p(x)}{q(x)}$ behaves this way near x_3?

19. Repeat the process for graphing just `Y3`. Use `trace` to try to determine what occurs on the graph at x_3. Can you see it?

Notes

Topics 1.8-1.10 (Day 1) Homework

1. Determine the value of each limit, if it exists, for f graphed below.

   (a) $\lim_{x \to -3} f(x)$

   (b) $\lim_{x \to 1^-} f(x)$

   (c) $\lim_{x \to 1^+} f(x)$

   (d) $\lim_{x \to 1} f(x)$

   (e) $\lim_{x \to -1^-} f(x)$

   (f) $\lim_{x \to -1^+} f(x)$

   (g) $\lim_{x \to -1} f(x)$

   (h) $\lim_{x \to \infty} f(x)$

   (i) $\lim_{x \to -\infty} f(x)$

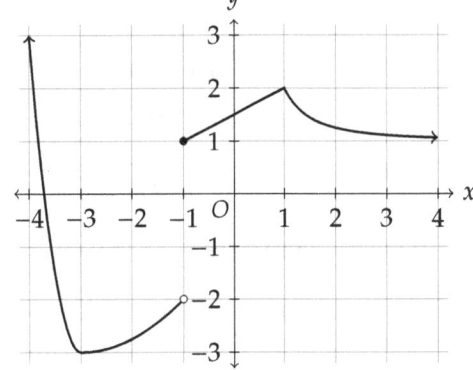

2. Working with polynomials and rational functions will require that you are able to *factor* - this includes factoring out greatest common factors (GCFs) and quadratic expressions. Factor each of the following.

   (a) $x^2 + 8x - 9$

   (b) $x^2 - 3x - 54$

   (c) $x^2 - 25$

   (d) $9x^4 - 16$

   (e) $4x^2 + 4x + 1$

   (f) $2x^3 - 50x$

   (g) $6x^2 - 13x + 5$

   (h) $-16t^2 + 80t$

   (i) $x^2 + 14x - 51$

   (j) $4y^2 + 12y + 9$

   (k) $y^2 - 14y + 24$

   (l) $x^4 - 1$

   (m) $2x^2 - 11x - 6$

   (n) $81 - 9y^2$

   (o) $x^2 - 64x - 65$

3. Find the following, if there are any, for $f(x) = \dfrac{x^3 - 4x}{x(x^2 - 5x + 6)}$.

   (a) Zero/s

   (b) Vertical asymptote/s

   (c) Hole/s (y-coordinate, too!)

4. Let $r(x) = \dfrac{x^2 - 8x + 15}{9 - x^2}$. Determine the following without using a calculator.

   (a) $\lim_{x \to -3^-} f(x)$

   (b) $\lim_{x \to -3^+} f(x)$

   (c) $\lim_{x \to 5} f(x)$

   (d) $\lim_{x \to 3} f(x)$

5. Construct a function that has a hole at $(2, 3)$, a vertical asymptote at $x = 4$, and a zero at $x = -\frac{3}{2}$.

Topics 1.8-1.10 (Day 1) Solutions/Notes

Activity

4. $x_1 = -4$ or $x_1 = -3$. The solutions to #7 include answers to each situation.

7. (a) Both values of x_1 give $p(x) \to 0$.
   (b) For $x_1 = -4$, $q(x) \to -210$. For $x_1 = -3$, $q(x) \to -96$.
   (c) Both values of x_1 give $\dfrac{p(x)}{q(x)} \to 0$.
   (d) It makes sense since $\dfrac{0}{-210} = 0$ (and $\dfrac{0}{-96} = 0$).

8. When the numerator of a rational function has a zero at a given input and the denominator does not, the rational function has a *zero* at that input.

9. $x_2 = 3$ or $x_2 = -1$. The solutions to #11 include answers to each situation.

11. (a) For $x_2 = 3$, $p(x) \to 84$. For $x_2 = -1$, $p(x) \to -12$.
    (b) Both values of x_2 give $q(x) \to 0$
    (c) For $x_2 = 3$, $\dfrac{p(x)}{q(x)} \to -524.7$ as $x \to x_2$ from the left and $\dfrac{p(x)}{q(x)} \to 525.3$ as $x \to x_2$ from the right.

    For $x_2 = -1$, $\dfrac{p(x)}{q(x)} \to 74.19$ as $x \to x_2$ from the left and $\dfrac{p(x)}{q(x)} \to -75.8$ as $x \to x_2$ from the right.
    (d) For $x_2 = 3$, $p(x)/q(x)$ is dividing by smaller and smaller positive/negative numbers, making the quotient reach arbitrarily large positive/negative values. For $x_2 = -1$, $p(x)/q(x)$ is dividing by smaller and smaller positive/negative numbers, making the quotient reach arbitrarily large negative/positive values.
    (e) Error

13. Vertical asymptote

14. When the denominator of a rational function has a zero at a given input and the numerator does not, the rational function has a *vertical asymptote* at that input.

15. $x_3 = 1$

18. (a) $p(x) \to 0$
    (b) $q(x) \to 0$
    (c) $\dfrac{p(x)}{q(x)} \to -2.5$
    (d) If $\dfrac{p(x)}{q(x)}$ is evaluated after dividing out the $(x-1)$ terms, the result is $\dfrac{(1+4)(1+3)}{2(1-3)(1+1)} = -\dfrac{20}{8} = -\dfrac{5}{2}$.

19. There appears to be nothing there!

Notes

The *zeros* of a rational function are precisely the zeros of the *numerator* that are not also zeros of the denominator.

The zeros of the *denominator* that are not a zero of the numerator are where *vertical asymptotes* occur. At vertical asymptotes, the numerator will approach a finite number as the denominator approaches 0; this will lead to their *ratio* approaching ∞ or $-\infty$. In other words, at any asymptote $x = a$,

$$\lim_{x \to a^-} \frac{p(x)}{q(x)} = \pm\infty \qquad \lim_{x \to a^+} \frac{p(x)}{q(x)} = \pm\infty$$

Examples are below.

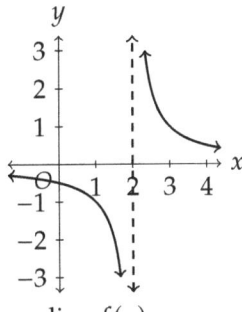 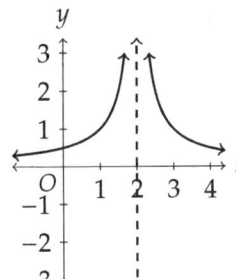 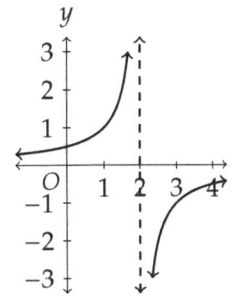 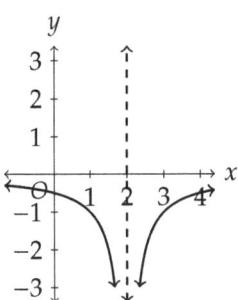

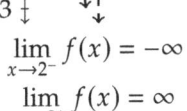 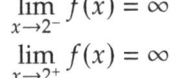 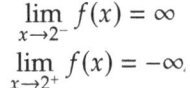

$\lim_{x \to 2^-} f(x) = -\infty$ $\lim_{x \to 2^-} f(x) = \infty$ $\lim_{x \to 2^-} f(x) = \infty$ $\lim_{x \to 2^-} f(x) = -\infty$

$\lim_{x \to 2^+} f(x) = \infty$ $\lim_{x \to 2^+} f(x) = \infty$ $\lim_{x \to 2^+} f(x) = -\infty$ $\lim_{x \to 2^+} f(x) = -\infty$

The $x \to a^-$ means "as x approaches *a from the left*" (below *a*) and $x \to a^+$ means "as x approaches *a from the right*" (above *a*).

If the numerator and denominator *share* a zero with the *same multiplicity*, the graph of $\frac{p(x)}{q(x)}$ will have a HOLE. Because the factors with these zeros can divide out, the function will still approach a finite output L: in limit notation, if a hole occurs at (h, L) on the graph of a function f, then

$$\lim_{x \to h^-} f(x) = \lim_{x \to h^+} f(x) = L$$

To compute L, simply evaluate the function at $x = h$ after dividing out the terms with h as a solution.

If the numerator and denominator share a zero, but the zero has a *greater multiplicity in the denominator*, then another *vertical asymptote* will occur.

All together, the zeros and vertical asymptotes are precisely the endpoints of the intervals for which $\frac{p(x)}{q(x)} \geq 0$ or $\frac{p(x)}{q(x)} \leq 0$. To determine each set of intervals, simply

- Identify all zeros and vertical asymptotes, then
- Test values *between* these zeros and vertical asymptotes to determine their sign

Homework

1. (a) −3   (b) 2   (c) 2   (d) 2   (e) −2   (f) 1   (g) DNE   (h) 1   (i) ∞

2. (a) $(x+9)(x-1)$ (f) $2x(x+5)(x-5)$ (k) $(y-12)(y-2)$
   (b) $(x-9)(x+6)$ (g) $(3x-5)(2x-1)$ (l) $(x+1)(x-1)(x^2+1)$
   (c) $(x+5)(x-5)$ (h) $-16t(t-5)$ (m) $(2x+1)(x-6)$
   (d) $(3x^2+4)(3x^2-4)$ (i) $(x+17)(x-3)$ (n) $9(3+y)(3-y)$
   (e) $(2x+1)^2$ (j) $(2y+3)^2$ (o) $(x-65)(x+1)$

3. $f(x)$ factors and simplifies into $f(x) = \dfrac{x(x^2-4)}{x(x-2)(x-3)} = \dfrac{x(x+2)(x-2)}{x(x-2)(x-3)} = \dfrac{x+2}{x-3}$.

   (a) $x = -2$
   (b) $x = 3$
   (c) $\left(0, -\tfrac{2}{3}\right)$ and $(2, -4)$

4. $r(x)$ factors into $r(x) = \dfrac{(x-3)(x-5)}{(3+x)(3-x)}$ and simplifies to $r(x) = -\dfrac{x-5}{x+3}$. There is a hole at $\left(3, -\tfrac{1}{3}\right)$.

   (a) As x approaches -3 from the left, $x - 3 < 0$ and $x - 5 < 0$, so the numerator will be positive. Similarly, $3 + x < 0$ and $3 - x > 0$, so the denominator will be negative. Therefore, the function will be approaching $-\infty$, which is the value of the limit.
   (b) Using similar logic to (a), it can be found that $\lim\limits_{x \to -3^+} r(x) = \infty$.
   (c) 0
   (d) $-\tfrac{1}{3}$

5. A vertical asymptote can be created with a $x - 4$ factor in the denominator and the zero can be created with a $2x - 3$ factor in the numerator. To introduce the hole, factors of $x - 2$ can be introduced in both the numerator and denominator, which gives

$$f(x) = \dfrac{(2x-3)(x-2)}{(x-4)(x-2)}.$$

Reflection

Suggested Class Time. 45 minutes

Prerequisites. Students need to be comfortable finding zeros given linear factors. They also should have some familiarity with vertical asymptotes and holes. Additionally, students should have an understanding that $\tfrac{k}{0}$ is undefined for all k, including $k = 0$.

Instructional Strategies. In this activity, it all really came together in the Notes section. Students got through the activity reasonably easily, though some students needed help with the notation: some misread what x_1 was supposed to be (numerator equal to 0, denominator not equal to 0). Others needed some assistance with "from each side," but this was cleared up quickly.

In the Notes section, free from all of the notational rigor of the activity, students wrapped their heads around the conditions for the existence of zeros, vertical asymptotes, and holes very quickly. Discussing left- and right-hand limits was very effective, and I even sketched some sample diagrams and had students try to write the correct one-sided limits for each. For holes, I used the following Geogebra sketch that worked to perfection: https://www.geogebra.org/m/mnyyggp6. The fact that the point disappears right at $x = 1$ really made students "see" (ironically) the hole.

Technology. Students used the TI-84 for this activity.

Topics 1.8-1.10 ~ Zeros, Vertical Asymptotes, and Holes (Day 2)

Learning Objectives

1. (1.8.A) Determine the zeros of rational functions.
2. (1.9.A) Determine vertical asymptotes of graphs of rational functions.
3. (1.10.A) Determine holes in graphs of rational functions.

Success Criteria

1. I can find zeros, vertical asymptotes, and holes of a rational function.
2. I can explain the difference between a zero, vertical asymptote, and hole.
3. I can express the behavior of a function near a hole or vertical asymptote using limit notation.

Today, you'll practice working with rational functions and their end behavior, zeros, vertical asymptotes, and holes.

1. Let $f(x) = \dfrac{(x-3)^2(5-4x)}{x^2 - 8x + 15}$.

    (a) For what value/s of x is $f(x) = 0$?
    (b) For what value/s of x is $f(x)$ undefined?
    (c) At the value/s identified in (b), classify each as a hole or vertical asymptote.

2. Let r be defined by $r(x) = \dfrac{6}{3x+1} - \dfrac{4}{x+2}$.

    (a) Rewrite $r(x)$ in the form $\dfrac{p(x)}{q(x)}$.
    (b) Compute the zero/s of $r(x)$.

3. Below are the graphs of the functions $f(x)$ and $g(x)$, which are polynomials with degree less than 5.

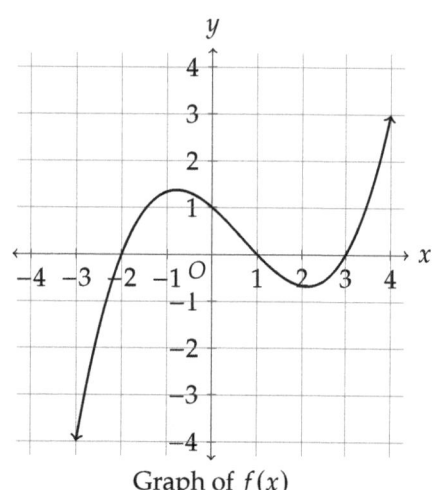

Graph of $f(x)$

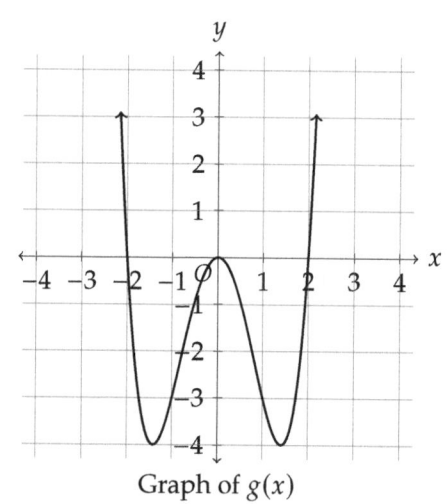

Graph of $g(x)$

Consider the function $r(x) = \dfrac{f(x)}{g(x)}$.

    (a) Find the x-intercepts, vertical asymptotes, and x-coordinates of the holes of $r(x)$.

    (b) What are the degrees of $f(x)$ and $g(x)$?

    (c) Does $r(x)$ have a horizontal asymptote, slant asymptote, or neither? Explain.

4. (a) Determine whether $f(x) = \dfrac{(x+3)^4}{(x+3)^5}$ has a hole or a vertical asymptote at $x = -3$.

    (b) Explain how f behaves as x gets arbitrarily close to $x = -3$ in words.

    (c) Express what you wrote in (b) with limits.

5. Create a function f that satisfies the following.

- $\lim\limits_{x \to 1/2^+} f(x) = \infty$
- $\lim\limits_{x \to 1/2^-} f(x) = -\infty$
- $\lim\limits_{x \to 4} f(x) = 0$

6. Suppose $f(x)$ is a polynomial function with zeros at $x = 2$, $x = 3$ with multiplicity 3, and $x = -4$. The polynomial function $g(x)$ has outputs displayed in the table below for selected values of x.

x	−4	−3	−2	−1	0	1	2	3
$g(x)$	2	0	1	0	−5	3	−2	0

Cconsider the function $q(x) = \dfrac{g(x)}{f(x)}$.

    (a) Find a value of x for which q must have a vertical asymptote.

    (b) Find a value of x for which q must have a zero.

    (c) Find a value of x for which q *could* have a hole. Under what condition/s would q have a hole at this x-value?

7. Find all zeros, holes, and vertical asymptotes of $r(x) = \dfrac{2x^3 - 32x}{3x^2 + 11x - 4}$.

Topics 1.8-1.10 (Day 2) Homework

1. Let $f(x) = \dfrac{x^3 + 8x^2 - 20x}{4x(2x-1)(x-2)}$. Find each of the following.

    (a) Zero/s

    (b) Vertical asymptote/s

    (c) Hole/s

    (d) Horizontal or slant asymptote

    (e) $\lim\limits_{x \to \infty} f(x)$

    (f) $\lim\limits_{x \to -\infty} f(x)$

2. Write a function that could produce the graph below.

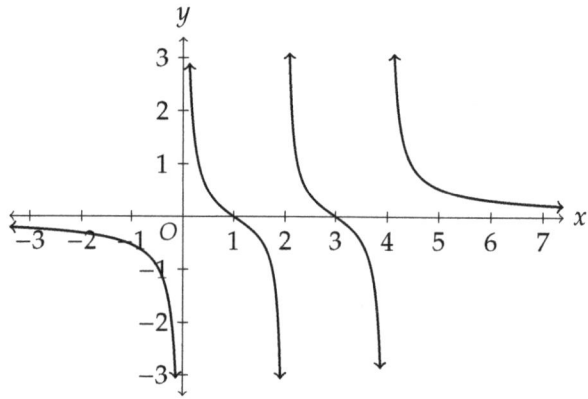

3. Let $f(x) = \dfrac{x^2 - 4}{4x + 8}$.

    (a) Show that $f(x)$ has a hole at $x = -2$.

    (b) Evaluate $\lim\limits_{x \to -2} f(x)$.

    (c) Explain why the limit from (b) exists even though f is undefined at $x = -2$

4. Let $f(x) = x^2 - 2x - 24$ and let $g(x) = x^2 + bx + 12$, where b is an integer. For what value/s of b will the graph of $r(x) = \dfrac{f(x)}{g(x)}$ have a hole?

5. (▣) Let $f(x) = x^3 + 5x^2 - 23x + 18$ and $g(x)$ be the function graphed below. Find all x-intercepts of $r(x) = \dfrac{f(x)}{g(x)}$.

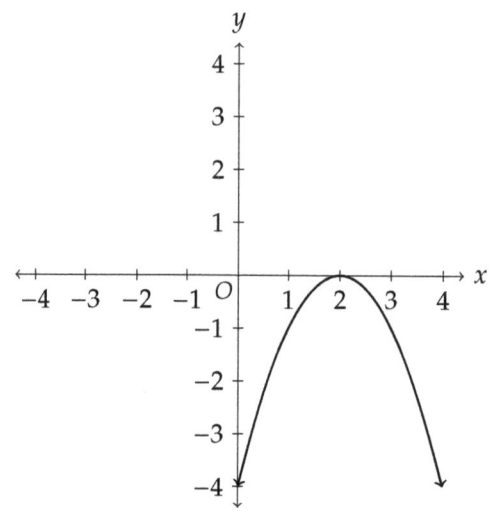

Topics 1.8-1.10 (Day 2) Solutions/Notes

Activity

1. f can be rewritten as $f(x) = \dfrac{(x-3)^2(5-4x)}{(x-3)(x-5)} = \dfrac{(x-3)(5-4x)}{x-5}$ for $x \neq 3, 5$.

   (a) $f(x) = 0$ when $(x-3)(5-4x) = 0$, or at $x = 3$ and $x = \frac{5}{4}$. However, f is not defined for $x = 3$, so the only zero is $x = \frac{5}{4}$.

   (b) f is undefined for $x = 3$ and $x = 5$.

   (c) A hole occurs when $x = 3$, and $x = 5$ is a vertical asymptote.

2. (a) $r(x) = \dfrac{6}{3x+1}\left(\dfrac{x+2}{x+2}\right) - \dfrac{4}{x+2}\left(\dfrac{3x+1}{3x+1}\right) = \dfrac{6x+12-(12x+4)}{(3x+1)(x+2)} = \dfrac{-6x+8}{(3x+1)(x+2)}$

   (b) The only zero of r is when the numerator equals 0, so $-6x + 8 = 0$ gives $x = \frac{4}{3}$.

3. (a) f has zeros at $x = -2, x = 1$, and $x = 3$. g has zeros at $x = -2, x = 0$, and $x = 2$. Therefore, f has a hole at $x = -2$, x-intercepts at $x = 1$ and $x = 3$, and vertical asymptotes at $x = 0$ and $x = 2$.

   (b) The degrees are both stated to be less than 5. f has three zeros, so it's degree is at least 3; its end behavior indicates that clearly the degree is odd, so the degree of f is 3. The degree of g must be even based on its end behavior, but it must also be at least 3 due to the presence of 3 zeros. Therefore, the degree of g is 4.

   (c) Because the degree of the denominator is higher than the degree of the numerator, the function $r(x)$ has a horizontal asymptote at $y = 0$.

4. (a) The function equals $f(x) = 1/(x+3)$ for x arbitrarily close to $x = -3$. This still results in division by 0 when substituting in $x = -3$, so $x = -3$ is a vertical asymptote.

   (b) As x approaches -3 from the left, the function will output negative values with increasing magnitude. As x approaches -3 from the right, the function will output positive values with increasing magnitude.

   (c) $\lim\limits_{x \to -3^-} f(x) = -\infty$, $\lim\limits_{x \to -3^+} f(x) = \infty$

5. Many can work, but the simplest is $f(x) = -\dfrac{x-4}{2x-1} = \dfrac{4-x}{2x-1}$.

6. (a) The denominator of q has a zero at $x = 2$, while the numerator does not share this zero. Therefore, there must be a vertical asymptote at $x = 2$. The same can be said for $x = -4$.

   (b) The numerator of q has a zero at $x = -1$, while $x = -1$ is not a zero of $f(x)$. Since f is a polynomial, $f(-1)$ must be defined at $x = -1$ and therefore q has a zero at $x = -1$. This also applies to $x = -3$.

   (c) q could have a hole if the numerator and denominator share a zero and the multiplicity of the zero in the numerator is greater than or equal to the multiplicity of the zero in the denominator. The only zero that fits the bill occurs at $x = 3$, so there will be a hole here if the multiplicity of $x = 3$ for $g(x)$ is greater than or equal to 3.

7. First, r factors into
$$r(x) = \dfrac{2x(x^2-16)}{(3x-1)(x+4)} = \dfrac{2x(x+4)(x-4)}{(3x-1)(x+4)}.$$

   There is a hole when $x = -4$, a vertical asymptote at $x = \frac{1}{3}$, and zeros at $x = 0$ and $x = 4$. The y-coordinate of the hole is
$$\dfrac{2(-4)(-4-4)}{3(-4)-1} = -\dfrac{64}{13}.$$

Homework

1. The function factors into $f(x) = \dfrac{x(x+10)(x-2)}{4x(2x-1)(x-2)}$, which equals $f(x) = \dfrac{x+10}{8x-4}$ for $x \neq 0, 2$.

   (a) $x = -10$

   (b) $x = \frac{1}{2}$

   (c) Holes at $\left(0, -\frac{5}{2}\right)$ and $(2, 1)$

   (d) Horizontal asymptote at $y = \frac{1}{4}$

   (e) $\lim\limits_{x \to \infty} f(x) = \frac{1}{8}$

   (f) $\lim\limits_{x \to -\infty} f(x) = \frac{1}{8}$

2. There are zeros at $x = 1$ and $x = 3$ and vertical asymptotes at $x = 0$, $x = 2$, and $x = 4$. Therefore,
$$f(x) = \frac{(x-1)(x-3)}{x(x-2)(x-4)}.$$

3. (a) $f(x) = \dfrac{(x+2)(x-2)}{4(x+2)}$, so there is a hole where $x + 2 = 0$, or when $x = -2$.

   (b) $\lim\limits_{x \to -2} f(x) = \dfrac{-2-2}{4} = -1$

   (c) Even though f is undefined at -2, the values of $f(x)$ get arbitrarily close to -1 as x gets very close to $x = -2$.

4. $r(x) = \dfrac{(x-6)(x+4)}{x^2+bx+12}$ will have a hole only if $(x-6)$ or $(x+4)$ is a factor of $g(x)$. These factorizations would be
$$x^2 + bx + 12 = (x-6)(x-2) = x^2 - 8x + 12,$$
or
$$x^2 + bx + 12 = (x+4)(x+3) = x^2 + 7x + 12$$
so $b = -8$ or $b = 7$.

5. Using the calculator, the numerator f has zeros of $x = -8.110$, $x = 1.110$, and $x = 2$. The first two of these are the x-intercepts of $r(x)$; a hole occurs at $x = 2$ because this is also a zero of $g(x)$.

Reflection

Suggested Class Time. 80 minutes

Prerequisites. Students should be proficient with the Day 1 activity material. Additionally, as always, students should be fluent with factoring by this point.

Instructional Strategies.
By and large, students struggled with the same things: getting zeros, holes, and vertical asymptotes mixed up. In particular, I noticed that a *lot* of students still didn't have a good enough grasp of $\frac{0}{k}$, $\frac{k}{0}$, and $\frac{0}{0}$. I tried a few different tactics to help students understand. The first was an approach based on an analytical definition of division: $\frac{a}{b} = c \Leftrightarrow a = bc$. I used this definition along with a constant, say 12, (to not overload students with variables) and we discussed the following.

$$\frac{0}{12} = x \Leftrightarrow 0 = 12x \Leftrightarrow x = 0$$

$$\frac{12}{0} = x \Leftrightarrow 12 = 0x \to \text{no such value exists!}$$

$$\frac{0}{0} = x \Leftrightarrow 0 = 0x \to \text{literally every } x \text{ satisfies this, hence it "lacks definition"}$$

For problems like #3 in which zeros were provided separately for the numerator and denominator, I had struggling groups make a table of all of the zeros of each function, including the value of the *other* function at that zero, like that shown below.

Zero	$f(x)$	$g(x)$	$\dfrac{f(x)}{g(x)}$
-2	0	0	undefined (hole)
1	0	-3	0 (zero)
2	-0.6	0	undefined (vert. asym.)
3	0	50	0 (zero)

This helped students keep track of things while reinforcing the rules for identifying zeros/holes/vertical asymptotes.

Technology. Students were *not* supposed to use a calculator for any of these problems, but some still tried!

Problems.

1. (■) The function $g(x)$ is a quadratic function with zeros of $x = 2$ and $x = 4$. If
$$r(x) = \frac{2x^3 + 4x^2 - 13x - 6}{g(x)},$$
then $r(x)$ has zeros at which of the following value/s?

   (A) $x = 2$ and $x = 4$
   (B) $x = 2$ only
   (C) $x = -3.581$ and $x = -0.419$
   (D) $x = -3.581, x = -0.419$, and $x = 2$

2. The functions f and g have been graphed below.

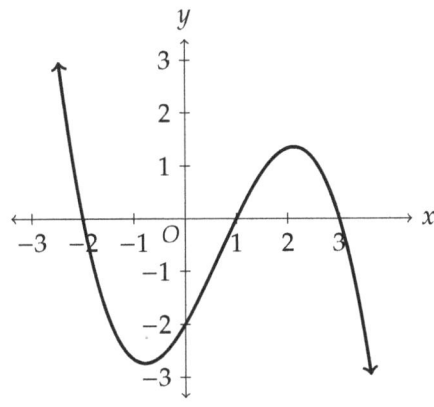
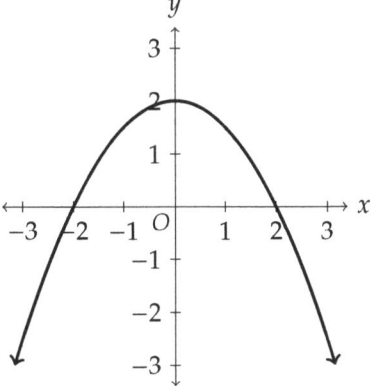

   Graph of $f(x)$ Graph of $g(x)$

   If $h(x) = \dfrac{g(x)}{f(x)}$, for which of the following x-values does $h(x) = 0$?

   (A) $x = 2$ only
   (B) $x = 2$ and $x = -2$
   (C) $x = -2, x = 1$, and $x = 3$
   (D) $x = -2, x = 1, x = 2$, and $x = 3$

3. Which of the following functions $f(x)$ would have a vertical asymptote at $x = -3$?

(A) $f(x) = \dfrac{(x+3)(x-2)^2}{(x+1)(x+3)}$

(B) $f(x) = \dfrac{(x+3)^3(x-2)^2}{(x+1)^2(x+3)}$

(C) $f(x) = \dfrac{(x+3)^2(x-2)}{(x+1)(x+3)^2}$

(D) $f(x) = \dfrac{(x+3)^2(x-2)^2}{(x+1)^2(x+3)^3}$

4. The function g is given by $g(x) = 2x(x-1)(x+3)$. For another function $f(x)$, the table below shows values of the function for certain x-values. It is known that zero of f has a multiplicity of 1.

x	-3	-2	-1	0	1	2
$f(x)$	0	3	-2	-1	0	4

Let h be given by $h(x) = \dfrac{f(x)}{g(x)}$. Which of the following is true?

(A) h has three holes and no vertical asymptotes.

(B) h has two holes and one vertical asymptote.

(C) h has one hole and two vertical asymptotes.

(D) h has no holes and three vertical asymptotes.

5. The function $f(x) = -\dfrac{1}{2x+4}$ is graphed below. Which of the following is true?

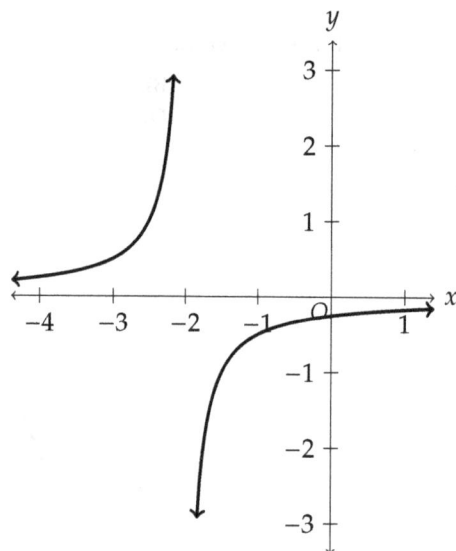

(A) $\lim\limits_{x \to \infty} f(x) = -2$ (B) $\lim\limits_{x \to -\infty} f(x) = -2$ (C) $\lim\limits_{x \to -2^-} f(x) = -\infty$ (D) $\lim\limits_{x \to -2^-} f(x) = \infty$

6. The function $h(x) = \dfrac{x^2 - 3x - 18}{x^2 + 3x}$ has a hole at (a, b). What is the value of b?

(A) $b = -3$ (B) $b = -1$ (C) $b = 1$ (D) $b = 3$

Solutions.

1. Graphing $y = 2x^3 + 4x^2 - 13x - 6$ reveals that the polynomial has three zeros at $x = -3.581$, $x = -0.419$, and $x = 2$. Thus, $x = 2$ is a zero of both the numerator and denominator so it is not a zero of $r(x)$. The only zeros of $r(x)$ are $x = -3.581$ and $x = -0.419$. The correct answer is therefore **(C)**.

2. The zeros of $g(x)$ are $x = -2$ and $x = 2$. The zeros of $f(x)$ are $x = -2$, $x = 1$, and $x = 3$. Thus, the zero of $h(x)$ is only $x = 2$ since $g(2) = 0$ but $f(2) \neq 0$. The correct answer is therefore **(A)**.

3. Note that each answer choice results in a zero numerator and a zero denominator when x is -3. However, the function in choice D can be simplified to
$$f(x) = \frac{(x+3)^2(x-2)^2}{(x+1)^2(x+3)^3} = \frac{(x-2)^2}{(x+1)^2(x+3)}$$
which gives numerator that is not zero and a denominator that is when $x = -3$. This indicates a vertical asymptote at $x = -3$. The correct answer is therefore **(D)**.

4. The zeros of $f(x)$ are $x = -3$ and $x = 1$. The zeros of $g(x)$ are $x = 0$, $x = 1$, and $x = -3$. Two of the zeros ($x = -3$ and $x = 1$) are shared by the numerator and denominator of h with the same multiplicity. Hence, these are holes. The other zero ($x = 0$) is only a zero of the denominator of h, so this leads to a vertical asymptote. The correct answer is therefore **(B)**.

5. As x increases without bound, the function tends to 0, so choice (A) is incorrect. As x decreases without bound, the function tends to 0, so choice (B) is incorrect. As x goes to -2 from the left, the function increases without bound. The correct answer is therefore **(D)**.

6. The function factors as
$$h(x) = \frac{x^2 - 3x - 18}{x^2 + 3x} = \frac{(x-6)(x+3)}{x(x+3)}.$$
Since $x = -3$ is a zero of both numerator and denominator with the same multiplicity, the hole occurs at $x = -3$. To get the y-value of the hole, we simplify further to get $h(x) = (x-6)/x$. Then $h(-3) = (-3-6)/(-3) = -9/(-3) = 3$. The correct answer is therefore **(D)**.

Topic 1.11 ~ Equivalent Representations

Learning Objectives

1. (1.11.A) Rewrite polynomial and rational expressions in equivalent forms.
2. (1.11.B) Determine the quotient of two polynomial functions using long division.
3. (1.11.C) Rewrite the repeated product of binomials using the binomial theorem.

Success Criteria

1. I can use or create factored forms of rational expressions to identify characteristics of their graphs.
2. I can divide two polynomials using long division and use this to find slant asymptotes of rational functions.
3. I can expand a repeated product of binomials using the binomial theorem.

The first Learning Objective concerns itself with factoring polynomials and rational expressions to identify zeros, holes, and vertical asymptotes. As we have already focused on this in Topics 1.8-1.10, we will instead focus on the second and third learning objectives.

You learned in Topic 1.7 that a rational function has a *slant asymptote* when the degree of the numerator is one greater than the degree of the denominator. Now, we discuss how to actually compute these.

To compute a slant asymptote, we use *polynomial long division*. Here's an example.

Example 1. Find the slant asymptote of $h(x) = \dfrac{f(x)}{g(x)} = \dfrac{3x^3 - x^2 + 4}{(x+3)(x-1)}$.

Step 0: Both the numerator and denominator need to be in standard form, so we multiply out the denominator.
$$(x+3)(x-1) = x^2 + 2x - 3$$
This is called Step 0 because it may be unnecessary if both numerator and denominator are already in standard form.

Step 1: Set up the long division structure, making sure to put a 0 as a placeholder for any "missing" x^k terms. In our case, $3x^3 - x^2 + 4$ has no x term, so we place a $0x$ in its position.

$$x^2 + 2x - 3 \overline{\smash{\big)}\, 3x^3 - x^2 + 0x + 4}$$

Step 2: Determine what multiplies by the *leading coefficient* of the divisor (what you're dividing by) to get the *leading coefficient* of the dividend (what is being divided). In this case, we determine what times x^2 equals $3x^3$, which is clearly $3x$.

$$\begin{array}{r} 3x \phantom{{}+3x^3-x^2+0x+4} \\ x^2 + 2x - 3 \overline{\smash{\big)}\, 3x^3 - x^2 + 0x + 4} \end{array}$$

Step 3: Multiply this new factor by the *entirety* of the divisor and place this beneath the divisor in parentheses.

$$\begin{array}{r} 3x \\ x^2+2x-3 \overline{\smash{\big)}\, 3x^3 - x^2 + 0x + 4} \\ -(3x^3 + 6x^2 - 9x) \end{array}$$

Step 4: Subtract downward. **Don't forget to distribute this subtraction.**

$$\begin{array}{r} 3x \\ x^2+2x-3 \overline{\smash{\big)}\, 3x^3 - x^2 + 0x + 4} \\ \underline{-(3x^3 + 6x^2 - 9x)} \\ -7x^2 + 9x + 4 \end{array}$$

Step 5: Repeat this process until the divisor won't multiply into the resulting dividend. What is left is the remainder. Here, we can continue:

$$\begin{array}{r} 3x - 7 \\ x^2+2x-3 \overline{\smash{\big)}\, 3x^3 - x^2 + 0x + 4} \\ \underline{-(3x^3 + 6x^2 - 9x)} \\ -7x^2 + 9x + 4 \\ \underline{-(-7x^2 - 14x + 21)} \\ 23x - 17 \end{array}$$

Step 6: Rewrite the final quotient $\frac{f(x)}{g(x)}$ in the form $q(x) + \frac{r(x)}{g(x)}$, where r is the remainder. For our quotient, we get

$$h(x) = \frac{3x^3 - x^2 + 4}{x^2 + 2x - 3} = 3x - 7 + \frac{23x - 17}{x^2 + 2x - 3}$$

Long division can help us simplify end behavior for rational functions. Think about it: the remainder of long division will by definition be *bottom heavy*, which means that the remainder will go to 0 as the inputs increase without bound. In short, if a rational function $f(x)$ can be rewritten as $f(x) = q(x) + r(x)$, where r is the remainder, then

$$\lim_{x \to \infty} f(x) = \lim_{x \to \infty} q(x) + r(x)$$
$$= \lim_{x \to \infty} q(x) + \lim_{x \to \infty} r(x)$$
$$= \lim_{x \to \infty} q(x) + 0$$
$$= \lim_{x \to \infty} q(x)$$

Therefore, for x of large magnitude (the same applies for $x \to -\infty$), we can see that $f(x)$ and $q(x)$ will be nearly identical ($f(x) \approx q(x)$). When this function q is linear, it is called a SLANT ASYMPTOTE.

In the case of the function $h(x) = \frac{3x^3-x^2+4}{(x+3)(x-1)}$, we can say that h will behave very similarly to the function $q(x) = 3x - 7$. Because this is linear, the graph of h has a slant asymptote of $y = 3x - 7$.

Practice 1. Do each of the following for the functions below.

(i) Evaluate each quotient using polynomial long division.

(ii) Identify a polynomial function that the graph of the rational function will resemble for x of large magnitude.

(iii) Determine if there is a slant asymptote; if there is, what is its equation?

(a) $f(x) = \dfrac{2x^3 + x^2 - 4x + 1}{x + 2}$

(b) $g(x) = \dfrac{2x^4 + x^2 + 5}{x^3 + 2x^2 + x + 1}$

Now, what happens when we get a remainder of *0*? Well, this means that the divisor is a <u>factor</u> of the dividend. This can then be used to actually factor polynomials!

Practice 2. Show that $x + 3$ is a factor of $f(x) = 3x^3 + 16x^2 + 23x + 6$. Then, find all zeros of $r(x) = \dfrac{f(x)}{x + 3}$.

Polynomials can be rewritten not only using division, but also *exponentiation*. For polynomials with more than 2 terms, this can be incredibly cumbersome. For just binomials, though, there is a remarkable theorem.

The Binomial Theorem. For any binomial $x + y$,

$$(x+y)^n = \sum_{k=0}^{k=n} \binom{n}{k} x^{n-k} y^k$$

where n is a positive integer.

The coefficient $\binom{n}{k}$ is called the BINOMIAL COEFFICIENT and is read as "n choose k." Values of the binomial coefficient are given by Pascal's Triangle, which is displayed below.

```
           1
          1 1
         1 2 1
        1 3 3 1
       1 4 6 4 1
      1 5 10 10 5 1
```

For $\binom{n}{k}$, n is the row and k is the entry in the row, *keeping in mind that there is a 0th row and a 0th entry*. For instance, take the row with entries 1, 4, 6, 4, 1. This is *row 4* (not 5!), and the entries read

$$1 \quad 4 \quad 6 \quad 4 \quad 1$$
$$\binom{4}{0} \quad \binom{4}{1} \quad \binom{4}{2} \quad \binom{4}{3} \quad \binom{4}{4}$$

Practice 3. Determine the values of $\binom{3}{2}, \binom{5}{1}, \binom{6}{2},$ and $\binom{3}{3}$.

Example 2. To expand a binomial using the Binomial Theorem, we simply utilize the structure of the expression $\binom{n}{k} x^{n-k} y^k$. For example, to expand $(x+3)^4$, we have that "y" is 3, so we get

Binomial Coefficient $\binom{n}{k}$	1	4	6	4	1
x^n	x^4	x^3	x^2	x	$x^0 = 1$
y^n	$3^0 = 1$	$3^1 = 3$	$3^2 = 9$	$3^3 = 27$	$3^4 = 81$
Final Sum	x^4	$+12x^3$	$+54x^2$	$+108x$	81

Therefore, $(x+3)^4 = x^4 + 12x^3 + 54x^2 + 108x + 81$.

Practice 4. Expand each of the following.

(a) $(x-1)^3$

(b) $(2x+5)^3$

(c) $(2x-3)^4$

Notes

Topic 1.11 Homework

1. Compute the following quotients.

    (a) $\dfrac{3x^2 - 9x + 4}{x - 4}$

    (b) $\dfrac{2x^3 - 7x^2 - 5}{2x + 1}$

    (c) $\dfrac{8 - x^3}{x^2 + x + 1}$

2. Determine whether $2x + 3$ is a factor of $8x^3 + 10x^2 - 13x - 15$.

3. Expand the following expressions.
    (a) $(x - 1)^4$ (b) $(x + 3)^3$ (c) $(2x - 1)^3$ (d) $(3x + 2)^4$ (e) $(x + 1)^6$

4. Consider the binomial $px + q$, where p, q are integers. If $(px + q)^3 = ax^3 - 108x^2 + 144x - d$, what are the values of p and q?

5. Find the holes, vertical asymptotes, and zeros of $f(x) = \dfrac{x^3 + 6x^2 + 12x + 8}{x^3 + 7x^2 + 10x}$.

Topic 1.11 Solutions/Notes

Activity
Practice 1.

(a) (i) The quotient is $2x^2 - 3x + 2 - \dfrac{3}{x+2}$.

$$\begin{array}{r}
2x^2 - 3x + 2 \\
x+2\,\overline{\smash{\big)}\,2x^3 + x^2 - 4x + 1}\\
-(2x^3 + 4x^2)\\
\hline
-3x^2 - 4x + 1\\
-(-3x^2 - 6x)\\
\hline
2x + 1\\
-(2x + 4)\\
\hline
-3
\end{array}$$

(ii) $q(x) = 2x^2 - 3x + 2$

(iii) No, as q is not linear.

(b) (i) The quotient is $2x - 4 + \dfrac{7x^2 + 2x + 9}{x^3 + 2x^2 + x + 1}$. The line $y = 2x - 4$ is a slant asymptote of the graph of $\dfrac{2x^4 + x^2 + 5}{x^3 + 2x^2 + x + 1}$.

$$\begin{array}{r}
2x - 4 \\
x^3 + 2x^2 + x + 1\,\overline{\smash{\big)}\,2x^4 + 0x^3 + x^2 + 0x + 5}\\
-(2x^4 + 4x^3 + 2x^2 + 2x)\\
\hline
-4x^3 - x^2 - 2x + 5\\
-(-4x^3 - 8x^2 - 4x - 4)\\
\hline
7x^2 + 2x + 9
\end{array}$$

(ii) $q(x) = 2x - 4$

(iii) Yes; $y = 2x - 4$

Practice 2. First, perform the long division.

$$\begin{array}{r}
3x^2 + 7x + 2 \\
x+3\,\overline{\smash{\big)}\,3x^3 + 16x^2 + 23x + 6}\\
-(3x^3 + 9x^2)\\
\hline
7x^2 + 23x + 6\\
-(7x^2 + 21x)\\
\hline
2x + 6\\
-(2x + 6)\\
\hline
0
\end{array}$$

Because the remainder is 0, $x + 3$ must be a factor of $3x^3 + 16x^2 + 23x + 6$. In fact, we can write $f(x) = (x + 3)(3x^2 + 7x + 2)$, which means $r(x) = \dfrac{f(x)}{x+3} = 3x^2 + 7x + 2$ for $x \neq -3$. This factors into $r(x) = (3x + 1)(x + 2)$, so the zeros of $r(x)$ are $x = -\frac{1}{3}$ and $x = -2$.

Practice 3. $\binom{3}{2} = 3$, $\binom{5}{1} = 5$, $\binom{6}{2} = 15$, $\binom{3}{3} = 1$

Practice 4.

(a) The result is $x^3 - 3x^2 + 3x - 1$.

Binomial coefficient	1	3	3	1
x^n	x^3	x^2	x	1
y^n	1	$(-1)^1 = -1$	$(-1)^2 = 1$	$(-1)^3 = -1$
Final Sum	x^3	$-3x^2$	$+3x$	-1

(b) The result is $8x^3 + 60x^2 + 150x + 125$.

Binomial coefficient	1	3	3	1
x^n	$(2x)^3 = 8x^3$	$(2x)^2 = 4x^2$	$(2x)^1 = 2x$	1
y^n	1	$5^1 = 5$	$5^2 = 25$	$5^3 = 125$
Final Sum	$8x^3$	$+60x^2$	$+150x$	$+125$

(c) The result is $16x^4 - 96x^3 + 216x^2 - 216x + 81$.

Binomial coefficient	1	4	6	4	1
x^n	$(2x)^4 = 16x^4$	$(2x)^3 = 8x^3$	$(2x)^2 = 4x^2$	$(2x)^1 = 2x$	1
y^n	1	$(-3)^1 = -3$	$(-3)^2 = 9$	$(-3)^3 = -27$	$(-3)^4 = 81$
Final Sum	$16x^4$	$-96x^3$	$+216x^2$	$-216x$	$+81$

Notes

These aren't notes, but tips and things to keep in mind.

For polynomial long division...

- Don't forget to use 0 placeholders in case of missing terms.
- When subtracting a row, *distribute the negative.*

For the Binomial Theorem...

- The powers of "x" and "y" will always sum to the power of the binomial.
- Signs will either be all positive or will *alternate.*
- Don't forget your binomial coefficients!
- *Use parentheses!!!*
- The value at the bottom of a column is the *product* of the terms in the column.

Homework

1. (a) The quotient is $3x + 3 + \dfrac{16}{x-4}$.

$$
\begin{array}{r}
3x + 3 \\
x-4 \overline{\smash{)}\ 3x^2 - 9x + 4} \\
\underline{-3x^2 + 12x } \\
3x + 4 \\
\underline{-3x + 12} \\
16
\end{array}
$$

(b) The quotient is $x^2 - 4x + 2 - \dfrac{7}{2x+1}$.

$$
\begin{array}{r}
x^2 - 4x + 2 \\
2x+1 \overline{\smash{)}\ 2x^3 - 7x^2 - 5} \\
\underline{-2x^3 - x^2 } \\
-8x^2 \\
\underline{8x^2 + 4x } \\
4x - 5 \\
\underline{-4x - 2} \\
-7
\end{array}
$$

(c) The quotient is $-x + 1 + \dfrac{7}{x^2 + x + 1}$.

$$
\begin{array}{r}
-x + 1 \\
x^2+x+1 \overline{\smash{)}\ -x^3 + 8} \\
\underline{x^3 + x^2 + x } \\
x^2 + x + 8 \\
\underline{-x^2 - x - 1} \\
7
\end{array}
$$

2. Yes, it is. Long division results in a remainder of 0.

$$
\begin{array}{r}
4x^2 - x - 5 \\
2x+3 \overline{\smash{)}\ 8x^3 + 10x^2 - 13x - 15} \\
\underline{-8x^3 - 12x^2 } \\
-2x^2 - 13x \\
\underline{2x^2 + 3x } \\
-10x - 15 \\
\underline{10x + 15} \\
0
\end{array}
$$

3. (a) $x^4 - 4x^3 + 6x^2 - 4x + 1$

(b) $x^3 + 9x^2 + 27x + 27$

(c) $8x^3 - 12x^2 + 6x - 1$

(d) $81x^4 + 216x^3 + 216x^2 + 96x + 16$

(e) $x^6 + 6x^5 + 15x^4 + 20x^3 + 15x^2 + 6x + 1$

4. From the binomial theorem, we know $-108 = 3p^2q$ and $144 = 3pq^2$. We can divide each equation by 3 to get $p^2q = -36$ and $pq^2 = 48$. Since $p^2q < 0$ and $p^2 > 0$, we know $q < 0$. Dividing the two equations, we get $\frac{p}{q} = -\frac{3}{4}$. We test out $p = 3$ and $q = -4$ and find that both equations $p^2q = -36$ and $pq^2 = 48$ are satisfied.

5. The numerator factors as $(x+2)^3$ and the denominator factors into $x(x+2)(x+5)$, so f simplifies into

$$f(x) = \frac{(x+2)^2}{x(x+5)} \text{ for } x \neq -2.$$

There is a hole at $(-2, 0)$, vertical asymptotes at $x = 0$ and $x = -5$, and no zeros.

Reflection

Suggested Class Time. 60 to 70 minutes

Prerequisites. Students should be comfortable with multiplying polynomials and binomials.

Instructional Strategies. Students mostly flew through this activity, but my students had seen both polynomial long division and the Binomial Theorem before in Advanced Algebra (Algebra II). That said, many were obviously rusty, so the examples helped tremendously.

That said, some students weren't sure of what to "do" with the remainder. For these students, I went back to good ol' mixed numbers: I showed how 18 divided by 5 is 3 with a remainder of 3, but that this, as a mixed number, gets written as $3\frac{3}{5}$ or "3 *and* 3 fifths." In the same way, I told students, the remainder is added on and still divided by the divisor.

When it came to the Binomial Theorem, literally every student ran into the same problem: not using parentheses. When expanding $(2x+5)^3$, every single student's polynomial started with $2x^3$ as opposed to the correct $8x^3$. I talked with each individual group about this, but then made sure to clear it up. I emphasized that, when in doubt, students should check at least their first term by considering the true expansion, in this case $(2x+5)(2x+5)(2x+5)$, which would obviously result in a leading term of $8x^3$.

Additionally, when we discussed the activity, I pulled up Geogebra and graphed the function from Practice 1 (a), $\frac{2x^3+x^2-4x+1}{x+2}$. When I asked students what they saw, they just said things like "it looks like a paper clip?" Then, though, I zoomed out (asfter scaling the x-axis up to a ratio of 5:1 with the y-axis since the function grows so quickly). After zooming out, the graph clearly resembles a quadratic, which students picked up on. When I asked, "What quadratic do you think this is?," I heard crickets, but after saying - " guess what... it's $2x^2 - 3x + 2$," there were some ooh's and aah's. Similarly, I graphed the rational function from Practice 1 (b) and zoomed out, and this time students were quick to see that it resembled a line. We discussed that this "zooming out" is the effect of looking only at x of large magnitude, and that the remainders are by construction *bottom heavy*, which means they approach 0. This was a nice opportunity to revisit bottom heavy rational functions.

Technology. Students did not need any technology, though some wanted their calculators for subtraction purposes.

Problems.

1. When $(3x+1)^4$ expanded, the term with the greatest leading coefficient is

   (A) the x^4 term. (B) the x^3 term. (C) the x^2 term. (D) the x term.

2. In the expansion of $(2x-1)^4$, the coefficient of the x^3 term is

   (A) -32 (B) -8 (C) 8 (D) 32

3. Which of the following is true concerning the function $f(x) = \dfrac{6x^2 - 4x + 3}{3x - 2}$?

   (A) f has a horizontal asymptote of $y = 2$.
   (B) f has a horizontal asymptote of $y = 0$.
   (C) f has a slant asymptote of $y = 2x$.
   (D) f has a slant asymptote of $y = 2x + \frac{8}{3}$.

4. The function $f(x) = \dfrac{6x^2 - 5x + 2}{ax - b}$ is shown below.

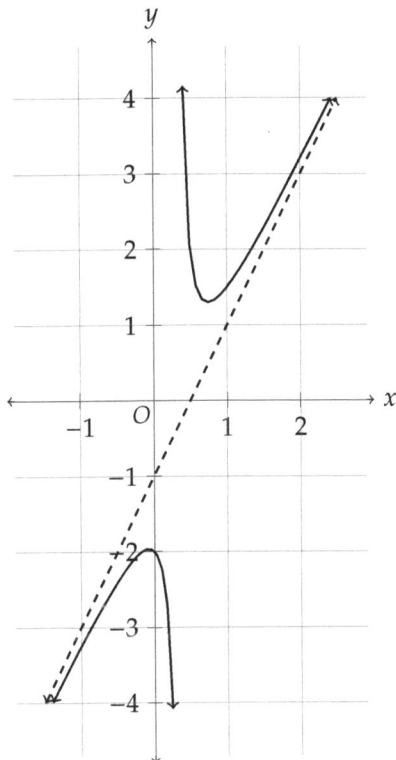

   What is the value of a?
   (A) $a = 1$ (B) $a = 2$ (C) $a = 3$ (D) $a = 6$

5. Which of the following functions has the same end behavior as the function g defined by $g(x) = \dfrac{3x^3 + 8x + 1}{x^2 - 2x + 3}$?
   (A) $f(x) = 3x - 6$ (B) $h(x) = 3x - 4$ (C) $j(x) = 3x + 4$ (D) $k(x) = 3x + 6$

Solutions.

1. By the Binomial Theorem, the expansion is

Binomial Coefficient $\binom{n}{k}$	1	4	6	4	1
x^n	$(3x)^4 = 81x^4$	$(3x)^3 = 27x^3$	$(3x)^2 = 9x^2$	$3x$	$x^0 = 1$
y^n	$1^0 = 1$	$1^1 = 1$	$1^2 = 1$	$1^3 = 1$	$1^4 = 1$
Final Sum	$81x^4$	$+108x^3$	$+54x^2$	$+12x$	1

The greatest leading coefficient is 108 on the x^3 term. The correct answer is therefore **(B)**.

2. The x^3 term is
$$\binom{4}{1}(2x)^{4-1}(-1)^1 = 4 \cdot (2x)^3(-1) = -4 \cdot 8x^3 = -32x^3.$$
The correct answer is therefore **(A)**.

3. The quotient is $2x + \dfrac{3}{3x-2}$.

$$\begin{array}{r} 2x \\ 3x-2 \overline{\smash{\big)}\, 6x^2 - 4x + 3} \\ \underline{-6x^2 + 4x} \\ 3 \end{array}$$

So the function f has a slant asymptote of $y = 2x$. The correct answer is therefore **(C)**.

4. The graph indicates that the slant asymptote is $y = 2x - 1$. Then $2x$ must be the quotient upon dividing $6x^2 - 5x + 2$ by $ax - b$. Thus, $(2x)(ax) = 6x^2$ which implies that $2a = 6$. Then $a = 3$. The correct answer is therefore **(C)**.

5. The degree of the numerator is 1 greater than the degree of the denominator, so the quotient is the end behavior.

$$\begin{array}{r} 3x + 6 \\ x^2 - 2x + 3 \overline{\smash{\big)}\, 3x^3 + 8x + 1} \\ \underline{-3x^3 + 6x^2 - 9x} \\ 6x^2 - x + 1 \\ \underline{-6x^2 + 12x - 18} \\ 11x - 17 \end{array}$$

The end behavior is $y = 3x + 6$. The correct answer is therefore **(D)**.

Topics 1.7-1.11 ~ Review

Reconnaissance!

Your mission is as follows: obtain any and all information you can on the following rogue functions. Some information will be more detailed than other information, and some may not be attainable. Get a photo or sketch of each rogue - whatever you can find. Unfortunately, the enemy has blocked some of our communications, so calculators have been rendered useless.

Rogue 1: We have a known identity: $f(x) = \dfrac{x(2x+3)(x-5)}{3x^2 - 11x - 20}$	
End behavior	
Zeros/Multiplicity	
Hole/s	
Vertical asymptote/s	
Horizontal asymptote	
Slant asymptote	**Additional Information**

Rogue 2: We know the following. $g(x) = \frac{p(x)}{q(x)}$, where p is a polynomial with degree 4 and leading coefficient -3. $q(x)$ is a polynomial with integer coefficients that has a zero of $x = \frac{1}{2}$, a zero of $x = 3$ with multiplicity 3, and no other zeros. Last spotted was $q(2) = -3$. It is known p and q share no zeros.

End behavior	
Zeros/Multiplicity	
Hole/s	
Vertical asymptote/s	
Horizontal asymptote	
Slant asymptote	**Additional Information**

UNIT 1 TOPICS 1.7-1.11 ~ REVIEW 99

Rogue 3: $g(x) = -x(x+2)^3(3-2x)^2$	
End behavior	
Zeros/Multiplicity	
Hole/s	
Vertical asymptote/s	
Horizontal asymptote	
Slant asymptote	Additional Information

Rogue 4: Rogue 3 is a known associate. Rogue 4 uses Rogue 3 as a denominator while keeping its own identity hidden. Thus, Rogue 4 is given only only by $h(x) = \dfrac{a(x-k)^4}{g(x)}$, where g is Rogue 3. It is known that $a > 0$ and $k > 2$.

End behavior	
Zeros/Multiplicity	
Hole/s	
Vertical asymptote/s	
Horizontal asymptote	
Slant asymptote	**Additional Information**

UNIT 1   TOPICS 1.7–1.11 ~ REVIEW

Rogue 5: $f(x) = \dfrac{5}{2x-1} - \dfrac{9}{(2x-1)^2}$

End behavior	
Zeros/Multiplicity	
Hole/s	
Vertical asymptote/s	
Horizontal asymptote	
Slant asymptote	**Additional Information**

Solutions

Rogue 1: We have a known identity: $f(x) = \dfrac{x(2x+3)(x-5)}{3x^2 - 11x - 20}$

End behavior	
$\lim\limits_{x \to \infty} f(x) = \infty$ $\lim\limits_{x \to -\infty} f(x) = -\infty$	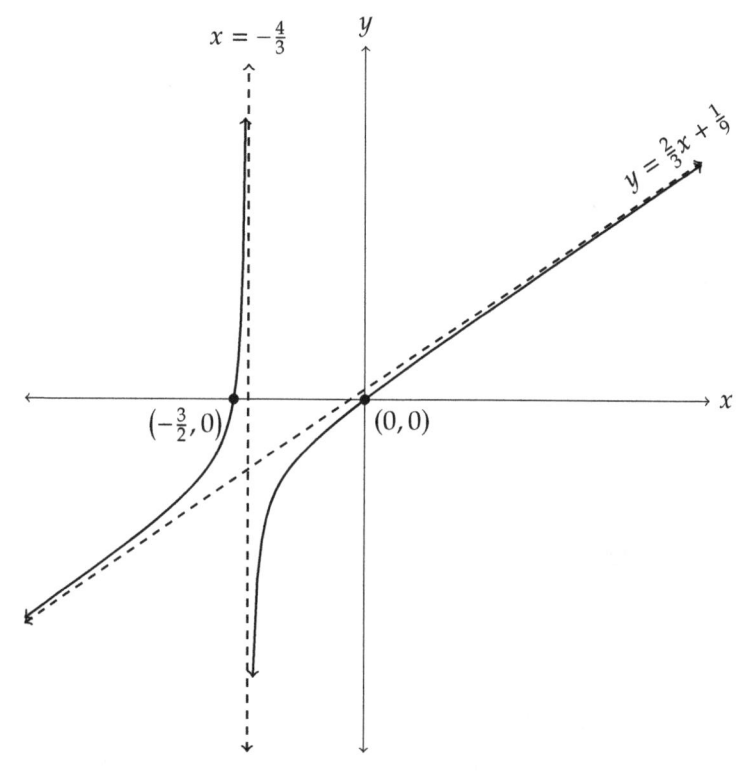
Zeros/Multiplicity	
$x = 0$ $x = -\frac{3}{2}$	
Hole/s	
$\lim\limits_{x \to 5} f(x) = \frac{5(13)}{19} = \frac{65}{19}$ so hole at $\left(5, \frac{65}{19}\right)$	
Vertical asymptote/s	
$x = -\frac{4}{3}$	
Horizontal asymptote	
None	
Slant asymptote	**Additional Information**
$y = \frac{2}{3}x + \frac{1}{9}$	The denominator factors into $(3x+4)(x-5)$ so that f simplifies to $f(x) = \frac{x(2x+3)}{3x+4}$.

Slant asymptote calculation:

$$
\begin{array}{r}
\frac{2}{3}x + \frac{1}{9} \\
3x+4 \overline{\smash{)}\ 2x^2 + 3x } \\
-2x^2 - \frac{8}{3}x \\
\hline
\frac{1}{3}x \\
-\frac{1}{3}x - \frac{4}{9} \\
\hline
-\frac{4}{9}
\end{array}
$$

Rogue 2: We know the following. $g(x) = \frac{p(x)}{q(x)}$, where p is a polynomial with degree 4 and leading coefficient -3. $q(x)$ is a polynomial with integer coefficients that has a zero of $x = \frac{1}{2}$, a zero of $x = 3$ with multiplicity 3, and no other zeros. Last spotted was $q(2) = -3$. It is known p and q share no zeros.

End behavior
$\lim\limits_{x \to \infty} g(x) = \frac{-3}{1} = -3$
$\lim\limits_{x \to -\infty} g(x) = \frac{-3}{1} = -3$

Zeros/Multiplicity
There are 4, but they are unknown...

Hole/s
None

Vertical asymptote/s
$x = \frac{1}{2}, x = 3$

Horizontal asymptote
$y = -\frac{3}{2}$

Slant asymptote
None

Additional Information

Rogue has the form $g(x) = \dfrac{-3x^4 + bx^3 + cx^2 + dx + e}{a(2x-1)(x-3)^3}$. Substituting $x = 2$ and solving $q(2) = -3$ gives $a = 1$.

Solutions

Rogue 3: $g(x) = -x(x+2)^3(3-2x)^2$

End behavior	
$\lim_{x \to \infty} g(x) = -\infty$ $\lim_{x \to -\infty} g(x) = -\infty$	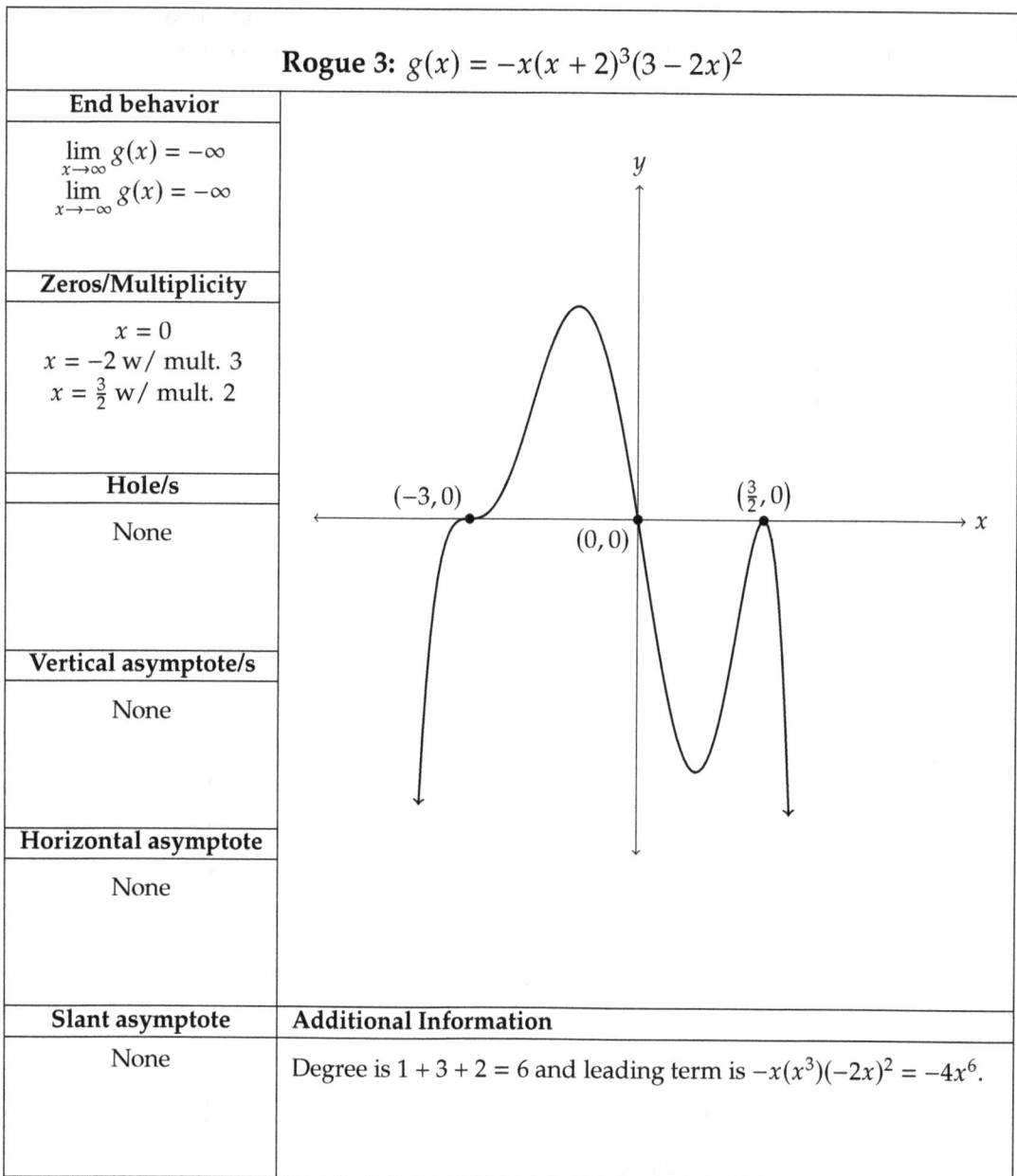
Zeros/Multiplicity	
$x = 0$ $x = -2$ w/ mult. 3 $x = \frac{3}{2}$ w/ mult. 2	
Hole/s	
None	
Vertical asymptote/s	
None	
Horizontal asymptote	
None	
Slant asymptote	**Additional Information**
None	Degree is $1 + 3 + 2 = 6$ and leading term is $-x(x^3)(-2x)^2 = -4x^6$.

Solutions

Rogue 4: Rogue 3 is a known associate. Rogue 4 uses Rogue 3 as a denominator while keeping its own identity hidden. Thus, Rogue 4 is given only by $h(x) = \dfrac{a(x-k)^4}{g(x)}$, where g is Rogue 3. It is known that $a > 0$ and $k > 2$.

End behavior	
$\lim_{x \to \infty} h(x) = 0$ $\lim_{x \to -\infty} h(x) = 0$	
Zeros/Multiplicity	
$x = k$ w/ mult. 4	
Hole/s	
None	
Vertical asymptote/s	
$x = 0, x = -2, x = \frac{3}{2}$	
Horizontal asymptote	
$y = 0$	
Slant asymptote	**Additional Information**
None	The numerator and g share no zeros since $k > 2$. For x with large magnitude, $h(x) \approx \dfrac{ax^4}{-4x^6} = -\dfrac{a}{4x^2}$. For all x with $x < 2$, $0 < x < \frac{3}{2}$, and $x > 2$, $h(x) > 0$. $h(x) < 0$ only for $-2 < x < 0$.

Solutions

Rogue 5: $f(x) = \dfrac{5}{2x-1} - \dfrac{9}{(2x-1)^2}$

End behavior
$\lim\limits_{x \to \infty} f(x) = 0$
$\lim\limits_{x \to -\infty} f(x) = 0$

Zeros/Multiplicity
$x = \frac{7}{5}$

Hole/s
None

Vertical asymptote/s
$x = \frac{1}{2}$

Horizontal asymptote
$y = 0$

Slant asymptote
None

Additional Information

f simplifies into $f(x) = \dfrac{5(2x-1) - 9}{(2x-1)^2} = \dfrac{10x - 14}{(2x-1)^2}$.

1.7-1.11 Review Reflection

Suggested Class Time. 80 minutes

Prerequisites. Students should be proficient or reasonably close to proficient with the material from Topics 1.7 to 1.11! Factoring is a must as well.

Instructional Strategies. I put the solution for each rogue in a separate folder and marked it accordingly. Then, I took these folders and hid them around the room (where they were still somewhat visible). Then, at the start of the activity, I assigned each group a different rogue to start on (I had 5 groups, so this worked perfectly, but it works all the same with different numbers of groups). Once a group was finished with a rogue, I would have them go check the solution. After they were done, I told them to go hide the folder in a new location. Students really had fun with the Easter egg nature of this.

Technology. No calculators!

Topic 1.12 ~ Transformations of Functions

Learning Objectives
1. (1.12.A) Construct a function that is an additive and/or multiplicative transformation of another function.

Success Criteria
1. I can construct a function that is a linear transformation of another function.
2. I can identify how linear transformations affect the domain and range of a function.

Up to now, you've studied polynomial functions (linear, quadratic, cubic, and of greater degree) and rational functions. Before we move into more function types, we want to discuss *transforming* functions.

1. Go to the link at https://www.geogebra.org/m/y8wq7zeq.

2. You will see the graph of a continuous piecewise function $f(x)$. By adjusting a, b, h, and k, we can *linearly transform* $f(x)$ into a new function $g(x)$. Click "Show Sliders & Transformed Function."

3. First, investigate a.

   (a) Move a to 2. How does the graph of g compare with the graph of f? Use one or more of the following terms: dilate, reflect, translate.

   (b) Move a to 0.5. How does the graph of g compare with the graph of f? Use one or more of the following terms: dilate, reflect, translate.

   (c) Move a to -1. How does the graph of g compare with the graph of f? Use one or more of the following terms: dilate, reflect, translate.

   (d) Move a to -3. How does the graph of g compare with the graph of f? Use one or more of the following terms: dilate, reflect, translate.

4. Fill in the blanks:

   The graph of $g(x) = af(x)$ is a _____ dilation of the graph of f by a factor of ____ with a reflection over the _____ if ____.

5. Next, investigate b. Return the other sliders to $a = 1, h = 0, k = 0$.

   (a) Move b to 2. How does the graph of g compare with the graph of f? Use one or more of the following terms: dilate, reflect, translate.

   (b) Move b to 0.5. How does the graph of g compare with the graph of f? Use one or more of the following terms: dilate, reflect, translate.

(c) Move b to -1. How does the graph of g compare with the graph of f? Use one or more of the following terms: dilate, reflect, translate.

(d) Move b to -2. How does the graph of g compare with the graph of f? Use one or more of the following terms: dilate, reflect, translate.

6. Fill in the blanks:

    The graph of $g(x) = f(bx)$ is a _____ dilation of the graph of f by a factor of ____ with a reflection over the _____ if ____.

7. Third time: investigate h. Return the other sliders to $a = 1, b = 1, k = 0$.

    (a) Move h to 2. How does the graph of g compare with the graph of f? Use one or more of the following terms: dilate, reflect, translate.

    (b) Move h to 4. How does the graph of g compare with the graph of f? Use one or more of the following terms: dilate, reflect, translate.

    (c) Move h to -3. How does the graph of g compare with the graph of f? Use one or more of the following terms: dilate, reflect, translate.

8. Fill in the blanks:

    The graph of $g(x) = f(x + h)$ is a _____ translation of the graph of f by _____.

9. Finally, investigate k. Return the other sliders to $a = 1, b = 1, h = 0$.

    (a) Move k to 2. How does the graph of g compare with the graph of f? Use one or more of the following terms: dilate, reflect, translate.

    (b) Move k to 5. How does the graph of g compare with the graph of f? Use one or more of the following terms: dilate, reflect, translate.

    (c) Move k to -3. How does the graph of g compare with the graph of f? Use one or more of the following terms: dilate, reflect, translate.

10. Fill in the blanks:

    The graph of $g(x) = f(x) + k$ is a _____ translation of the graph of f by _____.

11. Before moving to some practice, it's worth investigating the domain and range. Return $a = 1$, $b = 1, h = 0, k = 0$.

    (a) What are the domain and range of $f(x)$?

    (b) Change a to 2. Did $a = 2$ affect the domain, range, or both, and how did it affect them?

(c) Change a to -1. Did $a = -1$ affect the domain, range, or both, and how did it affect them?

(d) Return a to 1. Now change b to 2. Did $b = 2$ affect the domain, range, or both, and how did it affect them?

(e) Now change b to -0.5. Did $b = -0.5$ affect the domain, range, or both, and how did it affect them?

(f) Return b to 1. Now change h to 3. Did $h = 3$ affect the domain, range, or both, and how did it affect them?

(g) Return h to 0. Now change k to -2. Did $k = -2$ affect the domain, range, or both, and how did it affect them?

Try the following practice questions.

12. The function $h(x)$ is transformed to create the function $k(x) = -2h(3x)+1$. Describe the transformations – in order – that must occur to transform h into k.

13. The domain of the function f is $[-3, 4]$ and the range of f is $[-2, 9]$. What are the domain and range of $g(x) = -2f(x-4) + 1$?

14. The graphs of $f(x)$ and $g(x) = af(bx) + k$ are shown for integers a, b, k. What are $a, b,$ and k?

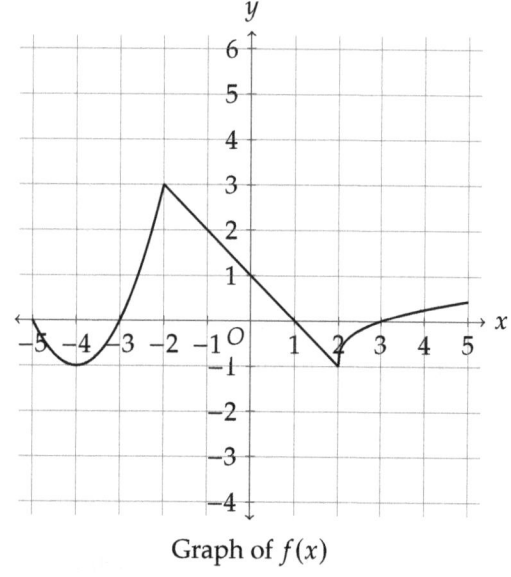

Graph of $f(x)$

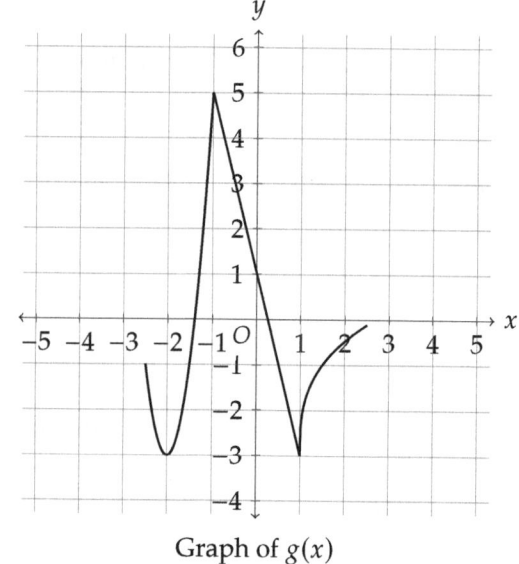

Graph of $g(x)$

Notes

Topic 1.12 Homework

1. The graph of $f(x) = x^2 - 3x + 5$ is to be translated left 4 units and down 4 units. Write, but do not simplify, an expression for this modified $f(x)$. (Note: Write your expression in terms of x, not simply $f(x)$.)

2. A teacher is going to use the function $S(x)$ to scale students' scores from a recent quiz she gave out of 50 points. She will scale a raw score of x (out of 50) to a scaled score of S (out of 50). After creating the function, the teacher would like to make some modifications. Firstly, she would like to increase all of the scaled scores by 10% (so a student whose scaled score was a 60 will increase to a 66). Next, she would like to add a buffer by adding 2 points to every student's scaled score. Write an expression for $M(x)$, the modified scaling function.

3. List the transformations that would transform the graph of a function $g(x)$ into the graph of the function $h(x) = -3g(2(x+3))$.

4. Shown below are values of the polynomial function f for given values of x.

x	-3	-2	-1	0	1	2	3
$f(x)$	-8	0	2	1	0	2	10

   (a) Determine the degree of f.
   (b) Sketch a graph of the function $g(x) = \frac{1}{2}g(\frac{1}{2}x)$.

5. Transformed functions of the form $g(x) = f(bx + c)$ can be tricky. Consider, for instance, the function $f(x) = x^2$ and the function $g(x) = f(2x + 4) = (2x + 4)^2$. The graphs of each are shown below. It may seem from $g(x) = f(2x + 4)$ that the graph of f is translated 4 units to the left, but it isn't. How much is the graph of f translated left by? Think about why and write your best explanation for this translation.

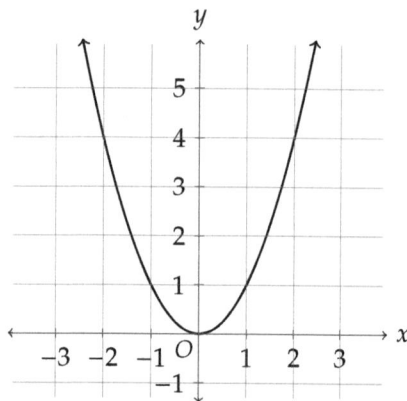 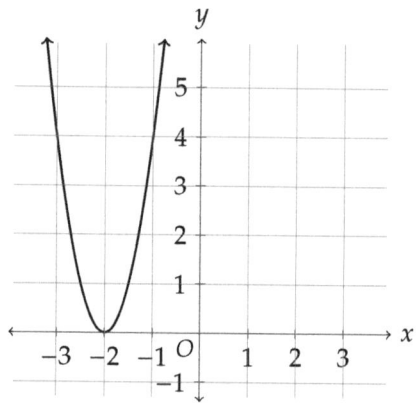

6. Describe the transformations that would transform $h(x)$ into $k(x) = 2h(3x - 1)$.

7. The graph of $f(x)$ includes the points $(2, 0)$, $(3, -1)$, $(4, 5)$, and $(6, 8)$. Let $g(x) = \frac{1}{2}f(2x) - 3$. Find a solution of $g(x) = 1$.

8. The domain of $f(x)$ is $[-4, 2]$ and the range is $[-3, 5]$. Find the domain and range of $g(x) = \frac{3}{2}f(2x) - 1$.

Topic 1.12 Solutions/Notes

Activity

3. (a) The graph of f is vertically dilated by a factor of 2.
    (b) The graph of f is vertically dilated by a factor of $\frac{1}{2}$.
    (c) The graph of f is reflected over the x-axis.
    (d) The graph of f is reflected over the x-axis and vertically dilated by a factor of 3.

4. The graph of $g(x) = af(x)$ is a <u>vertical</u> dilation of the graph of f by a factor of $|a|$ with a reflection over the x-axis if <u>$a < 0$</u>.

5. (a) The graph of f is horizontally dilated by a factor of $\frac{1}{2}$.
    (b) The graph of f is horizontally dilated by a factor of 2.
    (c) The graph of f is reflected over the y-axis.
    (d) The graph of f is reflected over the y-axis and horizontally dilated by a factor of $\frac{1}{2}$.

6. The graph of $g(x) = f(bx)$ is a <u>horizontal</u> dilation of the graph of f by a factor of <u>$|1/b|$</u> with a reflection over the <u>y-axis</u> if <u>$b < 0$</u>.

7. (a) The graph of f is horizontally translated by -2 units.
    (b) The graph of f is horizontally translated by -4 units.
    (c) The graph of f is horizontally translated by $+3$ units.

8. The graph of $g(x) = f(x + h)$ is a <u>horizontal</u> translation of the graph of f by <u>$-h$</u> units.

9. (a) The graph of f is vertically translated by $+2$ units.
    (b) The graph of f is vertically translated by $+5$ units.
    (c) The graph of f is vertically translated by -3 units.

10. The graph of $g(x) = f(x) + k$ is a <u>vertical</u> translation of the graph of f by <u>k</u> units.

11. (a) Domain: $[-5, 8]$, Range: $[-6, 5]$
    (b) The endpoints of the range are doubled; the range is $[-12, 10]$.
    (c) The endpoints of the range are each multiplied by -1, which reverses them to $[-5, 6]$.
    (d) The endpoints of the domain are halved; the domain is $\left[-\frac{5}{2}, 4\right]$.
    (e) The endpoints of the domain are each multiplied by -1, which reverses them, and then they are doubled; the new domain is $[-16, 10]$.
    (f) The endpoints of the domain each have 3 subtracted from them; the domain is $[-8, 5]$.
    (g) The endpoints of the range each have 2 subtracted from them; the range is $[-8, 3]$.
    (h) Horizontally dilated by a factor of $\frac{1}{3}$, then vertically dilated by a factor of 2 and reflected over the x-axis, then vertically translated 1 unit.
    (i) Domain: $[-3, 4] \to [-3 + 4, 4 + 4] \to [1, 8]$
        Range: $[-2, 9] \to [-9, 2] \to [-18, 4] \to [-17, 5]$
    (j) $f(6) = 4$ and $f\left(\frac{6}{2}\right) = f(3) = 4$. Since $g(3) = 3$, m must be -1.
    (k) $g(x) = 2f(2x)$.

12. Horizontal dilation by a factor of $\frac{1}{3}$, vertical dilation by a factor of 2 with a reflection over the x-axis, and a vertical translation by $+1$ (up one).

13. The domain will be translated 4 units to the right: $[1, 8]$. The range will be dilated by a factor of 2, reflected over the y-axis, and then translated up 1 unit. This results in $[-17, 5]$.

14. $a = 2$, $b = 2$, and $k = -1$ (so $g(x) = 2f(2x) - 1$).

Notes

For a function $f(x)$, any of the transformations a, b, h, k applied to create $g(x) = af(b(x + h)) + k$ is called a LINEAR TRANSFORMATION.

Multiplication of the output by a vertically dilates the graph of f by $|a|$ and reflects the graph of f over the x-axis if $a < 0$.

Multiplication of the input by b horizontally dilates the graph of f by $\left|\frac{1}{b}\right|$ and reflects the graph of f over the y-axis if $b < 0$.

Adding h to the input is a horizontal translation of the graph of f by $-h$ units.

Adding k to the output is a vertical translation of the graph of f by $+k$.

The domain and range can be affected by a transformation. It is useful to note that only b and h will affect the domain, while only a and k will affect the range.

Homework

1. $f(x) = (x + 4)^2 - 3(x + 4) + 1$ (The expanded and simplified form is $f(x) = x^2 + 5x + 15$.)

2. $M(x) = 1.1S(x) + 2$

3. In order: horizontally dilate by a factor of $\frac{1}{2}$; translate left by 3 units; vertically dilate by a factor of 3 and reflect over the x-axis.

4. (a) The 3rd differences are constant, so the degree of f is 3.

x	−3	−2	−1	0	1	2	3
$f(x)$	−8	0	2	1	0	2	10
1st diff.		8	2	−1	−1	2	8
2nd diff.			−6	−3	0	3	6
3rd diff.				3	3	3	3

(b) The graph of f is dilated horizontally by a factor of 2 and vertically by a factor of $\frac{1}{2}$, so we can create the following table of values of g. This allows us to sketch a graph of g.

x	−6	−4	−2	0	2	4	6
$g(x)$	−4	0	1	$\frac{1}{2}$	0	1	5

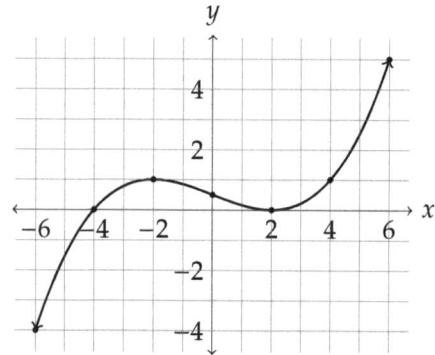

5. It has been translated 2 units to the left, because $f(2x + 4) = f(2(x + 2))$. Thus, $h = 2$.

6. We have $k(x) = 2h(3x - 1) = 2h(3(x - \frac{1}{3}))$. So h is dilated horizontally by a factor of $\frac{1}{3}$; translated horizontally by $\frac{1}{3}$ units to the right; and dilated vertically by a factor of 2.

7. We solve:
$$\frac{1}{2}f(2x) - 3 = 1$$
$$\frac{1}{2}f(2x) = 4$$
$$f(2x) = 8.$$

Since $(6, 8)$ lies on the graph of f, $f(6) = 8$. Hence, $2x = 6$, so $x = 3$.

8. Domain: $\left[-\frac{4}{2}, \frac{2}{2}\right] = [-2, 1]$; range $\left[\frac{3}{2}(-3) - 1, \frac{3}{2}(5) - 1\right] = \left[-\frac{11}{2}, \frac{13}{2}\right]$

Reflection

Suggested Class Time. 55-60 minutes

Prerequisites. Students should have proficiency with identifying domain and range and writing them as both intervals and inequalities.

Instructional Strategies. A few common student mistakes included:

- using the variable x in inequalities involving range;
- writing inequalities/intervals backwards or incorrectly (e.g. $[5, -3]$ or $5 \leq x \leq -3$); and
- not including absolute value when discussing the factor by which a function's graph dilates.

One major point of note: avoid saying "stretch" or "compress." Though these terms can be helpful if they're understood precisely (and correctly), it's actually best to simply stick to "dilate by a factor of..." in all multiplicative cases – this is also how it's written in the CED. This will sidestep any student confusion, especially for horizontal transformations. When it comes to translations, the CED also only uses "horizontal translation by $-h$" or "vertical translation by $+k$," but I didn't find any potential confusion for students in utilizing "left" or "right."

For helping students keep transformations straight, I went with an argument related to inputs/outputs:

Transformations affecting inputs will be *unintuitive*,
while transformations affecting outputs will be *intuitive*.

I wanted to avoid saying any loaded (or otherwise defined) terms like "opposite" or "inverse," so I went with "unintuitive." This (hopefully) helped students see that multiplication by b leads to a dilation by its reciprocal, and addition of h leads to a horizontal translation by $-h$.

In the discussion portion after the activity, we also talked a bit more at length about how to write intervals, as well as which transformations affect domain (b and h, which affect *inputs*) and range (a and k, which affect *outputs*). We spent a spare 5 minutes going over some odd-seeming transformations, including the following:

- The transformation of $f(x) = x^2$ to $g(x) = 9x^2$ can be considered a vertical dilation by a factor of 9 *or* a horizontal dilation by a factor of $\frac{1}{3}$.
- The transformation of $f(x) = x^2$ to $g(x) = (2x + 4)^2$ does not include a horizontal translation by -4, but just -2; students were somehow blown away, so I had to really emphasize the difference between $2(x + 4)$ and $2x + 4$ (and, more generally, $b(x + h)$ versus $bx + h$).

After grading the exit ticket questions I gave, though, I realized that it is going to take more practice for students to recognize transformations in functions expressed in terms of x, e.g. $f(x) = \frac{x^2}{4}$ or $g(x) = (x - 2)^2 + (x - 2) + 3$.

Technology. Students used their laptops to access a Geogebra link.

Problems.

1. If the range of $f(x)$ is $[-4, 6]$, what is the range of $g(x) = 3f(\frac{1}{2}x) - 2$?

   (A) $[-14, 16]$ (B) $[-10, 20]$ (C) $[-4, 1]$ (D) $[-10, 10]$

2. The graphs of the functions $f(x)$ and $g(x)$ are shown below.

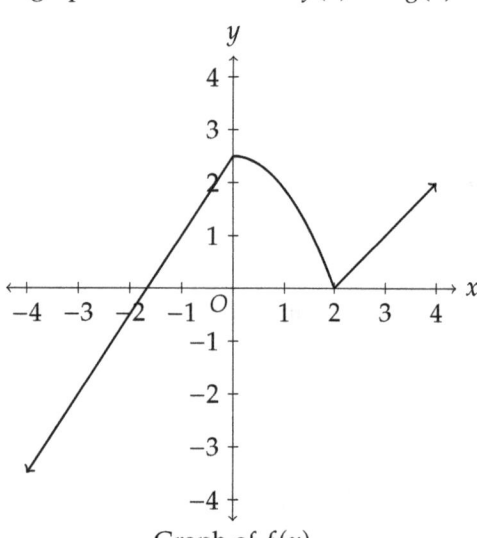
Graph of $f(x)$

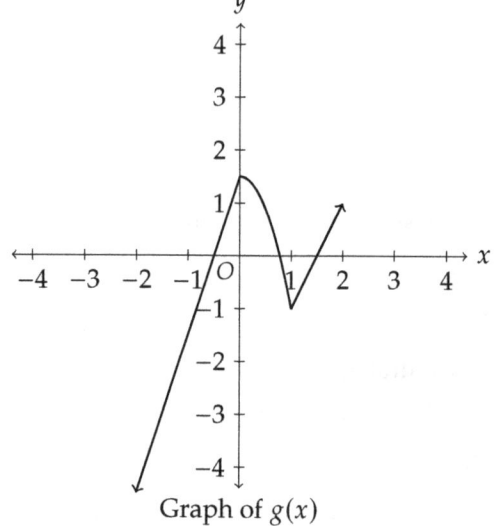
Graph of $g(x)$

   Which of the following expresses $g(x)$ in terms of $f(x)$?

   (A) $g(x) = f(\frac{4}{3}x) + 1$ (B) $g(x) = f(\frac{3}{4}x) - 1$ (C) $g(x) = f(2x) - 1$ (D) $g(x) = f(\frac{1}{2}x) - 1$

3. For the function g it is known that $g(2) = 0$ and $g(3) = -10$. The function f is given by $f(x) = g(x+3)$. Which of the following must be a solution to $f(x) = 0$?

   (A) $x = -3$ (B) $x = -1$ (C) $x = 0$ (D) $x = 2$

4. The domain of the function g is $-5 \leq x \leq 7$. The function h is given by $h(x) = 2g(\frac{1}{4}x) + 3$. What is the domain of h?

   (A) $-20 \leq x \leq 28$ (B) $-17 \leq x \leq 31$ (C) $-10 \leq x \leq 14$ (D) $-\frac{5}{4} \leq x \leq \frac{7}{4}$

5. The function f has a domain of all real numbers. Selected values of $f(x)$ are in the table below.

x	0	1	2	3	4	5
$f(x)$	-23	-6	1	2	3	10

   The function g is defined by $g(x) = 3f(2x - 1) + 4$. Which of the following must be a solution to $g(x) = 34$?

   (A) $x = 2$ (B) $x = 3$ (C) $x = 10$ (D) $x = \frac{33}{2}$

Solutions.

1. The function f undergoes a vertical dilation by a factor of 3 and a vertical translation of -2 units. The vertical dilation by a factor of 3 "multiplies" the range by 3: from $[-4, 6]$, we get $[-12, 18]$. Then the vertical translation of -2 decreases the range by 2 to become $[-14, 16]$. The correct answer is therefore **(A)**.

2. The graph of f has undergone a horizontal dilation by a factor of $\frac{1}{2}$ and a vertical translation of -1 units. Thus, the function g is $g(x) = f(2x) - 1$. The correct answer is therefore **(C)**.

3. Since $g(2) = 0$, and we want $g(x+3)$ to be zero, we need $x + 3 = 2$. Thus, $x = -1$. Hence, $f(-1) = g(-1+3) = g(2) = 0$. The correct answer is therefore **(B)**.

4. The function h is a horizontal dilation of the graph of g by a factor of 4, making the domain of h to be $-20 \leq x \leq 28$. The correct answer is therefore **(A)**.

5. We need $3f(2x-1) + 4 = 34$. Then $f(2x-1) = 10$. The value of x which gives $f(x) = 10$ is 5. Thus, $2x - 1 = 5$ which gives $x = 3$. The correct answer is therefore **(B)**.

Topic 1.13 ~ Selecting a Function Model

Learning Objectives

1. (1.13.A) Identify an appropriate function type to construct a function model for a given scenario.
2. (1.13.B) Describe assumptions and restrictions related to building a function model.

Success Criteria

1. I can determine whether a linear, quadratic, cubic, piecewise, or other function is appropriate to build a function model.
2. I can identify assumptions and restrictions that may apply to a given function model in context.

We have now studied a variety of mathematical functions, including polynomials and rational functions. These functions can often be used to *model* a given scenario. A model may exactly describe reality or may be used as a simpler means of approximating reality.

The most difficult part of modeling is not using a model, but *finding* one.

To pick a model, we often rely on the behavior of the zeros, extrema, and rates of change.

Model Types

1. *Linear function*
    - Roughly _____ rate of change

2. *Quadratic function*
    - Roughly _____ rate of change
    - Graphs are _____ with a single _____.
    - 2nd differences are roughly _____.
    - Geometric contexts involving _____

3. *Polynomial of degree n*
    - nth differences are roughly _____.
    - Multiple zeros and minima or maxima

4. *Piecewise function*
    - Exhibit different characteristics over nonoverlapping domain intervals

5. *Rational function*
    - Quantities that are _____
        Definition: Quantities a and b are _____ if $ab = k$ for some real k. This can also be rewrriten as $b = \frac{k}{a}$ or $a = \frac{k}{b}$.

It is important to note that all models come with *assumptions*. These may include:
- Domain restrictions: Are negative values allowed? Is there a fixed maximal input? Does the model only apply for certain inputs?
- Range restrictions: Do outputs have to be integers? Are outputs only reasonable in a certain range?
- Generalization: Is the model being used to generalize? Is this reasonable?

Examples

1. A wireless phone carrier allows customers to try their network for only $30 for 6 months. After this, the network costs $25 per month. However, if customers stay with the company for more than 2 years, this price drops to $15 per month. The company is always willing to "prorate," meaning that if a customer drops the network x days into a month, the company will only charge a fraction of the bill ($\frac{x}{\text{days in the month}}$). A model $C(x)$ is to be used modeling the cost C, in dollars, of being on this network for x months.

    (a) What are some observations you have about the relationship between x and C in context? Can you sketch a rough graph of what $C(x)$ should look like?

    (b) What kind of model could be used here?

    (c) Construct a function $C(x)$ that could be used to model the total cost of using this network for x months.

2. Rob has a chicken coop that is 10 feet wide by 6 feet long. He is going to extend both the width and length of the coop by x feet, and he would like to construct a model $A(x)$ that will give the increased amount of area A based on how much x he extends the coop by.

    (a) What are some observations you have about the relationship between x and A in context? Can you sketch a rough graph of what $A(x)$ should look like?

    (b) What kind of model could be used here?

    (c) Construct a function $A(x)$ that could be used to model the increase in area A based on extensions of x feet.

    (d) If Rob bought 40 feet of lumber to build the extended part of the coop (assuming the two sides that aren't being extended aren't reusable and that he may not use all of the lumber), what is a reasonable domain for $A(x)$?

3. The intensity I of light emitted from a camera's flash is inversely proportional to the square of the distance d that an object is from the camera. Suppose that a camera's flash has an intensity of 1,600 watt-seconds at a distance of 2 meters.

    (a) What are some observations you have about the relationship between d and I in context? Can you sketch a rough graph of what $I(d)$ should look like?

    (b) Construct a function $I(d)$ that could be used to model the intensity I in terms of the distance d in meters.

    (c) What would the intensity of the camera's flash be at a distance of 20 meters?

4. A streaming service in its infancy was tracking the number of subscribers S it had, in tens of thousands, after t months of launching the service. The table below shows values of s for $0 \leq t \leq 6$.

t	1	2	3	4	5	6
S	2.3	7.4	15.3	25.9	39.5	56.0

(a) What are some observations you have about the relationship between t and S in context? Can you sketch a rough graph of what $S(t)$ should look like?

(b) What kind of model could be used here?

Practice

1. In 2015, 1,800 bison were introduced to a protected wildlife reserve. From 2016 to 2022, the change in the number of bison were recorded yearly. The graph below shows the change in the population of bison (in hundreds) from the previous year, where $t = 0$ is 2016. The point $(1, -1)$ therefore represents a decrease in 100 bison from 2016 to 2017.

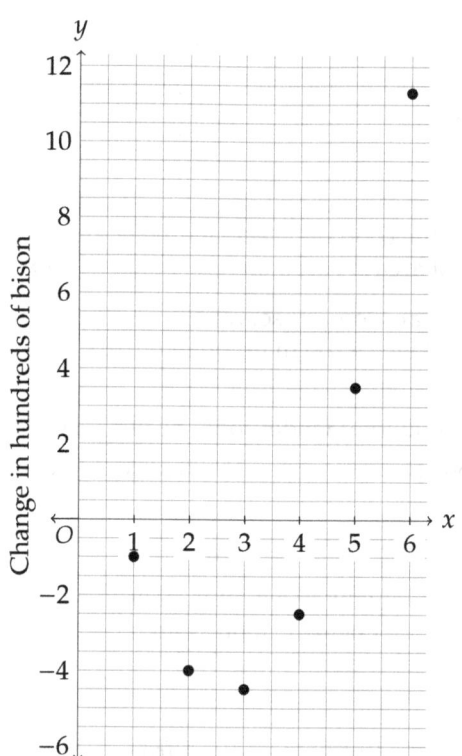

(a) Researchers at the organization running the preserve wish to create a function $P(t)$ that can be used to model the bison population based on the data provided in the graph. What kind of function should $P(t)$ be? Explain.

(b) One year, the researchers were concerned at the dropping bison population, so they decided to intervene over the next year, extending the roaming range for the bison and removing environmental factors that might be limiting population growth. The following year, the number of bison on the reserve increased for the first time. What year did the researchers intervene? Justify your answer.

(c) Is it reasonable to assume that the domain of $P(t)$ is $t \geq 0$?

Topic 1.13 Homework

1. The table below shows the domestic box office gross B, in billions of dollars, for each calendar year from $t = 0$, which represents 2012, to $t = 10$, which represents 2022.

Years since 2012	0	1	2	3	4	5	6	7	8	9	10
Box office (billions)	10.8	10.9	10.4	11.1	11.4	11.1	11.9	11.4	2.1	4.5	7.4

   (a) If a function model $B(t)$ were created for this data set, what type of function would be most appropriate? Explain.

   (b) Use your model to predict what the box office will be in 2023.

   (c) What, if any, assumptions are being made here about the domain and/or range? Are any asusmptions being made about the calculation in part (b)?

2. For each scenario described, determine what kind of function model may be appropriate to describe the relationship between the two variables outlined.

   (a) A cart is rolling down a hill. In the first second, it rolls 6 feet. Every second after that, the cart rolls 5 more feet than it did in the previous second. A model will be constructed to find the distance, in feet, the cart has traveled based on time, in seconds.

   (b) According to the Audubon Society's 2022 "State of the Birds" report, the population of ducks in the United States has varied quite a bit over the last 50 years.

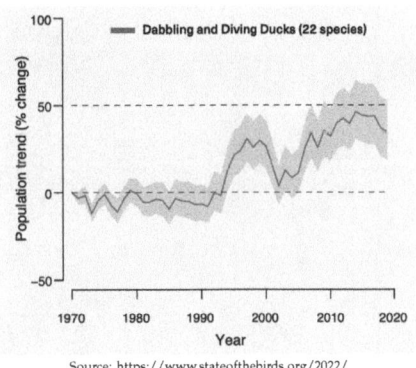

Source: https://www.stateofthebirds.org/2022/

   (c) The graph below shows the total number of broadband subscribers from 1998 to 2022. A function model could be used to create the upper-most graph.

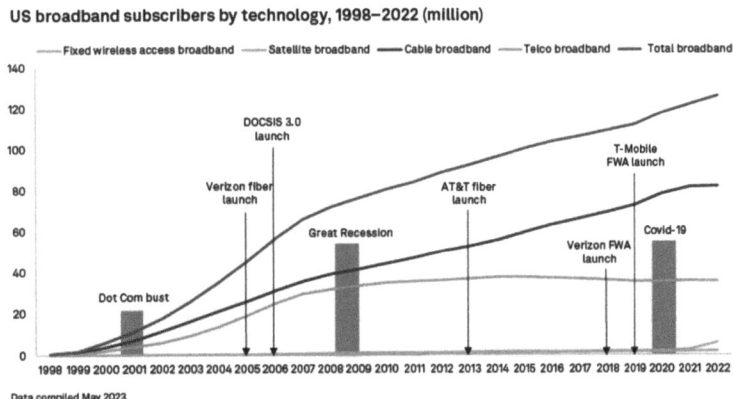

(d) A certain airplane can hold 20,000 pounds of luggage and can hold up to x bags, which varies from flight to flight. The airline that manages the airplane and schedules the flights is deciding how much each passenger is allowed to bring per bag on the flight. They will construct a model for the amount y (in pounds) each bag is limited to based on how many bags x they can fit on the airplane.

3. In the table below are values of the functions $f, g, h,$ and k for selected values of x.

x	1	2	3	4	5	6	7
$f(x)$	−1.4	−5	−7.8	−8	−4	5	21
$g(x)$	3.2	4.5	5.9	7.1	4.7	2.3	0
$h(x)$	12	8	6	4.8	4	3.42	3
$k(x)$	−8	−6	−2.1	4.2	12.2	22.3	34.1

Identify a type of appropriate function model with each function - you do not need to create the actual function model. (Assume that for x-values not shown in the table, the function and its rates of change don't change behavior.)

Topic 1.13 Solutions/Notes

Model Types

1. *Linear function*
   - Roughly <u>constant</u> rate of change

2. *Quadratic function*
   - Roughly <u>linear</u> rate of change
   - Graphs are <u>symmetric</u> with a single <u>min or max</u>.
   - 2nd differences are roughly <u>constant</u>.
   - Geometric contexts involving <u>area</u>

3. *Polynomial of degree n*
   - nth differences are roughly <u>constant</u>.
   - Multiple zeros and minima or maxima

4. *Piecewise function*
   - Exhibit different characteristics over nonoverlapping domain intervals

5. *Rational function*
   - Quantities that are <u>inversely proportional</u>
     Definition: Quantities a and b are INVERSELY PROPORTIONAL if $ab = k$ for some real k.
     This can also be rewrriten as $b = \frac{a}{k}$ or $a = \frac{b}{k}$.

Examples

1. (a) There are 3 separate constant rates of change (0, 25, and 15). The graph should look like 3 different line segments.
   (b) Piecewise linear
   (c) The function would be $C(x) = \begin{cases} 30 & 0 < x \leq 6 \\ 30 + 25(x - 6) & 6 < x \leq 24 \\ 30 + 25(18) + 15(x - 24) & x > 24 \end{cases}$

2. (a) This is an area problem, so a quadratic is likely involved. As x increases, the increase in area will involve the square of x, increasing at an increasing rate.
   (b) Quadratic
   (c) The increase in area is the total area of the coop minus the original area of the coop, or $A(x) = (x + 10)(x + 6) - 10(60) = x^2 + 16x$.
   (d) The diagram below will help.

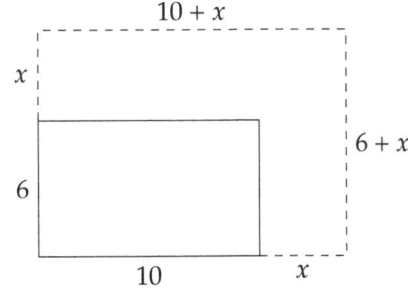

The new part of the perimeter requires $x + 10 + x + 6 + x + x = 16 + 4x$ feet of lumber, so we can solve $16 + 4x = 40$ to get $x = 6$ for the maximum x can be. Therefore, a reasonable domain is $0 < x \leq 6$.

3. (a) I and d^2 are inversely proportional, so as d increases, I decreases. A rational function will be appropriate.

   (b) The general form is $I(d) = \frac{k}{d^2}$, so we can solve $1600 = \frac{k}{2^2}$ to get $k = 6400$. Therefore, $I(d) = \frac{6400}{d^2}$.

   (c) $I(20) = \dfrac{6400}{20^2} = 16$ watt-seconds

4. (a) Looking at the first and second differences below, we see that the second differences are roughly constant (the first differences are increasing at a constant rate).

t	1	2	3	4	5	6
S	2.3	7.4	15.3	25.9	39.5	56.0
First diff.		5.1	7.9	10.6	13.6	16.5
Second diff.			2.8	2.7	3	2.9

   (b) A quadratic model would be appropriate.

Practice

1. (a) The graphed function is the first differences of $P(t)$. Continuing on by looking at the second and third differences, we get the table below.

t	1	2	3	4	5	6
First differences of $P(t)$	-1	-4	-4.5	-2.5	3.5	11.3
Second differences		-3	-0.5	2	5	7.8
Third differences			2.5	2.5	3	2.8

   Since the third differences are roughly constant, a cubic function best models this data.

   (b) The rate of change was first positive when $t = 5$, or in the year 2020 to 2021. Therefore, the intervention happened during 2020.

   (c) No. We have no reason to believe that the growth pattern will continue beyond $t = 6$.

Homework

1. (a) A piecewise function is appropriate as there are drastic differences in the rates of change for $t \leq 7$ and $t > 7$. For $t \in 0, 1, 2, \ldots, 7$, a linear model would be appropriate. For $t \in 8, 9, 10$, it is questionable whether the rate of change is constant, but we will use a linear model with a slope that averages the two rates of changes. We could write this function as

   $$B(t) = \begin{cases} 0.15t + 10.8 & t = 0, 1, 2, \ldots, 7 \\ 2.65(t - 8) + 2.1 & t = 8, 9, 10 \end{cases}$$

   where 0.15 is the average of the rates of change from $0 \leq t \leq 7$ and 2.65 is the average of the rates of change for $8 \leq t \leq 10$.

   (b) $B(11) = 10.05$, so 10.05 billion dollars.

   (c) The domain can only include integers, as these are *end-of-year* calculations. Any point with a decimal value would not make sense. In (b), there is an assumption that the "pattern" that has existed for the last 2 years will continue, which is quite a large assumption. (The domestic box office in 2023 was only 8.9 billion!)

2. (a) The rate of change is linear (slope of positive 5), so a quadratic function is appropriate.
   (b) A polynomial function of degree at least 3 seems applicable for years past 1990. Before 1990, you could argue for a constant function.
   (c) There appear to be two reasonably linear pieces of the function, one from 1998 to 2007 and one from 2007 to 2022, so a piecewise linear function seems appropriate.
   (d) The amount each passenger can bring and the total number of bags are inversely proportional, so a rational function is appropriate. ($40000 \approx xy \to y \approx \frac{40000}{x}$)

3. The 3rd differences of f are roughly constant, so a cubic function would be appropriate. For $g(x)$, there are two distinct intervals with roughly constant rates of change, so a piecewise linear function would be appropriate. For $h(x)$, the value decrease inversely proportionally with the inputs, so a rational function would be appropriate. For $k(x)$, the second differences are roughly constant, so a quadratic model could be used.

Reflection

Suggested Class Time. 80 minutes

Prerequisites. Basically all of Unit 1!

Instructional Strategies. If possible, try to spend two days on this lesson - my schedule did not allow for it, but many students found the sheer number of possible models overwhelming.

The biggest instructional strategy here is encouraging students to read closely, annotate, sketch pictures, or anything that will help them internalize the details of the scenario. Students were missing important details left and right.

On a stronger note, it really helped students to be able to memorize the inverse proportional relationship $u = \frac{k}{b}$. When it came to Topic 1.14, students were already fluent in utilizing inversely proportional relationships (now, *determining when this is applicable* was a slightly different story).

Technology. None was required.

Problems.

1. The director of the city aquarium wants to construct a tank for a shark and has determined that the function $V(x) = -\frac{1}{2}(x+2)(x-1)(2x-7)$ will model the volume V, in hundreds of gallons, based on the width x of the tank. Which of the following is a reasonable domain of $V(x)$?

   (A) $-2 < x < 3.5$ **(B)** $1 < x < 2$ **(C)** $x > 1$ **(D)** $1 < x < 3.5$

2. A population of wolves in a nature preserve approaches a value of 50 wolves as time increases. Which of the following could be an appropriate model for the population of wolves?

   (A) $P(t) = 50(t+1)$ **(B)** $P(t) = \dfrac{50}{t+1}$ **(C)** $P(t) = \dfrac{t+1}{50}$ **(D)** $P(t) = \dfrac{50t}{t+1}$

3. The table below gives the cost per gallon of oil, where n is the number of gallons. For example, the cost of 5 gallons is $30 per gallon, for a total of $150 for 5 gallons.

$0 < n \leq 4$	$4 < n \leq 10$	$n > 10$
$36	$30	$25

   Which of the function types is the best choice to model the total cost of n gallons of oil?

   (A) Piecewise linear   **(B)** Linear   **(C)** Quadratic   **(D)** Cubic

4. The table below describes values of the velocity $v(t)$ of a model train moving along its track, in meters per second, for each second over the first four seconds.

t	0	1	2	3	4
$v(t)$	0	5	13	24	38

A polynomial is used to model the velocity v based on time t in seconds. What is the most appropriate degree of that polynomial?

(A) A polynomial of degree 1 because the first differences are constant.

(B) A polynomial of degree 1 because the first differences are increasing at a constant rate

(C) A polynomial of degree 2 because the first differences are constant.

(D) A polynomial of degree 2 because the first differences are increasing at a constant rate.

5. A data set containing five points is graphed below.

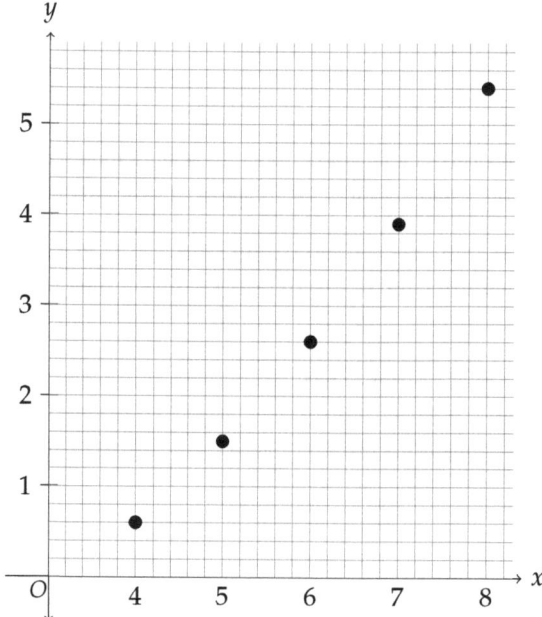

What type of function would be the most appropriate to model the relationship between x and y for this data set?

(A) A linear function because the rate of change is constant.

(B) A linear function because the points roughly appear to form a line.

(C) A quadratic function because the rate of rate of change is constant.

(D) A quadratic function because the points roughly appear to form a curve.

Solutions.

1. The zeros of the function are $x = -2$, $x = 1$, and $x = 3.5$. The function is a cubic with a negative leading coefficient, so $V > 0$ for $x < -2$ and $1 < x < 3.5$. However, x cannot be negative since this represents the width of the tank. Thus, $1 < x < 3.5$ is the domain. The correct answer is therefore **(D)**.

2. We seek a function whose values approach 50 as t increases without bound. This happens for $P(t) = \frac{50t}{t+1}$. The correct answer is therefore **(D)**.

3. The cost per gallon within each interval is constant, and the intervals do not overlap. Hence there is a constant rate of change in the price of n gallons over each interval. Thus, a piecewise linear function is the best model. The correct answer is therefore **(A)**.

4. We look at the differences. We get

t	0	1	2	3	4
$v(t)$	0	5	13	24	28
First differences of $v(t)$	5	8	11	14	
Second differences	3	3	3		

The second differences are constant, meaning the first differences are increasing at a constant rate. This indicates a polynomial of degree 2 would best model the velocity. The correct answer is therefore **(D)**.

5. Let's look at the differences in the outputs between successive equal-length intervals of the inputs.

x	4	5	6	7	8
y	0.6	1.5	2.6	3.9	5.4
First differences of $y(x)$	0.9	1.1	1.3	1.5	
Second differences	0.2	0.2	0.2		

The second differences are constant and positive, indicating the rate of the rate of change is constant. The correct answer is therefore **(C)**.

Topic 1.14 ~ Constructing a Function Model

> **Learning Objectives**
>
> 1. (1.14.A) Construct a linear, quadratic, cubic, quartic, polynomial of degree n, or related piecewise-defined function model.
> 2. (1.14.B) Construct a rational function model based on a context.
> 3. (1.14.C) Apply a function model to answer questions about a data set or contextual scenario.

> **Success Criteria**
>
> 1. I can construct a function model in context using linear function transformations.
> 2. I can construct a function model using technology and regressions.
> 3. I can use a function model to predict values and analyze rates of change with correct units.

In Topic 1.13, we discussed how to find a *type* of function that can be used to model the relationship between two variables. In this activity, we'll discuss how to actually *create* this function.

In general, you'll want to ask yourself three questions when determining what kind of function could be used to model data.

- Do the first, second, third, or nth differences form some noticeable pattern?
- If not... are there distinct patterns on distinct sub-intervals of the domain?
- If not... is there some giveaway from the context, like an inversely proportional relationship or a directly computable perimeter/area/volume model?

Once you've determined the kind of model that will be used, there are two possibilities.

- You'll directly compute the model.
- You'll use technology to perform *regression*.

REGRESSION is a statistical tool used to compute the "best" of a certain kind of function to model the data. A regression function will minimize the total amount of prediction error based on the data.

Example 1. Every year, more Americans are eligible to vote than the year prior. The table below shows the number N of voting-eligible Americans, in millions, for each presidential election year t from 1988 to 2008. (Source: https://www.presidency.ucsb.edu/statistics/data/voter-turnout-in-presidential-elections)

Year t	Voting-eligible population N
1988	173.6
1992	179.7
1996	186.3
2000	194.3
2004	203.5
2008	213.3

1. First, we can analyze the first differences. Press stat, then 1:Edit.... In L1, enter the years. In L2, enter the voting-eligible populations.

2. Go to the very top of L3 (where you're highlighting L3). Press 2nd, then stat, go over to OPS, then select 7:ΔList(.

3. Now, select L2 by typing 2nd then 2. Close the parentheses and hit enter. (To type the name of *any* list L_k, you press 2nd then the value of k, by the way).

4. You should now see the first differences of N in L3. You could argue these are *roughly* constant. If this is true, what kind of model would be appropriate?

5. We can visually check for a linear pattern by looking at a *scatterplot*. To create a scatterplot, do the following.

    - Press y= and clear any functions that are there.
    - Press 2nd y= (statplot). On 1:Plot1, press enter.
    - Highlight On and press enter. Then, for Type, select the first type of graph.
    - Input L1 and L2 into Xlist and Ylist, respectively. Then, press zoom and 9:ZoomStat.

6. Does the graph appear reasonably linear?

7. We can perform *linear regression* as follows.

    - So that we can view one particular statistic, press mode and go down to STAT DIAGNOSTICS. Turn them ON. (If you don't see a stat diagnostics option, let me know!)
    - Press 2nd mode to quit and return to the home screen.
    - Press stat, then go over to CALC. Select 4:LinReg(ax+b).
    - Enter L1 into Xlist and L2 into Ylist. Leave FreqList blank. Then, for Store RegEQ, we have the option of *storing* the resulting function into one of the calculator's y= functions. Enter Y1 by pressing alph trace and selecting Y1. Then, press Calculate.

What you should see now is a *linear regression model*. In this case, our model is a linear function $N(t)$.

8. Write an expression for $N(t)$ based on the regression.

$$N(t) = \underline{\hspace{3cm}}$$

9. The number r^2 is a measure of how well the function minimizes the prediction error. Values closer to 1 mean the function is a stronger predictor, while values closer to 0 mean the function is a weaker predictor. How strong of a predictor will $N(t)$ be?

10. We can visually see how closely the model fits our data by looking at the graph. Since you *already* stored this function into Y1, you can simply hit graph. What do you see?

11. What do you think is a reasonable domain for this function?

12. We can use $N(t)$ to predict future voting age populations. Use your model to predict how many Americans were voting-eligible in 2012. **Hint: Save yourself some typing and use Y1!**

13. The actual voting-eligible population in 2012 was 222.5 million. Was your model exactly correct? Did you expect it to be?

Now, we made the *assumption* that $N(t)$ was linear based on *reasonably* constant first differences. Perhaps, though, another function would do a better job.

14. Use L4 and L3 to find the *second* differences. Are they reasonably constant?

15. Using similar commands as before, perform *quadratic* regression on L1 and L2. Store the function in Y2.

    (a) Write the quadratic expression for $N(t)$: $N(t) =$ _____.

    (b) Do you think the quadratic $N(t)$ is a stronger predictor than the linear $N(t)$? Use both graphical and quantitative evidence to answer.

    (c) Use the quadratic $N(t)$ to predict the voting-eligible population in 2012. Does it do a better job than the linear $N(t)$?

Alongside linear and quadratic regression, your calculator can perform a variety of other regressions: cubic and quartic are the other polynomial types, as well as kinds we'll explore later in this course.

For a second, more brief, example, we'll look at an inversely proportional relationship.

Example 2. A project manager at a corporate software firm is in charge of assigning employees to work on rolling out software at various businesses. Over years of having managed rollouts, the manager noticed that the amount of employees he assigns and the work time required to complete the rollouts are inversely proportional. The table below displays some selected data the project manager has gathered.

Employees assigned	8	12	16	24	32
Hours to completion	150	100	75	50	37.5

16. Letting H be the number of work hours to completion and n be the number of employees assigned, write a function $H(n)$ that could model this data.

17. How long would a rollout take if the project manager assigned 20 employees to it?

18. The project manager has never assigned fewer than 6 employees to such a job and never plans to assign more than 50. What are a reasonable domain and range of $H(n)$?

19. Based on the constraints given, what is the least amount of time this project manager could get one of these rollouts complete in?

Notes

Topic 1.14 Homework

1. A teacher noticed over a few years that, the more students they were teaching in an AP course, the lower it seemed the pass rate (as a % of students passing the AP exam) was at the end of the year. They used 6 years' worth of data in the following table.

Number of students	Passing rate
4	100
17	76.4
19	96
23	73.9
38	71.1
72	65.2

   The teacher is interested in creating a function $P(n)$ that will model the pass rate P based on the number of students n.

   (a) Use technology to perform linear regression on the data. Write down the resulting function $P(n)$.
   (b) Use technology to perform quadratic regression on the data. Write down the resulting function.
   (c) Which function – linear or quadratic – better models the data? Explain.
   (d) Suppose the teacher taught 29 students in the course one year. How many students does the model you identified in (c) predict passed that year?
   (e) Explain why the y-intercept of your model makes no sense in context.
   (f) Based on the model, how many students would need to be in the class for the predicted pass rate to be 80%?
   (g) Identify a reasonable domain for $P(n)$.

2. A greeting card company's sales fluctuate throughout the year. The profits, in tens of thousands of dollars, of the company for the year of 2022 are displayed below.

Month of 2022	Profit (tens of thousands)
1	10
2	11
3	9.5
4	7
5	4.8
6	2.5
7	1
8	0.9
9	1.9
10	3.6
11	6
12	7

(a) Determine a kind of function that could be used to model this data. Justify your choice.

(b) Perform regression to create such a function $P(x)$ to model the profit P, in tens of thousands of dollars, x months into 2022.

3. A music producer charges a flat rate of $500 for people to use his studio for 3 or fewer hours. If people use the studio for longer than 3 hours, the producer simply charges a $100 fee and then charges $125 per hour. However, for individuals who are willing to commit to spending 20 or more hours in the studio, the producer drops the $100 fee and only charges $100 per hour. Assume that the producer will charge individuals pro rata (so 3.5 hours = 100 + 3.5(125) = $537.50). Write a function $C(t)$ for the cost C to rent this studio for t hours.

4. Boyle's Law states that the volume and pressure of a gas are inversely proportional. Suppose a gas with volume V, in liters, exerts pressure P, in kilopascals (kPa), on the walls of the container the gas is in. The table below gives some values of V and P.

V	22	28
P	2.100	1.650

Based on the information given, how many kilopascals of pressure would 15 liters of this gas exert on the walls of the container?

5. The graduating class of a high school is going to host a 10th reunion. As no funds are available for organizing the event, alumni are expected to pay equally to cover the costs of the reunion. The student council books a venue for $500, and it will cost $20 per attendee to eat. For their service in organizing the event, the alumni agree that the 10 student council members will get to attend for free.

(a) Construct a function $C(x)$ for the amount C it will cost each alumnus if x attend the reunion.

(b) Identify a reasonable domain of this function.

(c) The function $C(x)$ has a horizontal asymptote. Interpret it in context.

Topic 1.14 Solutions/Notes

Activity

4. Linear

6. Yes, it appears so, though there is a bit of curvature...

8. $N(t) = 1.985t - 3774.247$

9. r^2 is very close to 1, so $N(t)$ should be a very strong predictor.

10. The graph of $N(t)$ closely fits the given points.

11. $1988 \leq t \leq 2008$

12. $N(2012) \approx 219.573$ million people

13. The model isn't exactly correct, and this isn't surprising: it's a model!

14. Letting L4 equal Δlist(L3), the second differences are 0.5, 1.4, 1.2, and 0.6. These are all close to 1.

15. (a) $N(t) = 0.032t^2 - 126.904t + 124,984.199$

    (b) Since the r^2 is higher than that of the linear model, it appears the quadratic is a stronger predictor of voting-eligible population. Additionally, the graph of the quadratic $N(t)$ more closely follows the given points.

    (c) $N(2012) = 224.39$. This is indeed closer to the actual voting-eligible population by about 1 million.

16. We know $H(n) = \dfrac{k}{n}$ for some k. Plugging in $H = 150$ and $n = 8$, we get $k = 1,200$. Therefore, $H(n) = \dfrac{1200}{n}$.

17. $H(20) = \dfrac{1200}{20} = 60$ hours

18. Domain: $[6, 50]$, range: $\left[\dfrac{1200}{50}, \dfrac{1200}{6}\right] \to [24, 200]$

19. 24 work hours

Notes

REGRESSION is a means of finding a function that fits data in such a way that prediction errors are minimized. The measure of how well the model minimizes prediction error is r^2.

There are all sorts of function models, and some can be computed directly using information from a given scenario, like area models, rational functions from inversely proportional relationships, and piecewise functions.

Homework

1. (a) $P(n) = -0.457n + 93.618$

   (b) $P(n) = 0.011n^2 - 1.326n + 104.49$

   (c) The r^2 was substantially higher for the quadratic model ($0.725 > 0.591$) than the linear model, so a quadratic function is a better fit.

   (d) $P(29) \approx 75.181\%$

(e) First off, if the teacher taught 0 students, the class wouldn't exist! If it did, it certainly wouldn't have a pass rate of 104%!

(f) Storing the quadratic function in Y1 and setting Y2=80, we find the intersection occurs at $n = 23$.

(g) A negatively oriented quadratic will start to increase again, which makes no sense in this context. Additionally, a teacher can only have so many students. Perhaps the most reasonable domain is simply $0 < n < 72$ based on the teacher's existing data.

2. (a) There appear to be 3 distinct minima and maxima, so a quartic function should reasonably fit the data.

    (b) $P(x) = -0.01x^4 + 0.297x^3 - 2.680x^2 + 7.082x + 5.291$

3. $C(t) = \begin{cases} 500 & 0 < t \leq 3 \\ 125t + 100 & 3 < t < 20 \\ 100t & t \geq 20 \end{cases}$

4. Because V and P are inversely related, $P = \frac{k}{V}$. We can use either of the 2 points (P, V) given to compute k. Using $(22, 2.100)$, we get $2.1 = \frac{k}{22} \rightarrow k = 46.2$. Then, we can use this to compute that 15 liters of the gas would exert $\frac{46.2}{15} = 3.08$ kPa of pressure.

5. (a) The cost per person is the total cost divided by the number of paying alumni. There is a flat cost of \$500 plus \$20 per attendee x, but only $x - 10$ people will pay. Therefore,

$$C(x) = \frac{500 + 20x}{x - 10}.$$

   (b) There are 10 student council members at a minimum, so at least 10 people will attend. The highest the domain can go is the total number N of alumni, which we are not provided with. Therefore, the domain is $10 \leq x \leq N$, or $[10, N]$.

   (c) The degree of the numerator and denominator of $C(x)$ are equal, so the horizontal asymptote is the ratio of the leading coefficients, $y = \frac{20}{1} = 20$. This means that, as the number of people x who attend increases, the cost per person decreases, eventually approaching \$20.

Reflection

Suggested Class Time. 60-70 minutes

Prerequisites. Students need some familiarity with calculator skills (like using `alph trace` to access Y1). It helps if students have seen lines of best fit as well.

Instructional Strategies. I prefaced this activity with general discussion about *lines of best fit*, which students had a lot of familiarity with. I emphasized that these are more formally called *regression lines*, and that regression could be performed for a variety of functions, including most of the ones studied in the class thus far. To help students understand the need for this, I gave examples of where modeling could occur in actual, real life (not just the "real world" presented in contrived problems). [The main example I gave was about a bird feeder I'd just bought. I hadn't seen a single bird at it in a week, but I saw the volume of bird seed decreasing. I decided to start measuring the height of birdseed left in the cylindrical feeder, and to model the number of birds coming to the feeder, I could use regression (and some conversion, along with some major assumptions!) to create an estimate.]

Summarizing the activity was very brief. The main thing we went through was the practice example at the end. Many of my students actually created a *cubic* function to model the completion time based on the number of employees. Interestingly enough, the r^2 value was incredibly high for a cubic function.

We talked about whether this made the cubic model (as opposed to the inversely proportional function) "wrong," which was a rather fruitful discussion.

Technology. Students use the TI-84. It's important that students have the Stat Diagnostics turned on. For newer TI-84's, this is in the `mode` menu. For older calculators, it requires going to the `catalog` and finding `Diagnostic On`.

Problems.

1. A company that sells square glass sheets with side length x feet charges $20 per foot of length for sheets that are less than 3 feet long and adds an additional charge for larger sheets. This charge is equal to $2 per square foot of area of the sheet. Which of the following functions $C(x)$ gives the cost C, in dollars, of a square glass sheet with side length x?

   (A) $C(x) = \begin{cases} 20x & 0 \leq x < 3 \\ 2x & x \geq 3 \end{cases}$

   (B) $C(x) = \begin{cases} 20x & 0 \leq x < 3 \\ 2x^2 & x \geq 3 \end{cases}$

   (C) $C(x) = \begin{cases} 20x & 0 \leq x < 3 \\ 20x + 2x & x \geq 3 \end{cases}$

   (D) $C(x) = \begin{cases} 20x & 0 \leq x < 3 \\ 20x + 2x^2 & x \geq 3 \end{cases}$

2. The table below shows the population P of South Sudan, in millions of people, for each year since 2020.

t years since 2020	Population P
0	10.606
1	10.748
2	10.913
3	11.089

   A linear regression is used to construct a function $P(t)$ that will predict the population of South Sudan t years after 2020. What does $P(t)$ predict the population of South Sudan will be in 2030?
   (A) 11.265 million people
   (B) 11.727 million people
   (C) 12.211 million people
   (D) 12.343 million people

3. To save up for a car, Matilda takes a job working 10 hours a week at the library. For the first six weeks, the library pays Matilda $8 an hour. After that, she earns $11.50 an hour. Matilda puts all the money she earns each week in a savings account. On the day she starts work her savings account already holds $200. Which of the following functions S best models the amount of money in Matilda's savings account t weeks after her library job begins?

   (A) $S(t) = \begin{cases} 200 + 80t & 0 \leq t \leq 6 \\ 680 + 115(t-6) & t > 6 \end{cases}$

   (B) $S(t) = \begin{cases} 200 + 80t & 0 \leq t \leq 6 \\ 680 + 115t & t > 6 \end{cases}$

   (C) $S(t) = \begin{cases} 200 + 80t & 0 \leq t \leq 6 \\ 115(t-6) & t > 6 \end{cases}$

   (D) $S(t) = \begin{cases} 200 + 80t & 0 \leq t \leq 6 \\ 200 + 115(t-6) & t > 6 \end{cases}$

4. The function A, given by

   $$A(t) = \frac{1}{648}\left(-t^4 + 8t^3 + 90t^2 + 2808t + 22680\right),$$

   models the amount, in grams, of algae in a certain region of Lake Lanier over t days. The domain of A is $0 \leq t \leq 20$. Based on the model, approximately how many days did it take for the algae to increase from 60 grams to 80 grams?

   (A) 1.453 days (B) 3.375 days (C) 4.878 days (D) 8.252 days

5. The average price, in dollars per 1000 cubic feet, of natural gas used by residential households in the United States for selected years from 2002 to 2022 is given in the following table.

Year	2002	2004	2006	2008	2010	2012	2014	2016	2018	2020	2022
Price	7.89	10.75	13.73	13.89	11.39	10.65	10.97	10.05	10.50	10.78	14.75

A cubic polynomial is used to model the average price per 1000 cubic feet of natural gas based on the year. The model predicts an average price of $53.17 per 1000 cubic feet of natural gas used by residential households in the year 2030. Which of the following is true?

(A) The predicted price of $53.17 per 1000 cubic feet of natural gas in the year 2030 is not accurate but it allows us to predict that the price will be close to this predicted value.

(B) The predicted price of $53.17 per 1000 cubic feet of natural gas in the year 2030 is not accurate because a cubic polynomial model predicts that the price will continue to increase without bound.

(C) The predicted price of $53.17 per 1000 cubic feet of natural gas in the year 2030 is accurate because it is the value given by the cubic polynomial which models the data.

(D) The predicted price of $53.17 per 1000 cubic feet of natural gas in the year 2030 is accurate and any other polynomial used to model the data will also predict the price of $53.17.

Solutions.

1. For $0 \leq x < 3$, the company charges $20x$ dollars. For $x \geq 3$, the company charges the initial $20x$ dollars, plus another 2 dollars per area of the square. This is $20x + 2x^2$ dollars. The correct answer is therefore **(D)**.

2. Performing linear regression in the calculator yields $P(t) = 0.1614t + 10.5969$. The predicted population in 2030, or when $t = 10$, is $P(10) = 12.2109$. The correct answer is therefore **(C)**.

3. For weeks 0 through 6, the amount in the savings account is $200 + 80t$. For weeks greater than 6, we already have in the account all the money from the previous 6 weeks, which is a total of $200 + 80 \cdot 6 = 680$. We also have 10 hours each week at $11.50 per hour, for $115 per week. However, the number of weeks at this new rate does not include the first 6 weeks, so the additional salary is $115(t - 6)$. Hence, the savings for $t > 6$ is $680 + 115(t - 6)$. The correct answer is therefore **(A)**.

4. Graph the function $A(t)$ along with the lines $y = 60$ and $y = 80$. Use the intersection function to get the intersections of A with 60 (there are two such intersections) and the intersections of A with 80 (there are two). However, we want the intersections that indicate an increase of algae. These are for $t = 4.87764$ and $t = 8.25237$. Hence, it takes approximately $8.25237 - 4.87764 = 3.37473$ days. The correct answer is therefore **(B)**.

5. Using a cubic polynomial implies that as the inputs increase without bound, the outputs will increase without bound. This implies that the price will continue to increase without bound which is not realistic. The correct answer is therefore **(B)**.

Unit 1 Test ~ Polynomial and Rational Functions

> **Part A: 6 multiple choice and 1 free response**
> A calculator may be required for some questions on this section.

1. (Topic 1.1) An artisan who sells her crafts online has found that, if she spends t hours on a craft, the average consumer satisfaction rating out of 5 can be modeled by the polynomial function $S(t) = 0.11t^3 - 0.98t^2 + 2.81t + 1.5$. For $1 \leq t \leq 5$, which of the following is true?

   (A) The range of $S(t)$ is $[1.5, 4.8]$. This means the artisan typically spends between 1.5 and 4.8 hours on crafts.

   (B) The range of $S(t)$ is $[1.5, 4.8]$. This means the average satisfaction scores for this artisan's crafts are between 1.5 and 4.8 out of 5.

   (C) The range of $S(t)$ is $[3.44, 4.8]$. This means the artisan typically spends between 3.44 and 4.8 hours on crafts.

   (D) The range of $S(t)$ is $[3.44, 4.8]$. This means the average satisfaction scores for this artisan's crafts are between 3.44 and 4.8 out of 5.

2. (Topic 1.5) The polynomial function f has real coefficients and includes zeros of $x = \frac{1}{2}, x = 3$, and $x = 2i$. Which of the following could be an expression for $f(x)$?

   (A) $f(x) = (2x+1)(x+3)(x-2i)$
   (B) $f(x) = (2x+1)(x+3)(x-2i)(x+2i)$
   (C) $f(x) = (2x-1)(x-3)(x-2i)$
   (D) $f(x) = (2x-1)(x-3)(x-2i)(x+2i)$

3. (Topics 1.8-1.10) The function $f(x) = \dfrac{(3x-4)(x^2-x+2)}{(3x-4)(x-3)}$ has a hole where $x = \frac{4}{3}$. Which of the following is $\lim_{x \to \frac{4}{3}} f(x)$?

   (A) ∞
   (B) $-\dfrac{22}{15}$
   (C) 0
   (D) Because f has a hole at $x = \frac{4}{3}$, $\lim_{x \to \frac{4}{3}} f(x)$ does not exist.

4. (Topic 1.11) When $(2y - 1)^3$ is expanded, what is the sum of the coefficients?

   (A) 0   (B) 1   (C) 8   (D) 27

5. (Topic 1.14) A company that sells square glass sheets with side length x feet charges \$20 per foot of length for sheets that are less than 3 feet long and adds an additional charge for larger sheets. This charge is equal to \$2 per square foot of area of the sheet. Which of the following functions $C(x)$ gives the cost C, in dollars, of a square glass sheet with side length x?

   (A) $C(x) = \begin{cases} 20x & 0 \leq x < 3 \\ 2x & x \geq 3 \end{cases}$

   (B) $C(x) = \begin{cases} 20x & 0 \leq x < 3 \\ 2x^2 & x \geq 3 \end{cases}$

   (C) $C(x) = \begin{cases} 20x & 0 \leq x < 3 \\ 20x + 2x & x \geq 3 \end{cases}$

   (D) $C(x) = \begin{cases} 20x & 0 \leq x < 3 \\ 20x + 2x^2 & x \geq 3 \end{cases}$

6. (Topic 1.14) The table below shows the population P of South Sudan, in millions of people, for each year since 2020.

t years since 2020	Population P
0	10.606
1	10.748
2	10.913
3	11.089

A linear regression is used to construction a function $P(t)$ that will predict the population of South Sudan t years after 2020. What does $P(t)$ predict the population of South Sudan will be in 2030?

(A) 11.265 million people
(B) 11.727 million people
(C) 12.211 million people
(D) 12.343 million people

Free Response Part A: Instructions

- Show all of your work. Your work will be scored on the correctness and completeness of your responses as well as your answers. Answers without supporting work may not receive credit in cases where supporting work is requested.

- Unless otherwise specified, the domain of a function f is assumed to be the set of all real numbers x for which $f(x)$ is a real number.

- Avoid rounding intermediate computations on the way to the final result. Unless otherwise specified, any decimal approximations reported in your work should be accurate to three places after the decimal point.

- It may be helpful to use your graphing calculator to store information such as computed values for constants, functions you are working with, solutions to equations, and any intermediate values. Computations with the graphing calculator that use the stored information help to maintain as much precision as possible and ensure the desired accuracy in final answers.

Free Response #1. The function f is a quadratic function with zeros of $x = \frac{1}{2}$ and $x = 3$. The function g is given by $g(x) = x^3 - 2x^2 - 6x + 9$. The function r is given by $r(x) = \frac{f(x)}{g(x)}$.

(A) Find the x-values for which r is undefined.

(B) Classify what happens at each of the x-values from (a) as a hole or vertical asymptote in the graph of r. Explain your answer.

(C) Let $s(x) = 2r(x-3) - 1$. For what value/s of x is $s(x)$ undefined? Explain.

UNIT 1   UNIT 1 TEST ~ POLYNOMIAL AND RATIONAL FUNCTIONS   139

> **Part B: 14 multiple choice and 1 free response**
> A calculator is prohibited on this section.

7. (Topic 1.2) A car dealership has been tracking its sales since offering a new promotion starting on September 1. It is using a function $C(t)$ to model the number of cars C sold on day t, where $t = 0$ is September 1. The rate of change of C was approximately 6.2 on September 7. Assuming $C(t)$ is approximately linear, how many cars is the dealership expected to sell on September 12?

   (A) Approximately 6 cars.
   (B) Approximately 31 cars.
   (C) Approximately 6 more cars than it sold on September 7.
   (D) Approximately 31 more cars than it sold on September 7.

8. (Topic 1.3) For the function $y = f(x)$, it is known that $f(a) = b$ and $f(c) = d$. What is the average rate of change of f on the interval $a \leq x \leq c$?

   (A) $d - b$ **(B)** $b - d$ **(C)** $\dfrac{b-d}{a-c}$ **(D)** $\dfrac{b-d}{c-a}$

9. (Topic 1.4) Consider the function $f(x) = ax^2(x-3)(x-4)$ for some non-zero value of a. Which of the following *must* be true concerning f?

   (A) f has a local minimum at some x with $3 < x < 4$.
   (B) f has a local maximum or a local minimum, but not both, at some x with $0 < x < 3$.
   (C) f has both a global maximum and a global minimum.
   (D) f has neither a global maximum nor a global minimum.

10. (Topic 1.6) The function $g(x) = ax^n$ satisfies $\lim\limits_{x \to \infty} g(x) = -\infty$ and $\lim\limits_{x \to -\infty} g(x) = \infty$. Which of the following must be true?

    (A) $a < 0$ and n is odd   **(B)** $a < 0$ and n is even   **(C)** $a > 0$ and n is odd   **(D)** $a > 0$ and n is even

11. (Topic 1.7) As x increases without bound, the function $r(x) = \dfrac{3(x^2 - 2)(4x - 5)}{2x^2 - 5x + 3}$ will

    (A) approach the line $y = \frac{3}{2}$.
    (B) approach the line $y = 6$.
    (C) approach a line with a slope of $\frac{3}{2}$.
    (D) approach a line with a slope of 6.

12. (Topic 1.8) Let $f(x) = 2x^3 - 8x$ and $g(x) = 3x^2 - 11x + 10$. The function given by $q(x) = \dfrac{f(x)}{g(x)}$ has zeros at which x-values?

    (A) $x = \frac{5}{3}$ and $x = 2$ only
    (B) $x = 0$ and $x = 2$ only
    (C) $x = -2$ and $x = 0$ only
    (D) $x = -2, x = 0, x = \frac{5}{3}$, and $x = 2$

13. (Topic 1.9) The tables below show the zeros and their respective multiplicities for the functions f and g.

$f(x)$				
Zero	−2	0	4	5
Multiplicity	1	3	2	2

$g(x)$				
Zero	−3	−2	0	5
Multiplicity	1	2	2	1

    At which of the following x-values does the function $h(x) = \dfrac{f(x)}{g(x)}$ have a vertical asymptote?

    (A) $x = -3$ only   **(B)** $x = -2$ only   **(C)** $x = -3$ and $x = -2$ only   **(D)** $x = -3, x = -2, x = 0$, and $x = 5$

AP Precalculus: The Teacher's Compendium ~ © 2024 David Hornbeck and Chuck Garner

14. (Topic 1.10) The function $f(x) = \dfrac{x^2 + bx - 8}{5x^2 + 21x + 4}$ will have a hole for what integer value of b?

    (A) $b = -4$ (B) $b = -2$ (C) $b = 2$ (D) $b = 4$

15. (Topics 1.8-1.10) Which of the following could be an expression for the function $f(x)$ graphed below?

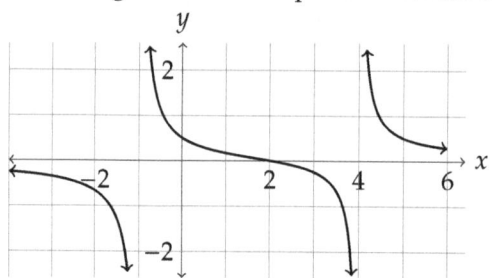

    (A) $f(x) = \dfrac{x - 2}{(x + 1)(x - 4)}$ (B) $f(x) = \dfrac{x + 2}{(x - 1)(x + 4)}$ (C) $f(x) = \dfrac{(x - 4)(x + 1)}{x - 2}$ (D) $f(x) = \dfrac{(x + 4)(x - 1)}{x + 2}$

16. (Topic 1.11) The quotient $\dfrac{4x^3 - x + 2}{x^2 + x + 3}$ is equivalent to $4x - 4 + r(x)$, where the degree of r is less than 2. Which expression is equivalent to $r(x)$?

    (A) $r(x) = -9x + 10$ (B) $r(x) = -9x + 14$ (C) $r(x) = -7x + 10$ (D) $r(x) = -7x + 14$

17. (Topic 1.12) If the range of $f(x)$ is $[-4, 6]$, what is the range of $g(x) = 3f\left(\tfrac{1}{2}x\right) - 2$?

    (A) $[-14, 16]$ (B) $[-10, 20]$ (C) $[-4, 1]$ (D) $[-10, 10]$

18. (Topic 1.12) The graphs of the functions $f(x)$ and $g(x)$ are shown below.

    Graph of $f(x)$

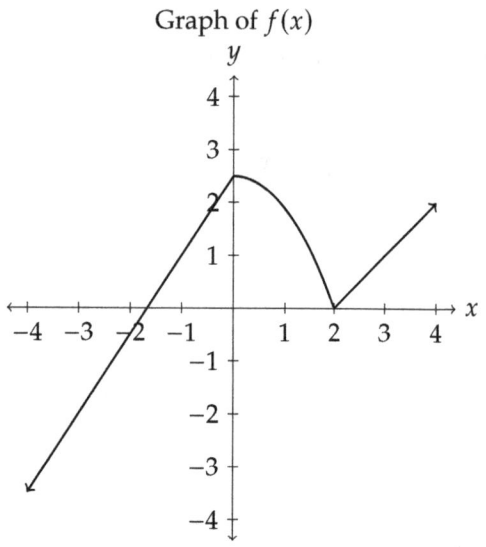

    Graph of $g(x)$

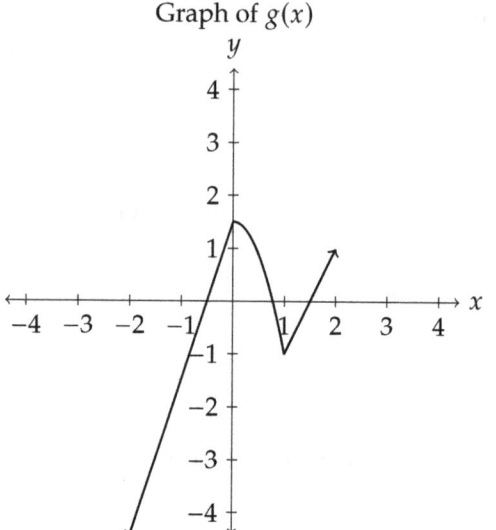

    Which of the following expresses $g(x)$ in terms of $f(x)$?

    (A) $g(x) = f\left(\tfrac{4}{3}x\right) + 1$ (B) $g(x) = f\left(\tfrac{3}{4}x\right) - 1$ (C) $g(x) = f(2x) - 1$ (D) $g(x) = f\left(\tfrac{1}{2}x\right) - 1$

19. (Topic 1.13) A data set containing five points is graphed below.

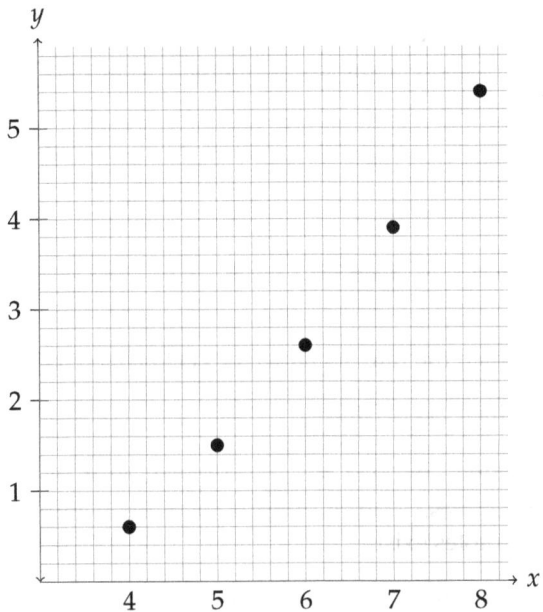

What type of function would be the most appropriate to model the relationship between x and y for this data set?

(A) A linear function because the rate of change is constant.

(B) A linear function because the points roughly appear to form a line.

(C) A quadratic function because the rate of rate of change is constant.

(D) A quadratic function because the points roughly appear to form a curve.

20. (Topic 1.13) An aquarium is building a tank for a shark and has determined that the function $V(x) = -\frac{1}{2}(x+2)(x-1)(2x-7)$ will model the volume V, in hundreds of gallons, based on the width x of the tank. Which of the following is a reasonable domain of $V(x)$?

(A) $-2 < x < 3.5$ (B) $1 < x < 2$ (C) $x > 1$ (D) $1 < x < 3.5$

Free Response Part B - Instructions

- Show all of your work. Your work will be scored on the correctness and completeness of your responses as well as your answers. Answers without supporting work may not receive credit in cases where supporting work is requested.

- Unless otherwise specified, the domain of a function f is assumed to be the set of all real numbers x for which $f(x)$ is a real number.

Free Response #2. The function f is given by $f(x) = -3(x-5)(x-2)$ and the function g is given by $g(x) = 2(x-4)(x-1)$.

(A) Find all values of x for which the rate of change of $f(x)$ is decreasing.

(B) Find all values of x for which $f(x) > 0$.

(C) Let k be the function defined by $k(x) = \frac{f(x)}{g(x)}$. Determine if the graph of k has any horizontal asymptotes or slant asymptotes. Find the slopes for those asymptotes, or state that no horizontal asymptotes or slant asymptotes exist.

(D) Evaluate $\lim\limits_{x \to 4^+} k(x)$. Provide a reason for your answer.

Unit 1 Test Solutions and Scoring

1. Graphing the function S on the interval $1 \leq t \leq 5$ and finding the minimum and maximum values of S on that interval gives the range. Since S models average consumer satisfaction, the average satisfaction scores are between 3.44 and 4.8. The correct answer is therefore **(D)**.

2. The other complex root is $x = -2i$. Since $x = 1/2$ makes $2x - 1$ zero, the polynomial is $f(x) = (2x - 1)(x - 3)(x - 2i)(x + 2i)$. The correct answer is therefore **(D)**.

3. The factors of $3x - 4$ in the numerator and denominator are equal, so they divide to become 1. Thus the function resembles $(x^2 - x + 2)/(x - 3)$ as x gets arbitrarily close to $4/3$. When $x = 4/3$, this becomes
$$\frac{(4/3)^2 - 4/3 + 2}{4/3 - 3} = \frac{16/9 - 4/3 + 2}{4/3 - 3} = \frac{16 - 12 + 18}{12 - 27} = -\frac{22}{15}.$$
The correct answer is therefore **(B)**.

4. By the Binomial Theorem, the expansion is

Binomial Coefficient $\binom{n}{k}$	1	3	3	1
x^n	$(2y)^3 = 8y^3$	$(2y)^2 = 4y^2$	$2y$	$(2y)^0 = 1$
y^n	$(-1)^0 = 1$	$(-1)^1 = -1$	$(-1)^2 = 1$	$(-1^3) = -1$
Final Sum	$8y^3$	$-12y^2$	$6y$	-1

   The sum of the coefficients is $8 - 12 + 6 - 1 = 1$. The correct answer is therefore **(B)**.

5. For $0 \leq x < 3$, the company charges $20x$ dollars. For $x \geq 3$, the company charges the initial $20x$ dollars, plus another 2 dollars per area of the square. This is $20x + 2x^2$ dollars. The correct answer is therefore **(D)**.

6. Performing linear regression in the calculator yields $P(t) = 0.1614t + 10.5969$. The predicted population in 2030, or when $t = 10$, is $P(10) = 12.2109$. The correct answer is therefore **(C)**.

7. On September 7, the rate of change of C is 6.2. Since C is approximately linear, this means that each day, another 6.2 cars are sold. September 12 is 5 days later, so we expect $5 \cdot 6.2 = 31$ more cars to be sold than on September 7. The correct answer is therefore **(D)**.

8. The average rate of change is $\dfrac{f(a) - f(c)}{a - c} = \dfrac{b - d}{a - c}$. The correct answer is therefore **(C)**.

9. This is a fourth-degree polynomial which could either open upwards or downwards, depending on the sign of a. Hence, we cannot say for sure that will always be a maximum or a minimum over any particular interval, but we can say that there will either be one or the other, but not both, on an interval between the zeros of the function. The correct answer is therefore **(B)**.

10. Since the function grows arbitrarily large as x decreases without bound, and the function decreases without bound as x grows arbitrarily large, the degree n must be odd. Moreover, a must be negative since arbitrarily large positive input values give negative output values of f. The correct answer is therefore **(A)**.

11. As x increases without bound, the function resembles $\dfrac{3(x^2)(4x)}{2x^2} = \dfrac{12x^3}{2x^2} = 6x$, which is a line with a slope of 6. The correct answer is therefore **(D)**.

12. Since $f(x) = 2x(x^2 - 4) = 2x(x - 2)(x + 2)$ and $g(x) = (3x - 5)(x - 2)$, then
$$q(x) = \frac{2x(x-2)(x+2)}{(3x-5)(x-2)} = \frac{2x(x+2)}{3x-5}$$
for values of $x \neq 2$. The zeros are therefore $x = 0$ and $x = -2$. The correct answer is therefore **(C)**.

13. There will be a vertical asymptote if there is a value of x that is a zero of the denominator zero but not of the numerator, or if there is a value of x that makes both numerator and denominator zero but the multiplicity of the zero in the denominator is greater than the multiplicity of the zero in the numerator. The values $x = -2$, $x = 0$, and $x = 5$ are zeros of the numerator and denominator, but only $x = -2$ has multiplicity of the denominator greater than the multiplcity of the numerator. The value $x = -3$ is a zero of the denominator but not the numerator. Hence, there are vertical asymptotes at both $x = -3$ and $x = -2$. The correct answer is therefore **(C)**.

14. The polynomial in the denominator factors as $(5x + 1)(x + 4)$. We need the polynomial in the numerator to factor where one of the factors is $x + 4$. Hence, $x^2 + bx - 8 = (x + 4)(x - n)$ for some integer n. Since $4 \cdot (-n) = -8$, $n = 2$. Then $(x + 4)(x - 2) = x^2 + 2x - 8$, and $b = 2$. The correct answer is therefore **(C)**.

15. The function has vertical asymptotes at $x = -1$ and $x = 4$, and a zero at $x = 2$. Thus, the denominator of f must have factors of $x + 1$ and $x - 4$ and the numerator of f must have a factor of $x - 2$. The correct answer is therefore **(A)**.

16. We use long division.

$$\begin{array}{r}
4x - 4 \\
x^2 + x + 3 \overline{\smash{)}\, 4x^3 - x + 2} \\
\underline{-4x^3 - 4x^2 - 12x } \\
-4x^2 - 13x + 2 \\
\underline{4x^2 + 4x + 12} \\
-9x + 14
\end{array}$$

The function r is $r(x) = -9x + 14$. The correct answer is therefore **(B)**.

17. The dilation by a factor of 3 makes the range $[-4 \cdot 3, 6 \cdot 3] = [-12, 18]$. The vertical translation by -2 units makes this range $[-12 - 2, 18 - 2] = [-14, 16]$. The correct answer is therefore **(A)**.

18. The graph of f has undergone a horizontal dilation by a factor of $\frac{1}{2}$ and a vertical translation of -1 units. Thus, the function g is $g(x) = f(2x) - 1$. The correct answer is therefore **(C)**.

19. The change between 4 and 5 is $1.5 - 0.6 = 0.9$, the change between 5 and 6 is $2.6 - 1.5 = 1.1$, the change between 6 and 7 is $3.9 - 2.6 = 1.3$, and the change between 7 and 8 is $5.4 - 3.9 = 1.5$. The second differences are a constant (0.2), so a quadratic function would be the most appropriate model. The correct answer is therefore **(C)**.

20. The zeros of the function are $x = -2$, $x = 1$, and $x = 3.5$. The function is a cubic with a negative leading coefficient, so $V > 0$ for $x < -2$ and $1 < x < 3.5$. However, x cannot be negative since this represents the width of the tank. Thus, $1 < x < 3.5$ is the domain. The correct answer is therefore **(D)**.

Question 1 Scoring Guidelines

	Model Solution	Scoring	
(A)	r is undefined when $g(x) = 0$, which occurs at $x = 3$, $x = -2.303$, and $x = 1.303$.	Indicates $g(x) = 0$ All 3 zeros	1 point 1 point
		Total for part (A)	2 points
(B)	Because both $f(3) = 0$ and $g(3) = 0$ (with same multiplicity), the graph of r has a hole when $x = 3$. At the other two inputs where $g(x) = 0$, $f(x) \neq 0$, so the graph of r has vertical asymptotes at $x = -2.303$ and $x = 1.303$.	Classification Justification	1 point 1 point
		Total for part (B)	2 points
(C)	$s(x)$ is a translation of $r(x)$ 3 units to the right, so the inputs at which s is defined is simply the inputs at which r is undefined translated 3 to the right. Therefore, s is undefined for $x = 6$, $x = 1.303$, and $x = 4.303$.	All three values Explanation	1 point 1 point
		Total for part (C)	2 points
		Total for Question 1	6 points

We propose the following time limits for each part of the test, assuming similar timings as for the AP exam. That means each multiple-choice should be allowed 3 minutes and each free-response 15 minutes. This could vary slightly, depending on the difficulty of the multiple-choice: it is reasonable to expect 2 minutes per problem for the easier problems. This creates the range of timings as given in the table below.

Suggested Time Limits	
Part A	30-35 minutes
Part B	45-55 minutes

We choose 20 multiple-choice and 2 free-response to include on the test in order to equal the same proportions of those two types of problems as on the AP Exam. That means such a test can be given an "AP score" based on the number of points earned. Each multiple-choice problem is worth 1 point and each free-response is worth 6 points, for a total of 32 points available on the test. The table below shows the conversions to an "AP score."

Raw score	AP score
23-32	5
18-22	4
14-17	3
8-13	2
0-7	1

Question 2 Scoring Guidelines

	Model Solution	Scoring	
(A)	The graph of f is a negatively oriented parabola, which is concave down for all x. Therefore, the rate of change of f is decreasing for all real x.	All real numbers	1 point
		Total for part (A)	**1 point**
(B)	$f(x) = 0$ for $x = 5$ and $x = 2$. Because the graph of f is a negatively oriented parabola, the graph of f will be above the x-axis for x between 2 and 5. Therefore, $f(x) > 0$ for $2 < x < 5$.	Computation of zeros Correct interval	1 point 1 point
		Total for part (B)	**2 points**
(C)	The degree of the numerator of k and the degree of the denominator of k are equal, so for inputs of large magnitude, the outputs $f(x)$ will approach a constant value. So $\frac{-3x^2}{2x^2} = -\frac{3}{2} \implies y = -\frac{3}{2}$ is the only horizontal asymptote. There are no slant asymptotes.	Horizontal asymptote No other horizontal or slant asymptotes	1 point 1 point
		Total for part (C)	**2 points**
(D)	As $x > 4$ gets arbitrarily close to 4, the graph of k will be positive and arbitrarily large. Therefore, $\lim_{x \to 4^+} k(x) = \infty$.	Limit with justification	1 point
		Total for part (D)	**1 point**
		Total for Question 2	**6 points**

Unit 2

Exponential and Logarithmic Functions

Part 2

Exponential and Logarithmic Functions

Topic 2.1 ~ Change in Arithmetic and Geometric Sequences

Learning Objectives
1. (2.1.A) Express arithmetic sequences found in mathematical and contextual scenarios as functions of the whole numbers.
2. (2.1.B) Express geometric sequences found in mathematical and contextual scenarios as functions of the whole numbers.

Success Criteria
1. I can describe rates of change in arithmetic and geometric sequences.
2. I can write arithmetic and geometric sequences as functions of the whole numbers.

This activity is about *sequences*.

In general, we call the terms of a sequence $a_0, a_1, a_2, \ldots$ and so on. Any arbitrary term a_n is often called the GENERAL TERM. One thing to look out for: because the first term is a_0, the term a_n is technically the $(n+1)$th term.

Sequences can actually be written as *functions* with a domain of the *whole numbers* $\{0, 1, 2, \ldots\}$. For instance, the sequence a_n given by 1, 4, 7, 10 can be written as

$$a_n = 1 + 3n \qquad (\star)$$

with domain $0, 1, 2, 3, \ldots$. To see this, complete the table below:

Input n	Output $a_n =$
0	$a_0 =$
1	$a_1 =$
2	$a_2 =$
3	$a_3 =$

We could also write the sequence from ($\star$) using something other than the *first* term. For instance, starting at the a_2 term, we could write

$$a_n = 7 + 3(n-2)$$

Can you see how $a_n = 7 + 3(n-2)$ and $a_n = 1 + 3n$ are the same sequence?

Two common types of sequences are *arithmetic* and *geometric*. ARITHMETIC SEQUENCES have a *constant difference* between two consecutive terms, while GEOMETRIC SEQUENCES have a *constant ratio* between two consecutive terms.

Consider the four sequences below.

Sequence A: $7, 12, 17, 22, \ldots$

Sequence B: $2, 6, 18, 54, \ldots$

Sequence C: $29, 23, 17, 11, \ldots$

Sequence D: $16, -8, 4, -2, \ldots$

For each of the sequences A, B, C, and D from above, try to determine the following.

	Sequence			
	A: $7, 12, 17, 22, \ldots$	**B**: $2, 6, 18, 54, \ldots$	**C**: $29, 23, 17, 11, \ldots$	**D**: $16, -8, 4, -2, \ldots$
Arithmetic or Geometric?				
Constant diff. d (or constant ratio r)				
Initial term a_0				
Next two terms a_4 and a_5				
Expression for general term a_n using a_0				
Expression for general term a_n using a different term				

Notes

Topic 2.1 Homework

1. Write the general term a_n of the sequence $6, -4, -14, -24, \ldots$.

2. Write the general term g_n of the sequence $12, 3, \frac{3}{4}, \ldots$.

3. Find the 8th term of the sequence with general term $g_n = 24\left(\frac{1}{2}\right)^{(n-3)}$.

4. What is the first term of the sequence with general term $a_n = 12 - 4(n-6)$?

5. Suppose a sequence s_n has $s_0 = 2$ and $s_4 = 32$.
    (a) If s_n is arithmetic, compute s_1.
    (b) If s_n is geometric, compute the two possible values of s_1.

6. The sequences s_n and t_n are graphed below.

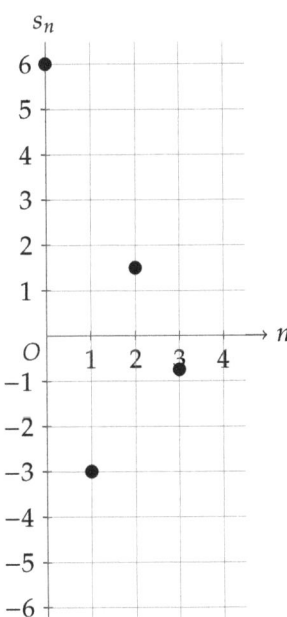 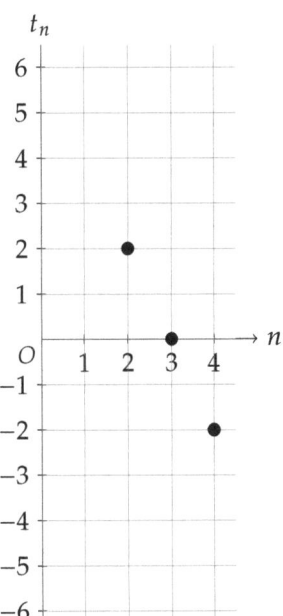

    (a) Write the general term of each of s_n and t_n.
    (b) Compute s_5 and t_5.

7. The initial term g_0 of an increasing geometric sequence g_n is 2. If the third term is $g_3 = 6$, what is g_6?

8. An arithmetic sequence has general term $a_n = 4 + 8(n-6)$.
    (a) Based on the general term, what is the 6th term?
    (b) Compute the difference $a_{2025} - a_{2023}$.

9. (■) The inflation per year in the United States has been on average 3.8% each year since 1960. This means that, each year, the prices of items generally cost 103.8% of what they cost the year prior. Let $C_0 = 50$ represent an item that costs \$50 this year (2023).
    (a) The cost of the item C_n for each year after 2023 can be approximated by a sequence. Is it a geometric or arithmetic sequence? Why?
    (b) Write the general term of C_n.
    (c) In what year is this item first expected to cost \$65 or more?

Topic 2.1 Solutions/Notes

Activity
Completed table:

Input n	Output $a_n =$
0	$a_0 = 1 + 3(0) = 1$
1	$a_1 = 1 + 3(1) = 4$
2	$a_2 = 1 + 3(2) = 7$
3	$a_3 = 1 + 3(3) = 10$

Answer to "Can you see how..."?: Algebraically, we can easily see that $7 + 3(n - 2) = 7 + 3n - 6 = 1 + 3n$. Conceptually, we can see that both expressions show a common difference of 3, but the form $7 + 3(n - 2)$ shows that, when $n = 2$, the term equals 7: this corresponds to $a_2 = 7$.

Sequences Information

	Sequence			
	A: $7, 12, 17, 22, \ldots$	**B**: $2, 6, 18, 54, \ldots$	**C**: $29, 23, 17, 11, \ldots$	**D**: $16, -8, 4, -2, \ldots$
Arithmetic or Geometric?	Arithmetic	Geometric	Arithmetic	Geometric
Constant diff. d (or constant ratio r)	$d = 5$	$r = 3$	$d = -6$	$r = -\frac{1}{2}$
Initial term a_0	7	2	29	16
Next two terms a_4 and a_5	$27, 32$	$162, 486$	$5, -1$	$1, -\frac{1}{2}$
Expression for general term a_n using a_0	$7 + 5n$	$2 \cdot 3^n$	$29 - 6d$	$16\left(-\frac{1}{2}\right)^n$
Expression for general term a_n using a different term	$12 + 5(n - 1)$ or $17 + 5(n - 2)$ or $22 + 5(n - 3)$ etc.	$6 \cdot 3^{(n-1)}$ or $18 \cdot 3^{(n-2)}$ or $54 \cdot 3^{(n-3)}$ etc.	$23 - 6(n - 1)$ or $17 - 6(n - 2)$ or $1 - 6(n - 3)$ etc.	$-8\left(-\frac{1}{2}\right)^{(n-1)}$ or $4\left(-\frac{1}{2}\right)^{(n-2)}$ or $-2\left(-\frac{1}{2}\right)^{(n-3)}$ etc.

Unit 2 — Topic 2.1 ~ Change in Arithmetic and Geometric Sequences

Notes

An ARITHMETIC SEQUENCE a_n has a common difference d between two consecutive terms. Its general term can be written as $a_n = a_0 + dn$ or $a_n = a_k + d(n-k)$, which is a function with a domain of the whole numbers $\{0, 1, 2, \ldots\}$. Arithmetic sequences increase or decrease at a *constant rate of change*.

A GEOMETRIC SEQUENCE g_n has a common ratio r between consecutive terms. Its general term can be written as $g_n = g_0 r^n$ or $g_n = g_k r^{(n-k)}$, which is a function with a domain of the whole numbers $\{0, 1, 2, \ldots\}$. Increasing geometric sequences increase by *increasing amounts* with each successive step.

Because sequences are functions, they can be graphed as a set of discrete points. For instance, $a_n = 1 + 3n$ from the activity could be graphed as

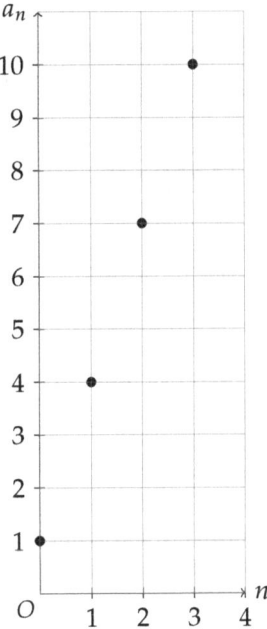

Arithmetic sequences will produce graphs that are *linear* when connected. Geometric sequences will produce graphs that are *exponential* when connected. These latter graphs will always maintain one kind of concavity (concave up or concave down) for all x.

Homework

1. $a_n = 6 - 10n$

2. $g_n = 12\left(\dfrac{1}{4}\right)^n$

3. The 8th term is g_7, so $g_7 = 24\left(\dfrac{1}{2}\right)^{(7-3)} = \dfrac{24}{16} = \dfrac{3}{2}$.

4. We rewrite a_n as $a_n = 12 - 4n + 24 = 36 - 4n$. The first term is 36.

5. (a) If s_n is arithmetic, then $s_n = 2 + dn$. Since $s_4 = 32$, then $32 = 2 + 4d$, or $d = 7.5$. This would make $s_1 = 2 + 7.5 = 9.5$.

   (b) If s_n is geometric, then $s_n = 2 \cdot r^n$. Since $s_4 = 32$, then $32 = 2 \cdot r^4$, or $r^4 = 16$. This gives $r = 2$ or $r = -2$. This would make $s_1 = 2(2) = 4$ or $s_1 = 2(-2) = -4$.

6. (a) $s_n = 6\left(-\dfrac{1}{2}\right)^n$, $t_n = 2 - 2(n-2)$.

   (b) $s_5 = 6\left(-\dfrac{1}{2}\right)^5 = -\dfrac{6}{32} = -\dfrac{3}{16}$, $t_5 = a_4 - 2 = -2 - 2 = -4$.

7. Let r be the common ratio. Then $g_0 = 2$ and $g_2 = 6$. Since g is geometric, $g_2 = 2r^3$, so $6 = 2r^3$ gives $r^3 = 3$. Therefore, $g_6 = g_3 \cdot r^3 = 6 \cdot 3 = 18$.

8. (a) The 6th term is 4.

   (b) We could plug in $n = 2025$ and $n = 2023$, but these are simply two terms apart, so the difference between them is just two times the constant difference, or $2(8) = 16$.

9. (a) Each year, the inflation rate increases by a factor of 1.038, so C_n is geometric.

   (b) $C_n = 50(1.038)^n$

   (c) We graph Y1=50(1.038)^X and Y2=65 and find the intersection. It occurs at $x = 7.04$, which is just beyond 7 years, so the first year in which the item would cost \$65 would be the eighth year after 2023, or 2031.

Reflection

Suggested Class Time. 40 minutes

Prerequisites. None, though it helps if students have seen arithmetic and geometric sequences before.

Instructional Strategies. Students flew through the first, third, and fourth rows of the table. There were a few hiccups.

Firstly, students wanted sequence D to have a constant ratio of -2. I reiterated this a couple of ways. I emphasized the repeated *multiplication*, and this seemed to work, but I felt dissatisfied with the quality of explanation. I then transitioned to an improved explanation, that of the constant ratio of consecutive terms. I showed students $\dfrac{-8}{16}, \dfrac{4}{-8}, \dfrac{-2}{4}$, etc., and this did the trick.

Students were systematically wrong at writing general terms of geometric sequences. For Sequence D, I saw a lot of $2 \cdot 3n$. What really seemed to help students was having them rewrite each term of the sequence in terms of the repeated operation, i.e.

$$a_0 = 2$$
$$a_1 = 2 \cdot 3$$
$$a_2 = 2 \cdot 3 \cdot 3$$
$$\vdots$$
$$a_n = ?$$

where I would have students fill in the last term. I had to repeat with many students, "Repeated addition is multiplication, but repeated multiplication is...".

One trick that helped students a lot with writing general terms based on something other than the initial term was rewriting $a_n = a_0 + dn$ as $a_n = a_0 + d(n-0)$, thereby highlighting that the *0th* term was given by a_n with $n - 0$. Similarly, when helping a few students with the last row for Sequence A, I wrote out the following:

$$a_n = 7 + 5(n - 0)$$
$$a_n = 12 + 5(n - 1)$$
$$a_n = 17 + 5(n - 2)$$

and let them infer more from there.

We finished going over the activity and taking notes with about 30 minutes of the 90-minute block remaining, so I decided to write up a few sample questions just for fun. In order, the problems I wrote were...

1. If the fourth term of an arithmetic sequence with a first term a_0 and a constant difference of -2 is 5, what is the 40th term?

2. The general term of a sequence is given by $s_n = 4 + 5(n-3)$ for $n = 0, 1, 2, \ldots$ Which is the first term of the sequence that will exceed 90?

    (A) the 19th term   **(B)** the 20th term   **(C)** the 21st term   **(D)** the 22nd term

3. The geometric sequence g_n is given by $g_n = 2375\pi \cdot 2^n$. What is the 94th term divided by the 92nd term?

4. If g_n is a geometric sequence with $g_{n+1} > g_n$ and $g_n > 0$ for all $n = 0, 1, 2, \ldots$, which of the following is true?

    (A) g is increasing at an increasing rate.

    (B) g is increasing at a decreasing rate.

    (C) g is decreasing at an increasing rate.

    (D) g is decreasing at a decreasing rate.

Students did great on Problem 1 (which is -67, by the way). For Problem 2, everyone figured out $n = 21$, but many students thought this was the *21st* term, not the 22nd. I highlighted the importance of paying attention to the subscripts of the terms (and whether s_0 is included). The third question flummoxed students - they were completely terrified of the 2375π and immediately reached for their calculators, clearly thinking they needed to compute the 94th and 92nd terms. I eventually talked them down and we discussed that the initial value was completely irrelevant. It helped them to see it written out algebraically:

$$\frac{g_{94}}{g_{92}} = \frac{g_0 \cdot 2^{94}}{g_0 \cdot 2^{92}} = 2^2 = 4.$$

Finally, the last question was one we discussed as a class (it's C).

Technology. No technology was required.

Problems.

1. The Mauna Loa Observatory in Hawaii began recording carbon dioxide levels in the atmosphere in 1958, when the atmosphere contained 315.71 parts per million (PPM) of carbon dioxide. There is an annual average increase of about 0.44% PPM of carbon dioxide over the year before. Which of the following is true?

    (A) The PPM in the atmosphere in the year 2050 is best approximated by an arithmetic sequence with first term 315.71 and constant difference 0.0044.

    (B) The PPM in the atmosphere in the year 2050 is best approximated by a geometric sequence with first term 315.71 and constant ratio 0.0044.

    (C) The PPM in the atmosphere in the year 2050 is best approximated by a geometric sequence with first term 315.71 and constant ratio 100.0044.

    (D) The PPM in the atmosphere in the year 2050 is best approximated by an arithmetic sequence with first term 315.71 and constant difference 100.0044.

2. An arithmetic sequence a_n with initial term a_0 has $a_2 = 6$ and $a_4 = 2$. Which of the following gives the general term of the sequence?

   (A) $a_n = 6 - 4(n-3)$ **(B)** $a_n = 6 - 2(n-3)$ **(C)** $a_n = 6 - 4(n-2)$ **(D)** $a_n = 6 - 2(n-2)$

3. A sequence s_n has the terms $s_0 = 6$ and $s_1 = -12$. If s_n is a geometric sequence, what is $\dfrac{s_{10}}{s_9}$?

   (A) -18 **(B)** -6 **(C)** -2 **(D)** $-\dfrac{1}{2}$

4. A geometric term with first term g_0 is given by the general term $g_n = 3 \cdot 2^{(n-4)}$. Which of the following does the general term reveal?

   (A) The first term in the sequence is 3.
   (B) The second term in the sequence is 3.
   (C) The fourth term in the sequence is 3.
   (D) The fifth term in the sequence is 3.

5. The table below shows the amount of milligrams of the antibiotic amoxicillin still found in the bloodstream of an average adult over an 8-hour period after the initial dose.

hours	0	1	2	3	4	5	6	7	8
milligrams	200	140	98	68.6	48.02	33.61	23.53	16.47	11.53

   A geometric series A_n is used to model the amount of amoxicillin A after n hours in the bloodstream. Which of the following is an expression for A_n?

   (A) $200(0.7)^{(n-1)}$ **(B)** $140(1.7)^{(n-1)}$ **(C)** $200(0.7)^n$ **(D)** $140(1.7)^n$

Solutions.

1. The increase is determined by a multiplicative factor, not an additive quantity, so the model should be geometric. Since there is an increase of 0.44%, the constant ratio of the geometric sequence is 100.0044. The correct answer is therefore **(C)**.

2. The general term of a_n can be written as $a_n = a_2 + d(n-2)$. Since $a_2 = 6$, $a_n = 6 + d(n-2)$. Also, since $a_4 = 2$, $a_4 = 2 = 6 + d(4-2)$. Solving for d gives $d = -2$. Thus, $a_n = 6 - 2(n-2)$. The correct answer is therefore **(D)**.

3. Since the sequence is geometric, the ratio of consecutive terms is the same. Hence,
$$\frac{s_{10}}{s_9} = \frac{s_1}{s_0} = \frac{-12}{6} = -2.$$
   The correct answer is therefore **(C)**.

4. When $n = 4$, we get $g_4 = 3 \cdot 2^0 = 3$, and this is the fifth term. The correct answer is therefore **(D)**.

5. We have that $140/200 = 0.7$ and the data is clearly decreasing. The initial value is 200, so the model is $200(0.7)^n$. The correct answer is therefore **(C)**.

Topics 2.2-2.3 ~ Change in Linear and Exponential Functions

> **Learning Objectives**
>
> 1. (2.2.A) Construct functions of the real numbers that are comparable to arithmetic and geometric sequences.
> 2. (2.2.B) Describe similarities and differences between linear and exponential functions.
> 3. (2.3.A) Identify key characteristics of exponential functions.

> **Success Criteria**
>
> 1. I can find a linear or exponential function to model a linear or geometric sequence, both in and out of context.
> 2. I can describe the rates of change of linear and exponential functions over successive equal-length intervals of the inputs.
> 3. I can construct or analyze an exponential function in terms of its base, initial value, and inputs.

In Topic 2.1, we discussed arithmetic and geometric sequences. We can actually express these sequences as *functions*. For instance, consider the sequence a_n given by $-4, -1, 2, 5, \ldots$.

The general term of this sequence is _____. We can consider the INDICES – the subscripts – as the *domain* and the actual sequence terms the *range* of a function. If we call our subscripts x and our outputs $f(x)$, then our function would be

$$f(x) = \text{\underline{\hspace{2cm}}}.$$

Because this was an arithmetic sequence, the rate of change was _____. Therefore, the function $f(x)$ was _____.

> **Fact:** Arithmetic sequences can be expressed as _____ functions.
>
> For just the sequence terms, the domain would be the _____. In a given context in which we might want to model more terms than just those in the sequence, the domain could be _____.

We can apply similar logic to a geometric sequence. For instance, consider the sequence g_n given by $100, 50, 25, 12.5, \ldots$.

The general term of the sequence is _____. Letting the subscripts be x and the outputs be $E(x)$, we can write the sequence as the function

$$E(x) = \text{\underline{\hspace{2cm}}}$$

This is called an _____.

> **Definition.** An _____ is a function $f(x) = ab^x$ where
>
> - the outputs have a *constant* _____ over successive equal-length intervals of inputs
> - an initial term a is repeatedly _____ by a constant _____ b (or r)

Why are we studying exponential functions alongside linear functions? Well,

- Linear functions can be expressed as an initial value _____ repeated _____ of a constant _____.

- Exponential functions can be expressed as an initial value _____ repeated _____ by a constant _____.

Because of these features, both exponential and linear functions can be completely determined by knowing only _____ input-output pairs.

Example 1.

(a) A linear function f contains the input-output pairs $(0, 4)$ and $(6, 7)$. Write an expression for the function $f(x)$.

(b) An exponential function g contains the input-output pairs $(0, 3)$ and $(3, 24)$. Write an expression for the function $g(x)$.

In the previous examples, we were given the initial term by the input-output pair with an input of 0. What if we don't have this input?

Example 2.

(a) A linear function f contains the input-output pairs $(5, 9)$ and $(11, 4)$. Write an expression for the function $f(x)$.

(b) An exponential function $g(x)$ contains the input-output pairs $(2, 36)$ and $(4, 16)$. Write an expression for the function $g(x)$.

> **Facts:**
>
> 1. A linear function $f(x)$ with known slope m and point (x_1, y_1) can be expressed in the form _____.
>
> 2. An exponential function $g(x)$ with a known ratio r and point (x_1, y_1) can be expressed in the form _____.

Exponential functions have a few notable characteristics. For $f(x) = ab^x$, we have the following.

- a is the INITIAL VALUE, while b is the BASE.
- If $a > 0$ and $b > 1$, the function demonstrates _____.
- If $a > 0$ and $0 < b < 1$, the function demonstrates _____.

UNIT 2 　　　　　　　　　TOPICS 2.2–2.3 ~ CHANGE IN LINEAR AND EXPONENTIAL FUNCTIONS　　　　159

- The graph of f is either always _____ or always _____ and is either always _____ or _____.

- For an exponential in general form, there are 3 possibilities as x increases/decreases without bound:

$$\lim_{x\to\pm\infty} f(x) = \underline{} \qquad \lim_{x\to\pm\infty} f(x) = \underline{} \qquad \text{or} \qquad \lim_{x\to\pm\infty} f(x) = \underline{}$$

Let's look through a couple of examples.

Example 3. A home was purchased at the beginning of 2023 that cost $225,000. The value of a home is expected to increase in value by 5% each year. Let $V(t)$ be a function that will model the value of the home V, in thousands of dollars, based on years t since 2023.

(a) Write an expression for $V(t)$.

(b) Sketch a graph of $V(t)$.

(c) (📱) In what year will the home first be worth more than $300,000?

Example 4. An exponential function $f(x) = ab^x$ passes through $(2, 3)$ and $(5, 1.5)$.

(a) (📱) Determine the values of a and b.

(b) For what x-values is $f(x)$ decreasing?

(c) For what x-values is the rate of change of f decreasing?

Practice 1. Which function/s below would contain the input-output pairs $(0, 64)$ and $(3, 8)$? Multiple answers may apply.

(A) $f(x) = 64 \cdot \left(\dfrac{1}{2}\right)^x$

(B) $g(x) = 64 \cdot \left(\dfrac{1}{2}\right)^{(x-1)}$

(C) $f(x) = 32 \cdot \left(\dfrac{1}{2}\right)^x$

(D) $g(x) = 32 \cdot \left(\dfrac{1}{2}\right)^{(x-1)}$

(E) $f(x) = 3 \cdot \left(\dfrac{1}{2}\right)^{(x-8)}$

(F) $g(x) = 8 \cdot \left(\dfrac{1}{2}\right)^{(x-3)}$

Practice 2. At time $t = 0$, a patient is administered 75 milligrams of a certain drug. Every hour, 30% of the drug that is in the bloodstream dissipates. The amount A of the drug remaining in the bloodstream after t hours can be modeled by $A(t)$.

(a) Write an expression for $A(t)$.

(b) Find $\lim\limits_{t\to\infty} A(t)$ and explain what this means in context.

Topics 2.2-2.3 Homework

1. Write a linear function $L(x)$ that could model an arithmetic sequence with a fifth term of 8 and a sixth term of 1.

2. Write an exponential function $E(x)$ that could model a geometric sequence with a fourth term of 4 and a fifth term of $\frac{2}{5}$.

3. Recall the following exponent rules.

   - For any positive x, $x^{1/n} = \sqrt[n]{x}$.
   - For any non-zero x, $x^{-a} = \dfrac{1}{x^a}$.

   (a) Use exponent rules to show that the functions $f(x) = 8 \cdot \sqrt{2}^x$ and $g(x) = 8 \cdot 2^{(x/2)}$ are the same function.

   (b) Use exponent rules to show that the functions $f(x) = 15 \cdot (\frac{1}{8})^x$, $g(x) = 15 \cdot 8^{(-x)}$, and $g(x) = 15 \cdot 2^{(-3x)}$ are the same function.

   (c) The sequence $g_n = 5 \cdot 16^{(-x/2)}$ can be rewritten in the form $g_n = 5b^x$. What is b?

4. (▤) A student is trying to create a function to model the relationship between x and y below.

x	2	4	6	8	10
y	3	3.75	4.7	5.86	7.3

   Explain why an exponential function would be appropriate.

5. Suppose the function $y = f(x)$, when graphed in the xy-plane, has the following characteristics.

   - The function has a domain of all real numbers.
   - The function is increasing for all x.
   - The function has an inflection point at $(3, 4)$.

   Could f be an exponential function? If yes, explain why. If not, explain why, and propose a possible type of function that f could be.

6. (▤) In Practice Problem 2, we looked at an example of a drug that diminished by 30% every hour. More generally, a drug is said to have a HALF-LIFE of t if it takes t (whatever units of time) for half of the substance to evaporate or otherwise disappear.

   (a) A certain isotope has a half-life of one day. If there are 500 micrograms of this substance at time $t = 0$, where t is in days, write a function $A_1(t)$ that would model the amount of the isotope left after t days.

   (b) Find the number of days, to the nearest half-day, after which there would be only 5% of the original isotope remaining.

   (c) Another isotope has a half-life of *three* days. If there are 500 micrograms of this substance at time $t = 0$, where t is in days, write a function $A_2(t)$ that would model the amount of this isotope left after t days.

   (d) Find after how many days, to the nearest half-day, there would be only 5% of the original amount of the second isotope remaining.

   (e) How do your answers from (d) and (b) relate to one another?

   (f) A researcher is trying to determine the initial amount of an isotope there was from a spill that occurred previously. Seven days after the spill, there were 42.5 micrograms of the isotope remaining. It is known that the half-life of this isotope is 4 days. How much isotope was there right before the original spill?

Topics 2.2-2.3 Solutions/Notes

Activity
The general term of this sequence is $a_n = -4 - 3n$.

... then our function would be
$$f(x) = -4 - 3x.$$

Because this was an arithmetic sequence, the rate of change was <u>constant</u>. Therefore, the function $f(x)$ was <u>linear</u>.

> **Fact:** Arithmetic sequences can be expressed as <u>linear</u> functions.
>
> For just the sequence terms, the domain would be the <u>whole numbers</u>. In a given context in which we might want to model more terms than just those in the sequence, the domain could be <u>all real numbers</u> (or any set!).

The general term of the sequence is $g_n = 100 \cdot (1/2)^n$. Letting the subscripts be x and the outputs be $E(x)$, we can write the sequence as the function

$$E(x) = 100 \cdot \left(\frac{1}{2}\right)^x$$

This is called an <u>exponential function</u>.

> **Definition.** An <u>exponential function</u> is a function $f(x) = ab^x$ where
> - the outputs have a *constant* <u>ratio</u> over successive equal-length intervals of inputs
> - an initial term a is repeatedly <u>multiplied</u> by a constant <u>ratio</u> b (or r)

Why are we studying exponential functions alongside linear functions? Well,

- Linear functions can be expressed as an initial value <u>plus</u> repeated <u>addition</u> of a constant <u>rate of change</u>.

- Exponential functions can be expressed as an initial value <u>times</u> repeated <u>multiplication</u> by a constant <u>ratio</u>.

Because of these features, both exponential and linear functions can be completely determined by knowing only _____ input-output pairs.

Example 1.

(a) The initial term is 4 and the constant rate of change must equal the average rate of change, which is $\frac{7-4}{6-0} = \frac{1}{2}$. Therefore, $f(x) = 4 + \frac{1}{2}x$.

(b) The initial term is 3. To obtain $(3, 24)$, the value 3 must be multiplied by a ratio r three times to obtain 24. Therefore, $3r^3 = 24$, so $r = 2$. We get $g(x) = 3 \cdot 2^x$.

Example 2.

(a) The average rate of change is $\frac{4-9}{11-5} = -\frac{5}{6}$. We can use the term $(5, 9)$ to generate $f(x) = 9 - \frac{5}{6}(x - 5)$. We could also use the term $(11, 4)$ to generate $f(x) = 4 - \frac{5}{6}(x - 11)$.

(b) We use "initial value" 36 to write $g(x) = 36b^{(x-2)}$. Plugging in the input 4 for x and 16 for $g(x)$, we get $16 = 36b^{(4-2)}$, or $\frac{4}{9} = b^2$. This gives $b = \frac{2}{3}$, so $g(x) = 36\left(\frac{2}{3}\right)^{(x-2)}$. We could also write $g(x) = 16\left(\frac{2}{3}\right)^{(x-4)}$.

> **Facts.**
>
> 1. A linear function $f(x)$ with known slope m and point (x_1, y_1) can be expressed in the form $f(x) = y_1 + m(x - x_1)$.
>
> 2. An exponential function $g(x)$ with a known ratio r and point (x_1, y_1) can be expressed in the form $g(x) = y_1 \cdot r^{(x-x_1)}$.

Exponential functions have a few notable characteristics. For $f(x) = ab^x$, we have the following.

- a is the INITIAL VALUE, while b is the BASE.
- If $a > 0$ and $b > 1$, the function demonstrates exponential growth.
- If $a > 0$ and $0 < b < 1$, the function demonstrates exponential decay.
- The graph of f is either always increasing or always decreasing and is either always concave up or concave down.
- For an exponential in general form, there are 3 possibilities as x increases/decreases without bound:

$$\lim_{x \to \pm\infty} f(x) = \infty \qquad \lim_{x \to \pm\infty} f(x) = -\infty \qquad \text{or} \qquad \lim_{x \to \pm\infty} f(x) = 0$$

Example 3:

(a) $V(t) = 225{,}000 \cdot 1.05^t$

(b)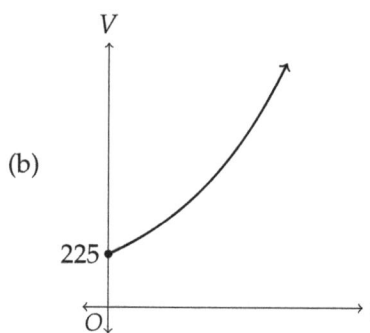

(c) Graphing $y = 225 \cdot 1.05^t$ and $y = 300$, we find an intersection point at 5.896. Therefore, the house will first be worth more than $300,000 in the year 2029.

Example 4.

(a) Substituting in $(2, 3)$ and $(5, 1.5)$ gives $3 = ab^2$ and $1.5 = ab^5$. Dividing the equations results in $\frac{1.5}{3} = \frac{ab^5}{ab^2}$, or $\frac{1}{2} = b^3$, which gives us $b = \left(\frac{1}{2}\right)^{1/3} \approx 0.7937$. Now, we can substitute in b to either equation and solve for $a \approx 4.762$.

(b) Because $a = 4.762 > 0$ and $b = 0.7937 < 1$, $f(x)$ is decreasing for all real x.

(c) The graph of $f(x)$ will always be concave up, so its rate of change will *never* be decreasing.

Practice 1. We solve $8 = 64b^3$ to get $b = \frac{1}{2}$. From here, the functions that are equivalent to $f(x) = 64 \cdot \left(\frac{1}{2}\right)^x$ are (A), (D), and (E).

Practice 2.

(a) $A(t) = 75 \cdot 0.7^t$

(b) $\lim_{t \to \infty} A(t) = 0$; as time goes on, eventually there will be none of the drug left in the bloodstream.

Homework

1. The constant difference is -7, so $L(x) = 1 - 7(x - 6)$ or $L(x) = 8 - 7(x - 5)$.

2. Note that $\frac{2}{5} = \frac{4}{10}$, so $b = \frac{1}{10}$. This gives $E(x) = 4\left(\frac{1}{10}\right)^{(x-4)}$ or $E(x) = \frac{2}{5}\left(\frac{1}{10}\right)^{(x-5)}$.

3. (a) $8 \cdot \sqrt{2}^x = 8 \cdot \left(2^{1/2}\right)^x = 8 \cdot 2^{(x/2)}$

   (b) $15 \cdot \left(\frac{1}{8}\right)^x = 15 \cdot \left(8^{-1}\right)^x = 15 \cdot \left(\left(2^3\right)^{-1}\right)^x = 15 \cdot 2^{-3x}$.

   Seeing $15 \cdot \left(\frac{1}{8}\right)^x = 15 \cdot 8^{(-x)}$ is obvious from the second expression.

   (c) $g_n = 5 \cdot 16^{(-x/2)} = 5 \cdot \left(16^{(-1/2)}\right)^x = 5 \cdot \left(\frac{1}{16^{1/2}}\right)^x = 5 \cdot \left(\frac{1}{\sqrt{16}}\right)^x = 5 \cdot \left(\frac{1}{4}\right)^x$. Therefore, $b = \frac{1}{4}$.

4. Evaluating the ratios of outputs over successive equal-length intervals of the inputs (of length 2) yields the ratios 1.25, 1.253, 1.247, and 1.246. These are roughly constant ratios, so an exponential function is appropriate.

5. f could not be an exponential function because exponential functions are always concave down or always concave up; therefore, they cannot have any inflection points. A possible solution would be a cubic function, though, like $f(x) = (x - 3)^3 + 4$.

6. (a) $A_1(t) = 500 \cdot \left(\frac{1}{2}\right)^t$

   (b) Graphing $y = A_t(t)$ and $y = 0.05(500) = 25$ gives an intersection at $t = 4.322$ days, or 4.5 days.

   (c) $A_2(t) = 500 \cdot \left(\frac{1}{2}\right)^{(t/3)}$

   (d) Graphing $y = A_2(t)$ and $y = 25$ gives an intersection at $t = 12.966$ days, or $t = 13.0$ days.

   (e) It took 3 times longer for the second isotope to reach this amount, which makes sense because the half-life is three times longer.

   (f) Letting $A(t)$ represent the amount remaining after the spill, where $t = 0$ is the day of the spill, we get $A(t) = a \cdot (1/2)^{(t/4)}$. Plugging in the input-output pair $(7, 42.5)$, we get $42.5 = a \cdot (1/2)^{7/4}$, or $a = 142.952$. This means there were originally 142.952 micrograms.

Reflection

Suggested Class Time. 60-70 minutes

Prerequisites. None beyond those covered in Topic 2.1

Instructional Strategies. Here are a few things that came up during the lecture.

Firstly, I think that the concept of using the indices as the domain of a function is rather abstract, so I started with a simpler example. I asked students whether they would believe that my 21-month old was already working with functions. They jokingly said yes, of course, because he's a math teacher's kid, but quickly they said no, babies can't do functions. I then told them that my son had learned to count to 3, and that the previous night he had been counting pieces of toast.

Students were enthralled by the notion that counting is a function: assigning numbers (1, 2, 3) to objects (pieces of toast, people, etc.) is in fact a function, as it assigns individual inputs to individual outputs. This really helped set the groundwork for using the indices of a sequence as the inputs for a function.

Most of the lesson past this was straightforward, but I did take a detour at one point. I wrote the function $f(x) = x^3$ on the board and asked students if it was exponential. Unsurprisingly, many students said yes. I had to really emphasize with them that an exponential function is not "one with an exponent." To hammer this home, I made the following chart (which we filled out from left to right).

Function	$f(5)$	$f(6)$
$f(x) = x^3$	$5 \cdot 5 \cdot 5$	$6 \cdot 6 \cdot 6$
$f(x) = 3 \cdot 2^x$	$3 \cdot 2 \cdot 2 \cdot 2 \cdot 2 \cdot 2$	$3 \cdot 2 \cdot 2 \cdot 2 \cdot 2 \cdot 2 \cdot 2$

We discussed the fact that an exponential function requires repeated multiplication by a *constant*. In the first function, the quantity being multiplied by changes, as it depends on the input. Rather, an exponential function's input simply tells how many of the base will be multiplied together (or, perhaps even better, how many terms into the geometric sequence it would be!).

Technology. We only used a calculator in part of Example 4.

Problems.

1. During flu season at one school, the number of students who had the flu increased exponentially. The function $F(t)$ can be used to model the number F of students who got the flu t weeks after the initial outbreak. The table below shows certain records of students with the flu.

Weeks after the initial outbreak	2	3
Number of students with the flu	50	60

   Which of the following is an expression for $F(t)$?
   (A) $F(t) = 50(10)^t$ **(B)** $F(t) = 50(10)^{(t-2)}$ **(C)** $F(t) = 50(1.2)^t$ **(D)** $F(t) = 50(1.2)^{(t-2)}$

2. Let the function f be given by $f(x) = 12 \cdot \left(\frac{3}{4}\right)^x$. Which of the following is true?

   (A) The rate of change of $f(x)$ is negative for all x.
   (B) The rate of change of $f(x)$ is positive for all x.
   (C) As x increases without bound, the outputs of $f(x)$ increase without bound.
   (D) As x decreases without bound, the outputs of $f(x)$ decrease without bound.

3. Which of the following describes the function $f(x) = 3(1.03)^x$?

   (A) f models exponential growth and $\lim_{x \to \infty} f(x) = 0$.
   (B) f models exponential growth and $\lim_{x \to \infty} f(x) = \infty$.
   (C) f models exponential decay and $\lim_{x \to \infty} f(x) = 0$.
   (D) f models exponential decay and $\lim_{x \to \infty} f(x) = \infty$.

4. An exponential function f contains the input-output pairs $(2, 4)$ and $(4, 36)$. What is the value of the output for the input 3?
   (A) 9   **(B)** 12   **(C)** 16.5   **(D)** 20

5. The function f has a domain of all real numbers, a range of positive real numbers, and is always decreasing. Which of the following could be true of the function f?

   (A) f is linear with a negative slope.

   (B) f is linear with a positive slope.

   (C) f is exponential with a negative rate of change.

   (D) f is exponential with a positive rate of change.

Solutions.

1. The ratio between consecutive terms is $60/50 = 1.2$. Then the function can be modeled by $F(t) = 50(1.2)^{(t-2)}$. The correct answer is therefore **(D)**.

2. Since $a = 12 > 0$ and $b = 3/4$ is positive and less than 1, this function models exponential decay. Hence, the rate of change is always negative. The correct answer is therefore **(A)**.

3. Since $a = 3 > 0$ and $b = 1.03 > 1$, this function models exponential growth. As x increases without bound, f also increases without bound. The correct answer is therefore **(B)**.

4. Using $(2,4)$, a model for the function is $f(x) = 4 \cdot b^{(x-2)}$. Using $(4, 36)$ in the model, we have $4 \cdot b^{(4-2)} = 36$, which simplifies to $b^2 = 9$. Thus, $b = 3$ and the model is $f(x) = 4 \cdot 3^{(x-2)}$. Then $f(3) = 4 \cdot 3^{(3-2)} = 4 \cdot 3 = 12$. The correct answer is therefore **(B)**.

5. The function cannot be linear since the domain is all reals and the range is only positives, so the function is exponential. Since the function is decreasing, the exponential has a negative rate of change. The correct answer is therefore **(C)**.

Topic 2.4 ~ Equivalent Exponential Forms

Learning Objectives

1. (2.4.A) Rewrite exponential expressions in equivalent forms.

Success Criteria

1. I can rewrite exponential expressions using exponent rules, including negative and rational exponents.

2. I can recognize that an exponential of the form b^{cx} is equivalent to $(b^c)^x$.

3. I can recognize and identify linear transformations of exponential functions from their analytical forms.

In Topics 2.2-2.3, we looked at exponential functions, particularly those written in the form $f(x) = ab^x$: we liked these forms, as they displayed the *initial value* and, as an added bonus, had a nice exponent of just x.

Often, we have to *transform* exponential functions into this form. For this, we use *exponent rules*. Recall the following rules.

Exponent Rules

1. $b^m \cdot b^n = b^{(m+n)}$	3. $b^{(1/k)} = \sqrt[k]{b}$
2. $(b^m)^n = b^{mn}$	4. $b^{-n} = \dfrac{1}{b^n}$

In the past, you used these rules almost exclusively for rewriting in *one direction*, like going from $x^2 \cdot x^3$ to the equivalent x^5. With exponential functions, though, we often want to carry this process out *backwards*. We very well may want to rewrite x^5 as x^{2+3}, which can then be split into $x^2 \cdot x^3$.

Try utilizing this for the following. Rewrite *each* of the following functions in the form $f(x) = ab^x$.

1. $f(x) = 4^{(x+2)}$

2. $f(x) = 9^{(\frac{1}{2}+x)}$

3. $f(x) = 5^{(x-2)}$

4. $f(x) = 8^{(x-\frac{1}{3})}$

5. (🔳) $f(x) = 6 \cdot 0.58^{(x-2.513)}$

We can use other exponent rules "backwards." Rather than *multiplying* exponents with an expression like $(x^2)^4$, we may be more interested in *pulling out* an exponential expression, like going from x^8 back to $(x^4)^2$.

Try rewriting each function in the form $f(x) = ab^x$.

6. $f(x) = 3^{2x}$

10. $f(x) = 16^{(x/4)}$

7. $f(x) = 3^{x/2}$

11. $f(x) = 8^{(2x/3)}$

8. $f(x) = 2 \cdot 4^{3x}$

12. $f(x) = 8 \cdot 5^{(-x/2)}$

9. $f(x) = 16^{(x/2)}$

13. $f(x) = -3 \cdot 12^{0.5x}$

We can even combine rules! Try to rewrite each of the following in the form $f(x) = ab^x$.

14. $f(x) = 2 \cdot 5^{(2x+1)}$

16. $f(x) = -3 \cdot (\sqrt{2})^{2x+6}$

15. $f(x) = \dfrac{8}{3} \cdot 16^{(2x-\frac{1}{2})}$

17. $f(x) = \dfrac{2}{3} \cdot 27^{(\frac{x+1}{3})}$

Lastly, here are two interesting features of exponential functions related to these exponent rules. We'll let $f(x) = b^x$ be a generic exponential function.

18. What kind of transformation transforms $f(x) = b^x$ into $g(x) = b^{x+c}$?

19. Show that $g(x) = b^{x+c}$ can actually be rewritten in the form $g(x) = ab^x$.

20. What kind of transformation transforms $f(x) = b^x$ into $g(x) = ab^x$?

21. The result is as follows: every _____ *translation* of an exponential function is equivalent to a _____ *dilation* of that function..

22. Similarly, what kind of transformation transforms $f(x) = b^x$ into $g(x) = b^{cx}$?

23. Show that $g(x) = b^{cx}$ is just an exponential function $f(x) = b^x$, but with a *different base* of that function.

24. The result is as follows: every _____ *dilation* of an exponential function is equivalent to a _____ of that function.

Notes

Topic 2.4 Homework

1. The function $f(x) = 49^{(-x/2)}$ can be rewritten as $f(x) = b^x$. What is b?

2. Let g be the function given by $g(x) = 4 \cdot 2^{(x/2-1)}$.

   (a) Compute $g(3)$.
   (b) Compute $g(1)$.
   (c) Rewrite g in the form $g(x) = ab^x$.

3. (Multiple Select) Which of the following are equivalent to $f(x) = 16^{(0.5x-1)}$? More than one may apply.

   (A) $f(x) = -16 \cdot 16^{0.5x}$
   (B) $f(x) = -16 \cdot 4^x$
   (C) $f(x) = 8 \cdot \left(\dfrac{1}{16}\right)^x$
   (D) $f(x) = \dfrac{1}{16} \cdot 16^{0.5x}$
   (E) $f(x) = 4 \cdot (-16)^x$
   (F) $f(x) = 4^{(x-2)}$
   (G) $f(x) = \dfrac{1}{16} \cdot 8^x$
   (H) $f(x) = \dfrac{1}{16} \cdot 4^x$

4. (🖩) The number of bacteria N in a culture after t days can be modeled by the function given by $N(t) = e^{(0.12t+6.215)}$.

   (a) Rewrite this function in the form $N(t) = ab^t$.
   (b) What was the initial number of bacteria in the culture?

5. The graph of $g(x) = e^{x+2}$ is a horizontal translation of the graph of $g(x) = e^x$ by -2. The graph of g is also a vertical dilation of f by what factor?

6. Identify two different transformation that could transform $f(x) = 3 \cdot 2^x$ into $g(x) = 6 \cdot 2^x$.

7. Identify two different transformations that could transform $f(x) = 16^x$ into $g(x) = 4 \cdot 16^x$.

8. (Multiple Select) Let $f(x) = 3^x$. Which of the following transformations would transform the graph of f into the graph of $g(x) = 9f(x)$? More than one may apply.

   (A) Horizontal dilation by 9
   (B) Horizontal dilation by -9
   (C) Horizontal dilation by 2
   (D) Horizontal dilation by -2
   (E) Horizontal translation by 9
   (F) Horizontal translation by -9
   (G) Horizontal translation by 2
   (H) Horizontal translation by -2
   (I) Vertical dilation by 9
   (J) Vertical dilation by -9
   (K) Vertical dilation by 2
   (L) Vertical dilation by -2
   (M) Vertical translation by 9
   (N) Vertical translation by -9
   (O) Vertical translation by 2
   (P) Vertical translation by -2

Topic 2.4 Solutions/Notes

Activity

1. $f(x) = 4^x \cdot 4^2 = 16 \cdot 4^x$

2. $f(x) = 9^{(1/2)} \cdot 9^x = 3 \cdot 9^x$

3. $f(x) = 5^{-2} \cdot 5^x = \frac{1}{25} \cdot 5^x$

4. $f(x) = 8^{(-1/3)} \cdot 8^x = \frac{1}{2} \cdot 8^x$

5. $f(x) = 6 \cdot 0.58^{-2.513} \cdot 0.58^x \approx 23.586 \cdot 0.58^x$

6. $f(x) = (3^2)^x = 9^x$

7. $f(x) = (3^{1/2})^x = \sqrt{3}^x$

8. $f(x) = 2 \cdot (4^3)^x = 2 \cdot 64^x$

9. $f(x) = (16^{1/2})^x = 4^x$

10. $f(x) = (16^{1/4})^x = 2^x$

11. $f(x) = (8^{2/3})^x = 4^x$ (because $8^{2/3} = (8^{1/3})^2 = 2^2 = 4$)

12. $f(x) = 8 \cdot (5^{(-1/2)})^x = 8 \cdot \left(\frac{1}{\sqrt{5}}\right)^x$

13. $f(x) = -3 \cdot (12^{0.5})^x = -3 \cdot (\sqrt{12})^x = -3 \cdot (2\sqrt{3})^x$

14. $f(x) = 2 \cdot 5^1 \cdot 5^{2x} = 10 \cdot (5^2)^x = 10 \cdot (25)^x$

15. $f(x) = \frac{8}{3} \cdot 16^{(-1/2)} \cdot 16^{2x} = \frac{8}{3} \cdot \frac{1}{4} \cdot (16^2)^x = \frac{2}{3} \cdot 256^x$

16. $f(x) = -3 \cdot (\sqrt{2})^6 \cdot (\sqrt{2})^{2x} = -3 \cdot 2^3 \cdot \left(\sqrt{2}^2\right)^x = -24 \cdot 2^x$

17. $f(x) = \frac{2}{3} \cdot 27^{x/3 + 1/3} = \frac{2}{3} \cdot 27^{1/3} \cdot 27^{x/3} = \frac{2}{3} \cdot 3 \cdot (27^{1/3})^x = 2 \cdot 3^x$

18. Horizontal translation by $-c$

19. $g(x) = b^{x+c} = b^c \cdot b^x = ab^x$, where $a = b^c$

20. Vertical dilation by $|a|$

21. Every <u>horizontal</u> translation of an exponential function is equivalent to a <u>vertical</u> dilation of that function.

22. Horizontal dilation by $\left|\frac{1}{c}\right|$

23. $g(x) = b^{cx} = (b^c)^x$, where the new base is b^c

24. Every <u>horizontal</u> dilation of an exponential function is equivalent to a <u>change of base</u> of that function.

> **Notes**
>
> Utilizing exponent rules, any function of the form $f(x) = (b_1)^{cx+d}$ can be turned into $f(x) = a \cdot (b_2)^x$.
>
> This results in the interesting facts that
>
> - any horizontal translation of an exponential function is equivalent to a *vertical dilation* of that same function
>
> - any horizontal dilation of an exponential function is equivalent to a *change of base* of that same function

Homework

1. $b = 49^{-1/2} = \dfrac{1}{\sqrt{49}} = \dfrac{1}{7}$

2. (a) $g(3) = 4 \cdot 2^{1/2} = 4\sqrt{2}$

   (b) $g(1) = 4 \cdot 2^{(-1/2)} = \dfrac{4}{\sqrt{2}} = \dfrac{4\sqrt{2}}{2} = 2\sqrt{2}$

   (c) $g(x) = 4 \cdot 2^{-1} \cdot (2^{1/2})^x = 4 \cdot \tfrac{1}{2} \cdot (\sqrt{2})^x = 2 \cdot (\sqrt{2})^x$

3. D, F, and H

4. (a) $N(t) = e^{6.215} e^{0.12t} \approx 500.196 e^{0.12t}$

   (b) Approximately 500

5. $g(x) = e^{x+2} = e^2 e^x$, so a vertical dilation by a factor of e^2

6. $g(x) = 6 \cdot 2^x = 3 \cdot (2 \cdot 2^x) = 3 \cdot 2^{x+1}$, so a vertical dilation by a factor of 2 or a horizontal translation of -1.

7. A vertical dilation by 4 or, since $g(x) = 4 \cdot 16^x = 16^{1/2} \cdot 16^x = 16^{x+1/2}$, a horizontal translation by $-\dfrac{1}{2}$.

8. H and I

Reflection

Suggested Class Time. 60 minutes

Prerequisites. Students should have some familiarity with exponent rules, particularly $b^m \cdot b^n = b^{(m+n)}$ and $(b^m)^n = b^{mn}$.

Instructional Strategies. By and large, my students flew through this activity. The two most common things I had to emphasize were:

- A number of students were not thorough with reading/finishing problems by writing their expressions in the form ab^x. I saw a lot of expressions like $8^{-1/2} \cdot 8^x$ (not simplified) and $2 \cdot 5^{2x}$. With students doing this, I tried to emphasize reading the problem, but I also tried to reinforce why ab^x is so convenient (initial value times common ratio).

- Many students struggled with Problem 17 because the exponent was not in the form $ax + b$. This was a bit frustrating, but unsurprising. As I walked around the room, I taught $\frac{x+1}{3} = \frac{1}{3}x + \frac{1}{3}$ in two different ways.

– Method 1: I call this "splitting" the fraction. I just emphasize to students that the division *must be distributed*. We went through quick examples like $\frac{a+b}{c} = \frac{a}{c} + \frac{b}{c}$ and $\frac{x+y+z}{z} = \frac{x}{z} + \frac{y}{z} + 1$.

– Method 2: It made more sense for some algebraically-minded students to see the aforementioned $\frac{x+1}{3}$ as $\frac{1}{3}(x+1)$. This helped them to *see* the need for distributing.

Personally, I additionally prefer Method 2, as it is more likely to result in students writing $\frac{1}{3}x$ than $\frac{x}{3}$, with the latter often obfuscating the slope/rate of change/coefficient for weaker students.

In the very short Notes portion after the activity, I designed a simple Geogebra sketch to help students visualize the multiple equivalent transformations of exponential functions. The sketch is online at https://www.geogebra.org/m/rqhb5bfu. I first showed how the transformations 2^{x+2} and $4 \cdot 2^x$ were equivalent, and we then played around with a few other combinations.

Technology. Students used the TI-84 for only 1 question. This question could easily be omitted.

Problems.

1. If $f(x) = 10 \cdot 16^x$, what is $f\left(-\frac{1}{4}\right)$?

    (A) $\frac{5}{2}$ (B) 5   (C) 20   (D) 40

2. Which of the following is true about the function $f(x) = 2^x$?

    (A) A vertical dilation of the graph of f by a factor of 4 is equivalent to a horizontal translation of the graph of f by -4 units.

    (B) A vertical dilation of the graph of f by a factor of 4 is equivalent to a horizontal translation of the graph of f by -2 units.

    (C) A vertical dilation of the graph of f by a factor of 4 is equivalent to a horizontal translation of the graph of f by $+2$ units.

    (D) A vertical dilation of the graph of f by a factor of 4 is equivalent to a horizontal translation of the graph of f by $+4$ units.

3. Which of the following is an expression equivalent to $f(x) = 4 \cdot 25^{(1+\frac{x}{2})}$?

    (A) $f(x) = 4 \cdot 5^x$ (B) $f(x) = 100 \cdot 5^x$ (C) $f(x) = 4 \cdot 12.5^x$ (D) $f(x) = 100 \cdot 12.5^x$

4. The graph of the function g is the image of the transformations of a horizontal translation by -2 units and a vertical dilation by a factor of 3 to the graph of the function f. If f is defined as $f(x) = 5(\frac{1}{4})^x$, which of the following is an expression for g?

    (A) $g(x) = 15\left(\frac{1}{4}\right)^{(x-2)}$ (B) $g(x) = 5\left(\frac{1}{4}\right)^{(x-1)}$ (C) $g(x) = \frac{15}{16}\left(\frac{1}{4}\right)^x$ (D) $g(x) = 15\left(\frac{1}{4}\right)^x$

5. The function h is defined by $h(x) = 13\left(\frac{5}{7}\right)^x$. Find the value of $\frac{h(24)}{h(23)}$.

    (A) $\frac{5}{7}$ (B) 1   (C) $\frac{7}{5}$ (D) $\frac{65}{7}$

Solutions.

1. $f(-1/4) = 10 \cdot 16^{-1/4} = 10 \cdot (2^4)^{-1/4} = 10 \cdot 2^{-1} = 10 \cdot 1/2 = 5$. The correct answer is therefore **(B)**.

2. A vertical dilation by a factor of 4 results in the expression $4 \cdot 2^x = 2^2 \cdot 2^x = 2^{(x+2)}$. This is also a horizontal translation by -2 units. The correct answer is therefore **(B)**.

3. We have $4 \cdot 25^{(1+\frac{x}{2})} = 4 \cdot 25^1 \cdot 25^{x/2} = 4 \cdot 25 \cdot (25^{1/2})^x = 4 \cdot 25 \cdot 5^x = 100 \cdot 5^x$. The correct answer is therefore **(B)**.

4. The translation makes the exponent $x + 2$ and the dilation multiplies the function by 3. Thus,
$$f(x) = 3 \cdot 5 \left(\frac{1}{4}\right)^{(x+2)} = 15 \left(\frac{1}{4}\right)^2 \cdot \left(\frac{1}{4}\right)^x = 15 \left(\frac{1}{16}\right) \cdot \left(\frac{1}{4}\right)^x = \frac{15}{16} \left(\frac{1}{4}\right)^x.$$

The correct answer is therefore **(C)**.

5. The ratio of two consecutive integer powers of an exponential function if simply the common ratio. The correct answer is therefore **(A)**.

Topics 2.5-2.6 ~ Exponential Function Modeling

Learning Objectives
1. (2.5.A) Construct a model for situations involving proportional output values over equal-length input-value intervals.
2. (2.5.B) Apply exponential models to answer questions about a data set or contextual scenario.
3. (2.6.A) Construct linear, quadratic, and exponential models based on a data set.
4. (2.6.B) Validate a model constructed from a data set.

Success Criteria
1. I can construct an exponential function model given 2 input-output pairs or an initial value and a reasonable proportion.
2. I can perform exponential regression and use it to answer questions in context.
3. I can use residuals to determine prediction errors and determine whether a function model is appropriate.

The table below shows the U.S. birth rate, in number of births per 1,000 people, for the years 2009 to 2018 (*National Student Clearinghouse Research Center*).

Year	Birth Rate
2009	13.558
2010	13.305
2011	13.051
2012	12.798
2013	12.544
2014	12.429
2015	12.317
2016	12.206
2017	12.096
2018	11.968

As of 2018, the birth rate had dropped for the 9th consecutive year. Researchers, policymakers, and other experts are all trying to figure out why. Among things that cause alarm bells are what might happen if this trend continues. The question is: what kind of trend, or *function model*, might that be?

The first differences, second differences, and ratios of consecutive terms have been shown below.

Year	Birth Rate	First Differences	Second Differences	Ratio
2009	13.558			
2010	13.305	−0.253		0.981
2011	13.051	−0.254	−0.001	0.981
2012	12.798	−0.253	0.001	0.981
2013	12.544	−0.254	−0.001	0.980
2014	12.429	−0.115	0.139	0.991
2015	12.317	−0.112	0.003	0.991
2016	12.206	−0.111	0.001	0.991
2017	12.096	−0.11	0.001	0.991
2018	11.968	−0.128	−0.018	0.989

1. Explain why any of a linear, quadratic, or exponential function might be appropriate for modeling the birth rate based on year.

2. Let's focus on the exponential model first. An exponential model can be created in one of two ways.

   - Method 1: Use an initial value a and an appropriate ratio b to construct $f(x) = ab^x$.
   - Method 2: Use two input-output pairs (x_1, y_1) and (x_2, y_2). Let y_1 be the initial value a and $\frac{y_2}{y_1}$ be the ratio b to construct $f(x) = ab^{(x-x_1)}$.

3. Use Method 1 to construct a model $B(t)$ to model the birth rate B based on t where t is *years since 2009*. Use 2009 as the initial year.

4. Use Method 2 to construct another model $B(t)$ to model the birth rate B based on t, where t is just the year. Just for practice, use the input-output pairs $(2012, 12.798)$ and $(2013, 12.544)$.

5. In what year does the model from Method 1 first predict that the birth rate will be below 10.5 per 1,000 people?

6. Rather than construct a model by hand, we can also perform *regression*. Input the year and birth rate data into L1 and L2 of your calculator and run `ExpReg` (you have to remember where to find this in the calculator!). *Make sure to store the regression equation in Y1*. Write the regression function $C(t)$ below.

$$C(t) = \underline{\hspace{2in}}$$

7. Use the regression model $C(t)$ to predict what the birth rate was in 2013.

8. The value from Problem 7 is not exactly equal to the actual birth rate in 2013, which was 12.544. The prediction error here is called a RESIDUAL. It is computed by the formula

$$\text{Residual} = \text{Actual output} - \text{Predicted output}$$

Compute the residual for the prediction you made in Problem 7.

9. The sign of a residual can help determine whether a prediction was an *overestimate* or *underestimate*. Use logic and the equation from Problem 8 to fill in the blanks.

   - When a residual is positive, the predicted y value is an _____ of the actual y value.

   - When a residual is negative, the predicted y value is an _____ of the actual y value.

10. Graphically, a residual appears as just a *signed vertical distance* from a point (x, y) to the regression function. Shown below is a scatterplot of the birth rate data, along with the exponential regression function. Draw the segment that represents the residual you calculated in Problem 8.

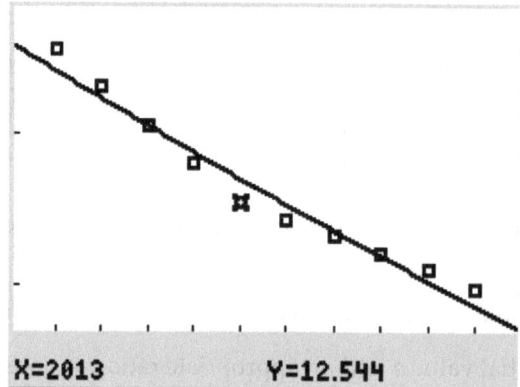

You have now used an exponential function to model the data - but what about a linear model or a quadratic model?

Linear and quadratic regression were computed for the birth rate data as well, and a scatterplot for each of the three models (including exponential) are shown below.

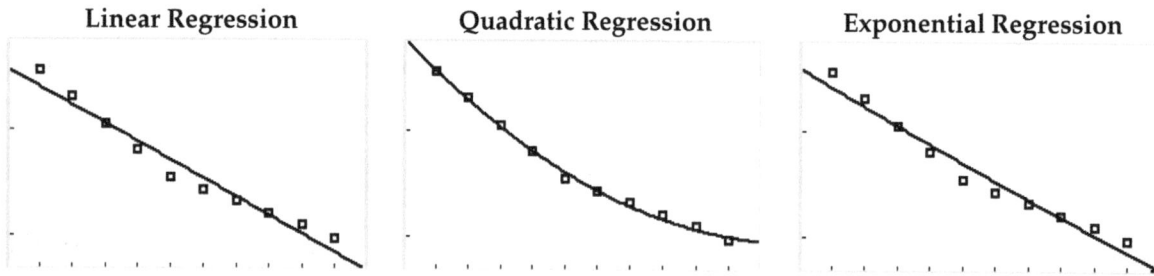

11. Based on the scatterplots, which of the three regression models do you think would best model the birth rate data? Why?

When choosing which of various regression models may be best - in this case, linear, quadratic, or exponential - it can often be hard to "eyeball" a scatterplot to see which model fits best. This leads to what is called a *residual plot*.

A RESIDUAL PLOT keeps the original input values along the x-axis, but changes the y-axis from the output values to the *residual* values. Essentially, what this does is *remove* the size of the *outputs*, which may be quite large, and help you to zoom in on *just the errors*.

The residual plots for each of the linear, quadratic, and exponential regressions are shown below.

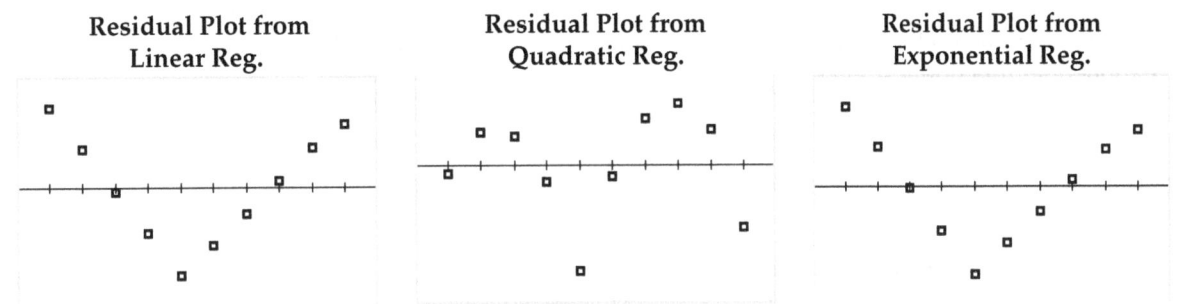

12. In Problem 11, you (hopefully!) identified the *quadratic* regression as the best fit. What do you notice about its residual plot compared to the residual plots from the linear and exponential regression models?

Notes

Topics 2.5-2.6 Homework

1. The relationship between the variables x and y, with values shown below, can be modeled by an exponential function $y = f(x)$.

x	2	5	8
y	3	4.5	6.75

   (a) Write an exponential function of the form $f(x) = ab^{(x-c)}$.
   (b) Find the value of $f(0)$.
   (c) Use (b) to rewrite f in the form $f(x) = ab^x$.

2. A telecommunications corporation recorded the average customer satisfaction score out of 5 (from a survey given at the end of any service call) for every call recorded each night. The corporation wanted to see how this was related to the number of service representatives answering service calls each night. The table below shows data from a week's worth of calls.

Number of representatives x	75	80	90	95	100	105	108
Average customer satisfaction score	3.2	3.5	3.7	3.9	4	4.3	4.4

   The corporation then performed regression to create an exponential function modeling this data. Based on this regression function, how many representatives would the corporation need answering service calls one night to get an average customer satisfaction score of 4.8?

3. A company's profits in April were reasonably high, but they decreased over the summer due to seasonal factors to their lowest monthly profit in August. The company's profits then increased back to April levels by December. Based on contextual clues, do you think a linear function, quadratic function, or exponential function would best model the company's monthly profits?

4. A student took the SAT (scored out of 1600) four times, each three months apart with the same amount of studying between each attempt, and their scores are recorded below.

Attempt a	1	2	3	4
Score S	1150	1200	1240	1300

   (a) Explain why a linear model could reasonably model the relationship between the score and the attempt.
   (b) Perform regression to find a linear function $S(a)$ to model this data.
   (c) Compute the residual for the student's 3rd attempt based on the model.
   (d) Did the linear model overestimate or underestimate the student's score?

5. Koi were introduced to a pond (that otherwise had no koi) with the goal of growing the koi population in the pond. The population of koi in the can be modeled by the function $P(t) = 6 \cdot 2^t$, where t is months from when the koi were initially introduced to the pond.

   (a) How many koi were initially placed in the pond?
   (b) How many koi does the model predict are the pond after 4 months?
   (c) Write an equivalent form of $P(t)$ that would model the koi population after t years.

6. A student conducted a research project in which they cooked the same kinds of meat at different temperatures and measured the change in the amount of bacteria from before cooking to after cooking for a set amount of time. The student then performed quadratic and exponential regression to find a model that would best model the relationship between temperature and change in bacteria. A graph of the residuals for each regression model are provided below.

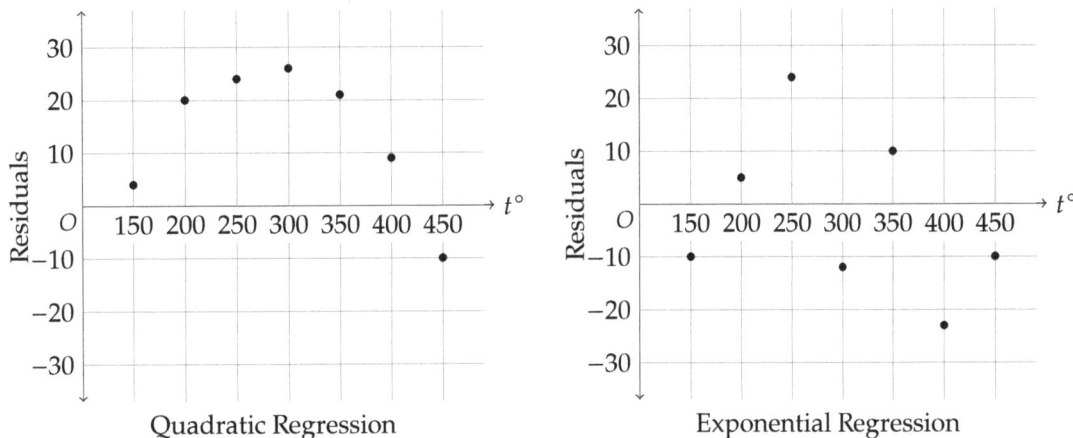

(a) Based on the graphs of the residual plots, which type of function – quadratic or exponential – is more appropriate for modeling the relationship between temperature and change in bacteria?

(b) For the piece of meat that was cooked at 350°, the predicted change in bacteria was 120 (using the regression identified in (a)). What was the actual change in bacteria for this piece of meat?

Topics 2.5-2.6 Solutions/Notes

Activity

1. The first differences and second differences are both roughly constant, so a linear or quadratic model would work. Additionally, the ratios of the outputs over successive equal-length intervals of the input are relatively constant.

3. $B(t) = 13.558(0.981)^t$

4. $B(t) = 12.798(0.981)^{t-2012}$

5. Graphing $y = 13.558(0.981)^x$ and $y = 10.5$, we get $t = 13.324$, or the year $2009 + 13.324 = 2022.324$. This would make the first year where the birth rate is below 10.5 per 1000 the year 2023.

6. $B(t) = (1.0387 \cdot 10^{13}) \cdot 0.9865^t$

7. Y1(2013) = 12.703

8. Residual = $12.544 - 12.703 = -0.159$

9.
   - When a residual is positive, the predicted y value is an underestimate of the actual y value.
   - When a residual is negative, the predicted y value is an overestimate of the actual y value.

10.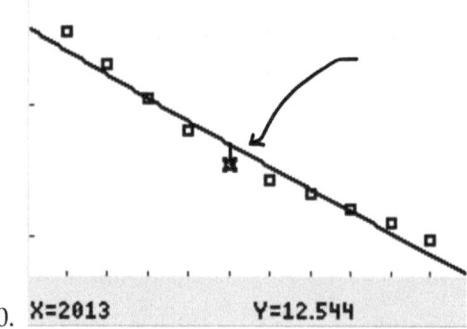

12. The residual plot for the quadratic regression appears like a random scatter of points, whereas the other two residual plots have clear patterns.

> **Notes**
>
> An exponential function can model growth patterns that have proportional outputs over successive equal-length intervals of inputs.
>
> An initial value (x_0, a) and a known ratio b can be used to create a model $f(x) = ab^x$.
>
> Similarly, two known input-output pairs (x_1, y_1) and (x_2, y_2) can be used to create a model
>
> $$f(x) = y_1 \left(\frac{y_2}{y_1}\right)^{(x-x_1)}.$$
>
> In both of the above models, b or $\frac{y_2}{y_1}$ can be understood as a *growth factor* in successive unit changes (changes by 1 unit).
>
> In many contextual scenarios, the base of an exponential function will be $e \approx 2.718$.
>
> Sometimes, exponential functions will be rewritten. For instance, if t represents days, then the function $f(t) = 3^t$ represents *tripling every day*. On the other hand, the equivalent form
>
> $$f(t) = \left(3^{1/24}\right)^{24t}$$
>
> represents growth by a factor of $3^{1/24}$ every *hour*.
>
> When determining which of various functions best model a data set, a *residual plot* can be used. If the residual plot for a regression shows *no pattern*, then the regression model is appropriate. If there is a *clear pattern* to the residual plot, then the regression model is *not appropriate*.

Homework

1. (a) The common ratio is $\sqrt[3]{\frac{4.5}{3}} = \sqrt[3]{1.5}$. There are 3 possible answers, but the simplest is $f(x) = 2 \cdot \sqrt[3]{1.5}^{(x-2)}$.

   (b) $f(0) = 2 \cdot \sqrt[3]{1.5}^{(-2)} \approx 2.289$

   (c) $f(x) = 2.289 \cdot \sqrt[3]{1.5}^x$

2. After performing regression, storing it into Y1, and then graphing this along with $y = 4.8$, the intersection occurs at $x = 118.2$. The model *predicts* that the company should have 118 or 119 representatives answering service calls to obtain an average customer satisfaction score of 4.8.

3. The context implies a decrease to a minimum and then a seemingly reasonably symmetric rise; this indicates a quadratic function might be an appropriate model.

4. (a) The first differences (50, 40, 60) are roughly constant, so a linear model is reasonable.

   (b) $S(a) = 49a + 1100$

   (c) The predicted score is $S(3) = 1247$, so the residual is $1240 - 1247 = -7$.

   (d) Because the residual is negative, the model overestimated the student's score by 7 points.

5. (a) 6

   (b) $P(4) = 6 \cdot 2^4 = 96$

   (c) $P(t) = 6 \cdot \left(2^{12}\right)^{(t/12)}$

182 TOPICS 2.5-2.6 ~ EXPONENTIAL FUNCTION MODELING UNIT 2

6. (a) The graph of the residuals from the quadratic regression shows a clear pattern, while the graph of the residuals from the exponential regression does not. Therefore, an exponential function is appropriate (while a quadratic function is not).

    (b) The residual for the piece of meat cooked at 350° was 10. Therefore,

    $$\text{Residual} = \text{Actual } y - \text{Predicted } y$$
    $$10 = y - 120$$
    $$y = 130$$

    The actual change in the number of bacteria was 130.

Reflection

Suggested Class Time. 60-70 minutes

Prerequisites. Students should be reasonably proficient with Topic 2.4 and should know how to compute regression in the calculator.

Instructional Strategies. The first half of the activity went quite well, particularly on the heels of the Topic 2.4 Activity. In the second half of the activity, students got through regressions well, but there was one main hiccup: students thought the quadratic residual plot in Problem 12 was correct because it either (a) was curved or (b) looked like a polynomial. Unfortunately, the residuals plot from this (real!) data for both linear and exponential regressions had forms that both looked the same and looked like "vees."

This was where the teaching had to come in. I teach residuals as essentially taking a model and straightening/flattening it out. I essentially tell students that we are removing the "blank space" in an xy-plane in order to really focus in on the prediction errors (residuals). We spent a good few minutes talking about what a residual plot would look like if a model was appropriate - the model should not consistently over- or under-estimate, and therefore the residuals should be a random scatter of positive and negative. On the other hand, a model that does consistently over- or under-estimate will result in a residual plot that shows some pattern (most likely curved). The applet at https://www.geogebra.org/m/hjtv2tSB is fantastic for displaying many residual plots in a short amount of time.

Technology. Students used TI-84s for the entirety of class.

Problems.

1. The function $f(x) = 4 \cdot b^x$ contains the input-output pair $(5, 8)$. The function f can be rewritten using the natural base e into the function $f(x) = 4 \cdot e^{(k \cdot x)}$ for some value of k. What is k?

    (A) 4.307   **(B)** 4.712   **(C)** 5.133   **(D)** 5.693

2. The table shows values for a function g at selected values of x.

x	-4	-2	2	4
$g(x)$	20	24	60	84

    An exponential regression $y = ab^x$ is used to model the data. What is the value of $g(3)$ predicted by the model?

    (A) 69.596   **(B)** 70.993   **(C)** 73.052   **(D)** 74.039

3. The graph of the exponential function $f(x)$ contains the input-output pairs $(0, 4)$ and $(2, \frac{4}{9})$. Which of the following could represent $f(x)$?

    (A) $f(x) = 2 \cdot \left(\frac{1}{9}\right)^x$ **(B)** $f(x) = 2 \cdot \left(\frac{1}{9}\right)^{(x-2)}$ **(C)** $f(x) = 4 \cdot \left(\frac{1}{3}\right)^x$ **(D)** $f(x) = 4 \cdot \left(\frac{1}{3}\right)^{(x-2)}$

4. A water tank is leaking, as shown in the table below.

Time in hours	0	10	20	25
Volume in gallons	20	18	15	7

   It is found that the transformed exponential function $V(t) = -2 \cdot 1.65^{(0.3t-4)} + 20$ can closely model the volume V of the tank after t hours. If the tank is empty after 30 hours, then which of the following is true about the model's prediction of the tank's volume after 30 hours?

   (A) The model underestimates the tank's volume after 30 hours by 1.340 gallons.

   (B) The model overestimates the tank's volume after 30 hours by 1.340 gallons.

   (C) The model underestimates the tank's volume after 30 hours by 4.460 gallons.

   (D) The model overestimates the tank's volume after 30 hours by 4.460 gallons.

5. Quadratic regression was used to model a data set, and a graph of the residuals from the regression is shown below.

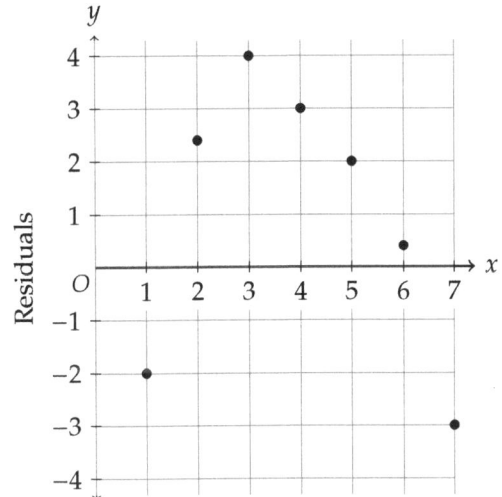

   Which of the following is true?

   (A) A quadratic model is appropriate, and the majority of the predictions are underestimates.

   (B) A quadratic model is appropriate, and the majority of the predictions are overestimates.

   (C) A quadratic model is not appropriate, and the majority of the predictions are underestimates.

   (D) A quadratic model is not appropriate, and the majority of the predictions are overestimates.

Solutions.

1. We have $8 = 4 \cdot e^{(k-5)}$ so that $2 = e^{(k-5)}$. Using the calculator to solve this equation, we get $k = 5.693$. The correct answer is therefore **(D)**.

2. The calculator gives an exponential regression of $y = 39.43827795 \cdot 1.2084312^x$. When $x = 3$, the regression curve gives $y = 69.596$. The correct answer is therefore **(A)**.

3. Let $f(x) = ab^x$ be the exponential function. Then from the input-output pair $(0, 4)$, we have $4 = ab^0 = a$. From the input-output pair $(2, \frac{4}{9})$, we have $\frac{4}{9} = ab^2$. However, $a = 4$, so $\frac{4}{9} = 4b^2$. Then $b^2 = \frac{1}{9}$ so that $b = \frac{1}{3}$. The correct answer is therefore **(C)**.

4. At $t = 30$ hours, the model gives

$$V(30) = -2 \cdot 1.65^{(0.3 \cdot 30 - 4)} + 20 = -2 \cdot 1.65^5 + 20 = -4.460.$$

This is an underestimate by 4.460 gallons since the tank is empty ($V = 0$ gallons) at $t = 30$ hours. The correct answer is therefore **(C)**.

5. The residual plot shows a clear pattern, and all but two of the residuals are positive. This indicates that the model is not appropriate and a majority of the predictions are underestimates. The correct answer is therefore **(C)**.

Topic 2.7 ~ Compositions of Functions

Learning Objectives

1. (2.7.A) Evaluate the composition of two or more functions for given values.
2. (2.7.B) Construct a representation of the composition of two or more functions.
3. (2.7.C) Rewrite a given function as a composition of two or more functions.

Success Criteria

1. I can evaluate the composition of two or more functions given analytically, graphically, tabularly, or verbally.
2. I can construct function compositions, including using algebra.
3. I can explain how certain function compositions correspond to graphical transformations.

Suppose you have an input x that you input into a function f to get an output $f(x)$. If you then take this output $f(x)$ and use it as an *input* in a function g to obtain the output $g(f(x))$, it is called the COMPOSITION of two functions. The resulting function $g(f(x))$ is read as "g of f of x" and can also be written as $(g \circ f)(x)$.

One general tip for compositions: they work "inside out." So, if you're given something like

$$f(h(g(x))),$$

we read this as...

$$f \text{ of } h \text{ of } g \text{ of } x,$$

and the first function evaluated is $g(x)$. A flow chart of sorts is shown below.

$$x \longrightarrow \boxed{g(\ )} \xrightarrow{g(x)} \boxed{h(\ )} \xrightarrow{h(g(x))} \boxed{f(\ )} \longrightarrow f(h(g(x)))$$

For the following problems, consider the following functions.

$$f(x) = x^2 + x - 3 \qquad h(x) = 2x \qquad k(x) = \frac{1}{x-3}$$
$$g(x) = \sqrt{2x-4} \qquad i(x) = x \qquad q(x) = x + 4$$

1. Compute $f(g(4))$.
2. Compute $g(f(4))$.
3. Are the answers to Problems 1 and 2 the same?
4. Fill in the following: *In general, function composition (as an operation) is not* _____.
5. Write an expression for $f(g(x))$.

6. Write an expression for $f(h(x))$.

7. Write an expression for $f(q(x))$.

8. Write an expression for $k(k(x))$.

9. Write an expression for each of the following compositions.
    (a) $f(i(x))$
    (b) $g(i(x))$
    (c) $h(i(x))$

10. What did you notice about composition with the function $i(x) = x$?

11. The i in $i(x)$ was not a coincidental choice: i was meant to stand for the *identity* function. Why do you think we call $i(x) = x$ the identity function?

12. With some compositions, domain restrictions will be involved.
    (a) What is the domain of $g(x)$?
    (b) What is the domain of $k(x)$?
    (c) Based on (a) and (b), determine the domain of $g(k(x))$. (Hint: You'll need to do some work!)
    (d) Based on (a) and (b), determine the domain of $k(g(x))$. (Hint: You'll need to do some work!)

13. Some function transformations correspond to *graphical transformations*.
    (a) Applying the function composition $f(q(x))$ is the same as applying what transformation to the graph of $f(x)$?
    (b) Applying the function composition $q(f(x))$ is the same as applying what transformation to the graph of $f(x)$?
    (c) Applying the function composition $g(h(x))$ is the same as applying what transformation to the graph of $g(x)$?
    (d) Applying the function composition $h(g(x))$ is the same as applying what transformation to the graph of $g(x)$?
    (e) Suppose the function composition $k(a(x))$ results in a horizontal translation of +2 and a horizontal dilation by a factor of $\frac{1}{4}$ to the graph of $k(x)$. What is the function $a(x)$?

14. One other skill that is often very useful in Calculus is to take an *already* composed function and to identify the "inner" and "outer" functions. This is known as DECOMPOSITION.

    For instance, the function $g(x) = \sqrt{2x-4}$ from above can be considered the composition of $a(x) = \sqrt{x}$ with $b(x) = 2x - 4$ so that $g(x) = a(b(x))$.

(a) Make sure you understand how $g(x) = a(b(x))$ in the example.

(b) Each of the functions below can be rewritten as $f(g(x))$ for some functions f and g. Identify each of f and g for each function. (Note: All function letters are now reset, so f, g, h, k, etc. do *not* refer to the functions on the first page.)

i. $h(x) = (3x + 4)^2$

ii. $h(x) = e^{3x+1}$

iii. $s(x) = -16(2x - 1)^2 + 25(2x - 1) + 1$

iv. $m(x) = \cos^2(x)$

v. $y = \dfrac{1}{3x^2 - 4x + 1}$

In the homework, you'll work on many of the exact same skills, but some of the representations will change: to find certain outputs, you'll need to rely on graphs, tables, and verbal representations.

Notes

Topic 2.7 Homework

1. Shown below is a graph of the function $f(x)$ and a table displaying values of the function $g(x)$.

    (a) Compute $f(g(2))$.

    (b) Compute $f(f(4))$.

    (c) Find a value $x = k$ such that $f(g(k)) = k$.

    (d) Compute $(f \circ g \circ f)(-2)$.

    (e) Let $h(x) = \frac{x}{2}$. How would the graph of $f(h(x))$ compare with the graph of $f(x)$?

    (f) How would the graph of $h(f(x))$ compare with the graph of $f(x)$?

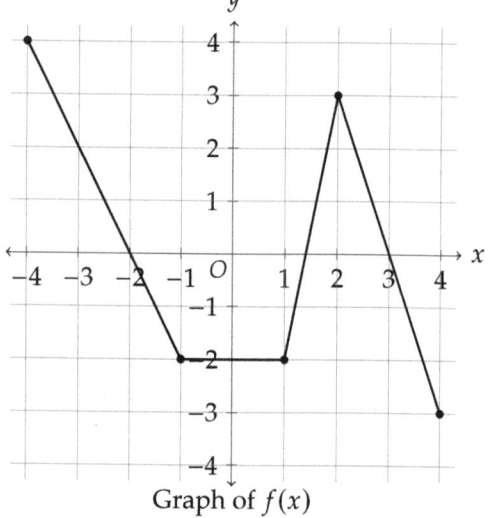

Graph of $f(x)$

Table of values for $g(x)$

x	-2	-1	0	1	2	3
$g(x)$	4	-2	3	-1	1	2

2. The graph of the function $y = f(x)$ contains the input-output pairs $(2, 3)$ and $(3, 4)$, and the graph of the function $y = g(x)$ contains the input output pairs $(2, 4), (3, 2)$ and $(4, 2)$. Compute the following if possible or state that not enough information is given.

    (a) $f(g(3))$

    (b) $f(f(3))$

    (c) $(g \circ f \circ f \circ g)(4)$

    (d) The symbol $g^{(10)}(x)$ means $g(g(g(\ldots(g(x))\ldots)))$, where there are 10 g's. Compute $g^{(99)}(2)$.

3. Let $f(x) = x^3 + 2x - 1$ and $g(x) = x - 3$. Write a simplified version of $f(g(x))$.

4. Let $f(x) = \sqrt{9 - x^2}$ and $h(x) = \dfrac{1}{2x^2}$.

    (a) Write and simplify an expression for $h(f(x))$.

    (b) Find the domain of $h(f(x))$.

5. The functions $f(x)$ and $g(x)$ are graphed below.

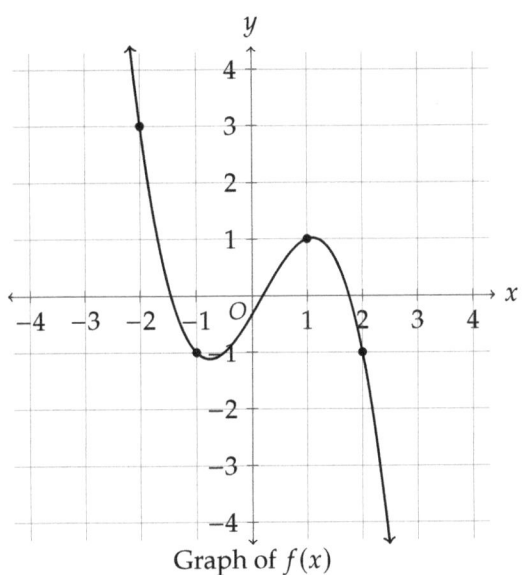
Graph of $f(x)$

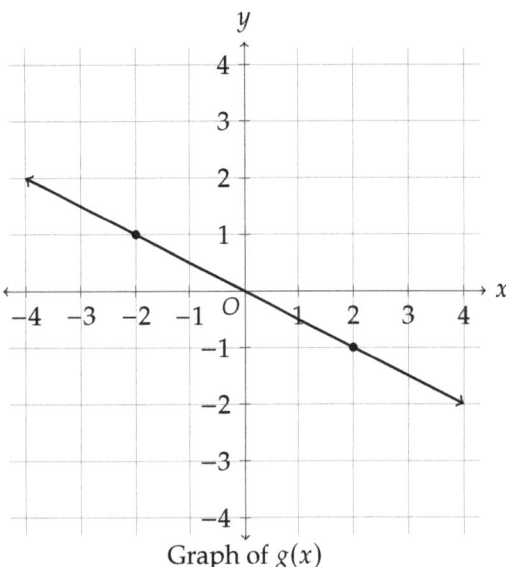
Graph of $g(x)$

On the graph below, sketch a graph of $f(g(x))$.

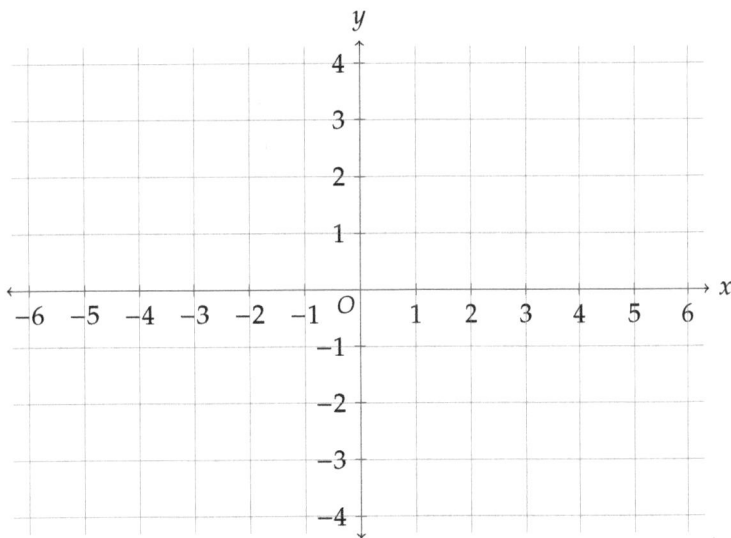

6. The functions below can be expressed in the form $f(g(x))$. Identify $f(x)$ and $g(x)$ for each.

   (a) $h(x) = 3 \cdot 2^{(1+x/2)}$

   (b) $y = \dfrac{1}{1+x^2}$

   (c) $k(x) = \sqrt{\dfrac{3x+2}{4}}$

   (d) $F(x) = \left(\dfrac{x-1}{2}\right)^2 - 2\left(\dfrac{x-1}{2}\right) + 3$

7. A teacher is going to take students' raw scores x out of 100 points and convert them into a scaled score using a "square root curve," where they take the square root of the percentage (as a decimal) of points the student got right. Right before applying the curve, the teacher realized each student's raw score should have been 2 points higher due to a typo on the test. If the student's final curved score as a percentage in decimal form is $C(x)$, identify an expression for $C(x)$.

Topic 2.7 Solutions/Notes

Activity

1. $f(g(4)) = f(\sqrt{2(4)-4}) = f(2) = 2^2 + 2 - 3 = 3$

2. $g(f(4)) = g(4^2 + 4 - 3) = g(17) = \sqrt{2(17)-4} = \sqrt{30}$

3. No.

4. In general, function composition is not *commutative*.

5. $f(g(x)) = (\sqrt{2x-4})^2 + (\sqrt{2x-4}) - 3 = 2x - 1 + \sqrt{2x-4}$

6. $f(h(x)) = (2x)^2 + (2x) - 3 = 4x^2 + 2x - 3$

7. $f(q(x)) = (x+4)^2 + (x+4) - 3 = x^2 + 8x + 16 + x + 4 - 3 = x^2 + 9x + 17$

8. $k(k(x)) = \dfrac{1}{\frac{1}{x-3} - 3} = \dfrac{1}{\frac{1}{x-3} - \frac{3x-9}{x-3}} = \dfrac{1}{\frac{10-3x}{x-3}} = \dfrac{x-3}{10-3x}$

9. (a) $f(i(x)) = x^2 + x - 3 = f(x)$
   (b) $g(i(x)) = \sqrt{2x-4} = g(x)$
   (c) $h(i(x)) = 2x = h(x)$

10. Composition with $i(x) = x$ leaves a function the same.

11. Composition with the identity function maintains a function's "identity."

12. (a) The radicand cannot be negative, so we write and solve $2x - 4 \geq 0$ to get a domain of $x \geq 2$.
    (b) All real numbers except $x = 3$
    (c) Anything that goes into $g(x)$ must be greater than or equal to 2. To determine what x will ensure this, we solve $\frac{1}{x-3} \geq 2$ to get $1 \geq 2x - 6$, or $x \leq \frac{7}{2}$. Therefore, the domain of $g(k(x))$ is $x \leq \frac{7}{2}$ with $x \neq 3$ (don't forget about $x \neq 3$!).
    (d) The domain of h was $x \geq 2$, but we also cannot input 3 into $k(x)$. Therefore, we must find the value of x such that $\sqrt{2x-4} = 3$. Solving gives us $2x - 4 = 9$, or $x = \frac{13}{2}$. Therefore, the domain of $k(g(x))$ is $x \geq 2$ with $x \neq \frac{13}{2}$.

13. (a) $f(q(x)) = f(x+4)$ is a horizontal translation of $f(x)$ by -4 units.
    (b) $q(f(x)) = f(x) + 4$ is a vertical translation of $f(x)$ by $+4$ units.
    (c) $g(h(x)) = g(2x)$ is a horizontal dilation by a factor of $\frac{1}{2}$.
    (d) $h(g(x)) = 2g(x)$ is a vertical dilation by a factor of 2.
    (e) $a(x) = 4(x-2)$

14. (b) $f(x) = x^2, g(x) = 3x + 4$
    (c) $f(x) = e^x, g(x) = 3x + 1$
    (d) $f(x) = -16x^2 + 25x + 1, g(x) = 2x - 1$
    (e) $f(x) = x^2, g(x) = \cos x$
    (f) $f(x) = \dfrac{1}{x}, g(x) = 3x^2 - 4x + 1$

> **Notes**
>
> (See the first part of the activity for the definition of function composition and a nice diagram.) When considering the domain of the composition of two functions, you must consider the domain of both the inner and outer functions. If the outer function has necessary restrictions, then you must determine what domain the *inner* function would have to ensure its *range* would not violate those restrictions for the outer function.
>
> Every graphical transformation is an example of a function composition. For a generic function f and a composition with a transformation function $t(x)$, the following transformations can be achieved.
>
> 1. A horizontal or vertical translation is composition of f with $t(x) = x + k$ where
>
>   (a) $f(t(x))$ is a horizontal translation of $-k$, and
>
>   (b) $t(f(x))$ is a vertical translation of $+k$.
>
> 2. A horizontal or vertical dilation is composition of f with $t(x) = kx$ where
>
>   (a) $f(t(x))$ is a horizontal dilation by $\left|\frac{1}{k}\right|$, and
>
>   (b) $t(f(x))$ is a vertical dilation by $|k|$.
>
> When decomposing functions, proper decomposition should replace every instance of the variable (x, t, etc.) with the "inner" function.

Homework

1. (a) $f(g(2)) = f(1) = -2$.
    (b) $f(f(4)) = f(-3) = 2$.
    (c) $k = 3$ works because $f(g(3)) = f(2) = 3$. Also, $k = 0$ works since $f(g(0)) = f(3) = 0$.
    (d) $(f \circ g \circ f)(-2) = (f \circ g)(0) = f(3) = 0$
    (e) The graph of $f(h(x)) = f\left(\frac{x}{2}\right)$ would be the graph of f horizontally dilated by a factor of 2 (so twice the distance from the y-axis).
    (f) The graph of $h(f(x)) = \frac{f(x)}{2}$ would be the graph of f vertically dilated by a factor of $\frac{1}{2}$ (so half the distance from the x-axis).

2. (a) $f(g(3)) = f(2) = 3$.
    (b) $f(f(3)) = f(4)$, and there is not enough information.
    (c) $(g \circ f \circ f \circ g)(4) = (g \circ f \circ f)(2) = (g \circ f)(3) = g(4) = 2$.
    (d) $g(g(2)) = 2$, so $g^{(2k)}(2) = 2$ for all $k = 1, 2, 3, \ldots$. Therefore, $g^{(99)}(2) = g(g^{(98)}(2)) = g(2) = 4$.

3. $f(g(x)) = (x-3)^3 + 2(x-3) - 1 = (x^3 - 9x^2 + 27x - 27) + 2x - 6 - 1 = x^3 - 9x^2 + 29x - 34$.

4. (a) $h(f(x)) = \dfrac{1}{2(\sqrt{9-x^2})^2} = \dfrac{1}{18 - 2x^2}$
    (b) $h(x)$ is undefined only for $x = 0$, so $h(f(x))$ is undefined when $f(x) = 0$, which occurs at $x = 3$ and $x = -3$. The domain of f is restricted to $-3 \leq x \leq 3$ since $9 - x^2$ must be greater than or equal to 0. Ruling out $x = 3$ and $x = -3$, we get that the domain of $h(f(x))$ is $-3 < x < 3$.

5. We notice that $g(x) = -\frac{1}{2}x$, so the graph of $f(g(x))$ is simply $f\left(-\frac{1}{2}x\right)$, which is a dilation of the graph of f by a factor of $\frac{1}{|-1/2|} = 2$ with a reflection over the y-axis. Therefore, the graph of $f(g(x))$ can be generated by simply taking the 4 known points and multiplying each x-coordinate by -2 (and connecting the dots!).

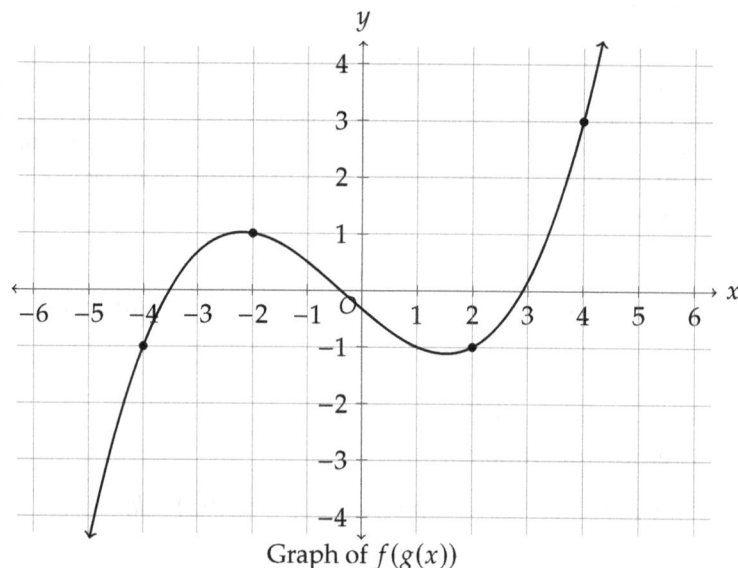
Graph of $f(g(x))$

6. (a) $f(x) = 3 \cdot 2^x$, $g(x) = 1 + \dfrac{x}{2}$

   (b) $f(x) = \dfrac{1}{x}$, $g(x) = 1 + x^2$

   (c) $f(x) = \sqrt{x}$, $g(x) = \dfrac{3x+2}{4}$

   (d) $f(x) = x^2 - 2x + 3$, $g(x) = \dfrac{x-1}{2}$

7. $C(x) = \sqrt{\dfrac{x+2}{100}}$

Reflection

Suggested Class Time. 50-60 minutes

Prerequisites. Students need to be familiar with transformations of functions.

Instructional Strategies. The first thing that comes to mind: emphasize the importance of parentheses, especially when it comes to expressions with exponents. At least five times during class, I had students write something like $2x^2$ or $x + 3^2$ instead of $(2x)^2$ or $(x+3)^2$. I have emphasized this with my students many times, but I had to do so again. This time, I tried to explain that there is a level of mathematical maturity they need to have (or gain) by the time they reach (insert more advanced class here). Basically, I told them, it's a matter of dotting i's and crossing t's: generally, just being precise, thorough, and careful.

Beyond this, some students just took time to get used to composition. They would be stuck on something as simple as Problem 1 for 5 minutes, but once they got the idea, they would be on Problem 12 within mere minutes. One thing I did notice, though, was that students were *not* simplifying expressions. For instance, answers to Problem 7 would often look like $f(q(x)) = (x+4)^2 + (x+4) - 3$. While I told students this was correct, I also warned them to look out for multiple choice or free response questions that might require or ask for simplification.

Problem 12 was definitely a challenge for students. I found many reaching for their calculators to try to determine the domain. For my students, at least, it is incredibly unnatural to actually write and solve an equation or inequality for what they want. They could not believe how simple writing and solving

$2x - 4 \geq 0$ was. For (c), similarly, setting up and solving $\sqrt{2x - 4} = 3$ seemed enlightening for students. We were also able to have nice conversations for writing things like "all real numbers greater than or equal to 2 except for $\frac{13}{2}$." In particular, I showed them

$$x \geq 2 \text{ with } x \neq \frac{13}{2} \quad \text{and} \quad \left[2, \frac{13}{2}\right) \cup \left(\frac{13}{2}, \infty\right).$$

Problem 13 was a great chance to review transformations with students. With the students who didn't necessarily get this the first time, it was a nice opportunity to revisit the more missed transformations (horizontal dilations especially).

Technology. As stated, some students used calculators, but you could also withhold calculators and make students problem-solve!

Problems.

1. Let f and g be the functions given by $f(x) = \sqrt{2x + 1}$ and $g(x) = \dfrac{x^2 + 3}{4}$. Which of the following is equivalent to $g(f(x))$ for $x \geq -\frac{1}{2}$?

    (A) $g(f(x)) = \sqrt{\dfrac{x^2 + 4}{2}}$ **(B)** $g(f(x)) = \sqrt{\dfrac{x^2 + 5}{2}}$ **(C)** $g(f(x)) = 2x + 1$ **(D)** $g(f(x)) = \dfrac{1}{2}x + 1$

2. The function f is given by $f(x) = 2x$. For any function $g(x)$, which of the following is true?

    (A) $g(f(x))$ is a vertical dilation of g by a factor of 2, and $f(g(x))$ is a horizontal dilation of g by a factor of 2.

    (B) $g(f(x))$ is a vertical dilation of g by a factor of 2, and $f(g(x))$ is a horizontal dilation of g by a factor of $\frac{1}{2}$.

    (C) $g(f(x))$ is a horizontal dilation of g by a factor of 2, and $f(g(x))$ is a vertical dilation of g by a factor of 2.

    (D) $g(f(x))$ is a horizontal dilation of g by a factor of $\frac{1}{2}$, and $f(g(x))$ is a vertical dilation of g by a factor of 2.

3. The function
$$h(x) = \dfrac{(x + 2)^2}{1 - (x + 2)^2}$$
can be written as the composition $f(g(x))$ for functions f and g. Which of the following could be f?

    (A) $f(x) = \dfrac{x}{1 - x}$ **(B)** $f(x) = \dfrac{1}{x}$ **(C)** $f(x) = x + 2$ **(D)** $f(x) = (x + 2)^2$

4. The table gives values of the function f for selected values of x.

x	2	12	17	20	24
$f(x)$	−1	10	20	15	6

    The function g is given by $g(x) = 4^x + \frac{1}{2}x^3$. What is the value of $f(g(2))$?

    (A) $-\dfrac{1}{4}$ **(B)** 10   **(C)** 15   **(D)** 20

5. The function f is given by $f(x) = 3^{(x-1)}$. The function g is given by $g(x) = x^2 - 18x$. The function h is given by $h = g(f(x))$. For what value of x does $h(x) = -81$?

    (A) −56   **(B)** 3   **(C)** 4   **(D)** $9 + \sqrt{76}$

Solutions.

1. The composition is
$$g(f(x)) = g(\sqrt{2x+1}) = \frac{(\sqrt{2x+1})^2 + 3}{4} = \frac{2x+4}{4} = \frac{x+2}{2} = \frac{1}{2}x + 1.$$
The correct answer is therefore **(D)**.

2. The composition $g(f(x)) = g(2x)$ is a horizontal dilation of g by a factor of $\frac{1}{2}$. The composition $f(g(x)) = 2g(x)$ is a vertical dilation of g by a factor of 2. The correct answer is therefore **(D)**.

3. Let $g(x) = (x+2)^2$. Then letting $f(x) = x/(1-x)$ will give $f(g(x)) = h(x)$. The correct answer is therefore **(A)**.

4. Since $g(2) = 4^2 + \frac{1}{2}(2^3) = 16 + 4 = 20$, we have $f(g(2)) = f(20) = 15$. The correct answer is therefore **(C)**.

5. We need to solve $g(f(x)) = -81$. Thus, we seek the input value for g that gives the output value -81. This is given by $x^2 - 18x = -81$, which can be written as $x^2 - 18x + 81 = (x-9)^2 = 0$. Hence, the input value for g is 9. This implies $f(x) = 9$, or $3^{(x-1)} = 9 = 3^2$. We conclude that $x = 3$. The correct answer is therefore **(B)**.

Topic 2.8 ~ Inverse Functions

Learning Objectives
1. (2.8.A) Determine the input-output pairs of the inverse of a function.
2. (2.8.B) Determine the inverse of a function on an invertible domain.

Success Criteria
1. I can find input-output pairs of functions and their inverses.
2. I can find the inverse of a function analytically and graphically.
3. I can describe when a function has an inverse analytically and graphically.

In Topic 2.7, you investigated function composition. One function that came up was the *identity function* $i(x) = x$. This leads to a definition.

Definition. The <u>INVERSE</u> of a function $f(x)$ is the function $f^{-1}(x)$ that

- Returns an output $f(x)$ back to x, and
- satisfies $f(f^{-1}(x)) = f^{-1}(f(x)) = x$.

This can be visualized as follows.

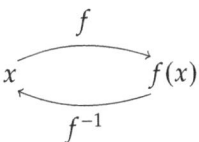

When a function has an inverse (on a certain domain), it is called *invertible*.

Try using the definition of an inverse function to answer the following.

1. Given an invertible function $f(x)$ with $f(2) = 3$, what is $f^{-1}(3)$?

2. The graph of the invertible function g contains the point $(4, -7)$. What point must the graph of $g^{-1}(x)$ contain?

3. Because an inverse function $f^{-1}(x)$ is defined strictly by its "undoing" of outputs of $f(x)$, we have a number of relationships that arise. Try completing the following.

Relationship 1	The inputs of $f(x)$ are the _____ of $f^{-1}(x)$, and the outputs of $f(x)$ are the _____ of $f^{-1}(x)$.
Relationship 2	The domain of $f(x)$ is the _____ of $f^{-1}(x)$, and the range of $f(x)$ is the _____ of $f^{-1}(x)$.*
Relationship 3	If a point (x, y) is on the graph of $f(x)$, then the point _____ is on the graph of $f^{-1}(x)$.

In Relationship 2, there was an asterisk. Let's investigate.

4. Consider $f(x) = x^2$.

    (a) What is $f(4)$? How about $f(-4)$?

    (b) Explain why you can't determine $f^{-1}(16)$.

5. Try to generalize the result of the previous question below.

$$\text{If a function } f \text{ has } \underline{\hspace{2cm}} \text{ with the } \underline{\hspace{2cm}},$$
$$\text{then the inverse function isn't } \underline{\hspace{2cm}}.$$

Problem 5 can be visualized graphically. Consider $f(x) = x^2$ below.

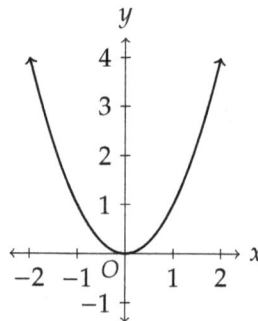

6. Visually, how can you see that multiple inputs have the same output?

7. This "problem" can be fixed with a *domain restriction*. What is the largest domain x that you could restrict $f(x)$ to such that *no* two inputs would have the exact same output?

So, in general, functions may only be invertible on a specific domain: this is often called the *invertible domain*. In this case, we have to modify Relationship 2 to state that

The INVERTIBLE DOMAIN of $f(x)$ is the range of $f^{-1}(x)$,
and the range of $f(x)$ *on its invertible domain* is the DOMAIN of $f^{-1}(x)$.

Now, you've looked at finding input-output pairs of an inverse function and even looked at when an inverse function exists (or doesn't). How do you actually *find* an inverse function, though? It comes down to Relationship 1. If $y = f(x)$, then the inputs of f are x and the outputs are y; for an inverse function, these inputs and outputs simply swap places, meaning x becomes y and y becomes x (analytically, this appears as $f^{-1}(y) = x$ – see how x and y have "swapped places"?). Let's try an example.

Use the box at the right to answer the following.

8. Let $f(x) = \dfrac{3x-1}{5}$. First, replace $f(x)$ with y.

9. Now, swap x and y – i.e., reverse the roles of the inputs and outputs.

10. Now, solve for y. This resulting y will actually be $f^{-1}(x)$.

> Finding the inverse of $f(x) = \dfrac{3x-1}{5}$

This process will work in general. Now, what about the *graph* of $f^{-1}(x)$?

12. We have stated that, for an inverse function, the inputs and outputs are reversed. If you were to graph $f(x)$ in the coordinate plane with x and y axes, how do you think you could graph $f^{-1}(x)$?

13. You have a separate piece of patty paper with the graph of $f(x) = \dfrac{3x-1}{5}$. To swap the x- and y-axes, you can actually *fold* the x-axis onto the y-axis.

14. Take the patty paper and fold the x-axis onto the y-axis. Press down to make a nice crease. Then, thickly trace over the graph of $f(x)$.

UNIT 2 TOPIC 2.8 ~ INVERSE FUNCTIONS **197**

15. Unfold the paper. You should now see a new line graphed (where you drew) – this line is the graph of $f^{-1}(x)$!

16. You should also see a line where your crease was made. What is the equation of this line?

17. Complete the following:

    For an invertible function f, the graph of $f^{-1}(x)$ is simply the graph of f _____.

Practice with the following.

18. Let g be the function given by $g(x) = \dfrac{1}{(x-2)^2}$, defined for $x > 2$.

    (a) Find $g^{-1}(x)$.

    (b) Find the domain and range of $g^{-1}(x)$.

19. The function $f(x)$ is graphed in the xy-plane below. On the graph to the right, sketch a graph of $f^{-1}(x)$.

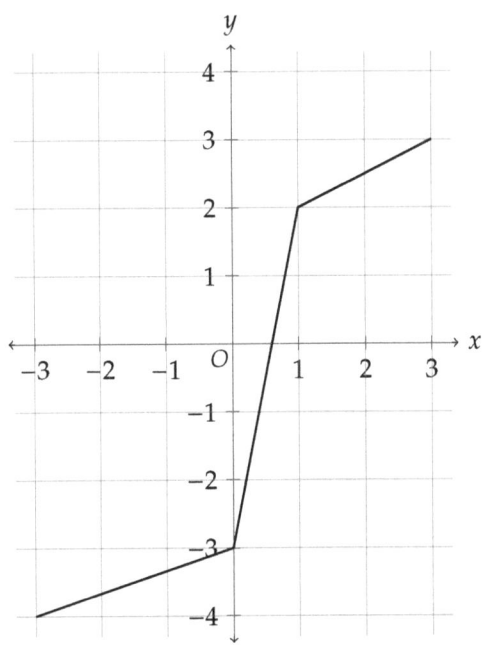 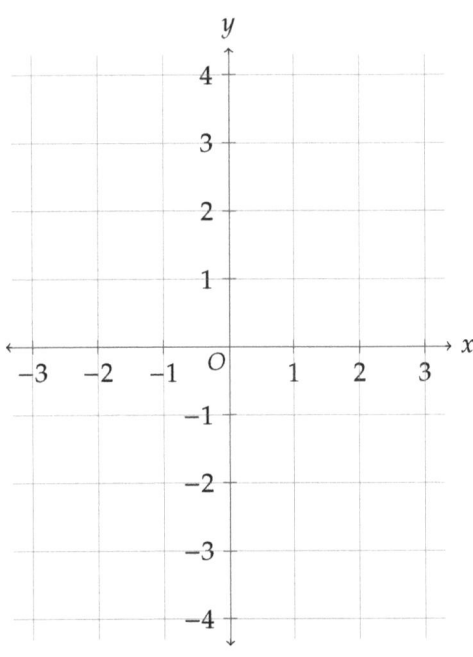

20. At a technologically primitive gas station, records are kept by using physical receipts, which are totaled at the end of each day. After a month's work of receipts were collected and totaled, the gas station owner realized that they had written down all 30 totals, but had accidentally forgotten to write down the dates of three of those totals.

The gas station manager was able to find an invertible function $R(d)$ that could model the receipt totals R for each day d of that month. The graph of $R(d)$ was increasing and concave down.

Shown below are statements about the inverse function $R^{-1}(d)$. For each underlined set of two words or phrases, circle the correct word or phrase.

(a) $R^{-1}(d)$ has receipt totals / dates on the x-axis and dates / receipt totals on the y-axis.

(b) $R^{-1}(d)$ is increasing / decreasing at an/a increasing / decreasing rate.

(c) Because $R(d)$ was invertible, there must have been / could not have been two or more days that had the same receipt totals.

(d) To find any of the missing dates, the manager should input the receipt totals into $R(d)$ / $R^{-1}(d)$.

Notes

Topic 2.8 Homework

This homework is a bit unique – it's "one problem."

1. Let f and g be functions given by

$$f(x) = \frac{6-2x}{7} \quad \text{and} \quad g(x) = (x-4)^2 - 9.$$

Additionally, output values of $h(x)$ are given in the table for select input values and the function $k(x)$ is graphed below.

x	−4	−3	−2	−1	0	1	2	3	4
$h(x)$	7	5	3	2	−1	0	3	4	5

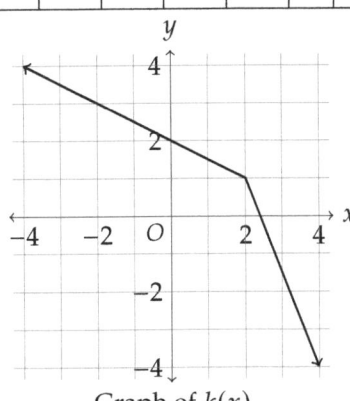

Graph of $k(x)$

(a) Compute $f^{-1}(3)$ without finding an expression for $f^{-1}(x)$.

(b) Find an expression for $f^{-1}(x)$.

(c) One of the invertible domains of g is $x \geq a$. What is a?

(d) Find an expression for $g^{-1}(x)$ for the invertible domain from part (c).

(e) Identify the domain and range of $g^{-1}(x)$ using the invertible domain from part (c).

(f) Explain why h is not invertible on the domain of all real numbers.

(g) Suppose h is invertible for $x \leq 1$. Complete what portions of the table below that you can.

x	1	2	3	4	5	6	7
$h^{-1}(x)$							

(h) If the range of h is $[-1, 10)$ for $x \leq 1$, find the domain and range of $h^{-1}(x)$.

(i) Sketch a graph of $k^{-1}(x)$ on the same graph as $k(x)$.

(j) Using characteristics of quadratics, determine if g is increasing/decreasing or concave up/down for $x \geq 3$.

(k) Determine if $g^{-1}(x)$ is increasing/decreasing or concave up/down for $x \geq 3$.

(l) Compute $f(f^{-1}(2023))$.

Image to print/trace onto patty paper

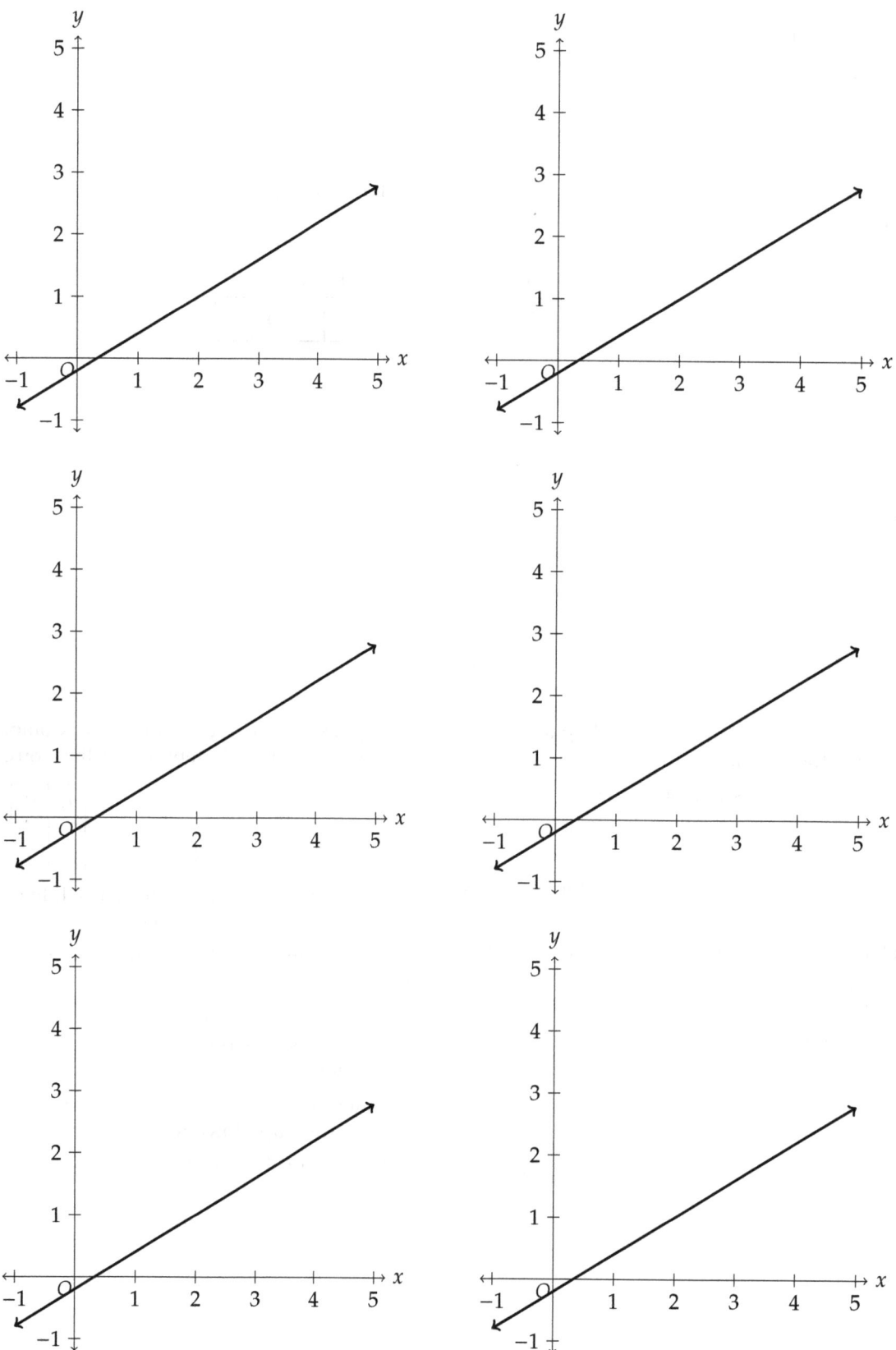

Topic 2.8 Solutions/Notes

Activity

1. $f^{-1}(3) = 2$

2. $(-7, 4)$

3.
Relationship 1	The inputs of $f(x)$ are the outputs of $f^{-1}(x)$, and the outputs of $f(x)$ are the inputs of $f^{-1}(x)$.
Relationship 2	The domain of $f(x)$ is the range of $f^{-1}(x)$, and the range of $f(x)$ is the domain of $f^{-1}(x)$.*
Relationship 3	If a point (x, y) is on the graph of $f(x)$, then the point (y, x) is on the graph of $f^{-1}(x)$.

4. (a) $f(4) = 16, f(-4) = 16$

   (b) Because it could be an output of 4 or −4!

5. If a function f has multiple inputs with the same output, then the inverse function isn't defined.

6. The graph of f fails the horizontal line test (a horizontal line intersects the function at multiple points).

7. $x \geq 0$ or $x \leq 0$

8. $y = \dfrac{3x - 1}{5}$

9. $x = \dfrac{3y - 1}{5}$

10. $x = \dfrac{3y - 1}{5} \Rightarrow 5x = 3y - 1 \Rightarrow 5x + 1 = 3y \Rightarrow y = \dfrac{5x + 1}{3} \Rightarrow f^{-1}(x) = \dfrac{5x + 1}{3}$

11. Answers will vary, but the x- and y-axes can be swapped.

16. $y = x$

17. For an invertible function f, the graph of $f^{-1}(x)$ is simply the graph of f reflected over the line $y = x$.

18. (a) $x = 1/(y-2)^2 \Rightarrow (y-2)^2 = 1/x \Rightarrow y - 2 = \pm\sqrt{1/x}$. Since $x \geq 2$, we only want the positive square root. Then $y - 2 = \sqrt{1/x} \Rightarrow y = 2 + \sqrt{1/x} \Rightarrow g^{-1}(x) = 2 + \sqrt{1/x}$.

    (b) The domain of $g^{-1}(x)$ is the range of $g(x)$, which is $x \geq 0$. The range of $g^{-1}(x)$ is the invertible domain of $g(x)$, which is $y \geq 2$.

19.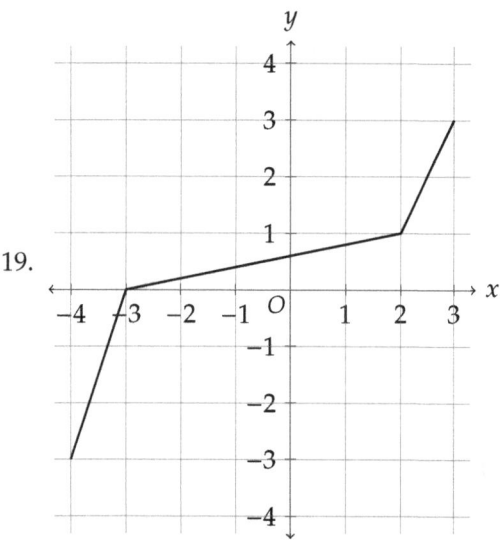

20. (a) $R^{-1}(d)$ has receipt totals on the x-axis and dates on the y-axis.
    (b) $R^{-1}(d)$ is increasing at an increasing rate.
    (c) Because $R(d)$ was invertible, there could not have been two or more days that had the same receipt totals.
    (d) To find any of the missing dates, the manager should input the receipt totals into $R^{-1}(d)$.

Notes

A function $f(x)$ is INVERTIBLE on any domain in which no two inputs have the same output (if its graph passes the Horizontal Line Test). *Note that the Horizontal Line Test is not in and of itself a sufficient explanation!*

To find an inverse function given an analytic expression of $y = f(x)$, reverse all x's for y's (and vice-versa) and solve for y.

Graphically, the graph of $f^{-1}(x)$ is the graph of $f(x)$ reflected over the line $y = x$. For any point (x, y) on the graph of $f(x)$, the point (y, x) is on the graph of $f^{-1}(x)$.

Once a function $f(x)$ has been restricted to its invertible domain (if necessary), the domain of $f^{-1}(x)$ is the range of $f(x)$ (on that invertible domain) and the range of $f^{-1}(x)$ is the invertible domain of $f(x)$.

The inverse of an increasing/decreasing function will still be increasing/decreasing. However, if the graph of a function f will have the opposite concavity of its inverse function $f^{-1}(x)$.

Homework

1. (a) $\dfrac{6-2x}{7} = 3 \Rightarrow 6-2x = 21 \Rightarrow -15 = 2x \Rightarrow x = -\dfrac{15}{2}$.
   (b) $\dfrac{6-2y}{7} = x \Rightarrow 6-2y = 7x \Rightarrow 6-7x = 2y \Rightarrow y = \dfrac{6-7x}{2} \Rightarrow f^{-1}(x) = \dfrac{6-7x}{2}$.
   (c) g is a quadratic with a vertex at $(4, -9)$, and due to the symmetry of quadratics, g must be restricted to $x \geq 4$.
   (d) $x = (y-4)^2 - 9 \Rightarrow x+9 = (y-4)^2 \Rightarrow \sqrt{x+9} = y-4 \Rightarrow y = 4 + \sqrt{x+9} \Rightarrow g^{-1}(x) = 4 + \sqrt{x+9}$.

(e) The range of $g^{-1}(x)$ is $y \geq 4$. The domain of $g^{-1}(x)$ is the range of $g(x)$. Since g is a quadratic with a positive leading coefficient and vertex of $(4, -9)$, its range is $[-9, \infty)$. Therefore, the domain of $g^{-1}(x)$ is $x \geq -9$.

(f) $h(-3) = h(4) = 5$ (two inputs have the same output).

(g)

x	1	2	3	4	5	6	7
$h^{-1}(x)$	?	−1	−2	?	−3	?	−4

(h) The domain is $[-1, 10)$ and the range is $y \leq 1$.

(i)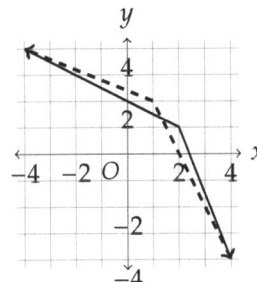

(j) g is increasing and and its graph is concave up.

(k) g^{-1} is increasing and its graph is concave down.

(l) 2023

Reflection

Suggested Class Time. 50-70 minutes

Prerequisites. Students could use some experience with inverse functions as well as function compositions. It helps if students have at least learned the Vertical Line Test (to contrast with the Horizontal Line Test).

Instructional Strategies. This activity was a lot for students - some flew through it, while others took quite a while. A few groups really struggled with Problems 6 and 7. For these students, I would sit down with them and go over the need for a function to be one-to-one in order to have an inverse. Showing students that $f(4) = f(-4) = 16$ with arrows and then *reversing* the arrows helped students to better see that, for functions with multiple inputs having the same output, the inverse would not be a function.

For the patty paper, a few students didn't trace over the graph of $f(x)$ after the fold, but instead *only* traced the crease and thought that $y = x$ was the graph of $f^{-1}(x)$. Make sure to look out for this! Also, encourage students to draw thickly. What may have confused students was that they felt they "weren't allowed" to trace on the back of the patty paper (which, of course, they were supposed to do!). To help students make sense of the reflection, I showed students how, informally, "swapping x and y" and "reflecting over $y = x$" were essentially the same thing.

Given how some students finished the entire activity in 45 minutes while others were around Problem 12 when I finished, I decided to wrap things up with about 20 minutes of class left. We first completed all of the notes and then ended up going through Problems 18 and 20 together. Problem 18 was a really good example to do as a class, because the domain restriction (and the need to check the $\pm\sqrt{1/x}$) led to really good conversations. At the end of class, on 3 relatively difficult Topic Questions that I gave students, they averaged 63% correct, or roughly 2 out of 3. The one that was most missed involved computing an inverse function *and* its domain and range. Students only seemed to miss the detail requiring them to use the *negative* square root. On a lengthy word problem, students either picked the correct option or the second best (forgetting to reverse x and y because they didn't actually identify the independent and dependent variables). Next year, I plan to have students annotate any similar question with explicit definitions of the variables.

Technology. None was required except for the patty paper.

Problems.

1. Let f be given by $f(x) = \dfrac{3}{2-x}$ on its maximal domain. Which of the following is true?

    (A) The inverse of f is $f^{-1}(x) = 2 - \dfrac{3}{x}$ and its range is all real numbers.

    (B) The inverse of f is $f^{-1}(x) = 2 - \dfrac{3}{x}$ and its range is all real numbers except 2.

    (C) The inverse of f is $f^{-1}(x) = \dfrac{3}{x} - 2$ and its range is all real numbers.

    (D) The inverse of f is $f^{-1}(x) = \dfrac{3}{x} - 2$ and its range is all real numbers except 2.

2. The function g is given by $g(x) = \dfrac{x-3}{3x+4}$. Which of the following is true of the inverse function $g^{-1}(x)$?

    (A) The domain of $g^{-1}(x)$ is all real numbers except $-\tfrac{4}{3}$.

    (B) The domain of $g^{-1}(x)$ is all real numbers except $-\tfrac{4}{3}$ and 3.

    (C) The range of $g^{-1}(x)$ is all real numbers except $-\tfrac{4}{3}$.

    (D) The range of $g^{-1}(x)$ is all real numbers except $-\tfrac{4}{3}$ and 3.

3. The graph of the piecewise linear function f is shown below.

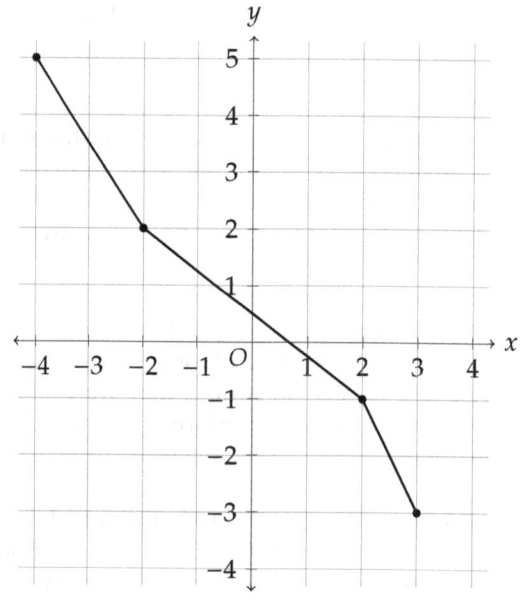

    Let g be the inverse function of f. What is the minimum value of g?

    (A) −5   (B) −4   (C) −3   (D) −2

4. The graph $y = g^{-1}(x)$ of the inverse of the function $g(x)$ is shown below.

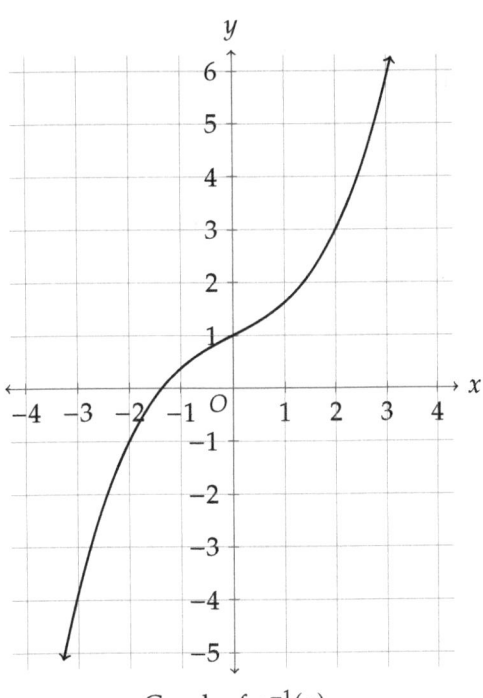

Graph of $g^{-1}(x)$

Which of the following points lies on the graph of $g(x)$?
(A) $(-2, -1)$ (B) $(1, 1)$ (C) $(2, 3)$ (D) $(3, 2)$

5. If f and g are invertible functions with outputs given for select inputs in the table below, what is $g^{-1}(f^{-1}(2))$?

x	0	1	2	3	4
$f(x)$	3	2	1	0	-2
$g(x)$	1	2	5	8	9

(A) 0 (B) 1 (C) 2 (D) 4

Solutions.

1. Let $y = f(x)$. Then interchange x and y. We get

$$x = \frac{3}{2-y}$$
$$x(2-y) = 3$$
$$2 - y = \frac{3}{x}$$
$$2 - \frac{3}{x} = y$$
$$y = f^{-1}(x) = 2 - \frac{3}{x}.$$

The range is the same as the domain of f, which is all real numbers except 2. The correct answer is therefore **(B)**.

2. The domain of g is all real numbers except $-\frac{4}{3}$, so the range of g^{-1} is all real numbers except $-\frac{4}{3}$. The correct answer is therefore **(C)**.

3. The minimum output value of the inverse function is the minimum input value of the function; this is -4. The correct answer is therefore **(B)**.

4. The point $(2,3)$ lies on the graph of g^{-1}, so $(3,2)$ lies on the graph of g. The correct answer is therefore **(D)**.

5. $g^{-1}(f^{-1}(2)) = g^{-1}(1) = 0$. The correct answer is therefore **(A)**.

UNIT 2   TOPICS 2.9-2.10 ~ LOGARITHMS, THE INVERSE OF EXPONENTIALS   207

Topics 2.9-2.10 ~ Logarithms, the Inverse of Exponentials

Learning Objectives

1. (2.9.A) Evaluate logarithmic expressions.
2. (2.10.A) Construct representations of the inverse of an exponential function with an initial value of 1.

Success Criteria

1. I can evaluate a logarithmic expression.
2. I can rewrite an exponential or logarithmic equation as a logarithmic or exponential equation, respectively.
3. I can describe the rates of change of logarithmic functions as inverses of exponential functions.
4. I can find an expression for the inverse of a logarithmic or exponential function.

In the first half of Unit 2, we investigated *exponential functions*. Exponential functions have the property that

As inputs increase linearly, outputs increase multiplicatively.

We then transitioned to studying *inverse functions*, in which the roles of x and y _____ . What, then, is the inverse of an exponential function?

Definition. The _____, written as $f^{-1}(x) = $ _____, is the inverse of the exponential function $f(x) = b^x$.

For example, consider the following exponential equations.

$$2^4 = 16 \qquad 5^{1/2} = \sqrt{5} \qquad 10^1 = 10 \qquad 9^{-2} = \frac{1}{81}$$

Each of these takes an *input* of an exponential function, like $f(4)$ for $f(x) = 2^x$, and outputs the base of the function to an exponent of the input. Logarithms, on the other hand, reverse this. Each of these expressions can be rewritten in terms of logarithms as follows.

$$\log_2(\underline{}) = \underline{} \qquad \log_5(\underline{}) = \underline{} \qquad \log_{10}(\underline{}) = \underline{} \qquad \log_9(\underline{}) = \underline{}$$

The logarithm with a base of 10 is called the _____ and is simply abbreviated _____. With a base of $e \approx 2.718$, the logarithm base e is called the _____ and is abbreviated _____.

Practice 1. Evaluate the following logarithmic expressions.

(a) $\log_3 9$

(b) $\log_2 \frac{1}{8}$

(c) $\log 1000$

(d) $\ln e^3$

(e) $\log_4 (16^3)$

(f) $\log_5 \sqrt[3]{5}$

(g) $\ln \frac{1}{\sqrt{e}}$

(h) $\log_2(\log_3 9)$

It's worthwhile also practicing writing the inverses of exponential functions for any input x.

Example 1. Write expressions for the inverses of $f(x) = 2^x$ and the function $g(x) = e^x$.

Because $f(x) = b^x$ and $g(x) = \log_b(x)$ are inverse functions, they must satisfy the identity _____.
Therefore, we can compute such difficult-looking expressions as follows quite easily.

Example 2. Compute the following.

(a) $\log_2\left(2^{23.5}\right)$
(b) $e^{\ln 1.55}$
(c) $2^{3+\log_2(6)}$

Additionally, we can graph any logarithm $\log_b(x)$ by simply _____
_____. (We can also just reverse the coordinates of any known ordered pair (x, b^x)!)

Example 3. Sketch a graph of $f(x) = \log_2 x$.

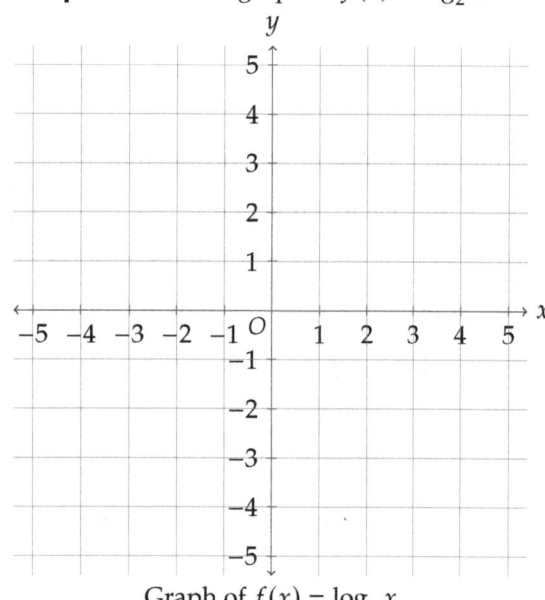
Graph of $f(x) = \log_2 x$

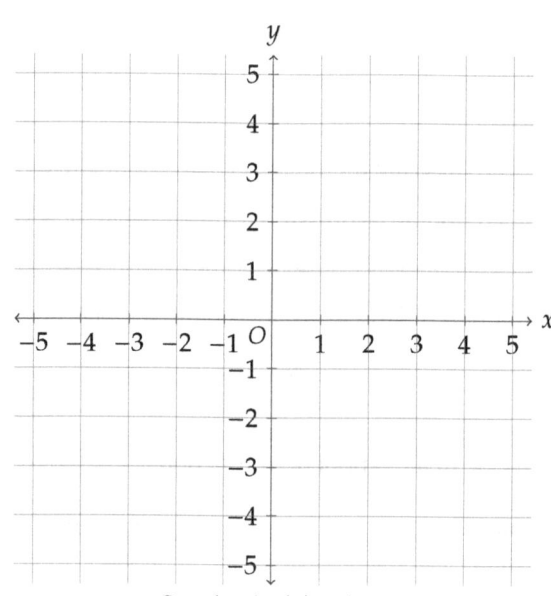
Graph of $g(x) = \ln x$

Practice 2. Sketch a graph of $g(x) = \ln x$ on the graph above. Identify the domain and range of g.

Now, let's consider a particular exponential function like $f(x) = 10^x$. Complete the table below.

x	0	1	2	3	4
$f(x) = 10^x$					

Observe how, for each increase of __ in the _____, the _____ increase by _____.

Because a logarithmic function is the *inverse* of an exponential function, what would we expect to see?

For each increase of __ in the _____, the _____ increase by _____.

We can find the inputs and outputs of $f^{-1}(x) = $ _____ by simply _____
_____. Complete the table below.

x					
$f^{-1}(x) = \log(x)$					

Practice 3. The function $f(x)$ is logarithmic for an unknown base b. If $f(2) = 5$ and $f(10) = 6$, then for what value of x does $f(x) = 7$?

Logarithms can also be used to *adjust the scale* for functions (or graphs) that quickly get into extreme values. For instance, the *power level* of a sound is actually measured as a *ratio* to a *reference power*. Sound power levels commonly have values like $1,000$, $1,000,000$, or larger.

For instance, the sound power level of a jet engine is 10^{14}, or a *hundred trillion*. This would make describing this sound very difficult, as the human brain struggles to comprehend such large numbers. Therefore, *logarithmic scales* can be used.

Definition. On a LOGARITHMIC SCALE, each unit represents a _____ change of the _____ _____.

The DECIBEL measures power level on a logarithmic scale of base $10^{1/10}$; therefore, an increase in power level by a factor of $10^{1/10} \approx 1.26$ corresponds to a linear increase of exactly *one* decibel. Similarly, an increase in power level by a factor of $10 = (10^{1/10})^{10}$ corresponds to a linear increase of exactly ____ decibels.

Example 4.

(a) Fill in the table below.

Sound power level	10	100	1,000	10,000	1,000,000
Decibels					

(b) Earlier, we said that a jet engine noise is roughly 140 decibels. On the other hand, a jackhammer makes a noise that is roughly 110 decibels. In terms of power level, how much more powerful is the sound from a jet engine than from a jackhammer?

(c) (■) The noise from an average vacuum cleaner is 70 decibels, while the sound from the average washing machine is 78 decibels. How many times more powerful is the sound level of a washing machine that a vacuum cleaner?

Topics 2.9-2.10 Homework

1. Evaluate the following logarithmic expressions or state that they are undefined.

   (a) $\log_3\left(\dfrac{1}{27}\right)$ (b) $\log_{\sqrt{2}} 8$ (c) $\ln \dfrac{1}{e}$ (d) $\log_b 1$ for any $b > 1$ (e) $\log_{1/2} 4$ (f) $\log 0.001$

2. Explain why the domain of $f(x) = \log_b(x)$ is $x > 0$ if $b \neq 1$ and $b > 0$.

3. Find the inverse of each of the following functions.

   (a) $f(x) = 10^x$ (b) $f(x) = \log_{\sqrt{3}} x$ (c) $g(x) = \log_2(\log_4 x)$

4. Sketch a graph of $f(x) = \log_3(x)$ with at least 3 labeled points.

5. Here is a table of values for the exponential function $E(x)$:

x	1	2	3	4
$E(x)$	3	6	12	24

   Consider the inverse function $E^{-1}(x)$. Describe the relationship between the inputs and outputs of $E^{-1}(x)$.

6. A student is creating a password for a website that allows users to create a password consisting of anywhere from 5 to 20 digits (where any digit can be the numbers 1 through 9 – no zeros are allowed). The student decides to create their own measure of security $S(d)$, which is the number S of possible passwords of length d digits.

   (a) If the student decides to measure S on a logarithmic scale, what base of the logarithm should they use?

   (b) Suppose the student uses the base you used in (a). How many times more possible passwords are there are of length 10 digits than possible passwords there are of length 7 digits?

7. (■) The Richter scale measures the magnitude of earthquakes. The Richter scale is logarithmic with base 10, so every increase of 1 on the Richter scale corresponds to a magnitude that is 10 times greater. In 2018, an earthquake measuring 4.4 on the Richter scale occurred outside of Atlanta. The previous Georgia record was an earthquake measuring 4.1 on the Richter scale. How much more powerful was the earthquake that occurred in 2018 than the previous record?

8. The graph of the exponential function $f(x)$ is given. Write an expression for the inverse function $f^{-1}(x)$.

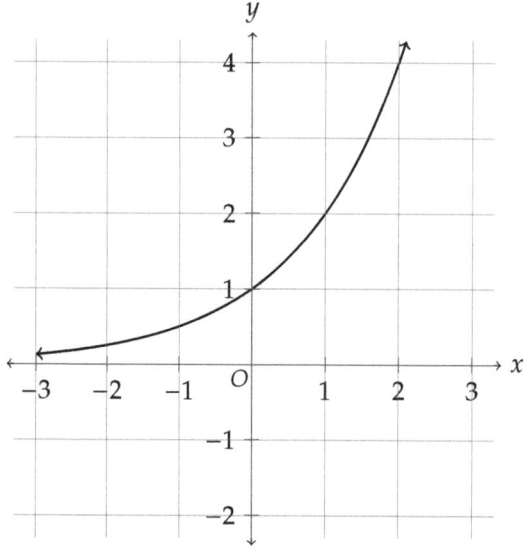

Topics 2.9-2.10 Solutions/Notes

Activity

... transitioned to studying <u>inverse functions</u>, in which the roles of x and y <u>are reversed</u>.

Definition. The <u>logarithm base b</u>, written as $f^{-1}(x) = \underline{\log_b(x)}$, is the inverse of the exponential function $f(x) = b^x$.

Each of these expressions can be rewritten in terms of logarithms as follows.

$$\log_2(16) = 4 \qquad \log_5(\sqrt{5}) = \frac{1}{2} \qquad \log_{10}(10) = 1 \qquad \log_9\left(\frac{1}{81}\right) = -2$$

The logarithm with a base of 10 is called the <u>common logarithm</u> and is simply abbreviated <u>log x</u>. With a base of $e \approx 2.718$, the logarithm base e is called the <u>natural logarithm</u> and is abbreviated <u>ln x</u>.

Practice 1. (a) 2 (b) -3 (c) 3 (d) 3 (e) 6 (f) $\frac{1}{3}$ (g) $-\frac{1}{2}$ (h) 1

Example 1. $f^{-1}(x) = \log_2 x$, $g^{-1}(x) = \ln x$

Because $f(x) = b^x$ and $g(x) = \log_b(x)$ are inverse functions, they must satisfy the identity $\underline{f(g(x)) = g(f(x))} \underline{= x}$.

Example 2. (a) 23.5 (b) 1.55 (c) $2^3 \cdot 2^{\log_2(6)} = 8 \cdot 6 = 48$

Additionally, we can graph any logarithm $\log_b(x)$ by simply <u>reflecting the graph of b^x over the line $y = x$</u>.

Example 3.

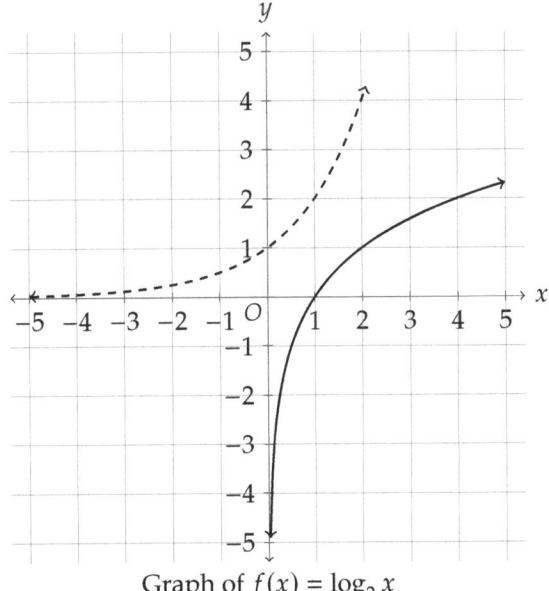
Graph of $f(x) = \log_2 x$

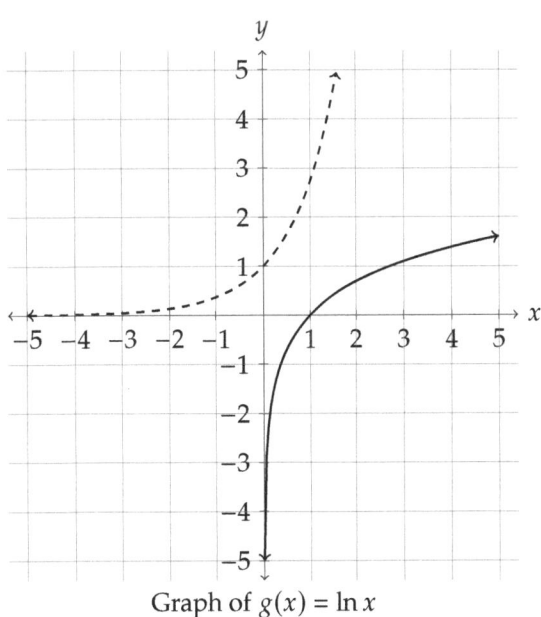
Graph of $g(x) = \ln x$

Practice 2. The graph of $g(x) = \ln x$ is on the rightmost graph above. The domain is $x > 0$ and the range is all real numbers.

x	0	1	2	3	4
$f(x) = 10^x$	1	10	100	1,000	10,000

Observe how, for each increase of 1 in the inputs, the outputs increase by a factor of 10.

Because a logarithmic function is the *inverse* of an exponential function, what would we expect to see?

For each increase of 1 in the outputs, the inputs increase by a factor of 10.

We can find the inputs and outputs of $f^{-1}(x) = \log x$ by simply reversing the inputs and outputs of $f(x) = 10^x$. Complete the table below.

x	1	10	100	1,000	10,000
$f^{-1}(x) = \log(x)$	0	1	2	3	4

Practice 3. An increase in the outputs by 1 corresponded to an increase in the inputs by a factor of $\frac{10}{2} = 5$. Therefore, another increase of 1 in the outputs would correspond to another increase by a factor of 5, so $f(x) = 7$ for $x = 10(5) = 50$.

Definition. On a LOGARITHMIC SCALE, each unit represents a multiplicative change of the base of the logarithm.

An increase in power level by a factor of $10 = (10^{1/10})^{10}$ corresponds to a linear increase of exactly 10 decibels.

Example 4.

(a)

Sound power level	10	100	1,000	10,000	1,000,000
Decibels	1	2	3	4	6

(b) 30 decibels corresponds to 3 linear increases by 10 decibels, or an increase by a factor of $10^3 = 1000$. Therefore, the sound from a jet engine is 1000 times more powerful than the sound from a jackhammer.

(c) An increase in 8 decibels corresponds to a sound level increase by a factor of $(10^{1/10})^8 \approx 6.310$. Therefore, the sound from a washing machine is about 6.310 times more powerful than the sound from a vacuum cleaner.

Homework

1. (a) -3 (b) 6   (c) -1 (d) 0   (e) -2 (f) -3

2. There is no exponent k such that $b^k \leq 0$ if $b > 0$. Otherwise, there exists a k such that $b^k = c$ for any $c > 0$.

3. (a) $f^{-1}(x) = \log x$ (b) $f^{-1}(x) = \sqrt{3}^x$ (c) $g^{-1}(x) = 2^{4^x}$

4.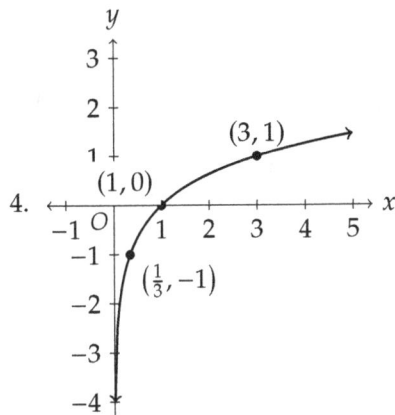

5. As the inputs increase by a factor of 2 (as the inputs double), the outputs will increase by 1.

6. (a) The number of combinations increases by a factor of 9 for each new digit added, so the student should use a base of 9.

    (b) A linear increase of 3 digits will increase the number of combinations by a factor of $9^3 = 729$. Therefore, there are 729 times more passwords of length 10 than there are of length 7.

7. $10^{4.4-4.1} = 10^{0.3} \approx 1.995$ times more powerful

8. This is the graph of $f(x) = 2^x$, so the inverse function is $f^{-1}(x) = \log_2(x)$.

Reflection

Suggested Class Time. 80 minutes

Prerequisites. Students need fluency with exponent rules, composition of functions, and the relationship between a function and its inverse. In short: most of Unit 2 up to this topic!

Instructional Strategies. For this part of the unit, I transitioned to a more teacher-centered approach. This was partially as a mini-experiment to see if it would affect student performance, but also because the amount of time I had to cover these rather challenging topics was much less than I would've wished. Using the guided notes, students were still equally engaged, if by the change of pace more than anything.

Throughout this lesson, the primary thing I would emphasize to anyone teaching this (and to my students) is the notions of inverses and, particularly, the relationship between their inputs and outputs. Too often, students come to my class thinking of an inverse as the "opposite" of a function, rather than what "undoes" a function. Even further, without a solid foundation in inverses, students will really struggle to understand what exactly $g(f(x)) = x$ is saying or how to apply it. Without that deep understanding, your students will assuredly do what my students have tried to do, which is to store "e is the inverse of ln" in their brains, rather than $f(x) = e^x$ and $g(x) = \ln x$ being inverse functions. I actually tried to bait my students with $\ln e^4$ and $\ln 4e$, and sure enough, they said 4 for both. We spent a solid extra 5 minutes going over *just* Example 2(b) to really hammer this home.

Most of the rest of the lesson is rather straightforward, and my students did not seem to have many issues. What worked really well for them was what I think College Board wants us emphasizing, which is *covariation*. Students should understand logarithms roughly as follows:

1. Exponential functions' outputs change multiplicatively while inputs change linearly/additively.

2. The inverse of an exponential function is a logarithmic function.

3. The inputs and outputs of an inverse function are the reverse of the original function.

4. Ergo, logarithmic functions' *inputs* change multiplicatively while *outputs* change linearly/additively.

To be honest, my students did struggle with logarithmic scales. My explanation seemed sufficient for them, and their answers in class were all spot on. Once tempted by a Topic Question, though, literally 95% of my students mistakenly said that a difference of 0.3 on a Richter scale indicated a difference of magnitude of a factor of 3: in other words, they simply multiplied 10 by 0.3, rather than computing $10^{0.3}$. I think this had something to do with decimals, but I would highly recommend doing multiple examples of logarithmic scales, even if they're simple and quick ones off to the side. I would also encourage using an analytical approach, for instance:

An increase of $+k$ on a logarithmic scale of base b corresponds to multiplication by b^k.

Technology. Students needed no technology at all, and it's not required for the teacher either, though I did pull up Geogebra to address the similarity of all functions of the form $\log_b x$ and to show the reflecting of an exponential over $y = x$.

Problems.

1. A graph of the exponential function $f(x)$ is shown.

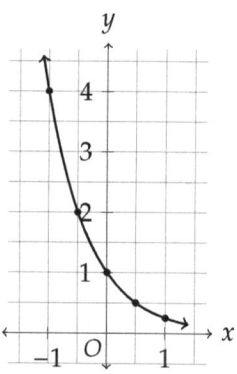

Which of the following points would lie on the graph of $f^{-1}(x)$?

(A) $(8, -2)$ (B) $(16, -2)$ (C) $(4, 2)$ (D) $(16, 2)$

2. The functions f and g are given by $f(x) = 10^x$ and $g(x) = \log_{10} x$, where x is nonzero. The function h is given by the composition $h(x) = f(g(x))$. Which of the following is true concerning h?

(A) h is a constant function.

(B) h is a linear function with a slope of 1.

(C) h is a linear function with a slope of 10.

(D) h is an exponential function with a base of 10.

3. The table below gives corresponding values of x and y. The data are modeled by the function g, where $y = g(x)$.

x	7	14	28	56	112
y	13	26	39	52	65

Which of the following could define $g(x)$?

(A) $g(x) = 13 \cdot 2^{(x/7-1)}$ (B) $g(x) = \dfrac{13}{2} \cdot 2^{(x-6)}$ (C) $g(x) = 13 \log_7(x)$ (D) $g(x) = 13 \log_2 \left(\dfrac{2x}{7} \right)$

4. A cybersecurity firm sells encryption packages that have security ratings that exist on a logarithmic scale of base 2. How many more times secure is a package with a rating of 7 than a package with a rating of 4?

   (A) 3   (B) 4   (C) 6   (D) 8

5. The function f is defined by $f(x) = 5 \cdot 3^x$. It is known that the function g satisfies the property that $f(g(x)) = x$ for all values of x. Which of the following is true?

   (A) g is increasing and the graph of g is concave down.
   (B) g is increasing and the graph of g is concave up.
   (C) g is decreasing and the graph of g is concave down.
   (D) g is decreasing and the graph of g is concave up.

Solutions.

1. By examining the graph we find that f has the input-output pairs $(0, 1)$ and $(-1, 4)$. Hence, each negative unit change in the inputs multiplies the output by 4. So when $x = -2$, the function must output $4 \cdot 4 = 16$, and $(-2, 16)$ is an input-output pair of f. Thus, $(16, -2)$ is such a pair on f^{-1}. The correct answer is therefore **(B)**.

2. The composition is $h(x) = f(g(x)) = f(\log_{10} x) = 10^{\log_{10} x} = x$. The correct answer is therefore **(B)**.

3. The inputs changes multiplicitavely while the outputs change additively, so the model should be logarithmic. Since the multiplicative factor is 2, we need a logarithmic function whose base is 2. The correct answer is therefore **(D)**.

4. Since the base is 2, the ratings of 4 and 7 correspond to the values 2^4 and 2^7. Thus, the rating of 7 larger than the rating of 4 by a factor of $2^7/2^4 = 2^{7-4} = 2^3 = 8$. The correct answer is therefore **(D)**.

5. The property that $f(g(x)) = x$ indicates that f and g are inverses. Therefore since f is an increasing exponential function, g is an increasing logarithmic function. Thus, the graph of g is concave down as well. The correct answer is therefore **(A)**.

Topic 2.11 ~ Logarithmic Functions

Learning Objectives
1. (2.11.A) Identify key characteristics of logarithmic functions.

Success Criteria
1. I can identify key characteristics of logarithmic functions like domain, end behavior, rate of change, concavity, and vertical asymptotes.

In Topics 2.9-2.10, we introduced the inverse of the exponential function, the LOGARITHM. We'll now identify a few of the major characteristics of a logarithmic function.

For this discussion, we will assume $b \neq 1$ and $b > 0$.

Characteristics of $\log_b(x)$		
Characteristic	Value/s	Explanation
Domain		
Range		
Asymptotic/ end behavior		
Rate of change		
Rate of rate of change/ concavity		

We will graph the general logarithmic function $f(x) = \log_b x$ on the axes on the next page.

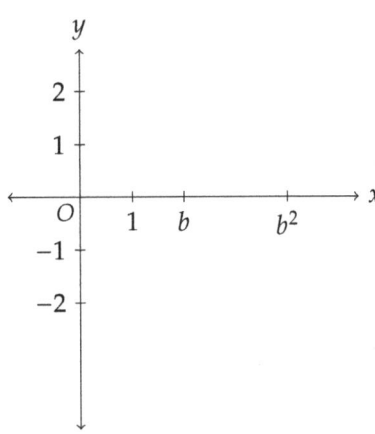

For the remainder of class, you will work some practice questions related to some previous topics.

Practice

1. Determine if the following are true or false for $f(x) = -\log x$. If they are false, correct them.

   (a) f is increasing for all x.

   (b) f is changing at an increasing rate for all x.

   (c) On the interval $0 < x \leq 10$, the maximum value is 0.

   (d) On the interval $0 < x \leq 10$, the minimum value is -1.

2. The function L is given by $L(x) = \log_b x$. Find the values of a, b, and c.

x	0.5	2	8	64
$L(x)$	a	0.5	1.5	c

3. For what value of k does $\log_6(k) = 3$?

4. Without a calculator, describe how you could transform the graph of $f(x) = \log_2 x$ into $g(x) = \log_{1/2} x$.

5. (**MAD LIB**) Complete the following Mad Lib.

   The function $f(x)$ is _____ at a/an _____ rate. As the inputs
   (increasing/decreasing)   (increasing/decreasing)
   increase _____ _____, the outputs increase _____
            (by a constant rate/by a factor of) (number)   (by a constant rate/by a factor of)
   _____. The graph of the function always has _____ concavity. Because of
   (number)   (positve/negative/zero)
   these characteristics, $f(x)$ must be a _____ function.
                                            (family type)

6. Find the inverse of the function $f(x) = 3\log_5 x$.

7. Find the domain of the function $g(x) = \dfrac{3x - 1}{\ln(2x - 4)}$.

8. Determine whether each of the statements below could apply to a linear function, exponential function, or logarithmic function. Any given statement might be able to apply to more than one of these. Assume that there are no domain restrictions other than $x > 0$ for logarithmic functions. Write some combination of **LIN**, **EXP**, and **LOG** for which function/s *could* possibly be described by each of the following statements.

    (a) As inputs increase by 1, the outputs increase by a factor of 6.
    (b) The rate of change of the function is positive for all x.
    (c) Over any successive interval of length 4 in the outputs, the inputs quadruple.
    (d) The graph of the function is always concave down.
    (e) The graph of the function is always concave down and $f(x) \geq 0$ for all x.
    (f) The function is increasing at a decreasing rate.
    (g) The function is decreasing at a decreasing rate.

9. An exponential function $E(x) = ab^x$, with $b > 0$, contains the points $(0, -10)$ and $(2, -90)$. Write an expression for E. Then, describe the rate of change of E (i.e. increasing at increasing rate for all x).

Topic 2.11 Homework

1. Sketch a graph of $f(x) = \log_2 x$. Then, on the same graph, sketch a graph of $g(x) = \log_2(2x) + 1$.

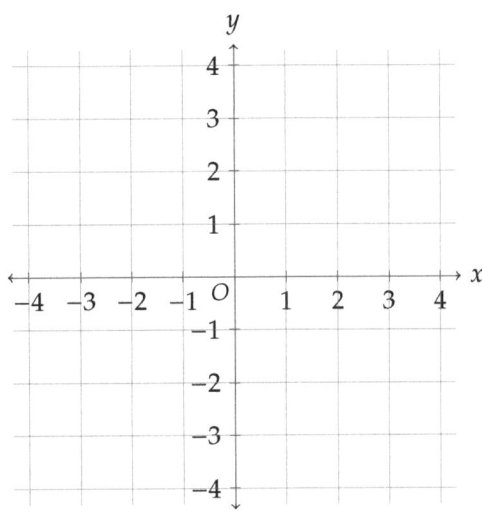

2. Determine the following characteristics of $g(x) = \log(x + 4) - 2$.

    (a) Domain

    (b) Range

    (c) $\lim\limits_{x \to \infty} g(x)$

    (d) Vertical asymptote

    (e) x-intercept

3. Consider the function $g(x)$ with values given in the table below. If $g(x) = a \log_b(x)$, find a and b.

x	5	25	125
$g(x)$	3	6	9

4. Let $f(x) = a \log_b x$. Determine whether each of the following statements about f are **always true**, **sometimes true**, or **never true**. Explain your answers.

    (a) The graph of f has an inflection point.

    (b) The graph of f has a global maximum.

    (c) f is increasing at an increasing rate.

    (d) f is decreasing at an increasing rate.

    (e) f is increasing at an decreasing rate.

    (f) $\lim\limits_{x \to \infty} f(x) = k$ for some constant k

    (g) As the inputs increase multiplicatively, the outputs increase (or decrease) linearly.

5. Write an expression for a function f that satisfies the following.

    - $\lim\limits_{x \to 3^+} f(x) = \infty$
    - $f(5) = 0$
    - $\lim\limits_{x \to \infty} f(x) = -\infty$

Topic 2.11 Solutions/Notes

Activity

Characteristics of $\log_b(x)$		
Characteristic	**Value/s**	**Explanation**
Domain	$x > 0$	$\log_b x = y$ means $b^y = x$. If $x < 0$ and $b > 0$, no such y exists.
Range	All real numbers	The domain of b^x is all real numbers, so the range of its inverse is all real numbers.
Asymptotic/ end behavior	$\lim_{x \to 0^+} f(x) = -\infty$, $\lim_{x \to \infty} f(x) = \infty$	As inputs continue to grow multiplicatively by a factor of b, inputs will continue to grow linearly/additively. However, as inputs decrease multiplicatively by a factor of b, they will approach 0, while outputs will decrease linearly/additively without bound.
Rate of change	Always positive or always negative	$f(x) = b^x$ is either growth or decay and cannot be both.
Rate of rate of change/ concavity	Always positive/concave up or always negative/concave down	Exponential growth is always speeding up or slowing down.

The function $f(x) = \log_b x$ is graphed below.

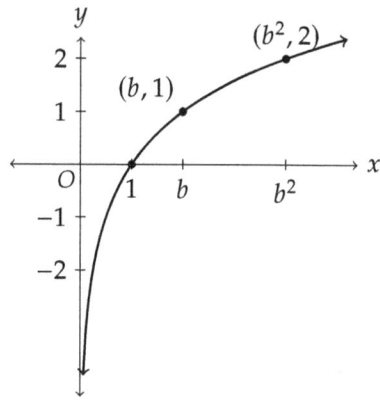

Practice.

1. The graph of $f(x) = -\log x$ is a reflection of the graph of $y = \log x$ over the x-axis, so:

    (a) False. f is decreasing for all x.
    (b) True. The graph is decreasing at an increasing rate.
    (c) False. Because $\lim_{x \to 0^+} f(x) = \infty$, f has no maximum value on this interval.
    (d) True. f is decreasing, so its minimum will be $f(10) = -\log(10) = -1$.

2. We have $\log_b(2) = 0.5$, so $b^{0.5} = \sqrt{b} = 2$. Therefore, $b = 4$. Now, $c = L(64) = \log_4(64) = 3$ and $a = \log_4\left(\frac{1}{2}\right) = -0.5$.

3. $k = 6^3 = 216$

4. If $g(x) = \log_{1/2} x$, then $\left(\frac{1}{2}\right)^{g(x)} = x$. Since $\frac{1}{2} = 2^{-1}$, we can write this as $\left(2^{-1}\right)^{g(x)} = x$, or more simply $2^{-g(x)} = x$. This can be rewritten as $-g(x) = \log_2 x$. Thus, $g(x) = -\log_2 x$. The graph of $g(x)$ is therefore the graph of $f(x)$ reflected over the x-axis.

5. Student answers will vary.

6. $y = 3\log_5 x \Rightarrow x = 3\log_5 y \Rightarrow \frac{x}{3} = \log_5 y \Rightarrow 5^{x/3} = y \Rightarrow f^{-1}(x) = 5^{x/3}$

7. The logarithm is defined only for positive inputs, so $2x - 4 > 0$ gives an initial domain of $x > 2$. However, if the logarithm outputs 0, the function g will be undefined. This occurs when $2x - 4 = 1$, or when $x = \frac{5}{2}$. Therefore, the domain of g is $x > 2$ with $x \neq \frac{5}{2}$.

8. (a) Exponential   (b) Linear, exponential, logarithmic   (c) Logarithmic   (d) Exponential, logarithmic
   (e) Exponential   (f) Logarithmic, exponential   (g) Exponential

9. Let $E(x) = ab^x$. Plugging in $(0, -10)$, we get $-10 = ab^0 = a$. Plugging in $(2, -90)$ gives $-90 = -10b^2$, or $b = 3$. Therefore, $E(x) = -10 \cdot 3^x$. The function E is decreasing at a decreasing rate for all x.

Homework

1.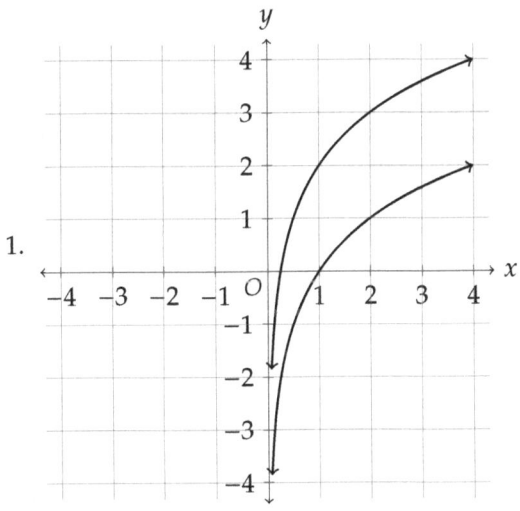

2. (a) $x + 4 > 0 \Rightarrow x > -4$ (b) All real numbers (c) ∞ (d) $x = -4$
   (e) $g(x) = 0 \Rightarrow \log(x+4) - 2 = 0 \Rightarrow \log(x+4) = 2 \Rightarrow 10^2 = x + 4 \Rightarrow x = 96$, so the x-intercept is $(96, 0)$.

3. As the outputs increase by a factor of 5, the inputs increase linearly, so $b = 5$. Therefore, $g(x) = a\log_5 x$. Plugging in $(5, 3)$, we get $3 = a\log_5 5$, or $a = 3$. Therefore, $g(x) = 3\log_5 x$.

4. (a) Never true. Logarithmic functions are always increasing or decreasing at an increasing or decreasing rate.

   (b) Sometimes true. Logarithmic functions have no global maxima or minima - unless on a restricted domain. If the domain is restricted, then it is possible for a logarithm to have a global maximum.

   (c) Never true. If a logarithmic function is increasing, then it is increasing at a decreasing rate.

   (d) Sometimes true. A logarithmic function could also be increasing at a decreasing rate.

   (e) Sometimes true. A logarithmic function could also be decreasing at an increasing rate.

   (f) Never true. Logarithmic functions either increase or decrease without bound.

   (g) Always true.

5. The first and third limits remind us of the vertical asymptote and end behavior of a logarithm. For $f(x) = \log_b x$, we usually have $\lim_{x \to 0^+} f(x) = -\infty$, though, so we need to shift right 3 units and reflect over the x-axis to get $f(x) = -\log_b(x-3)$. Now, we need $f(5) = 0$, or $-\log_b(5-3) = 0$. This leads to $b^0 = 2$, which is impossible. To adjust for this, we need $5 - 3 = 2$ to be scaled down to 1, so we multiply the inputs by $\frac{1}{2}$ (dilate horizontally by a factor of 2). This gives us $f(x) = -\log_b\left(\frac{1}{2}(x-3)\right)$, where b can be any number greater than 1. For convenience, we'll use the common log: $f(x) = -\log\left(\frac{1}{2}(x-3)\right)$.

Reflection

Suggested Class Time. 75 minutes

Prerequisites. We will revisit rates of change from Unit 1, and students should have some proficiency with logarithmic functions by this point.

Instructional Strategies. Honestly, this lesson went quite smoothly. A few things I noticed, though.

Firstly, I always taught the domain of a logarithm being $x > 0$ as being due to some version of "a positive number can't be raised to a power to equal a negative" or "a negative can't be a power of a positive." However, with the emphasis on inverse functions, and with the opportunity to emphasize graph characteristics, I transitioned this year to emphasizing that the domain and range of $f(x) = \log_b x$ are just the range and domain of $g(x) = b^x$, respectively.

In the graphing of a sample graph right below the table, I decided to do an additional example in class. I went with graphing $f(x) = -3\log_2(x-4) + 1$, which students helped me make up on the spot. I wanted to emphasize the importance of parent functions and graph transformations.

For the practice questions, rather than just give students the remaining 50 or so minutes to work on them, I broke it up into parts. Students worked the first 3 for about 15-20 minutes, and then we went over them. We repeated this for Problems 4-7, but I gave them longer. We finally went over Problem 8 as a class, as we were running out of time, unfortunately, and Problem 9 was omitted.

It's worth noting that the Mad Lib may very well be impossible – I had a different student give me their word for each different blank, and by the end, we had a description of a function that was impossible. It was still instructive, though, as it really had students thinking about characteristics of the parent functions.

Technology. Again, none needed for students, but I pulled up Geogebra a couple of times during the lesson.

Problems.

1. Which of the following is true about the graph of the function $f(x) = -2\log_3 x$ for all $x > 0$?
   - **(A)** f is increasing at an increasing rate.
   - **(B)** f is increasing at a decreasing rate.
   - **(C)** f is decreasing at an increasing rate.
   - **(D)** f is decreasing at a decreasing rate.

2. The exponential function $h(x)$ is decreasing and its graph is concave up for all x. Which of the following is true about the inverse function $h^{-1}(x)$?
   - **(A)** h^{-1} is increasing and its graph is concave up for all x in its domain.
   - **(B)** h^{-1} is increasing and its graph is concave down for all x in its domain.
   - **(C)** h^{-1} is decreasing and its graph is concave up for all x in its domain.
   - **(D)** h^{-1} is decreasing and its graph is concave down for all x in its domain.

3. The function f is defined by $f(x) = 3\log_5(2x - 1)$. Which of the following is true of f?

   (A) The domain of f is all positive real numbers and the range of f is all real numbers.
   (B) The domain of f is all real numbers greater than $\frac{1}{2}$ and the range of f is all real numbers.
   (C) The domain of f is all positive real numbers and the range of f is all positive real numbers.
   (D) The domain of f is all real numbers greater than $\frac{1}{2}$ and the range of f is all positive real numbers.

4. The function g defined by $g(x) = a\log_b(x)$ has values in the table below given by selected values of x.

x	4	16	64	256
$g(x)$	7	14	21	28

   What are the values of a and b?

   (A) $a = 14$ and $b = 16$ (B) $a = 7$ and $b = 4$ (C) $a = 4$ and $b = 7$ (D) $a = 16$ and $b = 14$

5. The function g has a domain of all real numbers and has values in the table below given by selected values of x.

x	3	6	9	12
$g(x)$	6	36	216	1296

   The function f is the inverse function of g. Which of the following is an expression for f?

   (A) $f(x) = 6^{(x/3)}$ (B) $f(x) = 6 \cdot 3^{(x-3)}$ (C) $f(x) = 3\log_6(x)$ (D) $f(x) = 6\log_3(x)$

Solutions.

1. The negative logarithm decreases at an increasing rate. The correct answer is therefore **(C)**.

2. The exponential function $h(x) = e^{-x}$ is decreasing and concave up for all x. Its inverse is $h^{-1}(x) = -\ln(x)$ which is also decreasing and concave up for all x in its domain. The correct answer is therefore **(C)**.

3. The logarithm requires that the argument be positive, so we need $2x - 1 > 0$. This implies the domain is all values of x greater than $1/2$. The range is all real numbers. The correct answer is therefore **(B)**.

4. As the inputs increase multiplicatively by a factor of 4, the outputs increase linearly by a constant rate of 7. This gives $a = 7$ and $b = 4$ as possibilities. Checking against the inputs and outputs gives $7\log_4 4 = 7$, $7\log_4 16 = 7 \cdot 2 = 14$, and so on, which matches what is in the table. The correct answer is therefore **(B)**.

5. We can write an expression for $g(x) = 6^{(x/3)}$ since the outputs increase by a factor of 6 for each increase of 3 in the inputs. Setting $y = g(x)$ and interchanging x and y gives $x = 6^{(y/3)}$. Then $\log_6 x = \frac{y}{3}$, and so $y = g^{-1}(x) = f(x) = 3\log_6(x)$. The correct answer is therefore **(C)**.

Topic 2.12 ~ Equivalent Logarithmic Forms

Learning Objectives

1. (2.12.A) Rewrite logarithmic expressions in equivalent forms.

Success Criteria

1. I can utilize the product, power, and change of base properties to rewrite logarithmic expressions.

Much how exponential expressions follow certain rules that enable us to rewrite them, logarithmic expressions come with an analogous set of rules.

\multicolumn{4}{c	}{**Properties of Logarithms**}
Name	Equation	Proof	Graphically
Product Property	$\log_b(xy) =$		
Power Property	$\log_b x^n =$		
Change of Base	$\log_b a =$		

Note: These rules apply for any positive base b! This means the above rules include $\log x$ and $\ln x$.

There will be no examples – just straight to problem solving.

Practice.

1. Compute $\log 2 + \log 20 + \log \frac{5}{2}$ without a calculator.

2. Rewrite $\ln x + 2 \ln y$ as a single logarithm.

3. If $\log_{10} 7 = a$ and $\log_{10} 2 = b$, write an expression for each of the following in terms of a and b.

   (a) $\log_{10} 49$ (b) $\log_{10} \frac{2}{7}$ (c) $\log_{10} \frac{\sqrt{7}}{8}$ (d) $\log_{10} 1.4$

4. Suppose $\log 9 \approx 0.954$. Compute the following without a calculator.

   (a) $\log 81$ (b) $\log \frac{10}{9}$ (c) $\log \frac{1}{9}$ (d) $\log 90$ (e) $\log 3$ (f) $\frac{\log_2 9}{\log_2 10}$

5. Let $f(x) = \log_3 x$. The function $g(x)$ has the graph of f, but translated vertically by $+2$. List two *different* expressions for $g(x)$.

6. Let $f(x) = \log_2 x$. The graph of the function $g(x) = \log_2(kx)$ is graphed below. What is k?

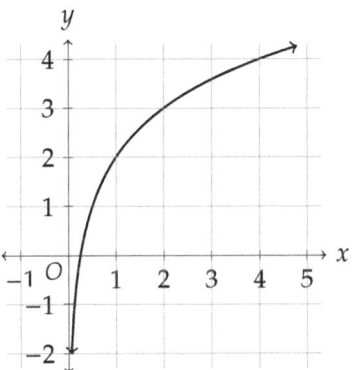

7. Complete the table for the function $\ln x$:

x	1					
$\ln x$	0	1	2	3	4	5

8. Let's go *old school*: using only the log button (no ln) on your calculator, compute $\log_3 10$, $\log_6 245$, and $\ln 27.3$.

9. For what value of b will $\log_b x = 3 \log_2 x$ for all $x > 0$?

10. If we want to go *really* old school, we would have to go back to before there were calculators: in those ancient times existed *log tables*. We won't replicate an entire log table here, we'll utilize some characteristics of a log table in the following problems.

    (a) In a table for $\log_{10} x$, one row would have the number 17. Then, the first column in would have the value 2304. What do you think this has to do with $\log 17$? Use your calculator to confirm your answer.

    (b) Suppose you wanted to compute $\log 1700$. The table only had rows that went from 10 to 50. How would you do it?

Topic 2.12 Homework

1. Let $f(x) = \log_5 x$ and $g(x) = x^4$. If $h(x) = f(g(x))$, then the graph of $h(x)$ is the graph of $f(x)$ under the image of what transformation?

2. Evaluate the following expressions (without a calculator!).

    (a) $\log 25 + \log 4$

    (b) $\ln \dfrac{1}{e^2}$

    (c) $\log_2 \dfrac{1}{8}$

    (d) $\log_9 2 - \log_9 6$

    (e) $\log 0.05 - \log 5$

3. What transformation would transform the graph of $f(x) = \log x$ into $g(x) = \log_2 x$?

4. Find a value of b such that $\log_b x = \dfrac{1}{2} \log_4 x$.

5. Rewrite each of the following as a single logarithmic expression of the form $\log_b(\text{expression})$.

    (a) $\ln x + \ln y - \ln 2x$

    (b) $3 \log x - 4 \log y + \log 2xy$

    (c) $\log 1 + \log 2 + \cdots + \log 6$

6. If $\log 2 = 0.301$ and $\log 3 = 0.477$, find the value of (c) from the previous question without a calculator.

7. Consider the functions $f(x) = \log_2 x$, $g(x) = 4 + \log_2 x$, and $h(x) = \log_8 x$.

    (a) Determine two different transformations that would transform the graph of f into the graph of g.

    (b) Determine a transformation that would transform the graph of f into the graph of h.

8. Let $f(x) = \log x$ and $g(x) = \log \dfrac{x}{1000}$. The graph of g is a translation of f by how much, and in what direction?

Topic 2.12 Solutions/Notes

Activity

Properties of Logarithms			
Name	Equation	Proof	Graphically
Product Property	$\log_b(xy)$ $= \log_b x + \log_b y$	Let $m = \log_b x$ and $n = \log_b y$. Then $b^m = x$ and $b^n = y$, so $b^m b^n = b^{m+n} = xy$. Therefore, $\log_b(xy) = m + n = \log_b x + \log_b y$.	A horizontal dilation by a constant is the same as a vertical translation. $\log_b(kx) = \log_b x + \log_b k$, where $\log_b k$ is a constant.
Power Property	$\log_b x^n$ $= n \log_b x$	Since $\log_b x^n = \log_b(x \cdot x \cdot x \cdots x)$, we can repeatedly use the Product Property to get $\log_b x^n = \log_b x + \log_b x + \cdots + \log_b x = n \log_b x$.	Raising the input to a power is the same as vertically dilating the graph of a logarithmic function.
Change of Base	$\log_b a$ $= \dfrac{\log_k a}{\log_k b}$	Let $y = \log_b a$. Then $b^y = a$. Taking $\log_k$ of both sides, we get $\log_k b^y = \log_k a$. Using the Power Property gives $y \log_k b = \log_k a$. Dividing by $\log_k b$ gives $y = \dfrac{\log_k a}{\log_k b}$.	All logarithmic functions are vertical dilations of each other: $\log_b x = \dfrac{1}{\log_a b} \cdot \log_a x$

Practice.

1. $\log 2 + \log 20 + \log \dfrac{5}{2} = \log\left(2 \cdot 20 \cdot \dfrac{5}{2}\right) = \log 100 = 2$

2. $\ln x + 2 \ln y = \ln x + \ln y^2 = \ln(xy^2)$

3. (a) $\log_{10} 49 = \log_{10} 7^2 = 2 \log_{10} 7 = 2a$

    (b) $\log_{10} \dfrac{2}{7} = \log_{10} 2 - \log_{10} 7 = b - a$

    (c) $\log_{10} \dfrac{\sqrt{7}}{8} = \log_{10} \sqrt{7} - \log_{10} 8 = \log_{10} 7^{1/2} - \log_{10} 2^3 = \dfrac{1}{2} \log_{10} 7 - 3 \log_{10} 2 = \dfrac{1}{2}a - 3b$

    (d) $\log_{10} 1.4 = \log_{10} \dfrac{14}{10} = \log_{10} 14 - \log_{10} 10 = \log_{10}(2 \cdot 7) - 1 = \log_{10} 2 + \log_{10} 7 - 1 = b + a - 1$

4. (a) $\log 81 = \log 9^2 = 2 \log 9 = 2(0.954) = 1.908$

    (b) $\log \dfrac{10}{9} = \log 10 - \log 9 = 1 - 0.954 = 0.046$

    (c) $\log \dfrac{1}{9} = \log 9^{-1} = -\log 9 = -0.954$

    (d) $\log 90 = \log(9 \cdot 10) = \log 9 + \log 10 = 0.954 + 1 = 1.954$

    (e) $\log 3 = \log 9^{1/2} = \dfrac{1}{2} \log 9 = \dfrac{1}{2}(0.954) = 0.477$

    (f) Reversing the change of base formula, this is simply $\log_{10} 9 = \log 9 = 0.954$.

5. Expression 1 is $f(x) = \log_3(x) + 2$. Expression 2 is $f(x) = \log_3 x + \log_3(9) = \log_3(9x)$.

6. The graph of $y = \log_2 x$ would pass through $(1,0)$ and $(2,1)$, so f is a vertical translation of $y = \log_2 x$ by 2, giving $f(x) = \log_2(x) + 2 = \log_2 x + \log_2 4 = \log_2(4x)$. Therefore, $k = 4$.

7.

x	1	e	e^2	e^3	e^4	e^5
$\ln x$	0	1	2	3	4	5

8. Using change of base formula, the values are 2.096, 3.070, and 3.307.

9. $\log_b x = \dfrac{\log_2 x}{\log_2 b} = 3\log_2 x$. This can be rewritten as $\dfrac{1}{3} = \log_2 b$, which gives $b = 2^{1/3} = \sqrt[3]{2}$.

10. (a) $\log 17 = 1.2304$

    (b) $\log 1700 = \log(17 \cdot 100) = \log 17 + \log 100 = 1.2304 + 2 = 3.2304$

Homework

1. $h(x) = f(g(x)) = \log_5 x^4 = 4\log_5 x$, so a vertical dilation by a factor of 4.

2. (a) $\log(25 \cdot 4) = \log 100 = 2$

   (b) -2

   (c) -3

   (d) $\log_9 \dfrac{2}{6} = \log_9 \dfrac{1}{3} = -\dfrac{1}{2}$

   (e) $\log \dfrac{0.05}{5} = \log 0.01 = \log \dfrac{1}{100} = -2$

3. $g(x) = \log_2 x = \dfrac{\log x}{\log 2}$, so a vertical dilation by a factor of $\dfrac{1}{\log 2}$.

4. $\log_b x = \dfrac{\log_4 x}{\log_4 b} = \dfrac{1}{2}\log_4 x$, which gives $\log_4 b = 2$, so $b = 16$.

5. (a) $\ln \dfrac{xy}{2x} = \ln \dfrac{y}{2}$

   (b) $\log x^3 - \log y^4 + \log 2xy = \log \dfrac{x^3(2xy)}{y^4} = \log \dfrac{2x^4}{y^3}$

   (c) $\log(1 \cdot 2 \cdot 3 \cdot 4 \cdot 5 \cdot 6) = \log 720$

6. We rearrange so that everything is in terms of log 2, log 3, or log 10.

$$\log 1 + \log 2 + \cdots + \log 6 = 0 + (\log 2 + \log 5) + \log 3 + \log 2^2 + \log(2 \cdot 3)$$
$$= \log 10 + \log 3 + 2\log 2 + \log 2 + \log 3$$
$$= 1 + 0.477 + 2(0.301) + 0.301 + 0.477$$
$$= 2.857$$

7. (a) One is a vertical translation by $+4$. Next, we can rewrite $g(x) = \log_2 16 + \log_2 x = \log_2(16x)$, so we also get a horizontal dilation by a factor of $\frac{1}{16}$.

   (b) $h(x) = \log_8 x = \dfrac{\log_2 x}{\log_2 8} = \dfrac{\log_2 x}{3} = \dfrac{1}{3}\log_2 x = \dfrac{1}{3}f(x)$, so a vertical dilation by a factor of $\dfrac{1}{3}$

8. $g(x) = \log \dfrac{x}{1000} = \log x - \log 1000 = \log(x) - 3$, so a vertical translation by -3.

UNIT 2   TOPIC 2.12 ~ EQUIVALENT LOGARITHMIC FORMS   229

Reflection

Suggested Class Time. 80 minutes

Prerequisites. The lesson does not assume prior knowledge of log rules, but students should of course know their exponent rules.

Instructional Strategies. For the proof of the Product Property, I actually decided to provide an alternative proof, and it seemed to resonate much more with students. It went like this.

Let $f(t) = b^t$. What is $f(x) \cdot f(y)$? According to exponent rules, we get

$$f(x) \cdot f(y) = b^x \cdot b^y \cdot b^{x+y} = f(x+y)$$

This means that, for an exponential function, *multiplying outputs* is the same as having *added inputs*.

For a logarithmic function, though, the inputs and outputs are reversed. Therefore, for $g(t) = \log_b t$, we must have that *multiplying* **inputs** is the same as having *added* **outputs**.

Therefore, $g(xy) = g(x) + g(y)$, or

$$\log_b(xy) = \log_b x + \log_b y$$

Throughout students working on the practice problems, this version came up again and again, and students seemed to understand it. Again, this emphasizes the relationship between inverse functions, so it pays off in multiple ways.

I did have students struggle with Problem 3. The most common thing was something like this: students saw $a = \log_{10} 7$ and $b = \log_{10} 2$, and when they saw $\log_{10} 14$, they immediately just saw $2 \cdot 7$ and wrote down ab. This was not an immediate fix, as some students were still doing later questions incorrectly with similar mistakes. What seemed to work the best was emphasizing that students should separate their terms (logarithms) first before worrying about anything else. When students immediately saw $8 = 2^3$ in (c), I told them to separate first, then use exponent and logarithm rules, and only *then* try to convert back to a and b.

For Problem 6, my recommendation is that you don't get too fancy with your explanation. Rather than try to discuss rates of change, I encourage simply plugging in points. It's what College Board did in a couple of Topic Questions; while I anticipated something deeper, I simply saw "this graph contains these points." My students tried out $k = 2$ and $k = \frac{1}{2}$, and while I did mention why $k = \frac{1}{2}$ would *slow down* the parent function (with the given graph clearly having a higher rate of change than the parent function), I just encouraged them to test out values of k, and they got it quickly.

Technology. No technology was used.

Problems.

1. Let $f(x) = \log_{10}(ax)$ and $g(x) = \log_{10}(bx^2)$ for positive numbers a and b. Let $k(x) = f(x) + g(x)$. Which of the following is expresses k as a single logarithmic expression?
    (A) $k(x) = \log_{10}(ax + bx^2)$ **(B)** $k(x) = 3\log_{10}(abx)$
    (C) $k(x) = \log_{10}(abx^3)$ **(D)** $k(x) = \log_{10}((a+b)x^2)$

2. The logarithmic function L can be expressed as $L(x) = A\log_3 x$ and also as $L(x) = B\log_2 x$. Which of the following is an expression for A in terms of B?

    (A) $A = B\log_2 3$ **(B)** $A = B \cdot \dfrac{\log_3 2}{\log_2 3}$ **(C)** $A = B \cdot \dfrac{\log_2 3}{\log_3 2}$ **(D)** $A = \dfrac{B}{\log_2 3}$

3. The function f defined by $f(x) = a + b \ln x$ is used to model a data set, where $a > 0$ and $b > 1$. The same model can be expressed as the function g defined by $g(x) = a + c \log_7 x$. What is the value of c in terms of b?

   (A) $\ln b$ (B) $\log_7 b$ (C) $b \ln 7$ (D) $\dfrac{b}{\ln 7}$

4. It is known that $\log_a b = 12$, where $a > 0$ and $b > 0$. What is the value of $\log_{a^3}(b^5)$?

   (A) 20 (B) $12^{3/2}$ (C) $12^{5/3}$ (D) $2 \cdot 12^3$

5. The function f is given by $f(x) = \log_5 x$. The function g is given by $g(x) = \log_{25} x$. Which of the following is true?

   (A) The graph of g is a horizontal dilation of the graph of f by a factor of $\ln 2$.
   (B) The graph of g is a horizontal dilation of the graph of f by a factor of $\frac{1}{2}$.
   (C) The graph of g is a vertical dilation of the graph of f by a factor of $\ln 2$.
   (D) The graph of g is a vertical dilation of the graph of f by a factor of $\frac{1}{2}$.

Solutions.

1. $f(x) + g(x) = \log_{10}(ax) + \log_{10}(bx^2) = \log_{10}(ax \cdot bx^2) = \log_{10}(abx^3)$. The correct answer is therefore **(C)**.

2. Use the change of base formula on the first expression to tuen the base of 3 into a base of 2. We get

$$L(x) = A \log_3 x = A \cdot \frac{\log_2 x}{\log_2 3} = \frac{A}{\log_2 3} \cdot \log_2 x.$$

   Since $L(x)$ also equals $B \cdot \log_2 x$, we must have $B = \frac{A}{\log_2 3}$, or $A = B \log_2 3$. The correct answer is therefore **(A)**.

3. We need f and g to be equal. This requires

$$b \ln x = c \log_7 x = c \cdot \frac{\ln x}{\ln 7}.$$

   Diving both sides by $\ln x$ and multiplying both sides by $\ln 7$ gives $c = b \ln 7$. The correct answer is therefore **(C)**.

4. We have

$$\log_{a^3}(b^5) = \frac{\log_a(b^5)}{\log_a(a^3)} = \frac{5 \log_a b}{3} = \frac{5 \cdot 12}{3} = 20.$$

   The correct answer is therefore **(A)**.

5. Since

$$\log_{25} x = \frac{\log_5 x}{\log_5 25} = \frac{\log_5 x}{2} = \frac{1}{2} \log_5 x,$$

   we see that g is a vertical dilation of f by a factor of $\dfrac{1}{2}$. The correct answer is therefore **(D)**.

Topic 2.13 ~ Exponential and Logarithmic Equations and Inequalities

Learning Objectives
1. (2.13.A) Solve exponential and logarithmic equations and inequalities.
2. (2.13.B) Construct the inverse function for exponential and logarithmic functions.

Success Criteria
1. I can utilize properties of exponentials and logarithms to solve inequations and inequalities, including analyzing the domains of logarithmic functions.

We will go through this Topic with worked examples.

Example 1. Find the inverse of $f(x) = 3 \cdot 5^{(x-2)} + 4$.

Example 2. Find the inverse of $g(x) = \ln(3x + 1) - 2$.

Example 3. Let $f(x) = \log_4(x^2 + x - 2)$ and $g(x) = \log_4(5x - 5)$.

(a) Find the domain of f and the domain of g.

(b) Find the value/s of x for which $f(x) = g(x)$.

Example 4. Let $f(x) = 3 \cdot 4^{(x-1)} - 2$. If $g(x)$ is the inverse of $f(x)$, find the solution/s to $g(x) = \frac{1}{2}x - 2$.

Example 5.

(a) Explain why $c^x = b^{x \log_b c}$ for any $b, c > 0$ with $b, c, \neq 1$.

(b) Use the relationship in (a) to solve $9^{(2x+3)} = 27^{(x-1)}$.

Example 6. Solve $\log_2(x-2) + \log_2(x-5) = 2$.

Example 7. Let $f(x) = \log(2x+1) - \log(x^2 + 2x - 3)$. Find all values of x for which $f(x) < 0$.

Topic 2.13 Homework

1. Find the value of x for which $\log_3(x^2 + 5x + 6) = \log_3(x + 3)$.

2. Find the inverse of each of the following.

    (a) $f(x) = 4e^{2x} - 3$
    (b) $g(x) = \frac{1}{4}\log_3(x - 1)$
    (c) $h(x) = \log_2(\log_4 x) - 1$

3. Find the value/s of x for which $1 = \log_{10}(x + 3) + \log_{10}(x - 6)$.

4. Solve the equation $4^{(2x+1)} = 8^{(1-2x)}$.

5. Show there are no solutions to $\ln(x^2 + 4x - 21) = \ln(x^2 - 3x)$.

6. The function $f(x) = \log_4 8^{2x}$ can be rewritten as a linear function. What is the slope of this line?

Topic 2.13 Solutions/Notes

Activity

Example 1. Let $f(x) = y$. Then we have

$$y = 3 \cdot 5^{(x-2)} + 4$$
$$x = 3 \cdot 5^{(y-2)} + 4 \qquad \text{(Interchange } x \text{ and } y\text{)}$$
$$x - 4 = 3 \cdot 5^{(y-2)} \qquad \text{(Subtract 4)}$$
$$\frac{x-4}{3} = 5^{(y-2)} \qquad \text{(Divide by 3)}$$
$$\log_5\left(\frac{x-4}{3}\right) = y - 2 \qquad \text{(Take } \log_5 \text{ of both sides)}$$
$$2 + \log_5\left(\frac{x-4}{3}\right) = y$$

Therefore $f^{-1}(x) = 2 + \log_5\left(\frac{x-4}{3}\right)$.

Example 2. Let $g(x) = y$. Then we have

$$y = \ln(3x + 1) - 2$$
$$x = \ln(3y + 1) - 2 \qquad \text{(Interchange } x \text{ and } y\text{)}$$
$$x + 2 = \ln(3y + 1) \qquad \text{(Add 2)}$$
$$e^{(x+2)} = 3y + 1 \qquad \text{(Rewrite as exponential)}$$
$$e^{(x+2)} - 1 = 3y \qquad \text{(Subtract 1)}$$
$$\frac{e^{(x+2)} - 1}{3} = y \qquad \text{(Divide by 3)}$$

Therefore $g^{-1}(x) = \frac{e^{(x+2)} - 1}{3}$.

Example 3.

(a) f is defined only if $x^2 + x - 2 > 0$. Factoring gives $(x+2)(x-1) > 0$. Sketching a graph reveals a quadratic that opens upward with zeros of $x = -2$ and $x = 1$, Therefore, $f(x) > 0$ on the intervals $(-\infty, -2)$ and $(1, \infty)$, which make up the domain of f. For $g(x)$, we get $5x - 5 > 0$, or $x > 1$ for the domain.

(b) We set the function expressions equal and solve. Because the bases of the logarithms are identical, we can simply set the arguments equal.

$$\log_4(x^2 + x - 2) = \log_4(5x - 5)$$
$$x^2 + x - 2 = 5x - 5 \qquad \text{(Set arguments equal)}$$
$$x^2 - 4x + 3 = 0 \qquad \text{(Subtract } 5x\text{, add 5)}$$
$$(x - 3)(x - 1) = 0 \qquad \text{(Factor)}$$

This gives solutions of $x = 3$ and $x = 1$. However, $x = 1$ is not in the domain of f or g, while $x = 3$ is in both domains, so our only solution is $x = 3$.

Example 4. First, we find an expression for $f^{-1}(x)$. Letting $y = f(x)$ and reversing x and y, we get

$$x = 3 \cdot 4^{(y-1)} - 2$$

$$\frac{x+2}{3} = 4^{(y-1)} \qquad \text{(Add 2 and divide by 3)}$$

$$\log_4\left(\frac{x+2}{3}\right) = y - 1 \qquad \text{(Take } \log_4 \text{ of both sides)}$$

$$1 + \log_4\left(\frac{x+2}{3}\right) = y \qquad \text{(Add 1)}$$

We get $f^{-1}(x) = 1 + \log_4\left(\frac{x+2}{3}\right)$, defined for $x > -2$. Inputting this to Y1 and $\frac{1}{2}x - 2$ into Y2, we graph and find the intersection points using 2nd trace (calc) and 5:intersect. The intersection points are $(-1.988, -2.994)$ and $(7.692, 1.846)$, so our solutions are $x = -1.988$ and $x = 7.692$.

Example 5.

(a) Because $f(x) = b^x$ and $g(x) = \log_b x$ are inverse functions, we have $f(g(x)) = b^{\log_b x} = x$. Therefore,

$$c^x = b^{\log_b c^x} = b^{x \log_b c}$$

where the last equality comes from the Power Property.

(b) Noticing that 9 and 27 are powers of 3, we can rewrite each side using a base of 3.

$$9^{(2x+3)} = 27^{(x-1)}$$

$$3^{\log_3 9^{(2x+3)}} = 3^{\log_3 27^{(x-1)}} \qquad \text{(Using relationship from (a))}$$

$$3^{(2x+3)\log_3 9} = 3^{(x-1)\log_3 27} \qquad \text{(Using Power Property)}$$

$$3^{(2x+3)\cdot 2} = 3^{(x-1)\cdot 3} \qquad \text{(Evaluating logarithms)}$$

$$3^{4x+6} = 3^{3x-3} \qquad \text{(Distribute)}$$

$$4x + 6 = 3x - 3 \qquad \text{(Setting exponents equal)}$$

$$x = -9$$

Example 6. On the left side, we can use the Product Property. On the right, we can rewrite $2 = \log_2 4$. This gives

$$\log_2((x-2)(x-5)) = \log_2 4$$

$$\log_2(x^2 - 7x + 10) = \log_2 4 \qquad \text{(Expand)}$$

$$x^2 - 7x + 10 = 4 \qquad \text{(Set arguments equal)}$$

$$x^2 - 7x + 6 = 0 \qquad \text{(Subtract 4)}$$

$$(x-6)(x-1) = 0 \qquad \text{(Factor)}$$

This gives solutions of $x = 6$ and $x = 1$. However, the domain of $\log_2(x-2)$ is $x > 2$ and the domain of $\log_2(x-5)$ is $x > 5$, and only $x = 6$ is within both of these domains. The only solution is $x = 6$.

Example 7. We write the inequality and solve.

$$\log(2x+1) - \log(x^2 + 2x - 3) < 0$$

$$\log(2x+1) < \log(x^2 + 2x - 3) \qquad \text{(Add } \log(x^2 - 2x - 3)\text{)}$$

$$2x + 1 < x^2 + 2x - 3 \qquad \text{(Utilize arguments since bases are the same)}$$

$$0 < x^2 - 4 \qquad \text{(Subtract } 2x \text{ and 1)}$$

$$0 < (x-2)(x+2)$$

The function $g(x) = x^2 - 4$ is an upward opening parabola with zeros of $x = 2$ and $x = -2$. Therefore, $g(x) > 0$ for $x < -2$ and $x > 2$. However, we must consider the domain of f. The first logarithm has a domain of $x > -\frac{1}{2}$, and the second logarithm has a domain of $x < -3$ and $x > 1$ (can you see why?). The domain of f is therefore $x > 1$. The only overlapping interval of our solution is $x > 2$.

Homework

1. We set the arguments equal to get $x^2 + 5x + 6 = x + 3$, which simplifies to $(x+3)(x+1) = 0$. This has solutions $x = -3$ and $x = -1$. The domain of $\log_3(x^2 + 5x + 6)$ is $x < -3$ and $x > -2$, and the domain of $\log_3(x+3)$ is $x > -3$. $x = -3$ is not contained in these domains, while $x = -1$ is contained within both, so the only solution is $x = -1$.

2. (a) $y = 4e^{2x} - 3 \Rightarrow x = 4e^{2y} - 3 \Rightarrow \frac{x+3}{4} = e^{2y} \Rightarrow \ln\left(\frac{x+3}{4}\right) = 2y \Rightarrow y = f^{-1}(x) = \frac{1}{2}\ln\frac{x+3}{4}$.

   (b) $y = \frac{1}{4}\log_3(x-1) \Rightarrow x = \frac{1}{4}\log_3(y-1) \Rightarrow 4x = \log_3(y-1) \Rightarrow 3^{4x} = y - 1 \Rightarrow y = g^{-1}(x) = 3^{4x} + 1$.

   (c) $y = \log_2(\log_4 x)) - 1 \Rightarrow x = \log_2(\log_4 y)) = 1 \Rightarrow x - 1 = \log_2(\log_4 y)) \Rightarrow 2^{(x-1)} = \log_4 y \Rightarrow y = h^{-1}(x) = 4^{(2^{(x-1)})}$.

3. Shown below.

$$\log_{10} 10 = \log_{10}((x+3)(x-6))$$
$$10 = x^2 - 3x - 18$$
$$0 = x^2 - 3x - 28$$
$$0 = (x-7)(x+4)$$

The solutions are $x = 7$ and $x = -4$, but the values of $\log_{10}(x+3)$ and $\log_{10}(x-6)$ are only both defined for $x = 7$, so this is the only solution.

4. Shown below.

$$4^{(2x+1)} = 8^{(1-2x)}$$
$$2^{\log_2 4^{(2x+1)}} = 2^{\log_2 8^{(1-2x)}}$$
$$2^{(2x+1)\log_2 4} + 2^{(1-2x)\log_2 8}$$
$$2^{(2x+1)\cdot 2} + 2^{(1-2x)\cdot 3}$$
$$4x + 2 = 3 - 6x$$
$$10x = 1$$
$$x = \frac{1}{10}$$

5. $x^2 + 4x - 21 = x^2 - 3x$ gives $7x - 21 = 0$, or $x = 3$. However, the values of $x^2 + 4x - 21$ and $x^2 - 3x$ are both 0 when $x = 3$, making both logarithmic expressions undefined at $x = 3$.

6. $f(x) = \dfrac{\log_2 8^{2x}}{\log_2 4} = \dfrac{2x \log_2 8}{2} = 3x$. The slope of the line is 3.

Reflection

Suggested Class Time. 70 minutes

Prerequisites. Most of Unit 2 is required here, and students also need to be comfortable solving polynomial inequalities (particularly quadratics).

UNIT 2   TOPIC 2.13 ~ EXPONENTIAL AND LOGARITHMIC EQUATIONS AND INEQUALITIES   237

Instructional Strategies. We spent a good while going over the Topic 2.12 homework, particularly numbers 4, 5, 6, and 8. What I really needed to emphasize with my students was a statement I make quite often in all of the classes I teach.

Math is about turning what you don't know how to do into what you do know how to do.

In particular, for questions like Problem 4 of the 2.12 homework, we discussed how solving such an equation for b involving turning what we *didn't know* – $\log_b$ – into what we *do know*, $\log_4 x$. We used change of base and quickly arrived at a solution.

This popped up again in Problem 5b, where we discussed what we didn't know – how to deal with exponentials of different bases – into what we did know, how to deal with exponentials with the same base. We ended up working through both versions of the solution to Problem 5b (using only traditional exponent rules, as well as the property from Problem 5a).

Most of the examples went smoothly enough, though it was admittedly a lot for students to handle. These questions are challenging, and a lot of considerations need to be made, especially with domains. I taught my students the term "argument" for a logarithm, so they were trained that *the argument must be greater than 0*. This helped them avoid the common traps of simply thinking that x needs to be greater than 0 or not thinking that they need to do any solving to find the domain.

Going back to Problem 5a, the property appears so convoluted that it's easy for its meaning and power to be lost on students. I wrote a simple example involving 5^x that showed $5^x = e^{x \ln 5}$, and we discussed that this enabled us to rewrite an exponential in terms of *another* exponential: in other words, this is the change of base formula *for exponentials*.

For Problem 6, one thing surprised me. Once we had used the product property to write $\log_2(x^2 - 7x + 10) = 2$, I thought students would most easily see that, by the definition of a logarithm, this simply meant that $x^2 - 7x + 10 = 2^2$. However, students were much keener to accept us *rewriting 2 as a log base 2 value*, i.e. $2 = \log_2 4$. I'm not sure if your students will find this easier as well, but I certainly did not see it coming.

Technology. Students need the TI-84 for one problem, and it helps if they have an operating system new enough to contain the LOGBASE(feature once they select math.

Problems.

1. A student is solving the equation $5^x = 10^{3x-1}$. They do not have their calculator, so they have to rely on a table of logarithms which provides values of $\log_{10} x$ for various x. Solving which of the following equations would enable the student to solve the original equation?

   (A) $\frac{1}{2}x = 3x - 1$ (B) $x \log_{10} \frac{1}{5} = 3x - 1$ (C) $x \log_{10} 5 = 3x - 1$ (D) $x + \log_{10} 5 = 3x - 1$

2. Which of the following is an expression for the inverse of $f(x) = 3^{2x} + 1$, along with its domain?

   (A) $f^{-1}(x) = \frac{1}{2} \log_3(x-1)$ for $x > 0$ (B) $f^{-1}(x) = \frac{1}{2} \log_3(x-1)$ for $x > 1$

   (C) $f^{-1}(x) = \log_3\left(\frac{x-1}{2}\right)$ for $x > 0$ (D) $f^{-1}(x) = \log_3\left(\frac{x-1}{2}\right)$ for $x > 1$

3. The functions f and g are given by $f(x) = \log_3(x-2)$ and $g(x) = \log_3(2x-1)$. Which of the following is all x for which $f(x) + g(x) > 2$?

   (A) $x > 3.5$ only   (B) $x < -1$ only   (C) $x > 3.5$ and $x < -1$ (D) $-1 < x < 3.5$

4. The functions f and g are given by $f(x) = \log_2(x-2) + \log_2(x+4)$ and $g(x) = \log_2(x+12)$, respectively. In the xy-plane, what are all x-coordinates of the points of intersection of the graphs of f and g?

   (A) $x = -10$ and $x = -5$ (B) $x = -5$ and $x = 4$ (C) $x = 4$ only   (D) $x = 10$ only

5. The function h is given by $h(x) = \log_5\left(\dfrac{125}{x^2}\right)$. Which of the following is an equation that gives the solution to $h(x) = 13$?

(A) $3 - \log_5\left(\dfrac{x}{2}\right) = 13$ 　 (B) $3 - 2\log_5 x = 13$ 　 (C) $\dfrac{2\ln x}{5} - 3 = 13$ 　 (D) $2\log_5 x - 3 = 13$

Solutions.

1. We take the logarithm base 10 of both sides. This yields $\log_{10} 5^x = 3x - 1$. Then by the Power Property, $x \log_{10} 5 = 3x - 1$. The correct answer is therefore **(C)**.

2. Let $y = f(x)$. Then interchanging x and y gives us $x = 3^{2y} + 1$. Subtracting 1 then taking the logarithm base 3 of both sides gives us $\log_3(x - 1) = 2y$. Finally, $f^{-1}(x) = \frac{1}{2}\log_3(x - 1)$. The domain of the inverse is the range of the f, which is all reals greater than 1. The correct answer is therefore **(B)**.

3. We have $f(x) + g(x) = \log_3((x - 2)(2x - 1)) = \log_3(2x^2 - 5x + 2)$. For this to be greater than 2, the argument of the logarithm must be greater than $3^2 = 9$. Hence, $2x^2 - 5x + 2 > 9$ implies $2x^2 - 5x - 7 > 0$. This factors as $(2x - 7)(x + 1) > 0$. This polynomial is positive for $x < -1$ and $x > 3.5$. However, only $x > 3.5$ satisfies the domains of both f or g. Hence, the solution is $x > 3.5$ only. The correct answer is therefore **(A)**.

4. We set f equal to g and solve. We have

$$\log_2(x - 2) + \log_2(x + 4) = \log_2(x + 12)$$
$$\log_2((x - 2)(x + 4)) = \log_2(x + 12)$$
$$x^2 + 2x - 8 = x + 12$$
$$x^2 + x - 20 = 0$$
$$(x + 5)(x - 4) = 0.$$

We reject $x = -5$ as a solution since it does not satisfy the domain of f. The only solution is $x = 4$. The correct answer is therefore **(C)**.

5. The function h can be rewritten as

$$h(x) = \log_5\left(\dfrac{125}{x^2}\right) = \log_5 125 - \log_5(x^2) = 3 - 2\log_5 x.$$

The correct answer is therefore **(B)**.

Topic 2.14 ~ Constructing a Logarithmic Model

Learning Objectives
1. (2.14.A) Construct a logarithmic function model.

Success Criteria
1. I can use properties of logarithms and their graphs, as well as contextual clues, to construct a logarithmic function model.
2. I can perform logarithmic regression and use it to make predictions.

Logarithmic functions appear in a variety of contexts, both on their own and as simply the inverses of exponential functions.

For starters, we'll begin by creating a logarithmic function to model a given data set. A logarithmic function can be generated by _____ points.

Example 1. Create a model $L(x) = a \log_b x + c$ that would pass through the points $(8, 5)$ and $(16, 6)$.

Solution. If we know the function is logarithmic, then we know that the outputs change _____ while the inputs change _____. Therefore, we could use the given points and notice that as the outputs change by ____, the inputs change by a factor of ____. Therefore, a reasonable base for our logarithm would be ____. To find the transformations a and c needed, we can consider the parent function $f(x) = \log_2 x$. For the input values of 8 and 16, we would have $\log_2 8 =$ ___ and $\log_2 16 =$ ___. Looking back at our original desired outputs, we can see that a vertical translation of ____ will suffice. Therefore, $L(x) =$ _____.

Practice 1. Find a logarithmic function $L(x) = a \log_b x + c$ that would contain the input-output pairs $\left(3, \frac{1}{2}\right)$ and $\left(27, \frac{3}{2}\right)$.

Practice 2. Find a logarithmic function $L(x) = a \log_b(x - c)$ that has a zero of $x = 4$ and such that $L(7) = 1$.

Oftentimes, the two (or more) points we are given are not so convenient that we can find a model by hand. For this, we rely on _____. In the calculator, we can actually perform *logarithmic regression*, but one thing worth noting is that it is *specifically* of the form $y = $ _____. Converting to any other base would require using the _____. To access the logarithmic regression, you press `stat`, `CALC`, and `9:LnReg`.

Practice 3. Matthew decided he wanted to lose weight, so he started on a new diet and exercise plan. He tracked how much total weight he had lost every two weeks for 24 weeks. His total weight loss W, in pounds, after t weeks on the diet and exercise plan are given below.

t	2	4	6	8	10	12	14	16	18	20	22	24
W	15	24.5	29	31	33	37	39	39.5	41	42	44	45

Logarithmic regression was performed to create a model $W(t) = a + b \ln t$. Based on this model, how much longer does Matthew need to be on diet and exercise plan to get his weight loss to 50 pounds?

Logarithmic functions can of course arise as the inverses of exponential functions.

Example 2. Suppose a factoring algorithm takes 800 hours to decode a 40-bit encrypted message, and that for every additional bit of encryption, the algorithm will take twice as long. Write a function $b(t)$ that will give the number of bits of encryption as a function of the time it takes to decode it.

Practice 4. Marina has two jobs. At one of them, she makes \$30,000 a year, while at the other, she makes \$45,000 a year. At the higher-paying job, Marina also gets a raise of 1.5% each year, starting this year. Write a function $t(s)$ that would give output values, in years t, for input values of Marina's salary s.

Topic 2.14 Homework

1. In a certain wildlife reserve, the number of acres populated by tule elk increased by 25% every year from 2018 to 2023. In 2018, the tule elk only populated roughly 1,000 acres of the wildlife reserve.

    (a) Write a function $y(a)$ for the year y since 2018 based on the amount of acres covered by tule elk.

    (b) (▤) Researchers measured at one point that the tule elk populated more than 1,500 acres. At a later point, the researchers measured that the number had grow to 2,000 acres. How long was it between these two measurements?

2. Construct a logarithmic function $L(x)$ passing through each pair of points. (To be clear, a different logarithmic function will be written for each part.)

    (a) $(4,3)$ and $(8,4)$

    (b) $(4,0)$ and $(12,2)$

    (c) $(e^2, 5)$ and $(e^4, 9)$

3. (▤) A teacher was analyzing the statewide end of course exam scores for their 8th grade students and comparing them to their 7th grade scores on the similar statewide end of course exam. The scores on each exam for each of 8 students are shown below.

7th grade score	68	72	76	81	86	91	97
8th grade score	87	89	91	93	95	97	99

    (a) If we were to make a function to model this data set, which variable would make the most sense to be the input and which would make the most sense to be the output? Explain.

    (b) Use the values in the table to explain why a logarithmic function might be an appropriate model for the variables you identified in (a).

    (c) Using your answer from (a), perform logarithmic regression to construct a function $f(x) = a + b \ln x$ that could be used to model the data.

    (d) Describe the rate of change and the rate of rate of change of f for $68 \le x \le 97$. Explain why these make sense in context.

    (e) Assuming the model is valid for all $0 < x < 100$, what would a student's 7th grade score need to have been to predict getting at least a 70 on the 8th grade exam?

Topic 2.14 Solutions/Notes

Activity
A logarithmic function can be generated by <u>two</u> points.

Example 1. If we know the function is logarithmic, then we know that the outputs change <u>multiplicatively</u> while the inputs change <u>additively</u>. Therefore, we could use the given points and notice that as the outputs change by <u>1</u>, the inputs change by a factor of 2. Therefore, a reasonable base for our logarithm would be <u>2</u>. To find the transformations a and c needed, we can consider the parent function $f(x) = \log_2 x$. For the input values of 8 and 16, we would have $\log_2 8 = \underline{3}$ and $\log_2 16 = \underline{4}$. Looking back at our original desired outputs, we can see that a vertical translation of $\underline{+2}$ will suffice. Therefore, $L(x) = \underline{\log_2 x + 2}$.

Practice 1. There are numerous correct answers. While we could utilize the fact that the inputs multiply by 9 as the outputs increase by 1, we notice that 3 and 27 both have convenient $\log_3$ values. In fact, $\log_3 3 = 1$ and $\log_3 27 = 3$. Looking at the outputs of $\frac{1}{2}$ and $\frac{3}{2}$, we notice these are just $\frac{1}{2}$ times the $\log_3$ outputs. Therefore, $f(x) = \frac{1}{2} \log_3 x$.

Practice 2. The function $f(x) = \log_b x$ will have a zero of $x = 1$ for any b ($b > 0, b \neq 1$), so we want to translate this horizontally by $+3$. This gives us $L(x) = \log_b(x - 3)$. Now, we want to find a suitable base. If we plug in 7, we need to get an output of 1. The simplest thing to do is solve $\log_b(7 - 3) = 1$, which gives $b = 4$. A suitable function is therefore $L(x) = \log_4(x - 3)$.

Oftentimes, the two (or more) points we are given are not so convenient that we can find a model by hand. For this, we rely on <u>regression</u>. In the calculator, we can actually perform *logarithmic regression*, but one thing worth noting is that it is *specifically* of the form $y = a + b \ln x$. Converting to any other base would require using the <u>change of base formula</u>.

Practice 3. We utilize `LnReg` and make sure to `Store RegEQ:Y`$_1$. Then, we graph Y$_2$=50, change the `window` accordingly, and find the intersection of Y$_1$ and Y$_2$. The intersection occurs at $x = 37.926$. Because the problem asks for how *much longer*, we would subtract 24 from this, giving us an answer of approximately 13.926 more weeks.

Example 2. The function $b(t)$ we need to find is the inverse of the function $t(b)$. From the context, we can construct this function $t(b)$ for the number of hours it would take to encode a b-bit encrypted message. We know $t(b)$ contains $(800, 40)$ and has a constant proportion of 2, so we can write the exponential function $t(b) = 800 \cdot 2^{b-40}$. To write the function $b(t)$, we simply need the inverse of $t(b)$, which is $t^{-1}(b)$. Hence, we interchange b and t, and solve for t.

$$b = 800 \cdot 2^{t-40}$$
$$\frac{b}{800} = 2^{t-40}$$
$$\log_2\left(\frac{b}{800}\right) = t - 40$$
$$40 + \log_2\left(\frac{b}{800}\right) = t$$

So $t^{-1}(b) = 40 + \log_2\left(\frac{b}{800}\right)$. Therefore $b(t) = 40 + \log_2\left(\frac{t}{800}\right)$.

Practice 4. Marina's overall yearly income in thousands of dollars can be modeled by $s(t) = 30 + 45 \cdot 1.015^t$. To write the function $t(s)$, we find the inverse of $s(t)$. Thus, we interchange s and t, and solve for s.

$$t = 30 + 45 \cdot 1.015^s$$
$$t - 30 = 45 \cdot 1.015^s$$
$$\frac{t-30}{45} = 1.015^s$$
$$\log_{1.015} \frac{t-30}{45} = s$$

So $s^{-1}(t) = \log_{1.015} \frac{t-30}{45}$. Therefore, $t(s) = \log_{1.015} \frac{s-30}{45}$ will output the year t based on Marina's overall yearly income s.

Homework

1. (a) The acreage based on the years since 2018 can be modeled by $a(y) = 1,000 \cdot 1.25^y$. The function $y(a)$ is simply the inverse of $a(y)$, which is $y(a) = \log_{1.25}\left(\frac{a}{1000}\right)$.
   (b) Ordinarily, we would solve this by graphing the function $a(y)$ and the horizontal lines at 1,000 and 2,000, then using the intersect feature of our calculators. Because we have the function $y(a)$ that will immediately output years, though, we can simply compute $y(2,000) - y(1,500) = 1.289$. The measurements occurred about 1.289 years apart.

2. (a) As x increases by a factor of 2, y increases by 1. A base of 2 seems reasonable, but $\log_2 4 = 2$ and $\log_2 8 = 3$. We can add a vertical translation of $+1$ to get a suitable function $L(x) = 1 + \log_2 x$.
   (b) Here, we have an x-intercept. The parent function $\log_b x$ has a zero of $(1,0)$, so a horizontal translation of 3 will have a zero of $(4,0)$. This starts us off with $\log_b(x-3)$. Now, it's a matter of finding a base such that $L(12) = 2$. Plugging in 12, we get $\log_b(12-3) = 2$, so we get $b^2 = 9$ or $b = 3$. The final function is $L(x) = \log_3(x-3)$.
   (c) As x increases by a factor of e^2, y increases by 4. Since L is logarithmic, we can deduce that an increase of x by a factor of e would lead to an increase of y by 2. The function $f(x) = 2\ln x$ would satisfy this property. However, to ensure the graph passes through $(e^2, 5)$ and $(e^4, 9)$, we must add 1 (try it for yourselves). The final function is $f(x) = 1 + 2\ln x$.

3. (a) Unless we're going back in time, it would make more sense to use the 7th grade scores as inputs to predict the 8th grade scores as outputs.
   (b) The ratios between consecutive terms over intervals of length 2 in the outputs are relatively constant, meaning the inputs are proportional over consecutive equal-length intervals of the outputs.

Ratio of 7th		1.059	1.056	1.066	1.062	1.058	1.066
7th grade score	68	72	76	81	86	91	97
8th grade score	87	89	91	93	95	97	99

   (c) $f(x) = -55.401 + 33.770 \ln x$
   (d) f is increasing, which makes sense because, generally, students who scored higher in 7th grade will likely score higher in 8th grade based on general mathematical ability, work ethic, prior education, etc. The graph is, however, increasing at a *decreasing rate*. This makes sense because there is less and less room for improvement the higher a student goes; the output values cannot exceed 100. On the other hand, students with lower 7th grade scores have much more room to improve, which explains some of the larger gains (and steeper part of the graph).
   (e) The graph of $f(x)$ and $y = 70$ intersect at $(40.992, 70)$. Therefore, a student would need approximately a 41 to have the model predict they will score at least a 70 in 8th grade.

Reflection

Suggested Class Time. 50-60 minutes

Prerequisites. Again, most of Topic 2 is used in this lesson.

Instructional Strategies. This lesson was very simple, and I have very few comments accordingly. It went about as expected.

Before we started the lesson, I made sure to help students zoom out a bit and gain some perspective about where we are in the course and what this Topic would be about: we are studying how variables change together, and up to this point, we had studied linear, quadratic, other polynomial, rational, and exponential functions. All of those have unique relationships between input and output variables and can be used to create regression functions to model a given relationship. We discussed common contexts for each of the aforementioned functions, and I asked students for some sample contexts that might be modeled by logarithmic functions. One that came up was population, which was interesting - later in the lesson, a few students mistakenly chose a *logistic* regression in the calculator, and I gave a very short lesson on what logistic functions look like and how they are somewhat of a combination of the exponential and logarithmic growth seen early on and much later, respectively in the growth of many populations. One that I brought up was Netflix, which had multiple quarters in a row where the number of subscribers added decreased; we talked about how this made sense, as there is a limited population from which to draw subscribers. (Of course, since this initial writing, Netflix has again been adding subscribers at an increasing rate!)

Otherwise, the lesson was pretty simple and seemed effective.

Technology. Students needed a graphing calculator for the third practice question.

Problems.

1. (■) Four equally rated professional weightlifters agreed to participate in an experiment where they would have different amounts of time to train for a deadlift competition. The amount of time trained, in months, and the maximum deadlifts, in pounds, are listed below.

Time training	1	3	6	10
Maximum deadlift	780	915	960	992

   Logarithmic regression was performed to create a function $f(x) = a + b \ln x$ to predict maximum deadlift based on months training. Based on the regression, how long would an equally rated weightlifter theoretically need to train to deadlift 1,100 pounds?

   (A) 28.161 months   **(B)** 145.573 months   **(C)** 1438.263 months   **(D)** 2,177.489 months

2. The table below gives values of x and y.

x	5	9	17
y	2	3	4

   Which of the following functions f could be used to model y based on x?

   (A) $f(x) = \log_2(x+1)$ **(B)** $f(x) = \log_2(x-1)$ **(C)** $f(x) = \log_2 x + 1$ **(D)** $f(x) = \log_2 x - 1$

3. (■) Two function models, k and p, are constructed to represent the population of king penguins on an island off the coast of Argentina. Both $k(t) = 0.62 + 3.68 \ln t$ and $p(t) = 0.62 + 0.28t$ represent the population of king penguins in hundreds, for $t \geq 1$. What is the first time t that the population predicted by the linear model will be 100 more than the population predicted by the logarithmic model?

   (A) 47.042   (B) 51.907   (C) 56.621   (D) 437.052

4. The Richter scale measures the magnitude of earthquakes, which is related to how much energy is released by the earthquake. The Richter scale uses a base 10 logarithm to model the magnitude. The highest Richter value ever recorded was 9.5 in 1960 in Chilé. In 2024, an earthquake struck Alaska with a Richter value of 4.7. Approximately how many times greater was the Chiléan earthquake than the Alaskan earthquake?

   (A) 2   (B) 5   (C) 100,000   (D) 1,000,000,000

5. (■) The table below shows the average life expectancies, in years, of Americans from 1900-2020.

Year	1900	1910	1920	1930	1940	1950	1960
Life Expectancy	47	50	54	60	63	68	70
Year	1970	1980	1990	2000	2010	2020	
Life Expectancy	71	74	75	77	79	77	

   A logarithmic regression $y = a + b \ln x$ is used to model the data, where $x = 1900$ corresponds to the year 1900. What is the life expectancy of Americans predicted to be in the year 2040 according to the model?

   (A) 82   (B) 84   (C) 86   (D) 88

Solutions.

1. Performing the logarithmic regression and storing it in the calculator, we find that the intersection with the line $y = 1100$ occurs when $x = 28.161$. The correct answer is therefore **(A)**.

2. Note that if we subtract 1 from the inputs, we get 4, 8, and 16. This suggests that the ouputs change by 1 and the inputs minus 1 change by factor of 2. So a reasonable base is 2, and the function that best models y is $f(x) = \log_2(x-1)$. The correct answer is therefore **(B)**.

3. Since we want k to be 1 hundred more than p, we need $k(t) + 1$ to be equal to $p(t)$. Graphing $k(t) + 1$ and $p(t)$ in the calculator, we find the intersection point given at $t = 56.621$. The correct answer is therefore **(C)**.

4. The Richter values are the results of base 10 logarithms, so to determine how many times larger, we take $10^{9.5}$ and $10^{4.7}$ and divide them. This gives $10^{9.5-4.7} = 10^{4.8} \approx 10^5 = 100,000$. The correct answer is therefore **(C)**.

5. Using the calculator, we enter the data and run a logarithmic regression. If the years are entered so that $x = 1$ corresponds to the year 1900, then the regression equation is $y = -3916.806 + 525.4713 \ln x$. The value $x = 2040$ then gives a predicted value of 87.655. The correct answer is therefore **(D)**.

Topic 2.15 ~ Semi-log Plots

Learning Objectives

1. (2.15.A) Determine if an exponential model is appropriate by examining a semi-log plot of a data set.
2. (2.15.B) Construct the linearization of exponential data.

Success Criteria

1. I can analyze a semi-log plot to determine the appropriate of an exponential function model or find an expression for an exponential function model.
2. I can linearize exponential data and find an exponential function given a known linearization of exponential data.

We already discussed in a previous topic the idea of a *logarithmic scale*. The entire purpose of a logarithmic scale is to make values that are increasing _____ increase instead at a more palatable, _____ pace. Let's investigate this analytically first.

Suppose we have the function $y = 2 \cdot 3^x$. If we take the logarithm base 10 of both sides, we get

And here it is! When we take an exponential function $y = a \cdot b^x$ and take the logarithm of both sides, what results is a function of the form _____ – which is _____!

Graphically, we are taking a logarithm of y while keeping x the same: this essentially _____ the rapidly growing y-values that make the function curved to begin with and slows them down to increase at a _____ rate.

When a graph plots x versus $\log_b y$, it is called a _____.

Here's a regular graph of $y = 2 \cdot 3^x$, along with a semi-log plot of the same function.

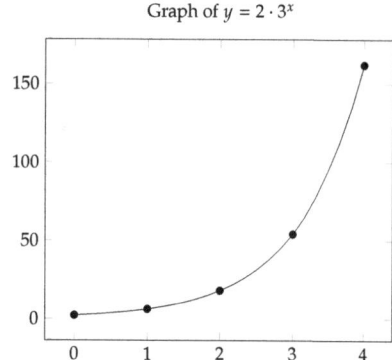
Graph of $y = 2 \cdot 3^x$

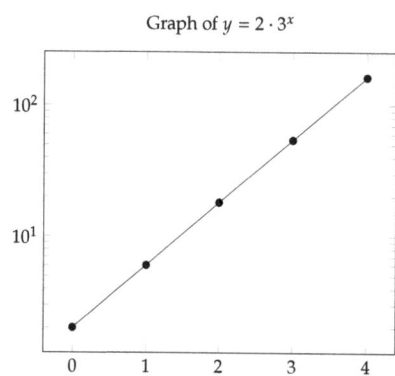
Graph of $y = 2 \cdot 3^x$

Fact: When the graph of an exponential function $E(x)$ is graphed in a semi-log plot, the resulting graph of $E(x)$ will be _____. As a result of taking the logarithm of an exponential function $y = f(x)$, both x and $\log_b y$ will form _____ for any $b > 0, b \neq 1$.

Reading semi-log plots takes a bit of getting used to, as the y-axis is on a _____ scale.

Example 1. The population of rabbits in a wildlife reserve are graphed in the semi-log plot below, where $t = 0$ represents when they were first released and $t = 2.5$ is two and a half years later.

(a) What does the semi-log plot reveal about how the population of rabbits was increasing?

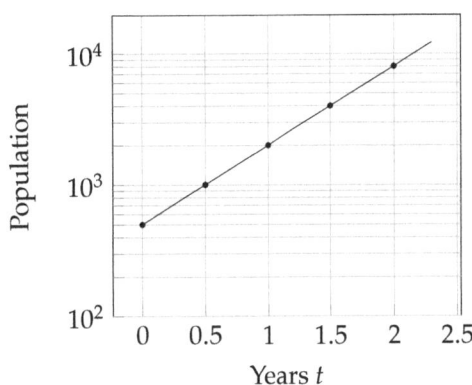

(b) Find an expression for $P(t)$, the population of rabbits t years after they were first released.

Example 2. (🖩) The function $g(x)$ can be used to model a country's GDP g, in billions of dollars, where x is in years since 2019 (so $x = 0$ represents the GDP in 2019). When graphed in a semi-log plot where the y-axis is on a logarithmic scale of base e, the function g forms the linear function $y = 0.693 + 0.0488x$.

(a) Interpret the slope of the line in the context of the country's GDP.

(b) What was the country's GDP in 2019?

(c) Write an expression for the function $g(x)$.

Practice 1. The function $y = f(x)$ is exponential. Values of x and $\ln y$ are given in the table below.

x	1	3	5	7	9
$\ln y$	4				10

Fill in the missing values in the table.

Practice 2. Let f be the function given by $f(x) = 25 \cdot 100^{3x}$.

(a) If f is graphed in a semi-log plot with the y-axis on a logarithmic scale of base 10, what is an expression for the transformed f?

(b) (🖩) If f were instead graphed in a semi-log plot with the y-axis on a logarithmic scale of base e, what is an expression for the transformed f?

Practice 3. Given the table of values for x and $\log_2(f(x))$ in the table below, sketch a graph of the function f in the xy-plane provided.

x	0	1	2	3
$\log_2(f(x))$	−1	0	1	2

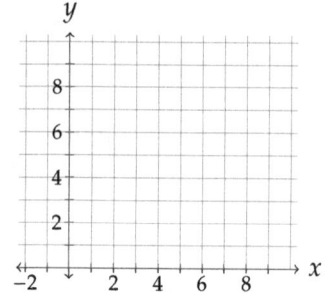

Topic 2.15 Homework

Work through the Unit 2B Progress Check in preparation for the Test Review.

Topic 2.15 Solutions/Notes

Activity

We already discussed in a previous topic the idea of a *logarithmic scale*. The entire purpose of a logarithmic scale is to make values that are increasing exponentially increase instead at a more palatable, linear pace. Let's investigate this analytically first.

Suppose we have the function $y = 2 \cdot 3^x$. If we take the logarithm base 10 of both sides, we get

$$\log_{10} y = \log_{10}(2 \cdot 3^x)$$
$$\log_{10} y = \log_{10} 2 + \log_{10} 3^x$$
$$\log_{10} y = \log_{10} 2 + x \cdot \log_{10} 3$$

And here it is! When we take an exponential function $y = a \cdot b^x$ and take the logarithm of both sides, what results is a function of the form $y = a + bx$ – which is linear!

Graphically, we are taking a logarithm of y while keeping x the same: this essentially slows down the rapidly growing y-values that make the function curved to begin with and slows them down to increase at a linear rate.

When a graph plots x versus $\log_b y$, it is called a semi-log plot.

Fact: When the graph of an exponential function $E(x)$ is graphed in a semi-log plot, the resulting graph of $E(x)$ will be linear. As a result of taking the logarithm of an exponential function $y = f(x)$, both x and $\log_b y$ will form arithmetic sequences for any $b > 0, b \neq 1$.

Reading semi-log plots takes a bit of getting used to, as the y-axis is on a logarithmic scale.

Example 1.

(a) Because the graph in the semi-log plot is linear, the population of rabbits was growing exponentially.

(b) From the points $(0, 500), (1, 2000),$ and $(2, 8000)$, we can see that the population of rabbits quadruples every year. Along with the initial value, we get $P(t) = 500 \cdot 4^t$.

Example 2.

(a) For every additional year, the country's GDP is expected to grow by a factor of $e^{0.0488} \approx 1.05$, or roughly 5%.

(b) $e^{0.693} \approx 2$, so the country's GDP is 2019 was about 2 billion dollars.

(c) $g(x) = 2 \cdot 1.05^x$

Practice 1. The values of $\ln y$ must form an arithmetic sequence with an average rate of change of $\frac{10-4}{9-1} = 0.75$. Over each successive interval of length 2 in the inputs, this corresponds to an increase of $2(0.75) = 1.5$. The missing values are therefore, in increasing order, 5.5, 7, and 8.5.

Practice 2.

(a) $\log f(x) = \log(25 \cdot 100^{3x}) \Rightarrow y = \log 25 + \log 100^{3x} \Rightarrow y = \log 25 + 3x \cdot \log 100 = \log 25 + 6x$

(b) $y = \ln 25 + 3 \ln 100 \cdot x$

Practice 3. We get $\log_2(f(0)) = -1$, so $f(0) = \frac{1}{2}$. From here, the inputs and outputs each form arithmetic sequences with a rate of change of 1, so f is exponential with a base equivalent to the base of the logarithm (2). The graph is below.

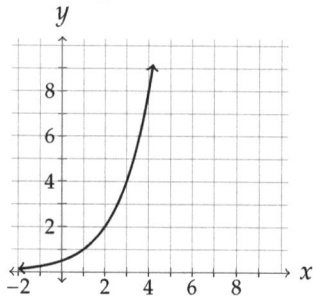

Reflection

Suggested Class Time. 60 minutes

Prerequisites. The primary prerequisite here is logarithmic scales.

Instructional Strategies. My way of teaching semi-log plots is to help students understand that they make exponential functions linear. We showed this analytically, but I think it really helps students to "see it." This is why I make a big deal of being animated, showing students how a semi-log plot is essentially "smushing" the y-axis so that what was exponential can be linear. (Smushing is not a technical term, and it does seem to imply that we're simply compressing, which is not what's happening, but the animated description of it is worthwhile all the same, as it helps students understand the purpose of the semi-log plot.)

When showing the first semi-log plot in the lesson (the graph of $y = 2 \cdot 3^x$), I made a point to talk through all the tick marks along the y-axis. We discussed why there were only 8 tick marks before 10 as opposed to 9 (because $\log_{10} 1 = 0$) and we notated each one (2, 3, 4, and so on). We did the same for the next 8 tick marks, labeling 200, 300, 400, and so on.

The examples are relatively straightforward, though there are a couple different ways of going through Example 2. One way is to refer back to the original linearization of $y = a \cdot b^x$ into $\ln y = \ln a + x \ln b$. Using this, you can get a one-to-one correspondence between the y-intercepts and the slopes to identify a and b. Another way, which is more informal, but also more intuitive, is to simply talk about the fact that we need to "re-exponentiate" to determine a and b. This skips the equation part and jumps straight to $a = e^{0.693}$ and $b = e^{0.0488}$.

I gave students about 20 minutes to do the 3 practice questions, and they struggled with both the first and the third. I expected the latter struggle, but I was really surprised with their struggle with the first. I think the analytical/verbal presentation just doesn't attract students' attention to the fact that the missing values are of a logarithm of an exponential function. I helped draw students' attention to this by writing the informal equation log exponential = linear. This seemed to help! When students were given a Topic Question that related to this exact same idea (and style of presentation) about 83% of my students it got it correct.

For Practice Problem 3, we first worked intuitively, using a base of 2 and then testing inputs and outputs to get the desired function. Instead, though, I realized that perhaps a better solution (which I showed students) was to just directly construct the linear function, which is $y = -1 + x = x - 1$, and then to solve as follows:

$$\log_2(f(x)) = x - 1$$
$$2^{\log_2(f(x))} = 2^{x-1}$$
$$f(x) = 2^{x-1}$$

Students seemed more comfortable with this method.

Overall, while people seem to be scared of semi-log plots, the logic behind their creation is quite intuitive: they are the graphical form of a logarithmic scale. If presented as such, students will find them just as intuitive.

Technology. We used a calculator for a couple of the questions.

Problems.

1. The function $g(x) = 3 \cdot 8^x$ is going to be graphed on a semi-log plot where the y-axis is on a logarithmic scale with base 10. On the semi-log plot, which of the following describes the graph of g?

    (A) g will be linear with a slope of $\log_{10} 3$.

    (B) g will be linear with a slope of $\log_{10} 8$.

    (C) g will be exponential with a constant ratio of 5.

    (D) g will be linear with a constant ratio of 8.

2. Since the early 1970s, the number of transistors in microprocessors has increased. A semi-log plot of T, the number of transistors, in year x since 1970 is shown below.

    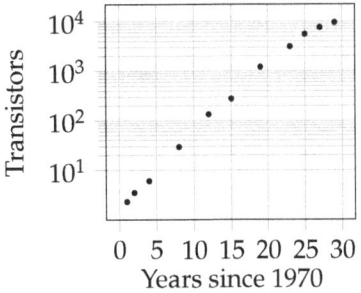

    Which of the following provides the correct kind of model that could be used to model $T(x)$ along with correct reasoning?

    (A) $T(x)$ could be modeled by a linear function because the semi-log plot is reasonably linear.

    (B) $T(x)$ could be modeled by an exponential function because the semi-log plot is reasonably linear.

    (C) $T(x)$ could be modeled by a logarithmic function because the semi-log plot is reasonably linear.

    (D) $T(x)$ could be modeled by a quadratic function because the semi-log plot is reasonably linear.

3. (🔳) The function h, given by $h(x) = 10^{(0.6x+7.2)}$, is graphed on a semi-log plot, where the y-axis is a logarithmic base 10 scale. On the semi-log plot, the graph of h is a line. The graph of the function g is also a line on the same semi-log plot. The slopes of the graph of h and of the graph of g are

the same. The y-intercept of the graph of h is $(0, c)$, and the y-intercept of the graph of g is $(0, \frac{1}{3}c)$. Which of the following values is closest to $g(3)$?

(A) 398   (B) 1995   (C) 15,849   (D) 63,095,734

4. Astronomers believe that the impact rate r of celestial bodies, assessed from craters on the moon, can be used to predict the age A of the craters. The function $A(r) = 12,000e^{-r}$ is to be used to predict the age of the craters, but the values of A get incredibly large for small impact rates r. Therefore, the astronomers decide to graph A on a semi-log plot where the y-axis is on a logarithmic scale of base e. Which of the following is true about the semi-log plot?

   (A) The semi-log plot will be exponential because $A(r)$ is exponential.
   (B) The semi-log plot will be logarithmic because the y-axis is on a logarithmic scale.
   (C) The semi-log plot will be linear because $A(r)$ is an exponential function.
   (D) The semi-log plot will not be linear because $A(r)$ is not an exponential function.

5. The graph shows the semi-log plot of an exponential function of the form $f(x) = a \cdot b^x$. The x-axis has a standard scale, and the y-axis has a logarithmic base 10 scale. What are the values of a and b?

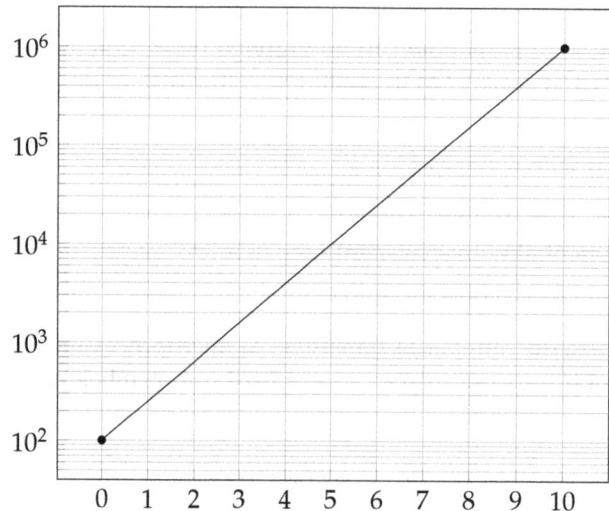

(A) $a = 2$ and $b = \frac{2}{5}$ (B) $a = 2$ and $b = 10^{(2/5)}$ (C) $a = 100$ and $b = \frac{2}{5}$ (D) $a = 100$ and $b = 10^{(2/5)}$

Solutions.

1. Taking the base 10 logarithm of the functions, we get $\log_{10} g(x) = \log_{10}(3 \cdot 8^x) = \log_{10} 3 + \log_{10}(8^x) = \log_{10} 3 + x \log_{10} 8$. Thus, on the semi-log plot, g will be linear with a slope of $\log_{10} 8$. The correct answer is therefore **(B)**.

2. A linear semi-log plot indicates an exponential model. The correct answer is therefore **(B)**.

3. We have that $c = 7.2$. Then $\frac{1}{3}c = 2.4$. Hence, $g(x) = 10^{(0.6x+2.4)}$, and so $g(3) = 10^{(1.8+2.4)} = 10^{4.2} \approx 15,849$. The correct answer is therefore **(C)**.

4. The function $A(r)$ is an exponential function with a base of e. Hence, the semi-log plot using a logarithm of base e will be linear. The correct answer is therefore **(C)**.

5. The line appears to have a slope of $\frac{4}{10} = \frac{2}{5}$ and a y-intercept of 2. On a semi-log plot, the graph of $f(x) = a \cdot b^x$ will resemble $\log_{10} f(x) = \log_{10} a + x \log_{10} b$. Therefore, $\log_{10} a = 2$ and $\log_{10} b = \frac{2}{5}$. This gives $a = 10^2 = 100$ and $b = 10^{2/5}$. The correct answer is therefore **(D)**.

Unit 2 Test ~ Exponential and Logarithmic Functions

> **Part A: 6 multiple choice and 1 free response**
> A calculator may be required for some questions on this section.

1. (Topic 2.3) The exponential function $f(x) = a \cdot b^x$ contains the input-output pairs $(0, 2.5)$ and $(2, 2.5 \cdot 0.96 \cdot 0.96)$. Which of the following is true?

   (A) $a = 2.5$, $b = 0.96$, and f demonstrates exponential growth.
   (B) $a = 2.5$, $b = 0.96$, and f demonstrates exponential decay.
   (C) $a = 0.96$, $b = 2.5$, and f demonstrates exponential growth.
   (D) $a = 0.96$, $b = 2.5$, and f demonstrates exponential decay.

2. (Topic 2.5) Josh has a salary of s thousand dollars this year and receives a raise of 3% of his salary every year. One of Josh's coworkers, Melody, has a salary that is $2,000 more than Josh's this year and also receives a bonus of $1,000 at the end of every year, which is not included as part of her salary. She also receives the 3% raise each year. Which of the following functions $M(t)$ models the amount of money Melody makes, in thousands of dollars, in year t, where $t = 0$ is this year?

   (A) $M(t) = s(1.03 + 1)^t + 2$
   (B) $M(t) = s(1.03 + 2)^t + 1$
   (C) $M(t) = (s + 2) \cdot 1.03^t + 1$
   (D) $M(t) = (s + 2) \cdot 1.03^{t+1}$

3. (Topic 2.13) Which of the following is an expression for the inverse of $f(x) = 3^{2x} + 1$, along with its domain?

   (A) $f^{-1}(x) = \frac{1}{2}\log_3(x - 1)$ for $x > 0$
   (B) $f^{-1}(x) = \frac{1}{2}\log_3(x - 1)$ for $x > 1$
   (C) $f^{-1}(x) = \log_3\left(\frac{x - 1}{2}\right)$ $x > 0$
   (D) $f^{-1}(x) = \log_3\left(\frac{x - 1}{2}\right)$ for $x > 1$

4. (Topic 2.14) Four equally rated professional weightlifters agreed to participate in an experiment where they would have different amounts of time to train for a deadlift competition. The amount of time trained, in months, and the maximum deadlifts, in pounds, are listed below.

Time training	1	3	6	10
Maximum deadlift	780	915	960	992

   Logarithmic regression was performed to create a function $f(x) = a + b \ln x$ to predict maximum deadlift based on months training. Based on the regression, how long would an equally rated weightlifter theoretically need to train to deadlift 1,100 pounds?

   (A) 28.161 months   **(B)** 145.573 months   **(C)** 1438.263 months   **(D)** 2,177.489 months

5. (Topic 2.6) In 2018, two function models – a quadratic function and an exponential function – are used to model the population of a country t years after 2018. Graphs of the residuals were created for both the quadratic and exponential regression functions.

Quadratic Model

Exponential Model

Which of the following is true based on the residual plots and the context?

   (A) The quadratic model has a smaller prediction error for the population of the country in 2023, but the exponential function is the more appropriate model overall.

   (B) The quadratic model has a smaller prediction error for the population of the country in 2023, and the quadratic function is the more appropriate model overall.

   (C) The exponential model has a smaller prediction error for the population of the country in 2023, but the quadratic function is still the more appropriate model.

   (D) The exponential model has a smaller prediction error for the population of the country in 2023, and the exponential function is the more appropriate model.

6. (Topic 2.15) Since the early 1970s, the number of transistors in microprocessors has increased. A semi-log plot of T, the number of transistors, in year x since 1970 is shown below.

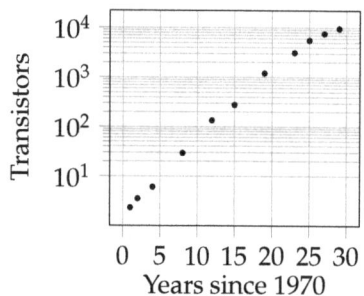

Which of the following provides the correct kind of model that could be used to model $T(x)$ along with correct reasoning?

   (A) $T(x)$ could be modeled by a linear function because the semi-log plot is reasonably linear.

   (B) $T(x)$ could be modeled by an exponential function because the semi-log plot is reasonably linear.

   (C) $T(x)$ could be modeled by a logarithmic function because the semi-log plot is reasonably linear.

   (D) $T(x)$ could be modeled by a quadratic function because the semi-log plot is reasonably linear.

Free Response Part A: Instructions

- Show all of your work. Your work will be scored on the correctness and completeness of your responses as well as your answers. Answers without supporting work may not receive credit in cases where supporting work is requested.
- Unless otherwise specified, the domain of a function f is assumed to be the set of all real numbers x for which $f(x)$ is a real number.
- Avoid rounding intermediate computations on the way to the final result. Unless otherwise specified, any decimal approximations reported in your work should be accurate to three places after the decimal point.
- It may be helpful to use your graphing calculator to store information such as computed values for constants, functions you are working with, solutions to equations, and any intermediate values. Computations with the graphing calculator that use the stored information help to maintain as much precision as possible and ensure the desired accuracy in final answers.

Free Response #1: Values of the invertible function f are provided in the table below at selected values of x.

x	1	3	9	27
$f(x)$	3	4	5	6

The function g is given by $g(x) = 2e^{0.5x} - 3$.

(A) (i) The function h is defined by $h(x) = (g \circ f)(x) = g(f(x))$. Find the value of $h(3)$ as a decimal approximation, or indicate that it is not defined.

(ii) Determine the x-values at which the inflection points of $g(x)$ occur, or indicate that there are none.

(B) (i) Determine the end behavior of g as x decreases without bound. Express your answer using the mathematical notation of a limit.

(ii) Find the zero/s of g as decimal approximations.

(C) (i) Use the table of values of $f(x)$ to determine if f is best modeled by a linear, quadratic, exponential, or logarithmic function.

(ii) Give a reason for your answer based on the relationship between the change in the output values of f and the change in the input values of f.

Part B: 14 multiple choice and 1 free response
A calculator is prohibited on this section.

7. (Topic 2.2) The geometric sequence s_n contains the terms $s_2 = 6r$ and $s_3 = 12r^2$, where $r \neq 0$. Which of the following is an expression for the general term of s_n?

   (A) $s_n = 6r^{(n-1)}$ (B) $s_n = 12r^{(n-1)}$ (C) $s_n = 3 \cdot (2r)^{(n-1)}$ (D) $s_n = 3 \cdot (2r)^n$

8. (Topic 2.4) Which of the following expressions is equivalent to $f(x) = 12 \cdot 6^{(x-2)}$?

   (A) $f(x) = 6^x$ (B) $f(x) = -1 \cdot 6^x$ (C) $f(x) = \frac{1}{3} \cdot 6^x$ (D) $f(x) = 12 \cdot \left(\frac{1}{36}\right)^x$

9. (Topic 2.7) The function f is given by $f(x) = 2(x-1)^2 + 1$, and values of the function g are provided in the table.

x	-1	1	2	3	4	5
$g(x)$	3	4	0	-2	-1	0

   What is the value of $g(f(2))$?

   (A) -2 (B) -1 (C) 0   (D) 3

10. (Topic 2.8) The function g is given by $g(x) = \dfrac{x-3}{x-2}$. Which of the following is true of the inverse function $g^{-1}(x)$?

    (A) The domain of $g^{-1}(x)$ is all real numbers except 2.
    (B) The domain of $g^{-1}(x)$ is all real numbers except 2 and 3.
    (C) The range of $g^{-1}(x)$ is all real numbers except 2.
    (D) The range of $g^{-1}(x)$ is all real numbers except 2 and 3.

11. (Topic 2.10) A graph of the exponential function $f(x)$ is shown.

    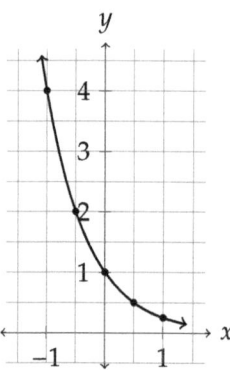

    Which of the following points would lie on the graph of $f^{-1}(x)$?

    (A) $(8, -2)$ (B) $(16, -2)$ (C) $(4, 2)$ (D) $(16, 2)$

12. (Topic 2.9) A cybersecurity firm sells encryption packages that have security ratings that exist on a logarithmic scale of base 2. How many more times secure is a package with a rating of 7 than a package with a rating of 4?

    (A) 3   (B) 4   (C) 6   (D) 8

13. (Topic 2.10) The functions f and g are given by $f(x) = 10^x$ and $g(x) = \log_{10} x$, where k is nonzero. The function h is given by the composition $h(x) = f(g(x))$. Which of the following is true concerning h?

    (A) h is a constant function.
    (B) h is a linear function with a slope of 1.
    (C) h is a linear function with a slope of 10.
    (D) h is an exponential function with a base of 10.

14. (Topic 2.11) The exponential function $h(x)$ is decreasing and its graph is concave up for all x. Which of the following is true about the inverse function $h^{-1}(x)$?

    (A) h^{-1} is increasing and its graph is concave up for all x.
    (B) h^{-1} is increasing and its graph is concave down for all x.
    (C) h^{-1} is decreasing and its graph is concave up for all x.
    (D) h^{-1} is decreasing and its graph is concave down for all x.

15. (Topic 2.12) Let $f(x) = \log_{10}(ax)$ and $g(x) = \log_{10}(bx^2)$ for positive numbers a and b. Let $k(x) = f(x) + g(x)$. Which of the following is expresses k as a single logarithmic expression?

    (A) $k(x) = \log_{10}(ax + bx^2)$ (B) $k(x) = 3\log_{10}(abx)$ (C) $k(x) = \log_{10}(abx^3)$ (D) $k(x) = \log_{10}((a+b)x^2)$

16. (Topic 2.12) The logarithmic function L can be expressed as $L(x) = A \log_3 x$ and also as $L(x) = B \log_2 x$. Which of the following is an expression for A in terms of B?

    (A) $A = B \log_2 3$ (B) $A = B \cdot \dfrac{\log_3 2}{\log_2 3}$ (C) $A = B \cdot \dfrac{\log_2 3}{\log_3 2}$ (D) $A = \dfrac{B}{\log_2 3}$

17. (Topic 2.13) The functions f and g are given by $f(x) = \log_3(x - 2)$ and $g(x) = \log_3(2x - 1)$. Which of the following is all x for which $f(x) + g(x) > 2$?

    (A) $x > 3.5$ only   (B) $x < -1$ only   (C) $x > 3.5$ and $x < -1$ (D) $-1 < x < 3.5$

18. (Topic 2.13) A student is solving the equation $5^x = 10^{3x-1}$. They do not have their calculator, so they have to rely on a table of logarithms which provides values of $\log_{10} x$ for various x. Solving which of the following equations would enable the student to solve the original equation?

    (A) $\dfrac{1}{2}x = 3x - 1$ (B) $x \log_{10} \dfrac{1}{5} = 3x - 1$ (C) $x \log_{10} 5 = 3x - 1$ (D) $x + \log_{10} 5 = 3x - 1$

19. (Topic 2.14) The table below gives values of x and y.

x	5	9	17
y	2	3	4

    Which of the following functions f could be used to model y based on x?

    (A) $f(x) = \log_2(x + 1)$ (B) $f(x) = \log_2(x - 1)$ (C) $f(x) = \log_2 x + 1$ (D) $f(x) = \log_2 x - 1$

20. (Topic 2.15) The function $g(x) = 3 \cdot 8^x$ is going to be graphed on a semi-log plot where the y-axis is on a logarithmic scale with base 10. On the semi-log plot, which of the following describes the graph of g?

   (A) g will be linear with a slope of $\log_{10} 3$.
   (B) g will be linear with a slope of $\log_{10} 8$.
   (C) g will be exponential with a constant ratio of 5.
   (D) g will be linear with a constant ratio of 8.

Free Response Part B - Instructions

- Show all of your work. Your work will be scored on the correctness and completeness of your responses as well as your answers. Answers without supporting work may not receive credit in cases where supporting work is requested.

- Unless otherwise specified, the domain of a function f is assumed to be the set of all real numbers x for which $f(x)$ is a real number.

Free Response #2:

(A) The functions f and g are given by

$$f(x) = 4 \ln x + \frac{1}{2} \ln x$$
$$g(x) = 5 \cdot 2^{(2x+3)}$$

   (i) Rewrite $f(x)$ as a single natural logarithm. Your result should be of the form ln(expression).
   (ii) Rewrite $g(x)$ as an exponential expression of the form $a \cdot b^x$.

(B) The functions h and j are given by

$$h(x) = 9 \cdot 3^{(x+1)}$$
$$j(x) = \log_2(3) + \log_2(x - 1)$$

   (i) Find the value of x for which $h(x) = 1$.
   (ii) Find the value of x for which $j(x) = 4$.

(C) The function m is given by

$$m(x) = 3e^{4x} + e$$

Find all values of x in the domain of m such that $m(x) > 2e$. Write your answer in the form $x > \frac{a - \log b}{c}$ for whole numbers a, b, and c.

Unit 2 Test Solutions and Scoring

1. The input $x = 0$ results in the output $a \cdot b^0 = 2.5$ so $a = 2.5$. The input $x = 2$ results in the output $2.5 \cdot b^2 = 2.5 \cdot 0.96^2$ so $b = 0.96$. Since $b = 0.96 < 1$, f denomonstrates exponential decay. The correct answer is therefore **(B)**.

2. Melody's current salary is $s + 2$. With a 3% raise each year, 3% of her current salary will be added to her current salary, giving her $(s + 2) \cdot 1.03^t$. Then she gets a bonus of 1 thousand dollars so that her salary is $(s + 2) \cdot 1.03^t + 1$. The correct answer is therefore **(C)**.

3. Let $y = f(x)$. Then interchanging x and y gives us $x = 3^{2y} + 1$. Subtracting 1 then taking the logarithm base 3 of both sides gives us $\log_3(x - 1) = 2y$. Finally, $f^{-1}(x) = \frac{1}{2}\log_3(x - 1)$. The domain of the inverse is the range of the f, which is all reals greater than 1. The correct answer is therefore **(B)**.

4. We need the inverse of f. Interchanging x and y, and then solving for y gives us the inverse:
$$f^{-1}(x) = e^{(x-a)/b}.$$
Performing the logarithmic regression in the calculator, we get $a = 791.9233965$ and $b = 92.29246995$. Then the number of months to deadlift 1100 pounds is
$$f^{-1}(1100) = e^{(1100-791.9233965)/92.29246995} = 28.16135305.$$
The correct answer is therefore **(A)**.

5. In 2023, the quadratic model has a smaller residual than the exponential model, so the prediction error is smaller in the quadratic model. However, the quadratic model has residuals in a pattern, while the exponential model does not. Hence, the exponential model is the more appropriate model overall. The correct answer is therefore **(A)**.

6. A linear semi-log plot indicates an exponential model. The correct answer is therefore **(B)**.

7. The ratio between consecutive terms is $12r^2/(6r) = 2r$, so $s_1 = \frac{6r}{2r} = 3$. The general term thus has the form $3 \cdot (2r)^{(n-1)}$. The correct answer is therefore **(C)**.

8. We have $12 \cdot 6^{(x-2)} = 12 \cdot 6^x \cdot 6^{-2} = 12 \cdot 6^x \cdot \frac{1}{6^2} = 12 \cdot 6^x \cdot \frac{1}{36} = \frac{12}{36} \cdot 6^x = \frac{1}{3} \cdot 6^x$. The correct answer is therefore **(C)**.

9. We have $f(2) = 2(2 - 1)^2 + 1 = 2 \cdot 1^2 + 1 = 2 + 1 = 3$, and $g(f(2)) = g(3) = -2$. The correct answer is therefore **(A)**.

10. The domain of g is all real numbers except 2, so this is the range of g^{-1}. The correct answer is therefore **(C)**.

11. By examining the graph we find that f has the input-output pairs $(0, 1)$ and $(-1, 4)$. Hence, each negative unit change in the inputs multiplies the output by 4. So when $x = -2$, the function must output $4 \cdot 4 = 16$, and $(-2, 16)$ is an input-output pair of f. Thus, $(16, -2)$ is such a pair on f^{-1}. The correct answer is therefore **(B)**.

12. Since the base is 2, the ratings of 4 and 7 correspond to the values 2^4 and 2^7. Thus, the rating of 7 larger than the rating of 4 by a factor of $2^7/2^4 = 2^{7-4} = 2^3 = 8$. The correct answer is therefore **(D)**.

13. The composition is $h(x) = f(g(x)) = f(\log_{10} x) = 10^{\log_{10} x} = x$. The correct answer is therefore **(B)**.

14. The exponential function $h(x) = e^{-x}$ is decreasing and concave up for all x. Its inverse is $h^{-1}(x) = -\ln(x)$ which is also decreasing and concave up for all x in its domain. The correct answer is therefore **(C)**.

15. $f(x) + g(x) = \log_{10}(ax) + \log_{10}(bx^2) = \log_{10}(ax \cdot bx^2) = \log_{10}(abx^3)$. The correct answer is therefore **(C)**.

16. Use the change of base formula on the first expression to tuen the base of 3 into a base of 2. We get
$$L(x) = A \log_3 x = A \cdot \frac{\log_2 x}{\log_2 3} = \frac{A}{\log_2 3} \cdot \log_2 x.$$
Since $L(x)$ also equals $B \cdot \log_2 x$, we must have $B = \frac{A}{\log_2 3}$, or $A = B \log_2 3$. The correct answer is therefore **(A)**.

17. We have $f(x) + g(x) = \log_3((x - 2)(2x - 1)) = \log_3(2x^2 - 5x + 2)$. For this to be greater than 2, the argument of the logarithm must be greater than $3^2 = 9$. Hence, $2x^2 - 5x + 2 > 9$ implies $2x^2 - 5x - 7 > 0$. This factors as $(2x - 7)(x + 1) > 0$. This polynomial is positive for $x < -1$ and $x > 3.5$. However, only $x > 3.5$ satisfies the domains of both f or g. Hence, the solution is $x > 3.5$ only. The correct answer is therefore **(A)**.

18. We take the logarithm base 10 of both sides. This yields $\log_{10} 5^x = 3x - 1$. Then by the Power Property, $x \log_{10} 5 = 3x - 1$. The correct answer is therefore **(C)**.

19. Note that if we subtract 1 from the inputs, we get 4, 8, and 16. This suggests that the ouputs change by 1 and the inputs minus 1 change by factor of 2. So a reasonable base is 2, and the function that best models y is $f(x) = \log_2(x - 1)$. The correct answer is therefore **(B)**.

20. Taking the base 10 logarithm of the functions, we get $\log_{10} g(x) = \log_{10}(3 \cdot 8^x) = \log_{10} 3 + \log_{10}(8^x) = \log_{10} 3 + x \log_{10} 8$. Thus, on the semi-log plot, g will be linear with a slope of $\log_{10} 8$. The correct answer is therefore **(B)**.

Question 1 Scoring Guidelines

		Model Solution	Scoring	
(A)	(i)	$g(f(3)) = g(4) = 11.778$	Value	1 point
	(ii)	Because g is always increasing at an increasing rate, there are no inflection points on the graph of g.	None with reason	1 point
			Total for part (A)	2 points
(B)	(i)	As x decreases without bound, the outputs of g approach -3. Therefore, $\lim_{x \to -\infty} g(x) = -3$.	-3 with limit notation	1 point
	(ii)	The only zero of g is $x = 0.811$.	Zero	1 point
			Total for part (B)	2 points
(C)	(i)	f is best modeled a logarithmic function	Logarithmic	1 point
	(ii)	The input values of f increase multiplicatively by a factor of 3 as the outputs increase additively by 1.	Reason	1 point
			Total for part (C)	2 points
			Total for Question 1	6 points

We propose the following time limits for each part of the test, assuming similar timings as for the AP exam. That means each multiple-choice should be allowed 3 minutes and each free-response 15 minutes. This could vary slightly, depending on the difficulty of the multiple-choice: it is reasonable to expect 2 minutes per problem for the easier problems. This creates the range of timings as given in the table below.

Suggested Time Limits

Part A	30-35 minutes
Part B	45-55 minutes

We choose 20 multiple-choice and 2 free-response to include on the test in order to equal the same proportions of those two types of problems as on the AP Exam. That means such a test can be given an "AP score" based on the number of points earned. Each multiple-choice problem is worth 1 point and each free-response is worth 6 points, for a total of 32 points available on the test. The table below shows the conversions to an "AP score."

Raw score	AP score
23-32	5
18-22	4
14-17	3
8-13	2
0-7	1

Question 2 Scoring Guidelines

		Model Solution	Scoring	
(A)	(i)	$f(x) = 4\ln x + \frac{1}{2}\ln x = \frac{9}{2}\ln x = \ln x^{(9/2)}, \; x > 0$	Expression	1 point
	(ii)	$g(x) = 5 \cdot 2^{(2x+3)} = 5 \cdot 2^{2x} \cdot 2^3 = 40 \cdot (2^2)^x = 40 \cdot 4^x$	Expression	1 point
			Total for part (A)	**2 points**
(B)	(i)	We solve: $$9 \cdot 3^{(x+1)} = 1$$ $$3^{(x+1)} = \frac{1}{9}$$ $$3^{(x+1)} = 3^{-2}$$ $$x + 1 = -2$$ $$x = -3$$	Value	1 point
	(ii)	We solve: $$\log_2 3 + \log_2(x-1) = 4$$ $$\log_2(3(x-1)) = \log_2 16$$ $$3x - 3 = 16$$ $$x = \frac{19}{3}$$	Value	1 point
			Total for part (B)	**2 points**
(C)		Again, we solve: $$3e^{4x} + e > 2e$$ $$3e^{4x} > e$$ $$e^{4x} > \frac{1}{3}e$$ $$\ln e^{4x} > \ln \frac{1}{3}e$$ $$4x > \ln \frac{1}{3} + \ln e$$ $$4x > -\ln 3 + 1$$ $$x > \frac{1 - \ln 3}{4}$$	Solution Form of $\dfrac{a - \log b}{c}$	1 point 1 point
			Total for part (C)	**2 points**
			Total for Question 2	**6 points**

Unit 3
Trigonometric and Polar Functions

Topics 3.1-3.2 ~ Periodic Phenomena and Radian Measure

Learning Objectives

1. (3.1.A) Construct graphs of periodic relationships based on verbal representations.
2. (3.1.B) Describe key characteristics of a periodic function based on a verbal representation.
3. (3.2.A) Determine the sine, cosine, and tangent of an angle using the unit circle (EK A.1 and A.2 only).

Success Criteria

1. I can sketch a periodic graph based on a description.
2. I can identify characteristics of a periodic function.
3. I can graph angles measured in radians in standard position.
4. I can compute arc lengths and angle measures in radians based on the definition of a radian.

In Units 1 and 2, we studied a variety of functions and their rates of change: we looked at whether functions increase or decrease and the rates at which they do so, reach maxima or minima, output zero, and more. Now, we move to a very special kind of function.

A PERIODIC function is one in which the output values demonstrate a *repeating pattern* over successive equal-length intervals of the inputs.

Below is data for what percentage of the moon was illuminated for a series of days in Baghdad, Iraq (roughly the location of ancient Babylon) in 2022.

Day	%	Day	%	Day	%	Day	%	Day	%	Day	%
1	0	11	82.3	21	73.2	31	0.3	41	73.6	51	81.1
2	1.2	12	89.5	22	64.4	32	3.2	42	82.3	52	73.2
3	5.2	13	94.9	23	54.9	33	8.8	43	89.5	53	64.4
4	11.7	14	98.4	24	45	34	16.8	44	94.9	54	54.9
5	20.4	15	99.9	25	35.1	35	26.4	45	98.4	55	45
6	30.6	16	99.5	26	24.5	36	36.9	46	99.9	56	35.1
7	41.6	17	97.3	27	16.6	37	47.8	47	99.5	57	25.4
8	52.8	18	93.3	28	9.1	38	58.5	48	97.3	58	16.6
9	63.6	19	87.9	29	3.5	39	52.8	49	93.3	59	9.1
10	73.6	20	81.1	30	0.5	40	63.6	50	87.9	60	3.5

On the graph below, sketch a scatterplot of the percentage of the moon that was illuminated for each of the 60 days.

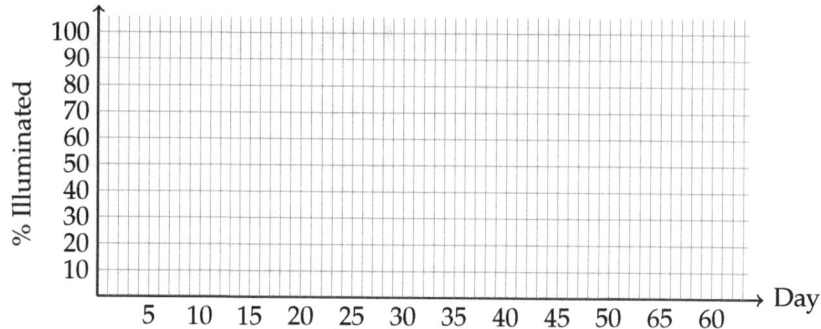

The percentage P can be predicted using the day x.

1. Sketch a graph of $P(x)$ by connecting the points with a curve.

2. Explain why P is (reasonably) periodic.

3. The PERIOD of a periodic function is the smallest interval length after which the function will begin to repeat itself. What do you think the period of P is?

4. Why does the value from Problem 3 make sense in context?

5. We can write the periodicity of a function with an analytical equation that shows that the output will repeat after an interval of length p (with p as the period). Fill in the blank below to show the periodicity of P.

$$P(x) = P(x\underline{})$$

Let's look at a verbal example.

A certain machine part exerts pressure, in Pascals (Pa), in a cyclic, periodically increasing and decreasing pattern. The machine exerts a minimum of 50 Pa of pressure and a maximum of 80 Pa of pressure. It takes about 2 minutes for the machine to make one cycle from 50 Pa to 50 Pa to occur again. Assume that the machine is producing 50 Pa at time $t = 0$ minutes.

6. Sketch a graph of P, the amount of pressure as a function of time t, in minutes, on the graph below.

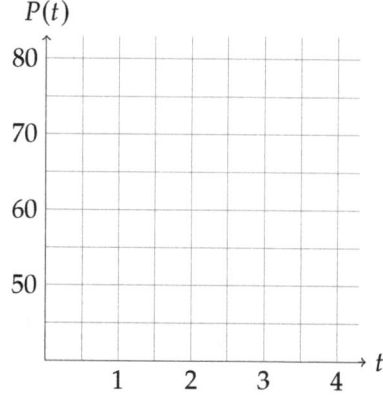

7. What is the period of the function $P(t)$?

8. In a span of one hour ($t = 0$ to $t = 60$), how many times will the machine be exerting its minimum amount of pressure?

9. In a span of one hour ($t = 0$ to $t = 60$), how many times will the machine be exerting its maximum amount of pressure?

Periodic relationships have been studied since the dawn of time... literally. Pun intended. Ancient astronomers and mathematicians studied the movement of the sun and stars in order to tell time, predict weather, and more. The oldest periodic relationships humans studied - and what we'll begin to study now - were related to *circles*.

There is some terminology when it comes to angles and the unit circle. An angle is comprised of an *initial ray*, which in *standard position* we will always set as the *positive x-axis*, and a *terminal ray*, or where the angle "ends." Positive angles will open from the positive x-axis in a *counterclockwise* direction, while negative angles will open from the positive x-axis in a *clockwise* direction.

Shown below are three angles.

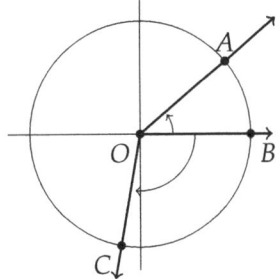

Try to fill in the blanks.

10. ∠AOB has terminal ray _____. Because it opened in a counterclockwise direction, ∠AOB is a _____ angle.

11. ∠BOC has terminal ray _____. Because it opened in a clockwise direction, ∠BOC is a _____ angle.

12. ∠AOC is // is not in standard position.

When we measure angles, we will use the unit RADIANS. These measure angles in terms of *radii*, which are the natural "language" of the size of circles. The radian is thus defined as the ratio of the *arc length* to the *radius*. For the angle below to the left, this means that an angle with an arc of length one radius measures $\frac{r}{r} = 1$ radian (which makes sense!).

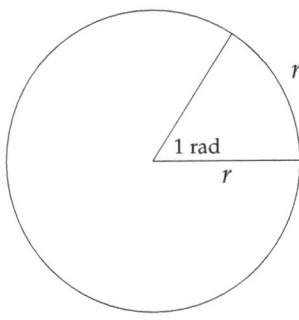

Angle: $\frac{r}{r} = 1$ radian

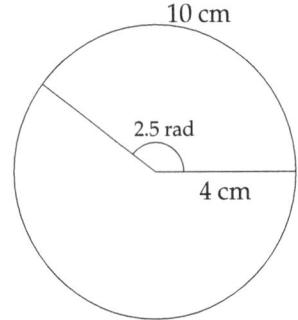

Angle: $\frac{10 \text{ cm}}{4 \text{ cm}} = 2.5$ radians

The CIRCUMFERENCE of a circle, or the distance around it, is $C = 2\pi r$, or 2π *radii*. Therefore, 2π *radians* form a complete circle, or 360°. This gives us the useful conversion equation

$$2\pi = 360°$$

13. Use the equation above to convert the following degree measures into radians.

    (a) 180°

    (b) 90°

    (c) 45°

    (d) 60°

    (e) 30°

    (f) 120°

    (g) 135°

    (h) 270°

14. Sketch each angle in standard position.

    (a) $\frac{\pi}{3}$ (b) $\frac{5\pi}{4}$ (c) $-\frac{\pi}{6}$ (d) $-\frac{\pi}{2}$

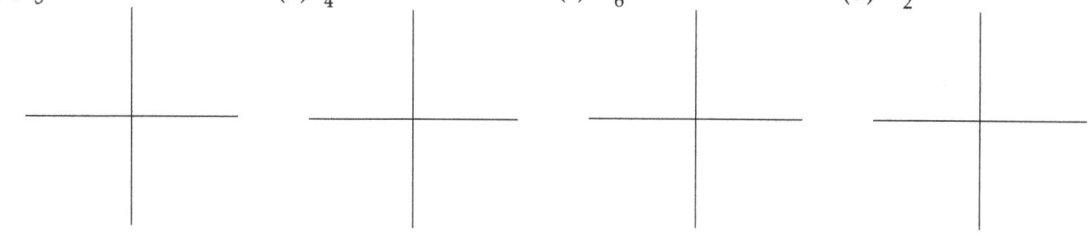

15. Each of the following intervals contains all angles in one or more specific quadrants. Identify the quadrant/s for each interval.

    (a) $0 \leq x \leq \pi$
    (b) $-\frac{\pi}{2} < x < \frac{\pi}{2}$
    (c) $0 \leq x < 2\pi$
    (d) $-\pi < x < -\frac{\pi}{2}$

16. Labeling angles can itself be pseudo-periodic. In the figures below, starting at 0, label the quadrantal angles as you rotate counterclockwise (left diagram) and counterclockwise (right diagram) for 3 full rotations. This should give you *six* different angles that all have the same terminal ray along each of the axes.

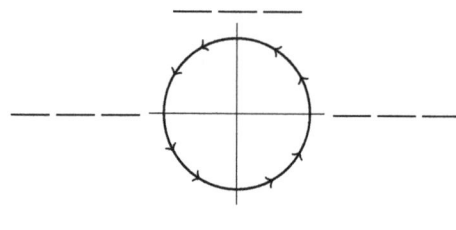 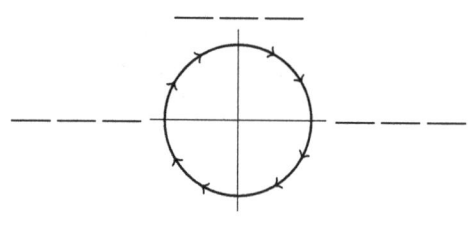

Notes

Topics 3.1-3.2 Homework

1. A salesperson has noticed that their sales increase and decrease in a periodic manner over time. Their sales peak around the 25th day of the year and reach their minimum around the 60th day of the year. The next peak in sales is around the 95th day of the year.

    (a) If $s(t)$ models the sales s on day t, what is the period of t?

    (b) Which day of a given year is the last day on which the sales will be at a minimum?

2. The blades on a wind turbine move in a circular path. The height h, in meters, of the end of one blade of the turbine t seconds after the turbine is started can be modeled using a periodic function $h(t)$ with a period of 6.

    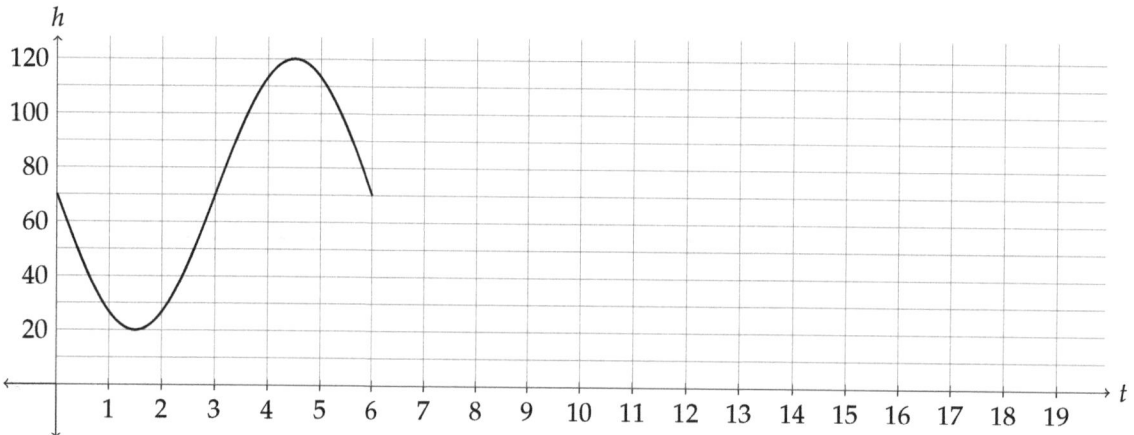

    (a) Sketch the rest of the graph of h for $0 \leq t \leq 19$.
    (b) What is the average height of the blade?
    (c) How long is the blade?
    (d) Can you determine if the blade is spinning clockwise or counterclockwise? If so, in which direction is it spinning?
    (e) What will the blade's height be 45 seconds after the turbine is turned on? How about 63 seconds?

3. Values of a periodic function f are given below.

x	3	6	9	12	15	18	21	24	27	30	33	36	39
$f(x)$	−1	4	3	−1	4	3	−1	4	3	−1	4	3	−1

    A student looks at this table and says the period of f is 18 because it appears that $f(x) = f(x + 18)$ for all given x. Explain why this likely is not the period of f, and identify a different estimate of the period.

4. Determine the missing angle measure (in radians) or arc length (in inches) for each figure below.

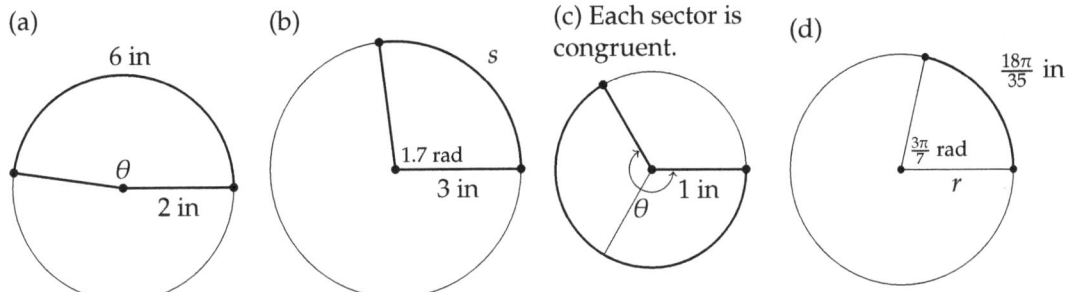

(a) 6 in, 2 in, θ

(b) s, 1.7 rad, 3 in

(c) Each sector is congruent. 1 in, θ

(d) $\frac{18\pi}{35}$ in, $\frac{3\pi}{7}$ rad, r

5. If the origin of the coordinate plane is O, sketch an angle COD with terminal ray $\overrightarrow{OC}$ in standard position such that $m \angle COD = \frac{2\pi}{3}$.

6. In preparation for the next lesson, let's explore some periodic behavior on a circle. First, you'll need to familiarize yourself with certain convenient angles: namely, those that are multiplies of $\frac{\pi}{6}$ or $\frac{\pi}{4}$ (which includes multiples of $\frac{\pi}{2}$ and $\frac{\pi}{3}$).

Complete the diagram below with the appropriate angle in radians.

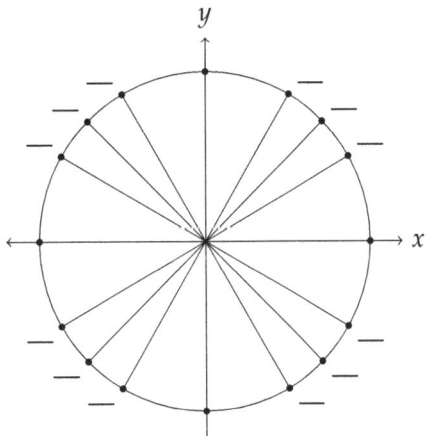

7. Now, the same grid as the one pictured above is given, but this time with a coordinate plane.

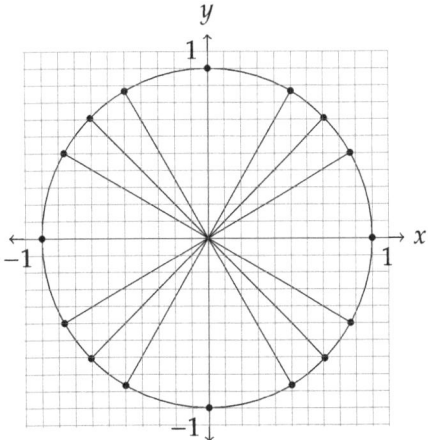

For each of the angles from Problem 7, the ray from the origin intersects the circle with radius 1 at a point. Estimate the *vertical displacement* of each point from the x-axis. For instance, the point where the angle $\frac{\pi}{6}$ meets the circle appears to have a y-coordinate of 0.5, so its vertical displacement is 0.5. Similarly, the angle $\frac{5\pi}{4}$ seems to have a y-coordinate of about -0.7, so its vertical displacement is -0.7.

Complete the table below, then sketch a graph of the relationship between the angle and the vertical displacement of the point.

Angle x	$\frac{\pi}{6}$	$\frac{\pi}{4}$	$\frac{\pi}{3}$	$\frac{\pi}{2}$	$\frac{2\pi}{3}$	$\frac{3\pi}{4}$	$\frac{5\pi}{6}$	π	$\frac{7\pi}{6}$	$\frac{5\pi}{4}$	$\frac{4\pi}{3}$	$\frac{3\pi}{2}$	$\frac{5\pi}{3}$	$\frac{7\pi}{4}$	$\frac{11\pi}{6}$	2π
Vert. disp. y	0.5									-0.7						

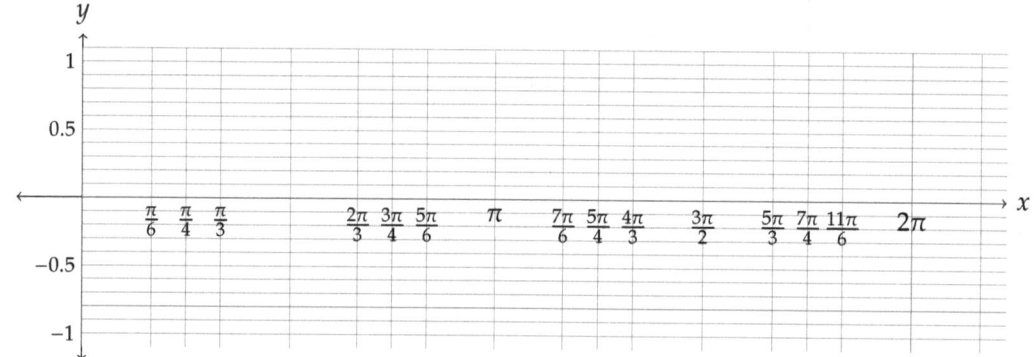

Topics 3.1-3.2 Solutions/Notes

Activity

1.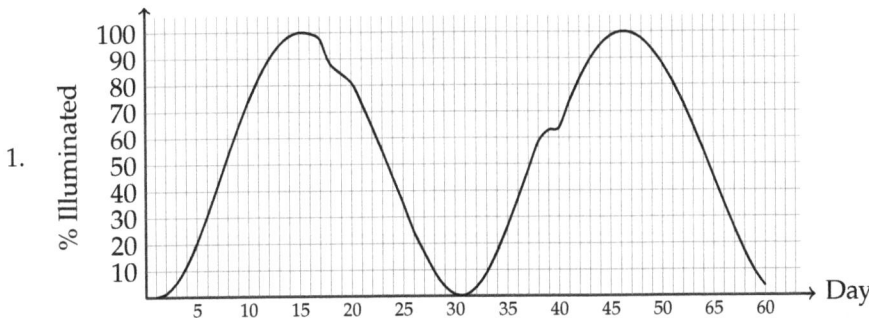

2. The percentage illuminated starts to repeat after about 30 days.

3. 30

4. This corresponds to roughly one month, which is roughly how long it takes the moon to orbit Earth.

5. $P(x) = P(x + 30)$

6.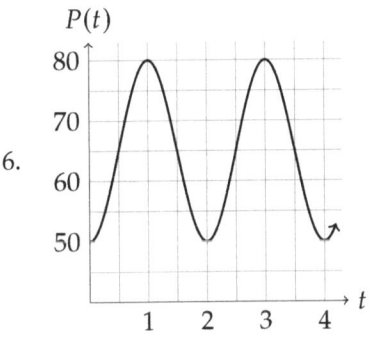

7. 2 minutes

8. The machine is exerting its minimum pressure at every even integer, so $t = 0, 2, 4, \ldots, 58, 60$. There are 31 even numbers from 0 to 60, so the minimum will occur 31 times.

9. The machine is exerting its maximum pressure at every odd integer, so $t = 1, 3, 5, \ldots, 57, 59$. There are 30 odd numbers from 0 to 60, so the maximum will occur 30 times.

10. $\angle AOB$ has terminal ray $\overrightarrow{OA}$. Because it opened in a counterclockwise direction, $\angle AOB$ is a underline{positive} angle.

11. $\angle BOC$ has terminal ray $\overrightarrow{OC}$. Because it opened in a clockwise direction, $\angle BOC$ is a underline{negative} angle.

12. $\angle AOC$ underline{is not} in standard position.

13. (a) π (b) $\frac{\pi}{2}$ (c) $\frac{\pi}{4}$ (d) $\frac{\pi}{3}$ (e) $\frac{\pi}{6}$ (f) $\frac{2\pi}{3}$ (g) $\frac{3\pi}{4}$ (h) $\frac{3\pi}{2}$

14.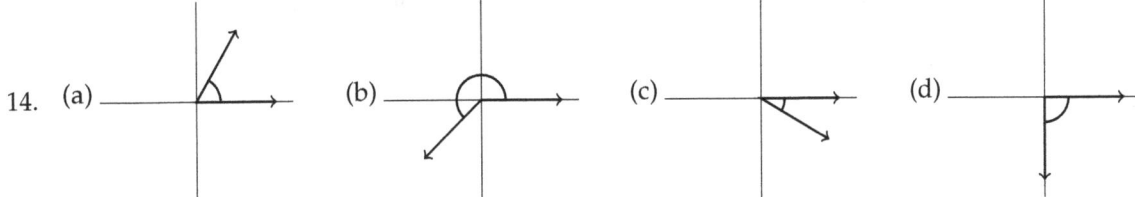

15. (a) Quadrants I and II (b) Quadrants I and IV (c) Quadrants I, II, III, and IV (d) Quadrant III

16.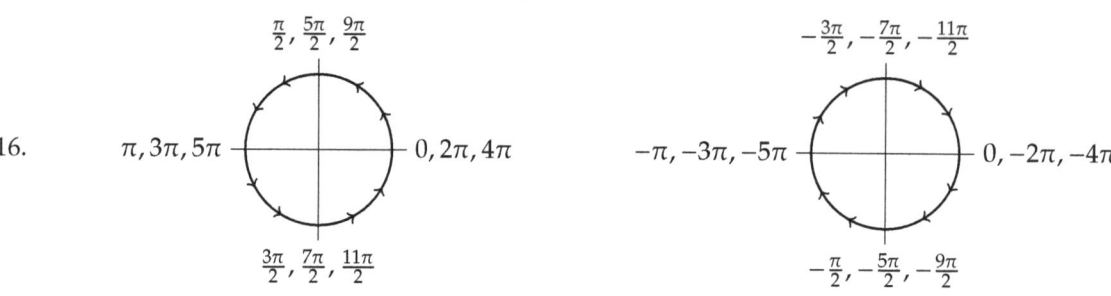

> **Notes**
>
> In a <u>PERIODIC</u> function, outputs will repeat over successive equal-length intervals of certain lengths of the inputs. The shortest of these lengths is called the <u>PERIOD</u>. If the period is p for a function $f(x)$, then
> $$f(x) = f(x+p) = f(x+2p) = \cdots = f(x+kp)$$
> for any integer k. Graphically, this looks like
>
>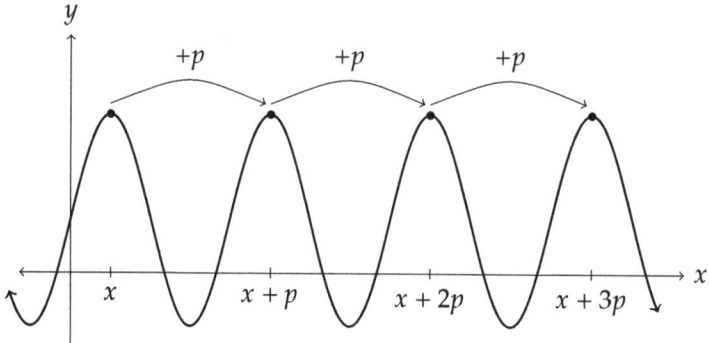
>
> The entire graph of a periodic function can be generated from one single period, or cycle.
>
> Most periodic behavior can be described using measurements related to angles graphed in a circle. Angles are measured using <u>RADIANS</u>, which measure the ratio of the arc length to the radius, or
> $$\theta = \frac{s}{r}$$
> where s is the arc length and r is the radius (θ is the Greek letter *theta*, which is a common letter used for arbitrary or unknown angles).
>
> Because the circumference of a circle is $C = 2\pi r$, a full angular rotation, or 360°, is 2π radians, which leads to the following useful conversions.
> $$\pi = 180° \qquad \tfrac{\pi}{2} = 90° \qquad \tfrac{\pi}{4} = 45° \qquad \tfrac{\pi}{3} = 60° \qquad \tfrac{\pi}{6} = 30° \qquad \tfrac{3\pi}{2} = 270°$$
> Angles are graphed in <u>STANDARD POSITION</u> by positioning their initial ray on the positive x-axis and then opening counterclockwise (for positive angles) or clockwise (for negative angles). The ending ray is called the <u>TERMINAL RAY</u>.

UNIT 3   TOPICS 3.1-3.2 ~ PERIODIC PHENOMENA AND RADIAN MEASURE   275

Homework

1. (a) 95 − 25 = 70 days
   (b) The minima will occur on the 60th, 130th, 200th, 270th, and 340th day of a single year. The last one that will occur is the 340th day.

2. (a)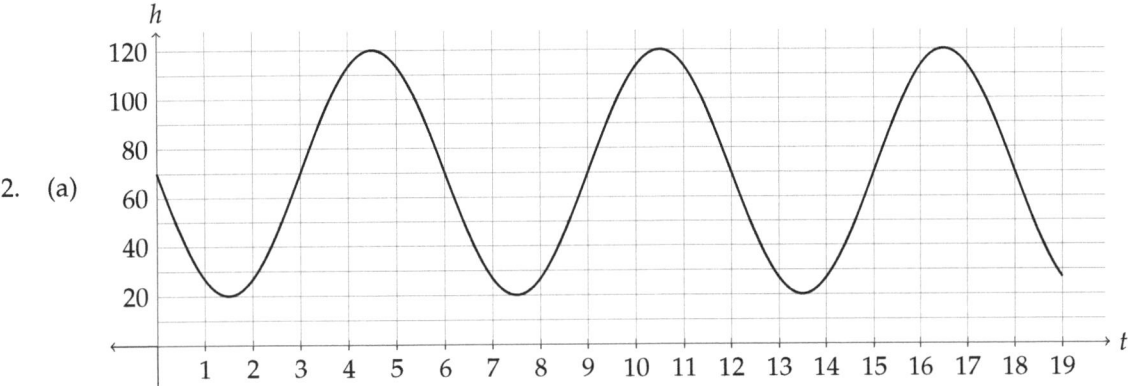

   (b) 70 meters
   (c) The height of the blade is the difference between the average height and the maximum (or minimum), which is 50 meters.
   (d) The blade's height begins dropping to start, so it must be spinning clockwise.
   (e) Because 45 = 7(6) + 3, the blade's height at 45 seconds will be the same as it was at 3 seconds, which is 70 meters. At 63 seconds, the blade's height will still be at 70 meters, because 63 is just 3 periods past 45 seconds.

3. The period is the *smallest* value of p for which $f(x) = f(x+p)$. From the table, it appears the period is 9, not 18.

4. (a) $\theta = \dfrac{6}{2} = 3$ radians

   (b) $1.7 = \dfrac{s}{3} \Rightarrow s = 5.1$ inches

   (c) θ is $\dfrac{2}{3}$ of a circle, so $\theta = \dfrac{2}{3}(2\pi) = \dfrac{4\pi}{3}$.

   (d) $\dfrac{3\pi}{7} = \dfrac{\frac{18\pi}{35}}{r} \Rightarrow r = \dfrac{18\pi}{35} \cdot \dfrac{7}{3\pi} = \dfrac{6}{5}$ inches

5.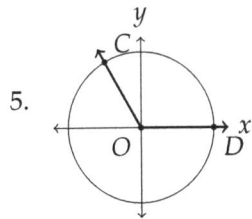

6.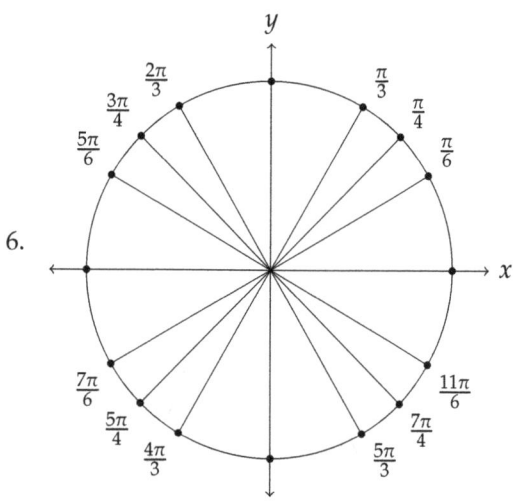

7.

x	$\frac{\pi}{6}$	$\frac{\pi}{4}$	$\frac{\pi}{3}$	$\frac{\pi}{2}$	$\frac{2\pi}{3}$	$\frac{3\pi}{4}$	$\frac{5\pi}{6}$	π	$\frac{7\pi}{6}$	$\frac{5\pi}{4}$	$\frac{4\pi}{3}$	$\frac{3\pi}{2}$	$\frac{5\pi}{3}$	$\frac{7\pi}{4}$	$\frac{11\pi}{6}$	2π
y	0.5	0.7	0.85	1	0.85	0.7	0.5	0	−0.5	−0.7	−0.85	−1	−0.85	−0.7	−0.5	0

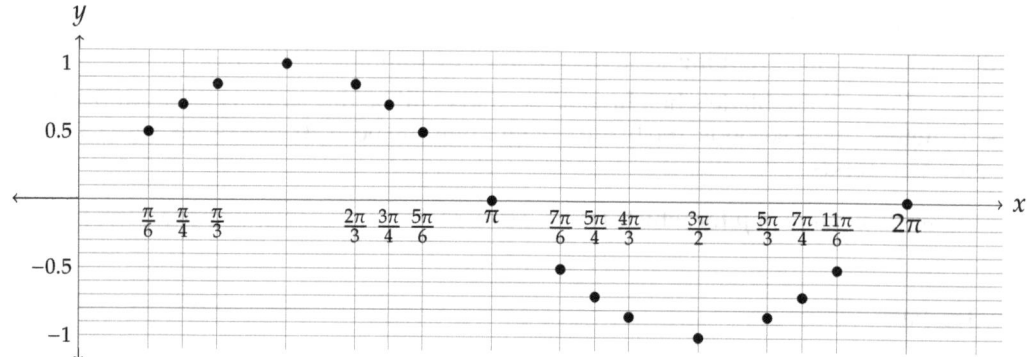

Reflection

Suggested Class Time. 60 minutes for activity, 20 for recap

Prerequisites. It didn't seem that students needed any previous exposure to radians or periodicity to get through the lesson.

Instructional Strategies. Students almost unanimously struggled at the same points in the activity. The first was on Problem 5. While students certainly got the idea, I saw 8-12 of my 19 students write the strangest thing (or a close version of it): $P(x) = P(x30)$. I found this flabbergasting (what the heck is "$x30$"?). Students all understood that the period was 30 or 31 days, but they did not express this as *addition* of inputs. As such, I covered this both on an individual group and a whole-class basis using a couple of tactics.

- I asked students what day had the same moon illumination percentage as day 1, day 2, and day 3, to which they promptly said 31, 32, and 33; I asked how they could generalize this to day x.

- I asked students how they could tell the graph was periodic, and they said something like "it's a copy of itself." I asked what transformation this corresponded to, and once we got to the fact that it is a translation, we were able to refer back to our transformation rules of $f(x+h)$ being a horizontal translation.

Students also struggled on PRoblems 10 and 11. At first, I thought this was my fault - maybe the notes weren't written well enough, or maybe there should be more examples, but I eventually decided against adding anything (not saying it's perfect!). To compensate for students struggling with all of the vocabulary during class, I just called everyone's attention to the board for literally 2 minutes. We went over standard position, initial ray, terminal ray, and positive and negative angles, and everyone had it down pat.

For radians, most students didn't have too much trouble, but I did have a lot of students writing $\frac{\pi}{2}$ as $\frac{1}{2}\pi$. I didn't want to discourage them from doing this, but some shoddy handwriting was leading towards $1/2\pi$, which could be mistaken for $\frac{1}{2\pi}$, or is simply non-standard. For the students who did struggle in any way, I really hammered using the equivalence $180° = \pi$ radians. To illustrate radians geometrically, I created the following Geogebra sketch: https://www.geogebra.org/m/hkeyfgj3

Also, everyone has their preferred way of teaching common degree and radian conversions, but one thing I've found very useful for students is the symmetry of the circle. When a student simply knows that $\frac{\pi}{4}$ is $\frac{1}{4}$ of a semicircle, they can identify where $\frac{5\pi}{4}$ is without ever having to compute a number. For those students that do prefer computation, I (and I believe College Board) would much prefer students think in terms of multiples of $\frac{\pi}{6}, \frac{\pi}{4}$, or $\frac{\pi}{3}$, rather than the old "multiply by $\frac{180}{\pi}$" formula. I show my students the following:

$$\frac{5\pi}{4} = 5 \cdot \left(\frac{\pi}{4}\right) \qquad \frac{7\pi}{6} = 7 \cdot \left(\frac{\pi}{6}\right)$$

and so on. Even then, for an angle like $\frac{7\pi}{6}$, I may not have students convert to degrees for sketching or computation. I'd rather them understand that $\frac{7\pi}{6}$ is simply $\frac{\pi}{6}$ beyond π.

Lastly, I emphasized with my students that degrees *will not* appear on the AP exam. To foster an environment where they will learn radians - and learn them quickly - we jokingly made up a tally on my board called "TPSD": Times People Say Degrees. Almost immediately after putting it on the board, a student referred to half of a circular rotation as 180°, so we got to put a tally on the board. I made a point of saying, "You may *think* in degrees, but you *never write them down.*" It's silly, and who knows if it works, but I think it will encourage faster retention and familiarity. I even encouraged students to rat each other out to me if they heard one of their classmates saying degrees.

Technology. No technology was required.

Problems.

1. The length of arc AB in the diagram is 4 centimeters long and angle θ has a measure of $\frac{2}{3}$ radians. What is the length of segment AO?

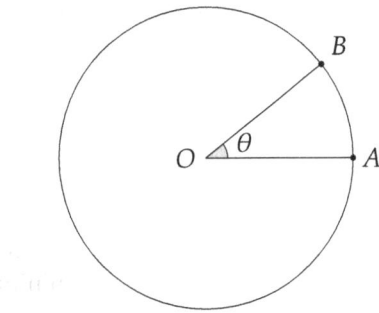

   (A) $\frac{8}{3}$ cm **(B)** 6 cm **(C)** $\frac{8}{3}\pi$ cm **(D)** 6π cm

2. The function f is periodic with a period of 4. Which of the following must be true for all x where $f(x)$ is defined?

   (A) $f(x) = f(4x)$ **(B)** $f(x) = f\left(\frac{1}{4}x\right)$ **(C)** $f(x) = f(x+2)$ **(D)** $f(x) = f(x+4)$

3. The function g shown in the graph is periodic, and g is defined for all real numbers.

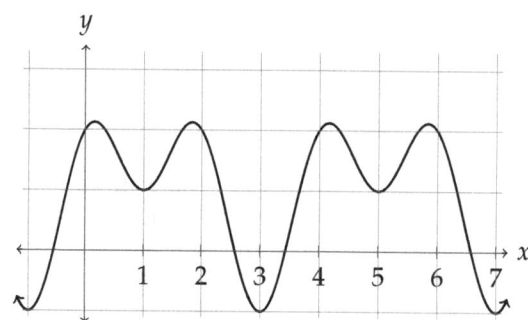

Which of the following is true of the graph of g on the interval $30 \leq x \leq 34$?

(A) The graph of g will obtain a local minimum at $x = 31$ only.
(B) The graph of g will obtain local minima at $x = 31$ and $x = 33$ only.
(C) The graph of g will obtain a local maximum at $x = 31$ only.
(D) The graph of g will obtain local maxima at $x = 31$ and $x = 33$ only.

4. Suppose an angle in standard position measuring θ radians, with $\frac{\pi}{2} < \theta < \pi$, has a terminal ray $\overrightarrow{OA}$. If another angle in standard position that measures α radians has the exact same terminal ray as $\overrightarrow{OA}$, which of the following could be true about α?

(A) $\alpha = \pi - \theta$ (B) $\alpha = \pi - \theta$ (C) $\alpha = 2\pi + \theta$ (D) $\alpha = 2\pi - \theta$

5. The periodic function f is defined for all real numbers. Values of the function f for selected values of x are shown in the table.

x	5	8	11	14	17	20	23	26	29	32	35
$f(x)$	16	15	4	−9	−16	−15	−4	9	16	15	4

What is the smallest possible period for f?

(A) 12 (B) 15 (C) 24 (D) 30

Solutions.

1. $\frac{4}{AO} = \frac{2}{3}$ implies $AO = \frac{4 \cdot 3}{2} = 6$. The correct answer is therefore **(B)**.

2. A function f is periodic if there is a value p such that $f(x) = f(x+p)$. Here, $p = 4$, so $f(x) = f(x+p)$. The correct answer is therefore **(D)**.

3. We see one period over $3 \leq x \leq 7$ and it has length 4. Thus, all other periods can found by adding multiples of 4 to 3 and 7. Therefore, substracting a multiple of 4 from both 30 and 34 can give us an interval pictured in the graph. Subtracting 28 yields $2 \leq x \leq 6$. So, the x-coordinates 31 and 33 correspond to the x-coordinates $31 - 28 = 3$ and $33 - 28 = 5$. At $x = 3$ and $x = 5$, the graph of g has local minima (recalling that global minima are also local minima). The correct answer is therefore **(B)**.

4. Different angles in standard position with the same terminal ray differ by complete circular rotations. Hence, it could be one rotation, so $\alpha = 2\pi + \theta$. The correct answer is therefore **(C)**.

5. The period is the smallest value of p for which $f(x) = f(x + p)$. From the table, $f(5) = f(29)$, and since $29 - 5 = 24$, the smallest possible period is 24. The correct answer is therefore **(C)**.

Topics 3.2-3.3 ~ Sine, Cosine, and Tangent

Learning Objectives

1. (3.2.A) Determine the sine, cosine, and tangent of an angle using the unit circle.
2. (3.3.A) Determine coordinates of points on a circle centered at the origin.

Success Criteria

1. I can compute sine, cosine, and tangent given an angle in standard position.
2. I can find the coordinates of a point on a circle given an angle in standard position and the radius of the circle.

Most periodic behavior observed in the real world can be modeled in terms of circular motion and, in particular, displacement from the x-axis, y-axis, and the origin. In particular, we're going to introduce the following *trigonometric functions*.

Sine, Cosine, and Tangent

Let x be an angle and let P be the point where the terminal ray of x intersects a circle centered at the origin.

- the SINE of an angle x, written $\sin(x)$, in standard position is the ratio of the *vertical displacement* of P from the x-axis to the *distance from the origin*.

- the COSINE of an angle x, written $\cos(x)$, in standard position is the ratio of the *horizontal displacement* of P from the y-axis to the *distance from the origin*.

- the TANGENT of an angle x, written $\tan(x)$, in standard position is the ratio of the *vertical displacement* of P from the x-axis to the *horizontal displacement* of P from the y-axis. This is equivalent to the *slope* of the terminal ray, as well as the ratio $\frac{\sin(x)}{\cos(x)}$.

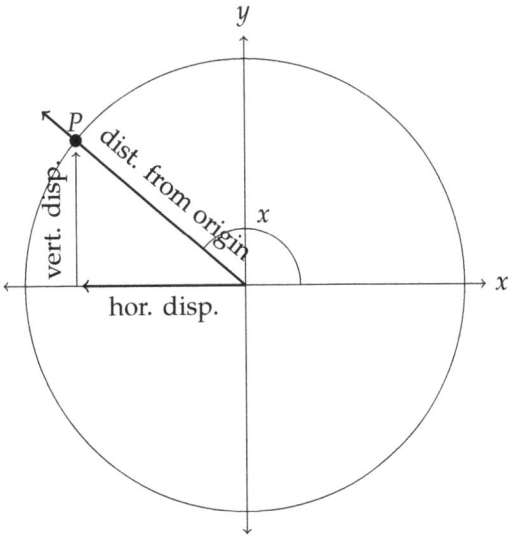

1. Consider the angle θ below.

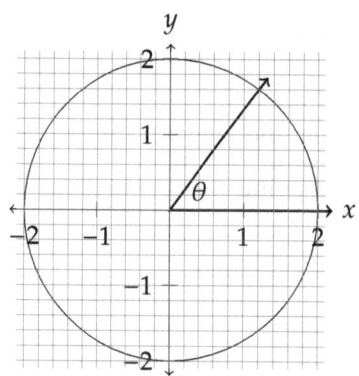

The terminal ray of θ intersects the circle at the point P.

   (a) Draw a segment that represents the vertical displacement of P. Find its length.
   (b) Draw a segment that represents the horizontal displacement of P. Find its length.
   (c) Draw a segment that represents the distance P is from the origin. Find its length.

2. Use the values from PRoblem 1 to compute the following.

   (a) $\sin(\theta)$
   (b) $\cos(\theta)$
   (c) $\tan(\theta)$

   Keep in mind, as you continue through these problems, that vertical and horizontal displacement can be *negative*.

3. (▣) Compute $\sin(x)$, $\cos(x)$, and $\tan(x)$ for the angle x below.

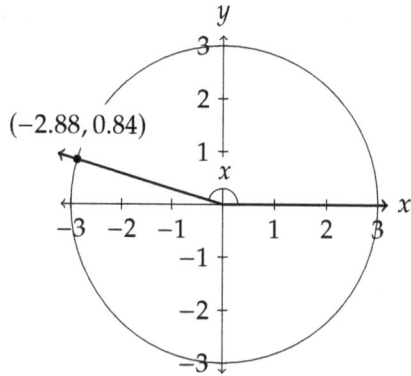

4. We can actually determine the sign of the sine, cosine, and tangent of an angle based solely on which quadrant the angle is in. Fill in the following with the correct inequality for an angle x.

$$
\begin{array}{c|c}
\textbf{Quad. II} \quad \begin{array}{l} \sin x \_\_ 0 \\ \cos x \_\_ 0 \\ \tan x \_\_ 0 \end{array} & \begin{array}{l} \sin x \_\_ 0 \\ \cos x \_\_ 0 \\ \tan x \_\_ 0 \end{array} \quad \textbf{Quad. I} \\
\hline
\textbf{Quad. III} \quad \begin{array}{l} \sin x \_\_ 0 \\ \cos x \_\_ 0 \\ \tan x \_\_ 0 \end{array} & \begin{array}{l} \sin x \_\_ 0 \\ \cos x \_\_ 0 \\ \tan x \_\_ 0 \end{array} \quad \textbf{Quad. IV}
\end{array}
$$

5. The values of $\sin(x)$ and $\cos(x)$ are quite easy to compute when the vertical displacement or horizontal displacement is nonexistent (0) or precisely equal to the distance from the origin.

   (a) The angles at which the vertical or horizontal displacement is equal to 0 or the distance from the origin are all multiples of what angle?

   (b) Complete the table.

x	0	$\frac{\pi}{2}$	π	$\frac{3\pi}{2}$	2π
$\sin(x)$					
$\cos(x)$					

6. For what value/s of x, with $0 \leq x \leq 2\pi$, will $\tan(x)$ equal 0? For what value/s of x, with $0 \leq x \leq 2\pi$, will $\tan(x)$ be undefined?

7. For what value/s of x, with $0 \leq x \leq 2\pi$, will $\tan(x) = 1$? How about $\tan(x) = -1$?

The definitions provided at the beginning of this lesson apply to *any* circle. One particular circle that makes sine and cosine even more accessible is the *unit circle*.

Definition: The UNIT CIRCLE is the circle centered at the origin with a radius of 1.

8. The unit circle doesn't change the definition of $\sin(x)$ or $\cos(x)$, but it does make for something very convenient. On the graph below, draw the point $P(0.6, 0.8)$. As you draw this, label the horizontal and vertical displacement.

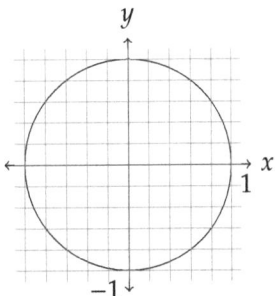

9. Draw a ray from the origin through P. Label the angle this ray makes with the x-axis x.

10. Find $\sin(x)$ and $\cos(x)$. What do you notice?

11. Can you use Problem 10 to compute $\tan(x)$?

12. Fill in the following fact:

> **Fact:** If an angle x is graphed in standard position and P is the point where the terminal ray intersects the *unit circle*, then
> - $\sin(x)$ is just _____
> - $\cos(x)$ is just _____
> - $\tan(x)$ is just _____

Let's look at an example.

13. The point $\left(-\frac{1}{4}, -\frac{\sqrt{15}}{4}\right)$ is the intersection of the unit circle and the terminal ray of the angle θ. Determine the values of $\sin(\theta)$, $\cos(\theta)$, and $\tan(\theta)$.

14. The line $y = mx$ intersects the unit circle at the point $\left(\frac{2}{3}, \frac{\sqrt{5}}{3}\right)$. It makes an angle of θ with the positive x-axis. Find $\sin(\theta)$, $\cos(\theta)$, $\tan(\theta)$, and m.

15. Some of the radicals from Problems 12 and 13 may have seemed odd, but they will become increasingly common. They are based on a very useful fact: for any point on a circle, the distance from the origin, the horizontal displacement, and the vertical displacement form a *right triangle*. This means that what theorem will apply?

16. Suppose for a certain angle x in Quadrant I that $\sin(x) = \frac{1}{2}$. Use the result of Problem 15 to compute $\cos(x)$. (A diagram will be helpful as you get used to this!)

We can use the fact from Problem 11 to actually find the coordinates of points on circles with *any* radii. Consider the diagrams below.

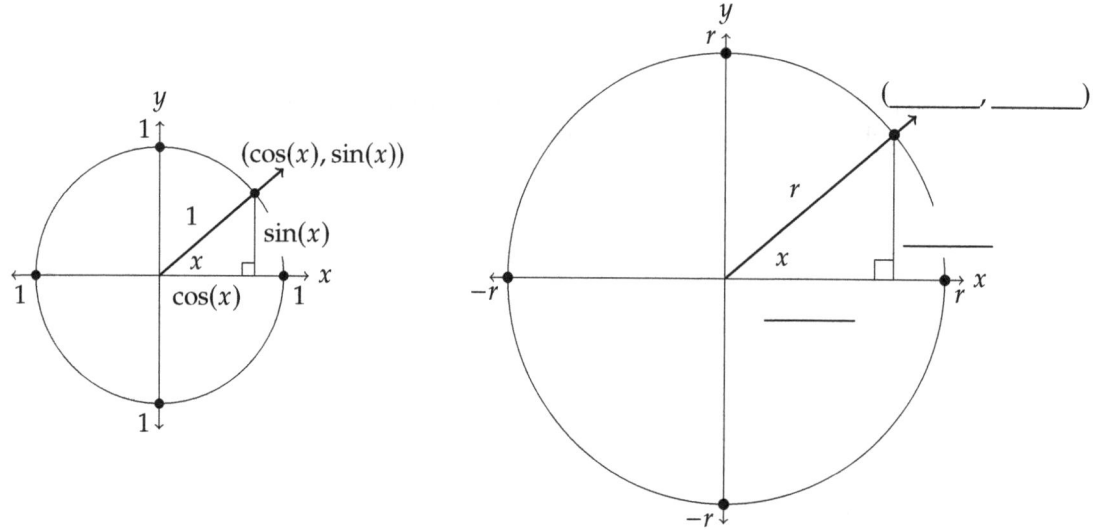

17. The circle and triangle on the right are simply a dilation of the circle and triangle on the left by what factor?

18. Use Problem 17 to fill in expressions for the horizontal displacement, vertical displacement, and coordinates of the point on the circle.

19. Fill in the fact below.

> **Fact:** If the terminal ray of an angle of measure θ in standard position intersects a circle of radius r at point P, then the coordinates of P are (_____, _____).

20. (▦) Find the coordinates of A in the circle below to 3 decimal places.

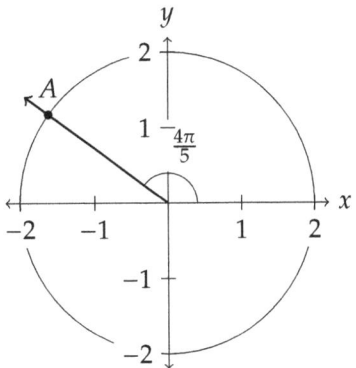

Note: One last thing concerning notation. You will have noticed that we use parentheses around the argument of sine, cosine, and tangent. That is, when we write $\sin(x)$ or $\sin(\theta)$, we put parentheses around the x or the θ. However, it is common that when the argument is only one term, we drop the parentheses. So it is common to see $\sin(x)$ as simply $\sin x$, or $\sin(\theta)$ as simply $\sin \theta$. So from here on out (including the homework), we will not use parentheses unless we need to make the argument of the trigonometric function clear.

Notes

Topics 3.2-3.3 Homework

1. The point $A\left(\frac{5}{13}, \frac{12}{13}\right)$ lies where the terminal ray of the angle x meets the unit circle. Compute $\sin x$, $\cos x$, and $\tan x$.

2. The point $(2, \sqrt{21})$ is the intersection of the terminal ray of an angle θ in standard position and a circle of radius r. Find r and compute $\sin \theta$, $\cos \theta$, and $\tan \theta$.

3. In the diagram, there are four angles in standard position: $\angle XOA$, $\angle XOB$, $\angle XOC$, and $\angle XOD$.

    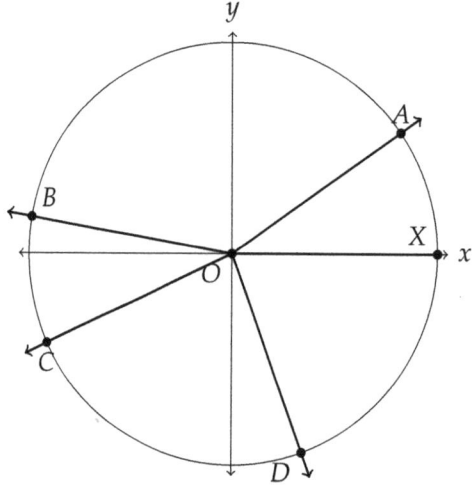

    Without any computation, fill in the 4 angles in each set of inequalities based on the diagram.

    $$\sin(\angle \underline{\quad}) < \sin(\angle \underline{\quad}) < \sin(\angle \underline{\quad}) < \sin(\angle \underline{\quad})$$
    $$\cos(\angle \underline{\quad}) < \cos(\angle \underline{\quad}) < \cos(\angle \underline{\quad}) < \cos(\angle \underline{\quad})$$
    $$\tan(\angle \underline{\quad}) < \tan(\angle \underline{\quad}) < \tan(\angle \underline{\quad}) < \tan(\angle \underline{\quad})$$

4. The terminal ray of the angle θ in standard position intersects a circle at the point $(x, 8)$. If $\sin \theta = \frac{1}{4}$, what is the radius of the circle?

5. For $0 \leq x < 2\pi$, there are exactly two solutions of $\sin x + \cos x = 1$. What are they?

6. The dashed rays in the figure below are at angles of $\frac{\pi}{6}, \frac{\pi}{4},$ and $\frac{\pi}{3}$ with the positive x- and y-axis, respectively. Given that $\sin \frac{\pi}{6} = \frac{1}{2}$ and $\cos \frac{\pi}{6} = \frac{\sqrt{3}}{2}$, determine the exact coordinates of the points $A, B,$ and C below.

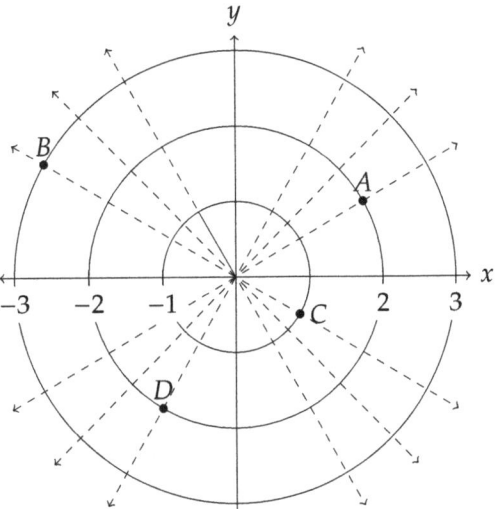

Topics 3.2-3.3 Solutions/Notes

Activity

1.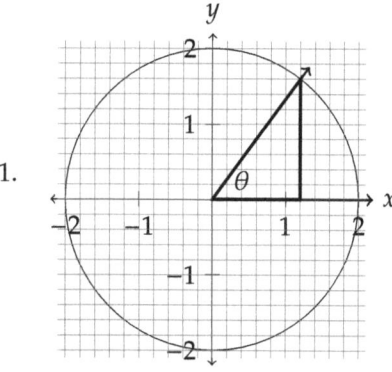

2. (a) $\sin(\theta) = \dfrac{1.6}{2} = \dfrac{4}{5}$ (b) $\cos(\theta) = \dfrac{1.2}{2} = \dfrac{3}{5}$ (c) $\tan(\theta) = \dfrac{\frac{4}{5}}{\frac{3}{5}} = \dfrac{4}{3}$

3. $\sin(x) = \dfrac{0.84}{3} = 0.28$, $\cos(x) = \dfrac{-2.88}{3} = 0.96$, $\tan(x) = \dfrac{0.28}{0.96} \approx 0.292$

4.
```
                    sin x > 0  |  sin x > 0
        Quad. II    cos x < 0  |  cos x > 0   Quad. I
                    tan x < 0  |  tan x > 0
        ───────────────────────┼─────────────────────── x
                    sin x < 0  |  sin x < 0
        Quad. III   cos x < 0  |  cos x > 0   Quad. IV
                    tan x > 0  |  tan x < 0
```

5.

x	0	$\frac{\pi}{2}$	π	$\frac{3\pi}{2}$	2π
$\sin(x)$	0	1	0	-1	0
$\cos(x)$	1	0	-1	0	1

6. $\tan(x) = \dfrac{\sin(x)}{\cos(x)}$ will equal 0 when $\sin(x) = 0$ and $\cos(x) \neq 0$. This will occur at $x = 0, \pi, 2\pi$, and any multiple of π. $\tan(x) = \dfrac{\sin(x)}{\cos(x)}$ will be undefined when $\cos(x) = 0$ and $\sin(x) \neq 0$. This will occur at $x = \frac{\pi}{2}, \frac{3\pi}{2}$, and any odd multiple of $\frac{\pi}{2}$.

7. $\tan(x) = \dfrac{\sin(x)}{\cos(x)} = 1$ when $\sin(x) = \cos(x)$. This will occur when $x = \frac{\pi}{4}$ and $x = \frac{5\pi}{4}$, or when the vertical and horizontal displacement are identical (with the same sign!). On the other hand, $\tan(x) = -1$ when the vertical and horizontal displacement have the same size but different sign, which occurs when $x = \frac{3\pi}{4}$ and $x = \frac{7\pi}{4}$.

8.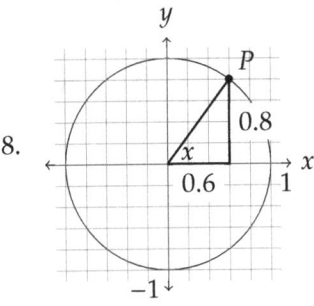

10. $\sin(x) = 0.8$, $\cos(x) = 0.6$

11. $\tan(x) = \dfrac{0.8}{0.6} = \dfrac{4}{3}$

12. **Fact:** If an angle x is graphed in standard position and P is the point where the terminal ray intersects the *unit circle*, then
    - $\sin(x)$ is just the y-coordinate of P
    - $\cos(x)$ is just the x-coordinate of P
    - $\tan(x)$ is just the ratio $\frac{y}{x}$ of the coordinates of P

13. $\sin(\theta) = -\dfrac{\sqrt{15}}{4}$, $\cos(\theta) = -\dfrac{1}{4}$, $\tan(\theta) = \dfrac{-\frac{\sqrt{15}}{4}}{-\frac{1}{4}} = \sqrt{15}$.

14. $\sin(\theta) = \dfrac{\sqrt{5}}{3}$, $\cos(\theta) = \dfrac{2}{3}$, $\tan(\theta) = m = \dfrac{\sqrt{5}}{2}$

15. Pythagorean

16. $\left(\dfrac{1}{2}\right)^2 + (\cos(x))^2 = 1 \;\Rightarrow\; \cos x = \sqrt{1 - \dfrac{1}{4}} = \sqrt{\dfrac{3}{4}} = \dfrac{\sqrt{3}}{2}$.

17. r

18. Horizontal: $r\cos(x)$, vertical: $r\sin(x)$, point: $(r\cos(x), r\sin(x))$

19. **Fact:** If the terminal ray of an angle of measure θ in standard position intersects a circle of radius r at point P, then the coordinates of P are $(r\cos(\theta), r\sin(\theta))$.

20. $A\left(2\cos\left(\dfrac{4\pi}{5}\right), 2\sin\left(\dfrac{4\pi}{5}\right)\right) \approx (-1.618, 1.176)$

Notes

The trigonometric functions $\sin(x)$, $\cos(x)$, and $\tan(x)$ measure ratios of displacement in the coordinate plane for angles x graphed in standard position.

When the circle has a radius of 1 (the unit circle), the coordinates of a point P at the intersection of the terminal ray of an angle x in standard position and the unit circle has coordinates that are exactly equal to the horizontal and vertical displacement: $P(x, y) = (\cos(x), \sin(x))$.

For a circle that doesn't have a radius of 1, then the intersection of the terminal ray and the circle is simply a dilation of $(\cos(x), \sin(x))$ by a factor equal to the radius. This means that any point P at the intersection of the terminal ray of an angle in standard position and a circle with radius r has coordinates $P(x, y) = P(r\cos(x), r\sin(x))$.

All angles that are multiples of $\frac{\pi}{2}$ will result in vertical or horizontal displacement of 0, 1, or -1, which makes computing $\sin\left(k \cdot \frac{\pi}{2}\right)$ and $\cos\left(k \cdot \frac{\pi}{2}\right)$ simple for any integer k.

For this reason, $\tan(x)$ will equal 0 when $\sin(x) = 0$, which occurs precisely when the terminal side coincides with the x-axis, or when $x = k\pi$. On the other hand, $\tan(x)$ will be undefined when $\cos(x) = 0$, which occurs precisely when the terminal side coincides with the y-axis, or when $x = (2k + 1) \cdot \frac{\pi}{2}$.

Homework

1. $\sin x = \frac{12}{13}$, $\cos x = \frac{5}{13}$, $\tan x = \frac{12}{5}$

2. From the Pythagorean Theorem, $r^2 = 2^2 + (\sqrt{21})^2$, or $r = 5$. Therefore, $\sin x = \frac{\sqrt{21}}{5}$, $\cos x = \frac{2}{5}$, and $\tan x = \frac{\sqrt{21}}{2}$.

3. Look at the size of the vertical displacement to order the values of sine, the horizontal displacement to order the values of cosine, and the size and sign of the ratio of the vertical to the horizontal displacement to order the values of tangent.

$$\sin(\angle XOD) < \sin(\angle XOC) < \sin(\angle XOB) < \sin(\angle XOA)$$
$$\cos(\angle XOB) < \cos(\angle XOC) < \cos(\angle XOD) < \cos(\angle XOA)$$
$$\tan(\angle XOD) < \tan(\angle XOB) < \tan(\angle XOC) < \tan(\angle XOA)$$

4. $\sin \theta = \frac{1}{4} = \frac{8}{r}$, so $r = 32$.

5. $\sin x + \cos x = 1$ only if $\sin x = 0$ and $\cos x = 1$ or if $\sin x = 1$ and $\cos x = 0$. These occur when $x = 0$ and $x = \frac{\pi}{2}$, respectively.

6. First, we get $A = (2\cos\frac{\pi}{6}, 2\sin\frac{\pi}{6}) = (1, \sqrt{3})$. Now, using symmetry, we can determine the coordinates of B and C because they are each within $\frac{\pi}{6}$ of the x-axis. Therefore,

$$B = \left(3\left(-\frac{\sqrt{3}}{2}\right), 3\left(\frac{1}{2}\right)\right) = \left(-\frac{3\sqrt{3}}{2}, \frac{3}{2}\right) \quad \text{and} \quad C = \left(\frac{\sqrt{3}}{2}, -\frac{1}{2}\right).$$

Now, D is $\frac{\pi}{3}$ away from the negative x-axis. However, closer inspection reveals that the triangle that D would form with the origin and negative x-axis is congruent to (a rotation of) the triangle A would form with the origin and positive x-axis, as shown below.

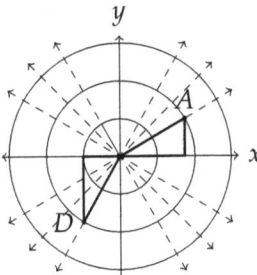

Therefore, we can determine that $D = \left(-1, -\sqrt{3}\right)$.

Reflection

Suggested Class Time. 70-80 minutes

Prerequisites. Students should be comfortable with radians. It also helps if students have had exposure to "SOH-CAH-TOA," as this is a nice way to introduce the new generalization of sine, cosine, and tangent to angles that are non-acute or negative.

Instructional Strategies. This lesson covers a *lot*. Originally, I intended for it to be a student-driven activity like most of the others in this book. However, it became clear pretty early on that students would need support throughout. As such, I decided to interrupt and bring the whole class together for guided instruction (we'd work through the problems together) at a couple points.

First, I put up answers to Problems 1-4 as students worked. While some flew through Problems 1-4, others were slower to catch on (mostly just second guessing themselves), so having the answers on the board helped them along.

The first interruption was when the entire class had gotten to or past Problem 5. The phrasing of Problem 5, which I stand by, was hard for them to visualize. As such, we worked through Problems 5-7 together. I made a quick Geogebra sketch (with a circle, a point on the circle, and a line going through this point and the origin) that would allow me to rotate the point around the circle so we could discuss the slope of the line. This drove our answering Problems 6 and 7.

After this, students resumed working, and I would slowly put up answers to Problems 8-12 and walk around the room again. Then, we came together again for Problems 13 and 14. We worked these as a class.

I then let students chew on Problems 15 and 16 again for a few minutes. They quickly got Problem 15, but many struggled to produce the diagram for Problem 16. After these few minutes, we came together again and worked Problem 16.

Finally, we worked Problems 17 and 18 as a class. I first asked students, "If I need to buy 3 chisels for my woodshop, how much will I spend?" A few students threw out nonsense answers, but pretty quickly, one student said, "Well, how much does one chisel cost?" I used this to introduce the idea of scaling the unit circle. We had gotten comfortable at this point with the idea that the x- and y-coordinates of a point on the unit circle represent the horizontal and vertical displacement, respectively, so this folded nicely into the chisel example: if we want the coordinates of a point on a larger circle, we can simply utilize $\cos(x)$ and $\sin(x)$ from the circle with radius 1 and scale up accordingly.

We actually ran out of time before Problems 19 and 20! I think we spent a bit too long going over the homework. I do not plan on shortening the activity next year; rather, I'll schedule the class ahead of time so that the class-wide discussions can occur at more prompt times.

One instructional strategy: when helping students with Problem 3 or any similar problem (when the vertical and horizontal displacement are not explicitly drawn), I always ask students to draw some random point, like $(-2, 3)$ or $(4, -1)$ in the coordinate plane. I then ask them to draw exactly what their brain was doing to generate this point - to draw the paths their eyes took as they constructed the point. This has always seemed to work well at helping students see that the x- and y-coordinates of any point in the plane represent precisely the horizontal and vertical displacement from the y- and x-axes, respectively.

For practice, it really helped students to be able to use this applet at home: https://www.geogebra.org/m/fdubatzx

Technology. Students used their calculators for a few computations.

Problems.

1. Ray $\overrightarrow{OA}$ lies on the terminal ray of angle α, and ray $\overrightarrow{OB}$ lies on the terminal ray of angle β.

   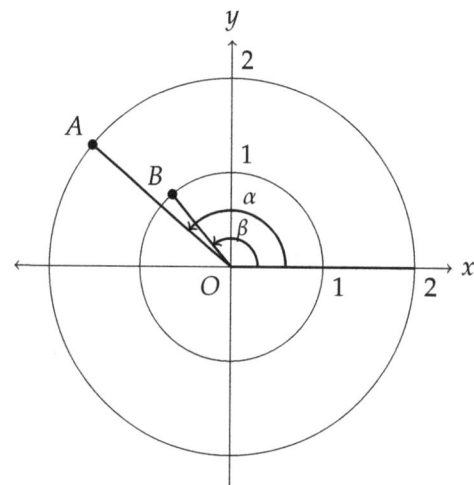

   Which of the following is true?

   (A) $\sin(\alpha) > \sin(\beta)$
   (B) $\sin(\alpha) = \sin(\beta)$
   (C) $\sin(\alpha) < \sin(\beta)$
   (D) It is impossible to determine without more precise labeling of the graph.

2. The function $f(x) = 2x$ passes through the origin. The graph of f makes an acute angle of θ with the positive x-axis. What is $\frac{\sin\theta}{\cos\theta}$?

   (A) -2 **(B)** $-\frac{1}{2}$ **(C)** $\frac{1}{2}$ **(D)** 2

3. An air traffic controller is working with software that specifies the location of planes in terms of an angle and the distance away from the control tower. The angle can be considered an angle in standard position in the coordinate plane if the control tower is at the origin. The air traffic controller spots a certain plane that is 5 miles away at an angle of 3 radians. Which of the following best describes the location of the plane respective to the control tower in terms of a cardinal direction?
   (A) $|3\sin 5|$ miles north of the tower
   (B) $|3\sin 5|$ miles south of the tower
   (C) $|5\cos 3|$ miles east of the tower
   (D) $|5\cos 3|$ miles west of the tower

4. The point $(5\sqrt{3}, -5)$ lies in the coordinate plane at the intersection of the terminal ray of an angle θ and a circle of radius r. What is the value of r?
   (A) $\frac{5}{2}$ **(B)** 5 **(C)** $5\sqrt{2}$ **(D)** 10

5. The point A lies at the intersection of the terminal ray of an angle in standard position measuring θ and the circle shown below.

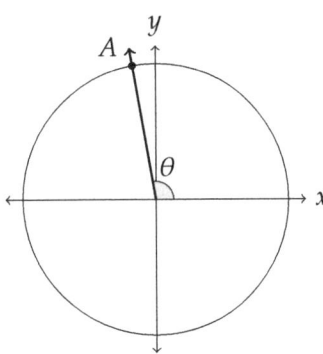

Which of the following is true?

(A) $\cos\theta < \sin\theta < \tan\theta$
(B) $\cos\theta < \tan\theta < \sin\theta$
(C) $\tan\theta < \sin\theta < \cos\theta$
(D) $\tan\theta < \cos\theta < \sin\theta$

Solutions.

1. Note that the ray $\overrightarrow{OA}$ intersects the unit circle at a point whose vertical displacement is less than the vertical displacement of B. Hence, $\sin\alpha < \sin\beta$. The correct answer is therefore **(C)**.

2. The ration of sine to cosine is the tangent, and the tangent of the angle is the slope of the line, which is 2. The correct answer is therefore **(D)**.

3. The distance from the origin horizontally is $5\cos 3$ and vertically is $5\sin 3$. Since 3 radians is close to π radians which is directly west of the tower, the distance must be $|5\cos 3|$ west of the tower. The correct answer is therefore **(D)**.

4. We have that $r\cos\theta = 5\sqrt{3}$ and $r\sin\theta = -5$. The presence of $\sqrt{3}$ implies that $\cos\theta = \frac{\sqrt{3}}{2}$ and $\sin\theta = -\frac{1}{2}$. Hence, $r = 10$. The correct answer is therefore **(D)**.

5. Note that $\cos\theta$ is small in magnitude and negative, and $\sin\theta$ is nearly 1. Hence, $\tan\theta$ is a number close to 1 divided by a negative number close to zero so $\tan\theta$ is negative and has larger magnitude than $\cos\theta$. Thus, $\tan\theta$ the smallest of the three, and $\sin\theta$ is the largest of the three. The correct answer is therefore **(D)**.

Topic 3.3 ~ Sine and Cosine Function Values

Learning Objectives
1. (3.3.A) Determine coordinates of points on a circle centered at the origin.

Success Criteria
1. I can compute the values of sine, cosine, and tangent for angles that are multiples of $\frac{\pi}{4}$ and $\frac{\pi}{6}$.
2. I can compute the coordinates of a point on a circle centered at the origin given the radius and angle.

It turns out that most values of $\sin x$, $\cos x$, and $\tan x$ are rather messy – in fact, if all the output values of these functions were an ocean, the amount of outputs that are rational or easily expressed as a radical wouldn't even make up a single drop.

Therefore, it is special when trigonometric functions do have "nice" outputs. It just so happens that this will occur for very convenient divisions of a circle.

We already discovered the fact that $\sin x$ and $\cos x$ equal either 0, 1, or −1 for any multiple of $\frac{\pi}{2}$, or any multiple of a quarter circle. What about other divisions of a circle?

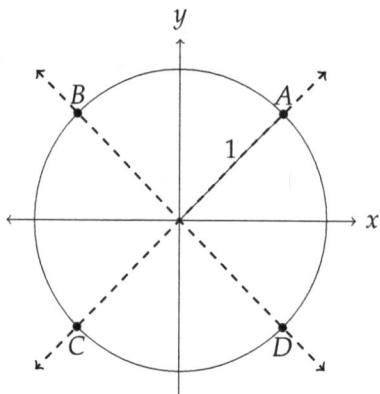

The circle above has been divided into 8 sectors, with rays drawn at each of the odd multiples of $\frac{\pi}{4}$. Using the geometry of an isosceles right triangle, we can compute $\sin \frac{\pi}{4}$ and $\cos \frac{\pi}{4}$.

Using symmetry, we can actually compute sine, cosine, and tangent for *any* multiple of $\frac{\pi}{4}$.

θ	$\frac{\pi}{4}$	$\frac{3\pi}{4}$	$\frac{5\pi}{4}$	$\frac{7\pi}{4}$
$\sin\theta$				
$\cos\theta$				
$\tan\theta$				

We can also compute the values of $\sin x$ and $\cos x$ for any multiples of $\frac{\pi}{6}$. Let's investigate. We can actually utilize the geometry of an _____ triangle, which we'll draw at the right, to determine the vertical and horizontal displacement of P.

$$\sin\frac{\pi}{6} = \underline{}$$

With the Pythagorean Theorem, we can establish

$$\cos\frac{\pi}{6} = \underline{} = \underline{} = \underline{}$$

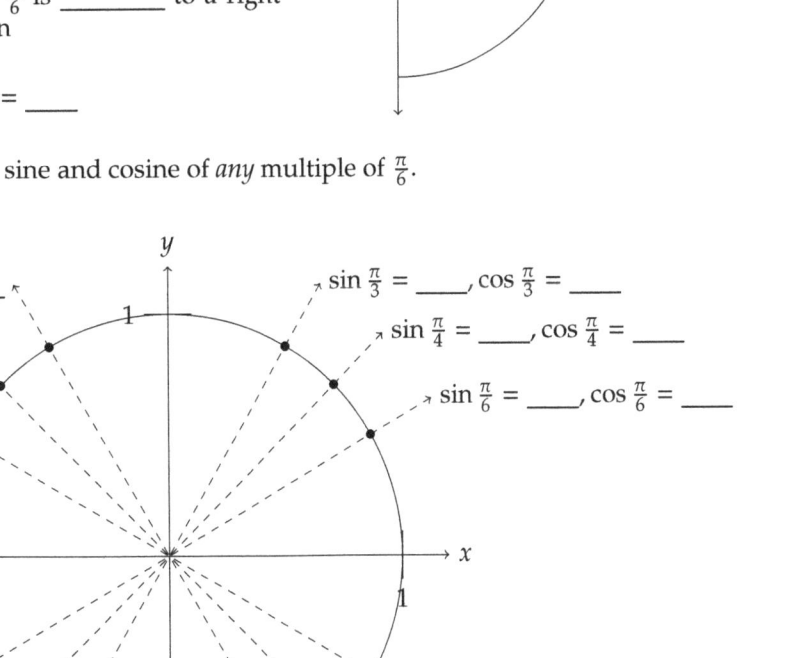

While this gave us $\sin\frac{\pi}{6}$ and $\cos\frac{\pi}{6}$, we can actually use the fact that a right triangle with an angle of $\frac{\pi}{6}$ is _____ to a right triangle with an angle of $\frac{\pi}{3}$ to ascertain

$$\sin\frac{\pi}{3} = \underline{}, \cos\frac{\pi}{3} = \underline{}$$

Using symmetry, we can compute the sine and cosine of *any* multiple of $\frac{\pi}{6}$.

$\sin\frac{2\pi}{3} = \underline{}, \cos\frac{2\pi}{3} = \underline{}$

$\sin\frac{3\pi}{4} = \underline{}, \cos\frac{3\pi}{4} = \underline{}$

$\sin\frac{5\pi}{6} = \underline{}, \cos\frac{5\pi}{6} = \underline{}$

$\sin\frac{\pi}{3} = \underline{}, \cos\frac{\pi}{3} = \underline{}$

$\sin\frac{\pi}{4} = \underline{}, \cos\frac{\pi}{4} = \underline{}$

$\sin\frac{\pi}{6} = \underline{}, \cos\frac{\pi}{6} = \underline{}$

$\sin\frac{7\pi}{6} = \underline{}, \cos\frac{7\pi}{6} = \underline{}$

$\sin\frac{5\pi}{4} = \underline{}, \cos\frac{5\pi}{4} = \underline{}$

$\sin\frac{4\pi}{3} = \underline{}, \cos\frac{4\pi}{3} = \underline{}$

$\sin\frac{11\pi}{6} = \underline{}, \cos\frac{11\pi}{6} = \underline{}$

$\sin\frac{7\pi}{4} = \underline{}, \cos\frac{7\pi}{4} = \underline{}$

$\sin\frac{5\pi}{3} = \underline{}, \cos\frac{5\pi}{3} = \underline{}$

Once we know values of the trigonometric functions for the _____, we can then dilate to find points that are ___ distance away from the origin for these special angles.

Fact: A point at the intersection of the terminal ray of an angle θ in standard position and a circle of radius r has coordinate (_____, _____).

Example 1: A line passes through the origin, Quadrant II, and Quadrant IV. The line makes an angle of $\frac{2\pi}{3}$ with the positive x-axis. At what point will this line intersect a circle with a radius of 6?

Example 2: The point $(-4, 4\sqrt{3})$ lies at the intersection of the terminal ray of an angle θ in standard position and a circle centered at the origin with radius r. What are θ and r?

Practice: In the figure, the smaller angle between the rays containing A and D is $\frac{7\pi}{12}$, and the rays containing B and C are perpendicular.

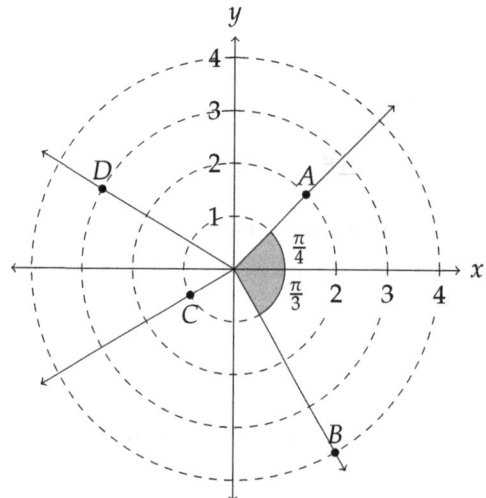

Find the coordinates of the points $A, B, C,$ and D.

Topic 3.3 Homework

1. For each angle θ and radius r given below, find the coordinates of the point at the intersection of the terminal ray of θ in standard position and the circle centered at the origin with radius r.

   (a) $\theta = \dfrac{\pi}{6}, r = 2$

   (b) $\theta = \dfrac{2\pi}{3}, r = 1$

   (c) $\theta = \dfrac{\pi}{3}, r = 5$

   (d) $\theta = \dfrac{3\pi}{4}, r = 12$

   (e) $\theta = -\dfrac{\pi}{4}, r = 4$

   (f) $\theta = -\dfrac{\pi}{2}, r = 6$

   (g) $\theta = \dfrac{4\pi}{3}, r = 10$

   (h) $\theta = \dfrac{7\pi}{4}, r = 5\sqrt{2}$

   (i) $\theta = 6\pi, r = 13$

2. Each point below is at the intersection of an angle in standard position measuring θ radians and a circle centered at the origin with radius r. Find θ, with $0 < \theta < 2\pi$, and r.

   (a) $(1, \sqrt{3})$

   (b) $\left(-5\sqrt{2}, -5\sqrt{2}\right)$

   (c) $(0, -8)$

   (d) $\left(\dfrac{5\sqrt{3}}{2}, \dfrac{5}{2}\right)$

   (e) $\left(-\dfrac{1}{4}, \dfrac{\sqrt{3}}{4}\right)$

   (f) $(2, -2)$

3. The perimeter of the rectangle below can be expressed as $a \sin \theta + b \cos \theta$ for integers a and b and an angle θ in radians. Find $a, b,$ and θ.

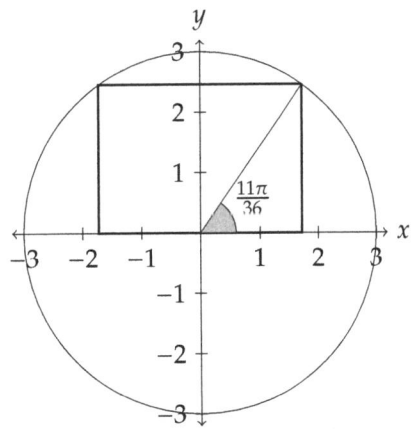

4. What is the midpoint of the segment AB with point $A\left(4\cos\frac{5\pi}{6}, 4\sin\frac{5\pi}{6}\right)$ and $B\left(4\cos\frac{7\pi}{6}, 4\sin\frac{7\pi}{6}\right)$?

Topic 3.3 Solutions/Notes

Activity

Derivation of $\sin\frac{\pi}{4}$ and $\cos\frac{\pi}{4}$: Let $x = \sin\frac{\pi}{4}$. Then $\cos\frac{\pi}{4} = x$ as well, as a right triangle with an angle of $\frac{\pi}{4}$ is isosceles.

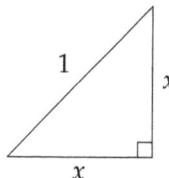

By the Pythagorean Theorem, we get

$$x^2 + x^2 = 1 \implies x = \sqrt{\frac{1}{2}} = \frac{\sqrt{2}}{2}$$

Therefore $\sin\frac{\pi}{4} = \cos\frac{\pi}{4} = \frac{\sqrt{2}}{2}$.

For the symmetry portion, if A is (x, y), then $B, C,$ and D can be seen to have coordinates of the form $(-x, y), (-x, -y),$ and $(x, -y)$. This can help to complete the table.

θ	$\frac{\pi}{4}$	$\frac{3\pi}{4}$	$\frac{5\pi}{4}$	$\frac{7\pi}{4}$
$\sin\theta$	$\frac{\sqrt{2}}{2}$	$\frac{\sqrt{2}}{2}$	$-\frac{\sqrt{2}}{2}$	$-\frac{\sqrt{2}}{2}$
$\cos\theta$	$\frac{\sqrt{2}}{2}$	$-\frac{\sqrt{2}}{2}$	$-\frac{\sqrt{2}}{2}$	$\frac{\sqrt{2}}{2}$
$\tan\theta$	1	-1	1	-1

We can also compute the values of $\sin x$ and $\cos x$ for any multiples of $\frac{\pi}{6}$. Let's investigate.

We can actually utilize the geometry of an equilateral triangle, which we'll draw at the right, to determine the vertical and horizontal displacement of P.

$$\sin\frac{\pi}{6} = \frac{1}{2}$$

With the Pythagorean Theorem, we can establish

$$\cos\frac{\pi}{6} = \sqrt{1 - \left(\frac{1}{2}\right)^2} = \sqrt{\frac{3}{4}} = \frac{\sqrt{3}}{2}$$

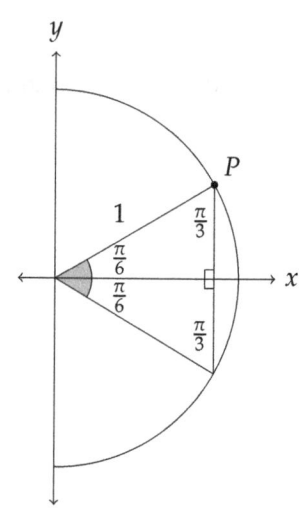

While this gave us $\sin\frac{\pi}{6}$ and $\cos\frac{\pi}{6}$, we can actually use the fact that a right triangle with an angle of $\frac{\pi}{6}$ is congruent to a right triangle with an angle of $\frac{\pi}{3}$ to ascertain

$$\sin\frac{\pi}{3} = \frac{\sqrt{3}}{2}, \cos\frac{\pi}{3} = \frac{1}{2}$$

Using symmetry, we can compute the sine and cosine of *any* multiple of $\frac{\pi}{6}$.

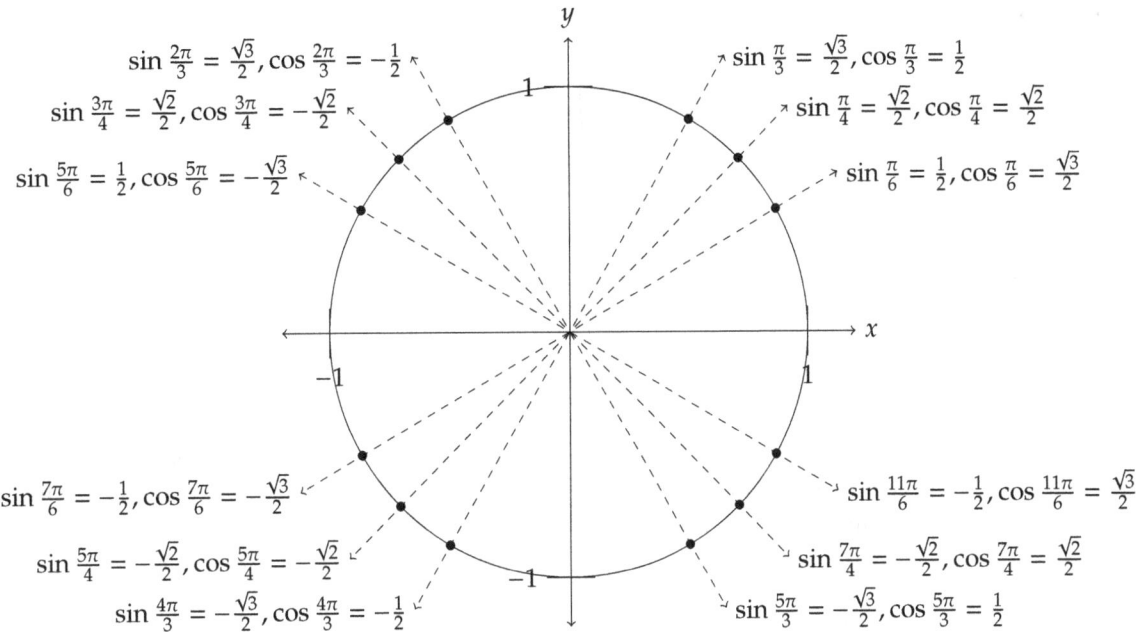

Once we know values of the trigonometric functions for the <u>unit circle</u>, we can then dilate to find points that are <u>any</u> distance away from the origin for these special angles.

Fact: A point at the intersection of the terminal ray of an angle θ in standard position and a circle of radius r has coordinates $(r\cos\theta, r\sin\theta)$.

Example 1 Solution: $\left(6\cos\frac{2\pi}{3}, 6\sin\frac{2\pi}{3}\right) \to \left(6\left(-\frac{1}{2}\right), 6\left(\frac{\sqrt{3}}{2}\right)\right) = (-3, 3\sqrt{3})$

Example 2 Solution: The y-coordinate involving $\sqrt{3}$ reminds us of a multiple of $\frac{\pi}{3}$. Along with the negative x-coordinate, we consider $\theta = \frac{2\pi}{3}$, which would intersect a unit circle at $\left(-\frac{1}{2}, \frac{\sqrt{3}}{2}\right)$. We can divide the coordinates of our desired point to find the factor of dilation and therefore the radius:

$$4\sqrt{3} \div \frac{\sqrt{3}}{2} = 4\sqrt{3} \cdot \frac{2}{\sqrt{3}} = 8$$

$$-4 \div -\frac{1}{2} = -4 \cdot -2 = 8$$

We have $\theta = \frac{2\pi}{3}$ and $r = 8$.

Solution to Practice:

$A = \left(2\cos\frac{\pi}{4}, 2\sin\frac{\pi}{4}\right) = \left(2 \cdot \frac{\sqrt{2}}{2}, 2 \cdot \frac{\sqrt{2}}{2}\right) = (\sqrt{2}, \sqrt{2})$

$B = \left(4\cos\left(-\frac{\pi}{3}\right), 4\sin\left(-\frac{\pi}{3}\right)\right) = \left(4 \cdot \frac{1}{2}, 4\left(-\frac{\sqrt{3}}{2}\right)\right) = (2, -2\sqrt{3})$

$C = \left(\cos\left(-\frac{\pi}{3} - \frac{\pi}{2}\right), \sin\left(-\frac{\pi}{3} - \frac{\pi}{2}\right)\right) = \left(\cos\left(-\frac{5\pi}{6}\right), \sin\left(-\frac{5\pi}{6}\right)\right) = \left(-\frac{\sqrt{3}}{2}, -\frac{1}{2}\right)$

$D = \left(3\cos\left(\frac{\pi}{4} + \frac{7\pi}{12}\right), 3\sin\left(\frac{\pi}{4} + \frac{7\pi}{12}\right)\right) = \left(3\cos\left(\frac{5\pi}{6}\right), 3\sin\left(\frac{5\pi}{6}\right)\right) = \left(3\left(-\frac{\sqrt{3}}{2}\right), 3 \cdot \frac{1}{2}\right) = \left(-\frac{3\sqrt{3}}{2}, \frac{3}{2}\right)$

Homework

1. (a) $\left(\sqrt{3}, 1\right)$

   (b) $\left(-\frac{1}{2}, \frac{\sqrt{3}}{2}\right)$

   (c) $\left(\frac{5}{2}, \frac{5\sqrt{3}}{2}\right)$

   (d) $\left(-6\sqrt{2}, 6\sqrt{2}\right)$

   (e) $\left(4\sqrt{2}, -4\sqrt{2}\right)$

   (f) $(0, -6)$

   (g) $\left(-5, -5\sqrt{3}\right)$

   (h) $(5, -5)$

   (i) $(13, 0)$

2. (a) $\theta = \frac{\pi}{3}, r = 2$

   (b) $\theta = \frac{5\pi}{4}, r = 10$

   (c) $\theta = \frac{3\pi}{2}, r = 8$

   (d) $\theta = \frac{\pi}{6}, r = 5$

   (e) $\theta = \frac{2\pi}{3}, r = \frac{1}{2}$

   (f) $\theta = \frac{7\pi}{4}, r = 2\sqrt{2}$

3. We have an angle in standard position and it intersects a circle of radius 3 at the topright corner of the rectangle. Therefore, the height of the rectangle is $3\sin\left(\frac{11\pi}{36}\right)$ and the width of the rectangle is double the horizontal displacement, or $2\left(3\cos\left(\frac{11\pi}{36}\right)\right) = 6\cos\left(\frac{11\pi}{36}\right)$. Therefore, the perimeter of the rectangle is $6\sin\left(\frac{11\pi}{36}\right) + 12\cos\left(\frac{11\pi}{36}\right)$, giving $a = 6, b = 12$, and $\theta = \frac{11\pi}{36}$.

4. A and B are the intersection points of two angles, $\frac{5\pi}{6}$ and $\frac{7\pi}{6}$, in standard position and a circle of radius 4. These points are reflections of one another over the x-axis. Therefore, the midpoint of AB is just $\left(4\cos\frac{5\pi}{6}, 0\right)$, or $(-2\sqrt{3}, 0)$.

Reflection

Suggested Class Time. 60-70 minutes

Prerequisites. Students should be very comfortable with radians, what $\sin x, \cos x$, and $\tan x$ measure, and the Pythagorean Theorem.

Instructional Strategies. This was a teacher-led lesson, and it went quite smoothly.

The derivation of $\sin\frac{\pi}{4}$ and $\cos\frac{\pi}{4}$ was straightforward. When we got to B in Quadrant II, I asked students what the angle measure for this ray was. A few quickly got $\frac{3\pi}{4}$, but I reiterated that the divisions of the circle made this easy. Rather than have to multiply 3 times 45° (or any other conversion formula), I showed that the angle of the ray through B was just three times the angle of the ray through B, hence $3 \cdot \frac{\pi}{4} = \frac{3\pi}{4}$. Students quickly enough caught on to the fact that C and D were the next odd multiples of $\frac{\pi}{4}$.

For teaching the equilateral triangle, I showed students that we actually were just taking half of a regular polygon again: for the $\frac{\pi}{4}$ angle, we looked at a triangle that was half of a square. This time, we were simply looking at half of an equilateral triangle. After we derived $\sin\frac{\pi}{6}$ and $\cos\frac{\pi}{6}$, I went to the TI-84 emulator and entered both values. I've found that giving students an intuitive idea of how large $\frac{\sqrt{3}}{2}$ and $\frac{\sqrt{2}}{2}$ are is useful for their generating unit circle values. I then asked students to predict the value of $\sin\frac{\pi}{3}$. Students had interesting responses - many predicted the correct $\frac{\sqrt{3}}{2}$, but others thought that perhaps the value of $\sin\frac{\pi}{6}$ should double. If we'd had more time, this could've led to a useful discussion about the growth rate of a sine function (and how it isn't linear!).

We did the top half of completing the unit circle together, and then students did the bottom half on their own. There were literally no questions!

Going into the polar coordinates, I made sure to emphasize the connection to the previous lesson, as we had somewhat rushed it (as mentioned in the previous reflection). I really wanted students to understand

UNIT 3　　　　　　　　　　　　　　　TOPIC 3.3 ~ SINE AND COSINE FUNCTION VALUES　　　299

the concept of dilating a known circle. Once we got to Example 1, I made a big deal about drawing a picture: most of the questions in AP Classroom are quite wordy and often come without diagrams, so I told my students it would be paramount that they be able to accurately translate words into pictures. Example 1 then went quite quickly.

For Example 2, I asked students what was "familiar" about the point $(-4, 4\sqrt{3})$, and I got surprisingly little response. Most likely, the $\frac{\sqrt{3}}{2}$ was still too new to them. I then mentioned how the $\sqrt{3}$ should generally make them think of $\frac{\pi}{6}$ and $\frac{\pi}{3}$. From there, our approach was to determine our quadrant, identify a known point with a similar structure (in this case, $\left(-\frac{1}{2}, \frac{\sqrt{3}}{2}\right)$), and then work backwards to determine the radius.

I gave students 7-8 minutes to do the practice question. Overall, it seemed like students were getting about 50-75% perfectly correctly, with some minor hiccups on identifying the precise angles that D and C corresponded to.

As an aside, I created a Delta Math assignment for identifying exact values of sine and cosine for my students as extra credit in addition to the homework. I gave them a week to complete it.

Technology. I used an emulator on the board, but students did not use any devices.

Problems.

1. In the xy-plane, the point A is the intersection of the terminal ray of an angle θ in standard position and a circle of radius r. Which of the following are the coordinates of A?

    (A) $(\sin \theta, \cos \theta)$　　**(B)** $(\cos \theta, \sin \theta)$　　**(C)** $(r \sin \theta, r \cos \theta)$　　**(D)** $(r \cos \theta, r \sin \theta)$

2. The point $(6, -6\sqrt{3})$ lies in the coordinate plane at the intersection of the terminal ray of an angle θ, $-\pi \leq x \leq \pi$, and a circle of radius r. What is the value of θ?

    (A) $-\dfrac{\pi}{6}$　　**(B)** $-\dfrac{\pi}{3}$　　**(C)** $\dfrac{2\pi}{3}$　　**(D)** $\dfrac{5\pi}{6}$

3. The point $(-4\sqrt{2}, 4\sqrt{2})$ lies in the coordinate plane at the intersection of the terminal ray of an angle $\theta, 0 \leq \theta \leq \pi$, and a circle of radius r. What is the value of $\tan \theta$?

    (A) -1　　**(B)** $-\dfrac{\sqrt{2}}{2}$　　**(C)** $\dfrac{\sqrt{2}}{2}$　　**(D)** 1

4. The point $A(3, 3\sqrt{3})$ lies in the coordinate plane at the intersection of the terminal ray of an angle α in standard position, $0 \leq \alpha \leq 2\pi$, and a circle of radius r. The point $B(-3\sqrt{3}, -3)$ lies in the coordinate plane at the intersection of the terminal ray of an angle β in standard position, $0 \leq \beta \leq 2\pi$, and the same circle of radius r. What is the length of the arc AB?

    (A) 5π　　**(B)** 7π　　**(C)** $3\sqrt{6} + 3\sqrt{2}$　　**(D)** $6\sqrt{6} + 6\sqrt{2}$

5. The line $y = x$ intersects a circle centered at the origin of radius 8. Let θ, for $0 \leq \theta \leq \pi$, be the angle the line makes with the positive x-axis. For this value of θ, what are the coordinates of the point of intersection of the line and the circle?

    (A) $(8, 8)$　　**(B)** $\left(8\sqrt{2}, 8\sqrt{2}\right)$　　**(C)** $\left(4\sqrt{2}, 4\sqrt{2}\right)$　　**(D)** $(4, 4)$

Solutions.

1. This is simply the definition of the point on the circle made by an angle. The correct answer is therefore **(D)**.

2. The y-coordinate involving $\sqrt{3}$ reminds us of a multiple of $\frac{\pi}{3}$. Since the y-coordinate is negative and the x-coordinate is positive, we use $-\pi/3$. The correct answer is therefore **(B)**.

3. Since $\tan\theta = \dfrac{\sin\theta}{\cos\theta} = \dfrac{r\sin\theta}{r\cos\theta}$, we have $\tan\theta = \dfrac{4\sqrt{2}}{-4\sqrt{2}} = -1$. The correct answer is therefore **(A)**.

4. The radius of the circle is 6, so $\alpha = \dfrac{\pi}{3}$ and $\beta = \dfrac{7\pi}{6}$. Then the measure of the arc is $6\left(\dfrac{7\pi}{6} - \dfrac{\pi}{3}\right) = 6 \cdot \dfrac{5\pi}{6} = 5\pi$. The correct answer is therefore **(A)**.

5. The slope of the line is $\tan\theta$. Since the slope is 1, the angle must be $\dfrac{\pi}{4}$. Then the coordinates are $(8\cos\frac{\pi}{4}, 8\sin\frac{\pi}{4}) = (4\sqrt{2}, 4\sqrt{2})$. The correct answer is therefore **(C)**.

Topics 3.4-3.5 ~ Graphs of Sine and Cosine (🖩)

Learning Objectives
1. (3.4.A) Construct representations of the sine and cosine functions using the unit circle.
2. (3.5.A) Identify key characteristics of the sine and cosine functions.

Success Criteria
1. I can construct a graph of sine and cosine and interpret how input-output pairs relate to the unit circle.
2. I can identify characteristics of the graph of a sinusoidal function given a graph or context.

We know that $\sin x$ is the ratio of the vertical displacement to the distance from the origin: we also know that, if the circle is a *unit circle*, this ratio can be considered *just* the vertical displacement. For $\cos x$, this amounts to being *just* the horizontal displacement. With that in mind, we can construct input-output pairs for $f(x) = \sin x$ and $g(x) = \cos x$ with the help of a unit circle.

To the left is a unit circle with 24 points - each one is labeled as the intersection of the terminal ray (not shown) of a given angle and the unit circle.

1. (🖩) For each angle given, draw a segment representing the vertical displacement. Then, **use your calculator** to fill in the table to the right with the corresponding vertical displacement.

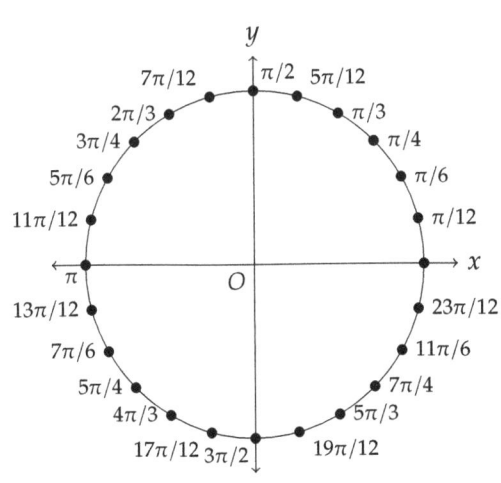

θ	$f(x) = \sin x$	θ	$f(x) = \sin x$
0		π	
$\pi/12$		$13\pi/12$	
$\pi/6$		$7\pi/6$	
$\pi/4$		$5\pi/4$	
$\pi/3$		$4\pi/3$	
$5\pi/12$		$17\pi/12$	
$\pi/2$		$3\pi/2$	
$7\pi/12$		$19\pi/12$	
$2\pi/3$		$5\pi/3$	
$3\pi/4$		$7\pi/4$	
$5\pi/6$		$11\pi/6$	
$11\pi/12$		$23\pi/12$	

2. Graph the input-output pairs from the table on the graph below.

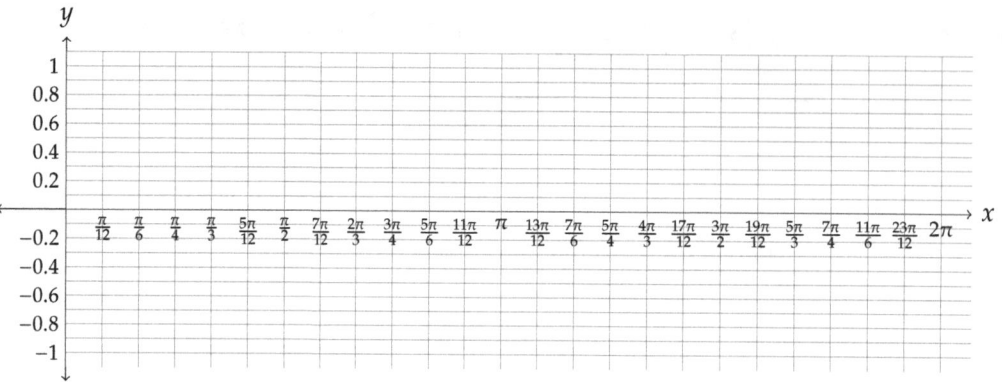

The graph of sine has a few notable characteristics.

- The domain of $f(x) = \sin x$ is all real numbers, and the range is _____.

- For any input-output pair $(x, \sin x)$, the output $\sin x$ represents the _____ of the intersection of the terminal ray of an angle measuring _____ and the _____. (Note: The inputs may be notated with a θ rather than an x to really emphasize the angular nature.)

- The MIDLINE is the average of the minimum and the maximum values. For $f(x) = \sin x$, the midline is the line _____.

- The AMPLITUDE of $f(x) = \sin x$ is the distance between the maximum and the midline (or the midline and the minimum). The amplitude of $f(x) = \sin x$ is ____.

- The PERIOD of $f(x) = \sin x$ is the shortest interval of inputs after which the outputs will begin to repeat. The period of $f(x) = \sin x$ is ____.

- The FREQUENCY of $f(x) = \sin x$ is the reciprocal of the period. The frequency of $f(x) = \sin x$ is ____.

- The graph of $f(x) = \sin x$ alternatively changes between increasing and decreasing and also between concave up and concave down.

The graph of $g(x) = \cos x$ is very similar to the graph of $f(x) = \sin x$.

3. For each angle given, draw a segment representing the horizontal displacement.

4. (🖩) **Use your calculator** to compute the horizontal displacements for *only the angles in Quadrant I*. Fill in the corresponding values in the table.

θ	$f(x) = \cos x$	θ	$g(x) = \cos x$
0		π	
$\pi/12$		$13\pi/12$	
$\pi/6$		$7\pi/6$	
$\pi/4$		$5\pi/4$	
$\pi/3$		$4\pi/3$	
$5\pi/12$		$17\pi/12$	
$\pi/2$		$3\pi/2$	
$7\pi/12$		$19\pi/12$	
$2\pi/3$		$5\pi/3$	
$3\pi/4$		$7\pi/4$	
$5\pi/6$		$11\pi/6$	
$11\pi/12$		$23\pi/12$	

5. Look at the unit circle with the horizontal displacement segments. How could you transform this diagram to look identical to the unit circle diagram for $\sin x$ in Problem 1?

6. Look at the outputs you got in the table - in order. How do they relate to the outputs in the table for $f(x) = \sin x$?

7. Fill in the rest of the values in the table *without* the calculator using the pattern you found in Problems 5-6.

8. Graph the input-output pairs from the table to generate the graph of $g(x) = \cos x$.

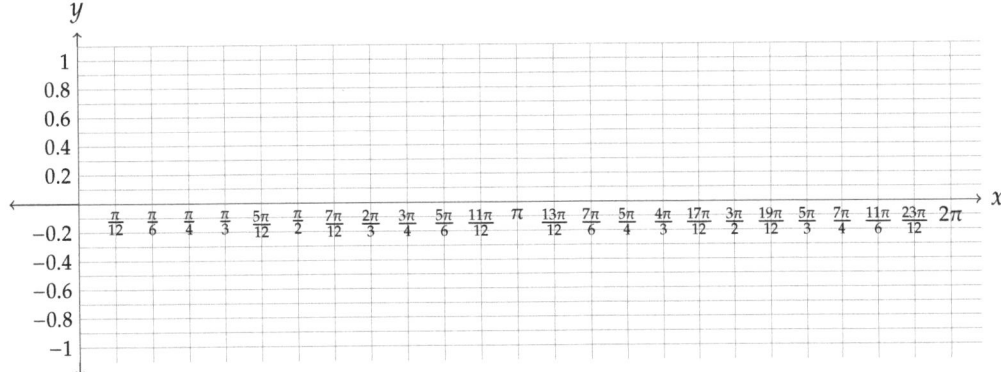

9. It would be a bit of a waste of time to spend time writing down all of the characteristics – domain, range, period, etc. – for $g(x) = \cos x$. Why?

10. It turns out that $g(x) = \cos x$ is just a horizontal _____ of $f(x) = \sin x$. Based on this, fill in the equation below:

For all x, $\cos(x) = \sin(\underline{\qquad})$

The functions $f(x) = \sin x$ and $g(x) = \cos x$ are called SINUSOIDAL functions, or functions that are additive or multiplicative transformations of $\sin x$. For instance, $\cos x$ is a sinusoidal function because $\cos x = \sin\left(x + \frac{\pi}{2}\right)$.

While $\sin x$ and $\cos x$ share many similarities, one major way they differ is their *symmetry*.

11. What kind of symmetry will the graph of $f(x) = \sin x$ have? Is $f(x) = \sin x$ an even function, an odd function, or neither?

12. What kind of symmetry will the graph of $g(x) = \cos x$ have? Is $g(x) = \cos x$ an even function, an odd function, or neither?

Sinusoidal functions appear frequently in natural phenomena (particularly in physics), and constructing sinusoidal models and answering questions about them in context can be aided by identifying features of the functions in context.

13. The daily average temperature in one city varied roughly sinusoidally over a period of 6 months. The high daily average temperature was 87° and the low daily average temperature was 52°. If a sinusoidal function were used to model the daily average temperature of this city over the six months, what would be the most reasonable estimates for the midline and amplitude of this function?

14. The voltage of current that a certain sinusoidal sound wave produces has a frequency of 440 hertz. What is the period of this wave?

15. The height of a seat on a circular amusement ride can be modeled by a sinusoidal function with a maximum of 25 feet. If the midline of this function is the line $y = 14$, what is the minimum height of the function?

Notes

Topics 3.4-3.5 Homework

1. The graph of $f(x) = \sin x$ contains the input-output pair $(0.6, 0.565)$. Interpret what the input and output represent in the context of the unit circle.

2. A student was working in their calculator and computed $\cos 3 = -0.9899924966$. They figured there must be a mistake because "cosine can't be negative." Explain why this value is indeed reasonable. Why is it so close to -1?

3. Identify the following characteristics of each function graphed below.

(a)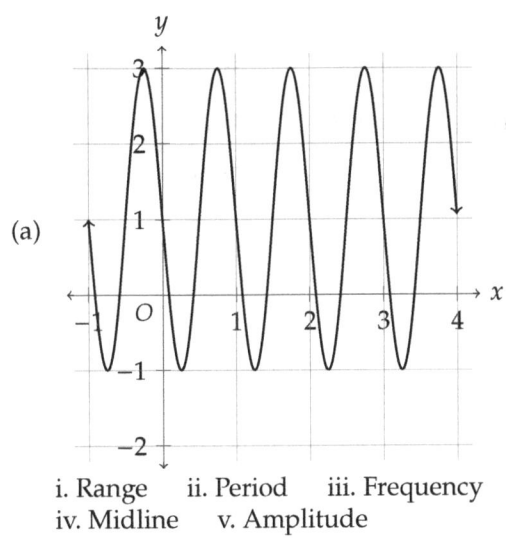

   i. Range   ii. Period   iii. Frequency
   iv. Midline   v. Amplitude

(b)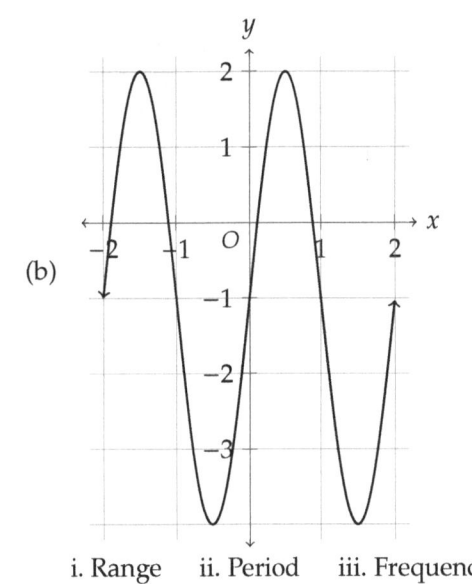

   i. Range   ii. Period   iii. Frequency
   iv. Midline   v. Amplitude

4. Rob and Dee were arguing over a homework question they had, which was to construct a function that could have the graph shown below.

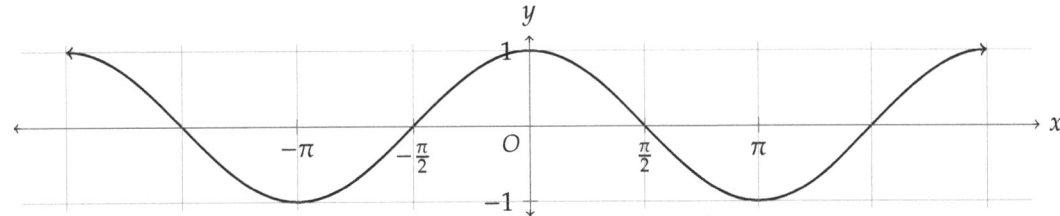

   Rob said this could be a sine function, and Dee argued it was a cosine function. Who is correct?

5. Suppose a sinusoidal function has a midline of $y = 8$ and an amplitude of 3. Come up with two variables in context that could be modeled by this function. What is the period of your function for these variables?

6. Suppose for angles α and β (alpha and beta, respectively) with $0 < \alpha < \frac{\pi}{2}$ and $\frac{\pi}{2} < \beta < \pi$ that the following values are known.
$$\sin \alpha = 0.42 \qquad \cos \beta = -0.85$$

   Determine the values of the following.

   (a) $\sin(-\alpha)$ (b) $\cos(-\beta)$ (c) $\sin\left(\frac{\pi}{2} + \beta\right)$ (d) $\sin(\pi - \alpha)$ (e) $\cos(\beta + \pi)$

Topics 3.4-3.5 Solutions/Notes

Activity

1.

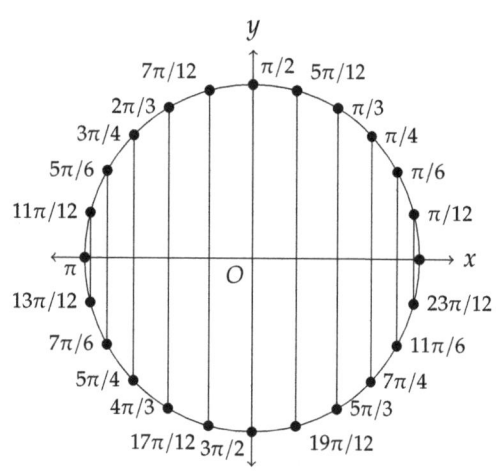

θ	$f(x) = \sin x$	θ	$f(x) = \sin x$
0	0	π	0
$\pi/12$	0.259	$13\pi/12$	-0.259
$\pi/6$	0.5	$7\pi/6$	-0.5
$\pi/4$	0.707	$5\pi/4$	-0.707
$\pi/3$	0.866	$4\pi/3$	-0.866
$5\pi/12$	0.966	$17\pi/12$	-0.966
$\pi/2$	1	$3\pi/2$	-1
$7\pi/12$	0.966	$19\pi/12$	-0.966
$2\pi/3$	0.866	$5\pi/3$	-0.866
$3\pi/4$	0.707	$7\pi/4$	-0.707
$5\pi/6$	0.5	$11\pi/6$	-0.5
$11\pi/12$	0.259	$23\pi/12$	-0.259

2.

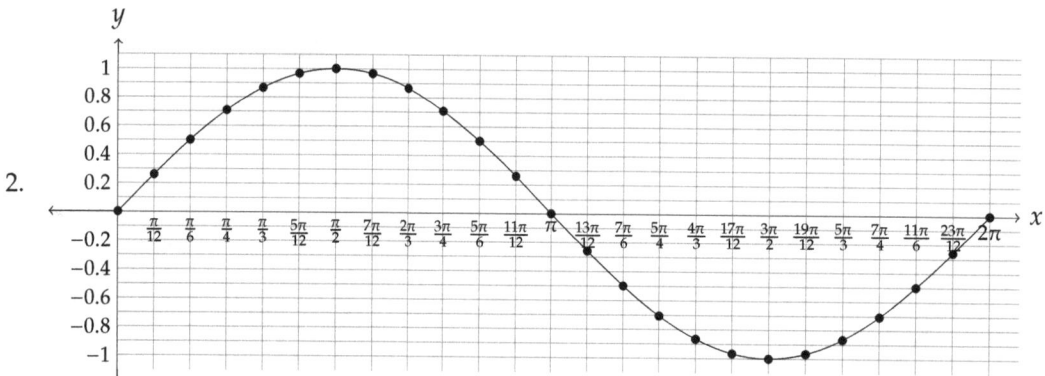

The graph of sine has a few notable characteristics.

- The domain of $f(x) = \sin x$ is all real numbers, and the range is $-1 \leq \sin x \leq 1$.

- For any input-output pair $(x, \sin x)$, the output $\sin x$ represents the vertical displacement from the x-axis of the intersection of the terminal ray of an angle measuring x radians and the unit circle. (Note: The inputs may be notated with a θ rather than an x to really emphasize the angular nature.)

- The MIDLINE is the average of the minimum and the maximum values. For $f(x) = \sin x$, the midline is the line $y = 0$.

- The AMPLITUDE of $f(x) = \sin x$ is the distance between the maximum and the midline (or the midline and the minimum). The amplitude of $f(x) = \sin x$ is 1.

- The PERIOD of $f(x) = \sin x$ is the shortest interval of inputs after which the outputs will begin to repeat. The period of $f(x) = \sin x$ is 2π.

- The FREQUENCY of $f(x) = \sin x$ is the reciprocal of the period. The frequency of $f(x) = \sin x$ is $\frac{1}{2\pi}$.
- The graph of $f(x) = \sin x$ alternatively changes between increasing and decreasing and also between concave up and concave down.

The graph of $g(x) = \cos x$ is very similar to the graph of $f(x) = \sin x$.

4.

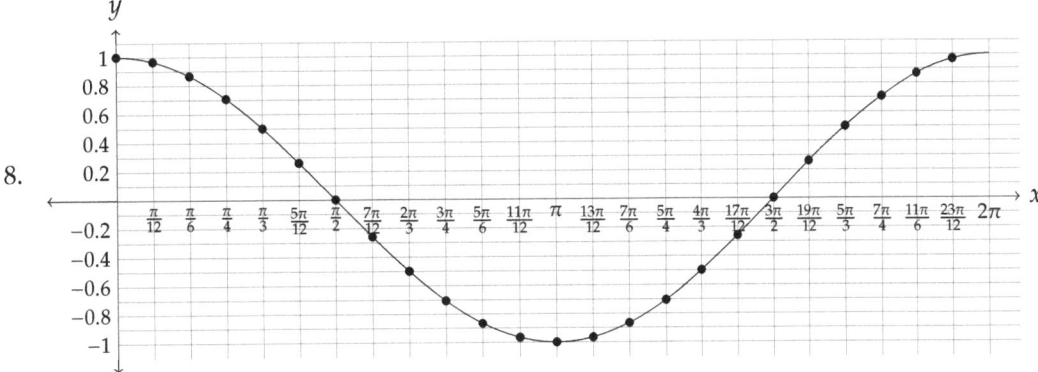

θ	$f(x) = \cos x$	θ	$g(x) = \cos x$
0	1	π	-1
$\pi/12$	0.966	$13\pi/12$	-0.966
$\pi/6$	0.866	$7\pi/6$	-0.866
$\pi/4$	0.707	$5\pi/4$	-0.707
$\pi/3$	0.5	$4\pi/3$	-0.5
$5\pi/12$	0.259	$17\pi/12$	-0.259
$\pi/2$	0	$3\pi/2$	0
$7\pi/12$	-0.259	$19\pi/12$	0.259
$2\pi/3$	-0.5	$5\pi/3$	0.5
$3\pi/4$	-0.707	$7\pi/4$	0.707
$5\pi/6$	-0.866	$11\pi/6$	0.866
$11\pi/12$	-0.966	$23\pi/12$	0.966

5. If the unit circle were rotated counterclockwise by $\frac{\pi}{2}$!

6. All of the outputs of $\cos x$ are just the outputs of $\sin x$, but for inputs shifted forward by $\frac{\pi}{2}$ ($\sin(x + \frac{\pi}{2})$).

8.

9. Most of the features are identical to those of $\sin x$.

10. $g(x) = \cos x$ is just a horizontal translation of $f(x) = \sin x$. For all x, $\cos(x) = \sin(x + \frac{\pi}{2})$.

11. The sign of vertical displacement is opposite for equivalent positive and negative angles, so $\sin x$ has rotational symmetry. It is an odd function.

12. The sign of horizontal displacement is the same for equivalent positive and negative angles, so $\cos x$ has reflectional symmetry. It is an even function.

13. The midline would be $y = \frac{87+52}{2} = 69.5$ and the amplitude would be $87 - 69.5 = 17.5$.

14. If the frequency is the reciprocal of the period, then the period is the reciprocal of the frequency. Therefore, the period is $\frac{1}{440}$.

15. The amplitude of the function is $25 - 14 = 11$, so the minimum height is the midline minus the amplitude, or $14 - 11 = 3$ feet.

> **Notes**
>
> The function $f(\theta) = \sin \theta$ takes an angle as an input and outputs the vertical displacement from the x-axis, while $g(\theta) = \cos \theta$ outputs the horizontal displacement (for θ in standard position and point where the ray intersects the unit circle).
>
> A function that involves additive or multiplicative transformations of $f(\theta) = \sin \theta$ is a SINUSOIDAL function. Cosine is a sinusoidal function because $\cos \theta = \sin(\theta + \frac{\pi}{2})$ for all θ.
>
> The functions $\sin x$ and $\cos x$ have the following characteristics.
>
> | | $\sin x$ | $\cos x$ |
> |---|---|---|
> | Symmetry | Rotational (Odd) | Reflectional (Even) |
> | Domain | All real numbers | |
> | Range | $[-1, 1]$ | |
> | Midline | $y = 0$ | |
> | Amplitude | 1 | |
> | Period | 2π | |
> | Frequency | $\frac{1}{\text{period}} = \frac{1}{2\pi}$ | |

Homework

1. For an angle measuring 0.6 radians in standard position in the xy-plane, the point where the angle's terminal ray intersects the unit circle is 0.565 above the x-axis.

2. The cosine of x measures horizontal displacement, which can be negative for any point in the plane to the left of the y-axis. For 3 radians, this is nearly π radians, which intersects the unit circle and the terminal ray of an angle measuring π in standard position at the point $(-1, 0)$. Therefore, the horizontal displacement will be very close to, but slightly less negative than, -1.

3. (a) i. $[-1, 3]$ ii. 1 iii. 1 iv. $y = 1$ v. 2
   (b) i. $[-3, 2]$ ii. 2 iii. $\frac{1}{2}$ iv. $y = -1$ v. 3

4. They could both be correct! This appears to be the graph of $\cos x$, which could also be written as $\sin(x + \frac{\pi}{2})$.

5. Student answers will vary.

6. (a) -0.42 (b) -0.85 (c) -0.85 (d) 0.42 (e) 0.85

Reflection

Suggested Class Time. 70 minutes

Prerequisites. Students need to know how to compute sine and cosine with their calculator in radian mode. They should also remember what sine and cosine measure, even and odd functions, and horizontal translations.

Instructional Strategies. Students were all over the place with filling out the table in Problem 1. It seemed like half the class was done with it (perfectly) in 5 minutes. Another chunk of the class had only part, or even very little of it complete. Some students were trying to fill in the ones they knew exact values for, while others had their calculators set in degrees. It was a learning experience for them, I'm sure, but it also meant I had to watch closely and make sure that groups that got significantly behind were still able to proceed in a timely manner. For next year, I will probably give everyone very clear verbal instructions on what to do for Problem 1, or I'll have one student in each group do all of the calculations. (Hopefully the calculator symbol and bold text will be clear enough!)

After about 20-30 minutes, we went over the answers to all of the "notable characteristics" as a class.

For the cosine graph, I had a ball with this. Students definitely did pick up on patterns, but they only picked up on the patterns *within* the table. In other words, they tried to use values of cosine and symmetry to predict other values of cosine. They certainly did not look to the table on the previous page for anything other than clues on symmetry.

What I did with each group was walk by and show them the two tables side by side - I would then shift the paper with the sine table down. This would show the tables lining up perfectly. Then, I'd take the unit circle graphs and show students how the horizontal displacements of cosine are just the vertical displacements of sine rotated precisely by $\frac{\pi}{2}$. It's hard to express just how fun this is to show students within the confines of a written reflection.

Once students got the cosine graph, it was smooth sailing. They were quite rusty on even and odd functions, but it didn't slow them down, as many skipped it. My way of teaching the even/odd when we got to the Notes was to show the following:

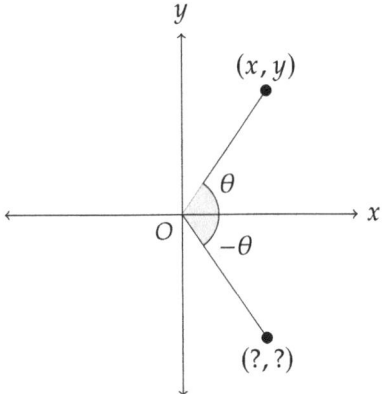

We did more formally write $\cos(x) = \cos(-x)$ and $\sin(-x) = -\sin(x)$, but the symmetry argument from above helped.

Technology. Students need their calculators for this lesson. Additionally, I pulled up a Geogebra sketch at the beginning of class to introduce the lesson: https://www.geogebra.org/m/fdubatzx.

Problems.

1. The point P lies at the intersection of the terminal ray of an angle θ, with $\pi < \theta < \frac{3\pi}{2}$, and a circle of radius 1.

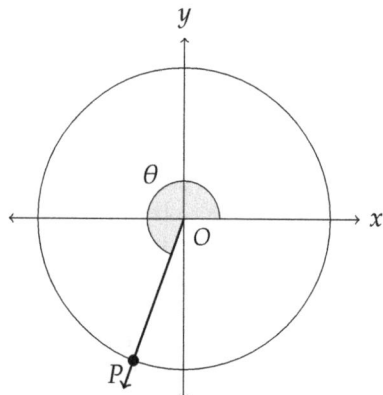

As the angle θ increases to $\frac{3\pi}{2}$, which of the following is true about the functions $\sin \theta$ and $\cos \theta$?

   (A) As θ increases to $\frac{3\pi}{2}$, $\sin \theta$ increases and $\cos \theta$ increases.

   (B) As θ increases to $\frac{3\pi}{2}$, $\sin \theta$ increases and $\cos \theta$ decreases.

   (C) As θ increases to $\frac{3\pi}{2}$, $\sin \theta$ decreases and $\cos \theta$ increases.

   (D) As θ increases to $\frac{3\pi}{2}$, $\sin \theta$ decreases and $\cos \theta$ decreases.

2. (▣) The function $g(x) = \cos x$ is graphed for $0 < x < \frac{\pi}{2}$. The point A lies at the intersection of the graph of g and the line $x = 0.3$.

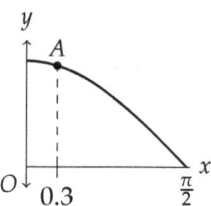

What is the y-coordinate of A and what does it represent in the context of the unit circle?

   (A) The y-coordinate of A is approximately 0.955. This means that the intersection of the unit circle and the terminal ray of an angle in standard position measuring 0.3 radian has a vertical displacement of 0.955 from the x-axis.

   (B) The y-coordinate of A is approximately 0.955. This means that the intersection of the unit circle and the terminal ray of an angle in standard position measuring 0.3 radian has a horizontal displacement of 0.955 from the y-axis.

   (C) The y-coordinate of A is approximately 0.955. This means that the intersection of the unit circle and the terminal ray of an angle in standard position measuring 0.3 radian has a vertical displacement of 0.045 from the x-axis.

   (D) The y-coordinate of A is approximately 0.955. This means that the intersection of the unit circle and the terminal ray of an angle in standard position measuring 0.3 radian has a horizontal displacement of 0.045 from the y-axis.

3. Let α be an angle in the interval $(\frac{\pi}{2}, \pi)$. Which of the following must be true?

    (A) $\sin \alpha > \cos \alpha$ **(B)** $\sin \alpha < \cos \alpha$ **(C)** $\sin \alpha < \tan \alpha$ **(D)** $\cos \alpha < \tan \alpha$

4. The daily average temperature D, in degrees Fahrenheit, in one city can be modeled by the function $D(t) = a \sin(b(t + c)) + d$, where t is the day of the year, with $t = 0$ representing January 1st. The maximum daily average temperature was $89°$ and the minimum daily average temperature was $35°$. Which of the following is the best estimate for the value of a?

    (A) 17.5   **(B)** 27   **(C)** 44.5   **(D)** 54

5. Let f be the function defined by $f(x) = \cos(x)$. Which of the following is equivalent to f for all x?

    (A) $g(x) = \sin(x - \pi)$ **(B)** $g(x) = \sin\left(x - \frac{\pi}{2}\right)$ **(C)** $g(x) = \sin\left(x + \frac{\pi}{2}\right)$ **(D)** $g(x) = \sin(x + \pi)$

Solutions.

1. The vertical displacement decreases to -1 and the horizontal displacement increases to 0. Hence, $\sin \theta$ decreases and $\cos \theta$ increases. The correct answer is therefore **(C)**.

2. A point on the graph of $\cos \theta$ measures the horizontal displacement from the y-axis to a point on the unit circle determined by θ. The correct answer is therefore **(B)**.

3. On the interval $(\frac{\pi}{2}, \pi)$, the cosine is negative and the sine is positive. Hence, the sine is larger than the cosine. The correct answer is therefore **(A)**.

4. Assuming $a > 0$, which is reasonable given the answer choices, the value of a is the amplitude, which is the distance from the midline to either the maximum or the minimum value. The midline is $\frac{35+89}{2} = 62$ so the amplitude is $89 - 62 = 27$. The correct answer is therefore **(B)**.

5. The graph of cosine is a horizontal translation of the graph of sine by $-\frac{\pi}{2}$ units. The correct answer is therefore **(C)**.

Topic 3.6 ~ Sinusoidal Transformations

Learning Objectives

1. (3.6.A) Identify the amplitude, vertical shift, period, and phase shift of a sinusoidal function.

Success Criteria

1. I can identify or choose a correct amplitude, vertical shift, period, and/or phase shift of a sinusoidal function given a graph or verbal description of that function.

In various previous topics, we have discussed function transformations. Refresh yourselves on the various transformations below. Consider a function $f(x)$ and a transformed function $T(x) = af(b(x+c)) + d$.

- *The effect of b*: The graph of $T(x)$ will be the graph of f _____ by a factor of ____.

- *The effect of c*: The graph of $T(x)$ will be the graph of f _____ by ____.

- *The effect of a*: The graph of $T(x)$ will be the graph of f _____ by a factor of ____.

- *The effect of d*: The graph of $T(x)$ will be the graph of f _____ by ____.

When we apply some combination of these transformations to sinusoidal functions, it ends up affecting the period, midline, amplitude, frequency, and so on. You may see sinusoidal functions written as $f(x) = a\sin(b(x+c)) + d$, but remember: *the specific letters used as variables are unimportant!*

As you practice various questions, it will be helpful to know some "key points" for each of $\sin x$ and $\cos x$.

1. Fill in the value of $\sin x$ at each given input. Then, sketch the input-output pairs on the graph to the right.

x	$\sin x$
0	
$\frac{\pi}{2}$	
π	
$\frac{3\pi}{2}$	
2π	

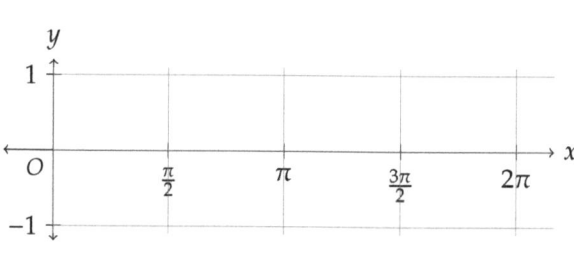

2. Fill in the value of $\cos x$ at each given input. Then, sketch the input-output pairs on the graph to the right.

x	$\cos x$
0	
$\frac{\pi}{2}$	
π	
$\frac{3\pi}{2}$	
2π	

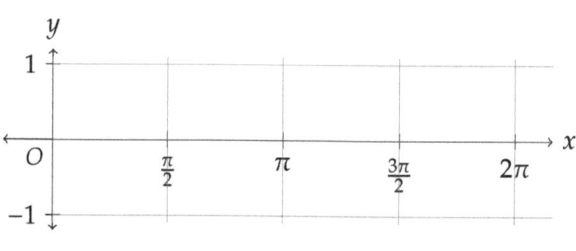

UNIT 3   TOPIC 3.6 ~ SINUSOIDAL TRANSFORMATIONS   313

Knowing these five input-output pairs, we can sketch graphs of more sinusoidal functions.

Just use the known points, transform them according to order of operations, and connect the points with a sinusoidal function.

3. On the xy-plane below, sketch a graph of $g(x) = 3\sin\left(x - \frac{\pi}{2}\right) + 1$. Use the table at the left to keep track of the five key points and how their outputs change with each successive transformation. (The columns stand for "Transformation 1," "Transformation 2," and so on.)

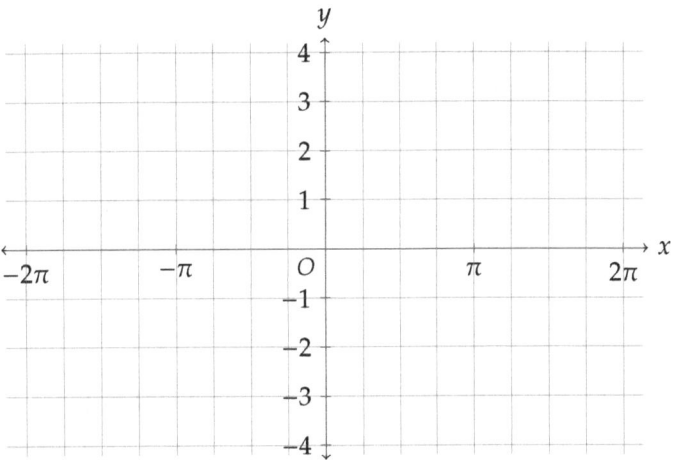

$(x, \sin x)$	T1	T2	T3
$(0, 0)$			
$\left(\frac{\pi}{2}, 1\right)$			
$(\pi, 0)$			
$\left(\frac{3\pi}{2}, -1\right)$			
$(2\pi, 0)$			

When a sinusoidal function has a horizontal translation, it is referred to as a PHASE SHIFT. For instance, the function g from Problem 3 had a phase shift of $+\frac{\pi}{2}$.

4. Look back at the graph from Problem 3. Suppose you wanted to work backwards to see why the function with this graph could be defined by $g(x) = 3\sin\left(x - \frac{\pi}{2}\right) + 1$.

   (a) How could you identify that $a = 3$ from the graph?

   (b) How could you identify that $d = 1$ from the graph?

5. Let's try another, this time with a horizontal dilation. Sketch a graph of $h(x) = 2\cos(2x)$.

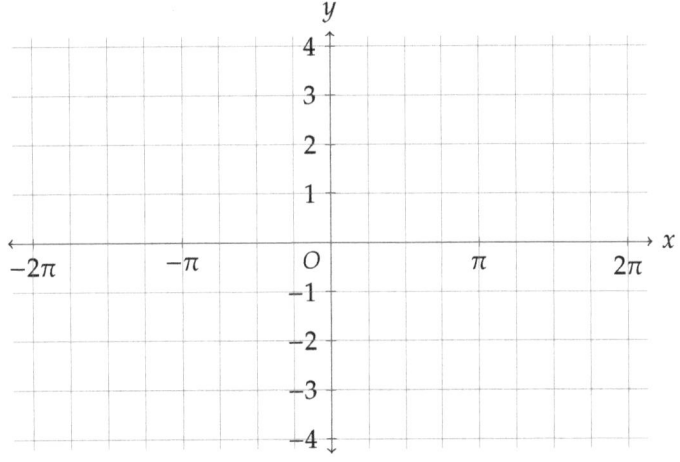

$(x, \cos x)$	T1	T2
$(0, 1)$		
$\left(\frac{\pi}{2}, 0\right)$		
$(\pi, -1)$		
$\left(\frac{3\pi}{2}, 0\right)$		
$(2\pi, 1)$		

6. What is the period of $h(x) = 2\cos(2x)$? How about the frequency?

7. On the same graph as the one from Problem 5, sketch a graph of $j(x) = 2\cos(-2x)$.

8. Use what you learned from the previous questions to try to complete the following fact.

   Fact: For a sinusoidal function $f(x) = a\sin(b(x+c)) + d$:
   - The amplitude is ____.
   - The midline is $y = $ ____.
   - There will be a phase shift of ____ units.
   - The period is ____ and the frequency is ____.

9. Now, try working backwards. Write a function of the form $f(x) = a\sin(b(x+c)) + d$ that could have the graph shown below.

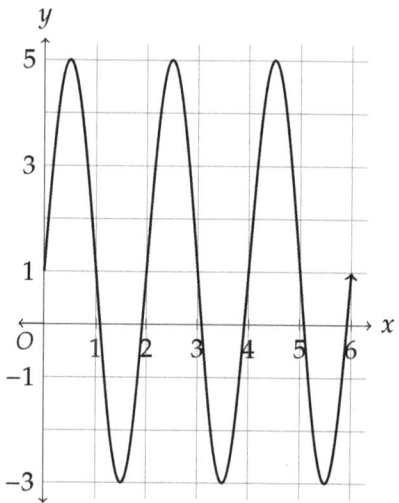

10. Try another one. Write a function of the form $f(x) = a\cos(bx) + d$ that could have the graph shown below.

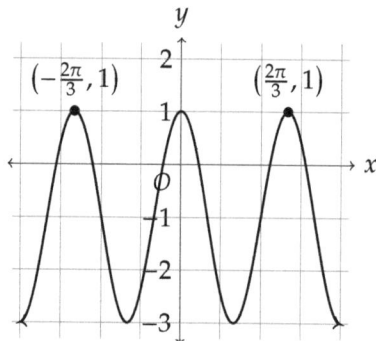

11. Suppose the graph of sinusoidal function f has a maximum at $\left(\frac{\pi}{4}, 3\right)$.

    (a) If $f(x) = a\sin(bx)$, what are possible values for a and b?

    (b) If $f(x) = \sin(x+c) + d$, what are possible values for c and d?

12. The graph of $g(\theta) = a\sin(b(\theta + c)) + d$ is shown. Find values of $a, b, c,$ and d.

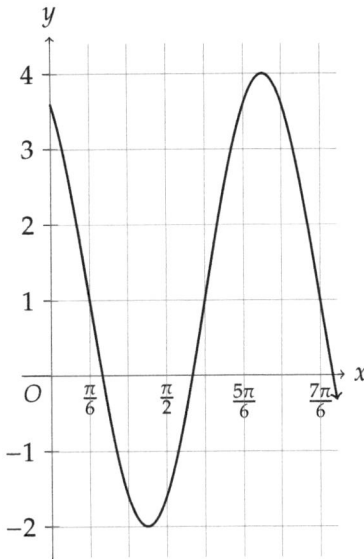

13. The graph of $g(x) = 4\sin\left(\frac{\pi}{4}(x + k)\right)$ has been graphed in the coordinate plane.

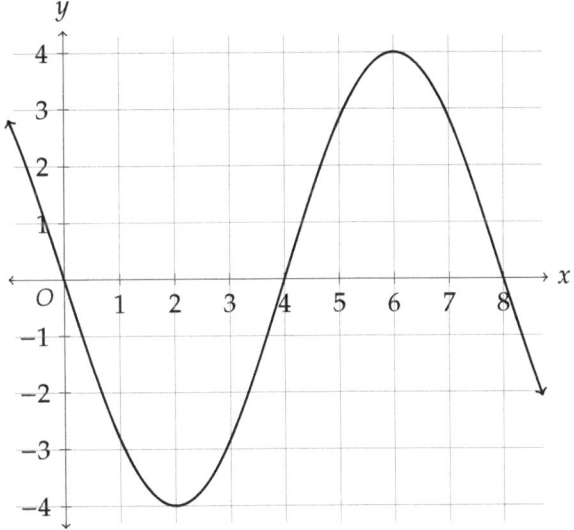

Find a possible value of k.

Notes

Topic 3.6 Homework

1. Shown below is a graph of $f(x) = a\sin(bx)$, where $a > 0$ and $b > 0$. Identify the values of a and b.

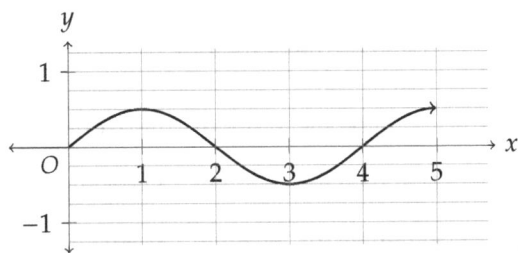

2. Sketch a graph of $g(\theta) = -2\cos\left(\theta + \frac{\pi}{2}\right) - 3$ for $-2\pi \leq \theta \leq 2\pi$. Draw your own xy-plane.

3. Identify the period of each the following functions.

   (a) $f(x) = \sin(2\pi x)$

   (b) $f(x) = 3\cos(4x)$

   (c) $g(x) = 5\cos\left(\frac{1}{3}x\right)$

   (d) $h(x) = 22.3\sin\left(\frac{45}{2\pi}(x - 30)\right) + 52$

   (e) $f(x) = \sin\left(-\frac{3}{4}x\right)$

   (f) $f(\theta) = -3\cos\left(\frac{\pi}{3} \cdot \theta\right) - 2$

4. A data set of input (x) and output (y) values includes a maximum y-value of 8. This y-value of 8 occurs for $x = 0, 2, 4$, and every other even integer. The minimum value is $y = 2$. It is known the relationship between x and y in this data set can be modeled by a trigonometric function $f(x) = a\sin(bx) + c$ or $f(x) = a\cos(bx) + c$. Identify which of the two models is better and find appropriate values of a, b, and c.

5. A darts player, Hugh, who has played for years, has found he hasn't improved much in recent years, but that he's okay with it. In a game of darts, the goal is to score exactly 501 in as few throws as possible, where 9 is the absolute minimum and the number of throws to get 501 is your "score." Hugh plays one game every day when he gets home from work, and he found that, over the last few years, his score T per day has fluctuated according to the function $T(d) = 24 + 6\cos(0.17592d)$, where d is days since January 1, 2018.

   (a) According to the function, what's the best score Hugh has achieved since 2018?

   (b) What is Hugh's average score, approximately, since 2018?

   (c) Hugh's scores fluctuate periodically every how many weeks, approximately?

   (d) When Hugh started recording the data on January 1, 2018, did his scores start improving or getting worse for the next few days?

Topic 3.6 Solutions/Notes

Activity

- *The effect of b*: The graph of $T(x)$ will be the graph of f <u>horizontally dilated</u> by a factor of $\left|\frac{1}{b}\right|$.

- *The effect of c*: The graph of $T(x)$ will be the graph of f <u>horizontally translated</u> by <u>$-c$</u>.

- *The effect of a*: The graph of $T(x)$ will be the graph of f <u>vertically dilated</u> by a factor of $|a|$.

- *The effect of d*: The graph of $T(x)$ will be the graph of f <u>vertically translated</u> by <u>$+d$</u>.

1. The table and graph are below.

x	$\sin x$
0	0
$\frac{\pi}{2}$	1
π	0
$\frac{3\pi}{2}$	-1
2π	0

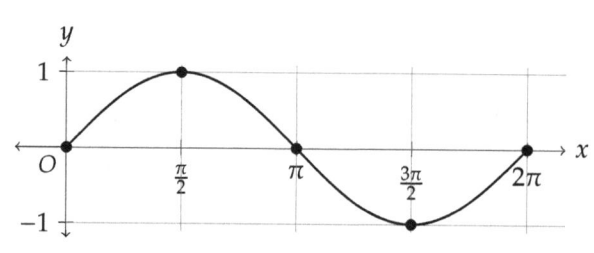

2. The table and graph are below.

x	$\cos x$
0	1
$\frac{\pi}{2}$	0
π	-1
$\frac{3\pi}{2}$	0
2π	1

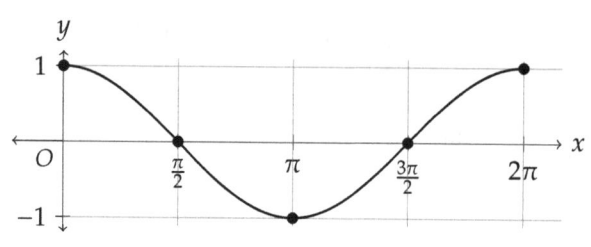

3. T1 is a horizontal translation of $+\frac{\pi}{2}$, T2 is a vertical dilation by a factor of 3, and T3 is a vertical translation of $+1$.

$(x, \sin x)$	T1	T2	T3
$(0, 0)$	$(\frac{\pi}{2}, 0)$	$(\frac{\pi}{2}, 0)$	$(\frac{\pi}{2}, 1)$
$(\frac{\pi}{2}, 1)$	$(\pi, 1)$	$(\pi, 3)$	$(\pi, 4)$
$(\pi, 0)$	$(\frac{3\pi}{2}, 0)$	$(\frac{3\pi}{2}, 0)$	$(\frac{3\pi}{2}, 1)$
$(\frac{3\pi}{2}, -1)$	$(2\pi, -1)$	$(2\pi, -3)$	$(2\pi, -2)$
$(2\pi, 0)$	$(\frac{5\pi}{2}, 0)$	$(\frac{5\pi}{2}, 0)$	$(\frac{5\pi}{2}, 1)$

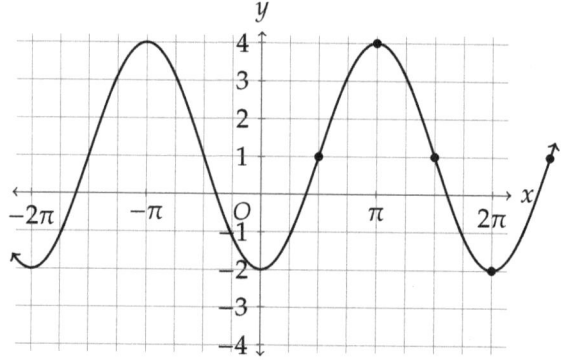

4. (a) The maximum is 4 and the minimum is -2. The difference between these is 6, so the amplitude is 3.

   (b) The midline is $y = 1$. This is the value of d.

5. T1 is a horizontal dilation by a factor of $\frac{1}{2}$ and T2 is a vertical dilation by a factor of 2.

$(x, \cos x)$	T1	T2
$(0, 1)$	$(0, 1)$	$(0, 2)$
$(\frac{\pi}{2}, 0)$	$(\frac{\pi}{4}, 0)$	$(\frac{\pi}{4}, 0)$
$(\pi, -1)$	$(\frac{\pi}{2}, -1)$	$(\frac{\pi}{2}, -2)$
$(\frac{3\pi}{2}, 0)$	$(\frac{3\pi}{4}, 0)$	$(\frac{3\pi}{4}, 0)$
$(2\pi, 1)$	$(\pi, 1)$	$(\pi, 2)$

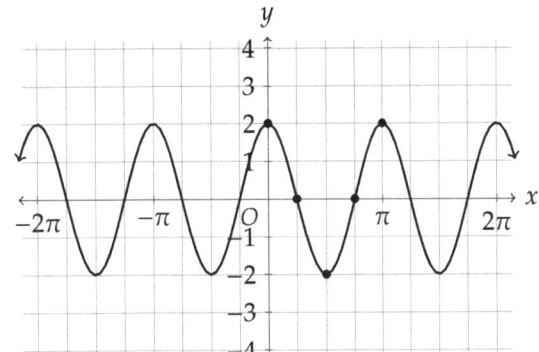

6. The period is π, and the frequency is $\frac{1}{\pi}$.

7. The graph is literally identical: $\cos x$ is an even function, and the transformations did not involve any horizontal translations. Therefore, $2\cos(2x) = 2\cos(-2x)$ for all x.

8. **Fact:** For a sinusoidal function $f(x) = a\sin(b(x+c)) + d$:
   - The amplitude is $|a|$.
   - The midline is $y = d$.
   - There will be a phase shift of $-c$ units.
   - The period is $\left|\frac{1}{b}\right| \cdot 2\pi$ and the frequency is $\frac{|b|}{2\pi}$.

9. The period is 2, so we can write and solve $\left|\frac{1}{b}\right| \cdot 2\pi = 2$ to get $b = \pm\pi$. The midline is $y = 1$, so $d = 1$. The difference between the maximum and the midline is $5 - 1 = 4$, so $a = \pm 4$. Because the graph is at its midline at $x = 0$ and iis immediately increasing for $x > 0$, it does not appear that there is a necessary phase shift. Therefore, one possible answer is $f(x) = 4\sin(\pi x) + 1$.

10. The period is $\frac{2\pi}{3}$, so $b = \pm 3$. The midline is $y = -1$, so $d = -1$. The difference between the maximum and midline is $1 - (-1) = 2$, so $a = \pm 2$. Therefore, a possible answer is $f(x) = 2\cos(3x) - 1$.

11. (a) The maximum value of our function will reach its maximum after it has completed $\frac{1}{4}$ of a period. Because the maximum occurs at $\frac{\pi}{4}$, we can make our period equal to π (since $\frac{\pi}{4} = \frac{1}{4}(\pi)$). This requires that $b = \pm 2$. To obtain a maximum value of 3 with no vertical translation, we need an amplitude of 3, which means $a = \pm 3$. Therefore, one possible combination is $a = 3$ and $b = 2$.

    (b) With no phase shift, the maximum occurs at $y = 1$ when $x = \frac{\pi}{2}$. To shift this over to $x = \frac{\pi}{4}$, we can apply a phase shift of $\frac{\pi}{4}$, making $c = -\frac{\pi}{4}$. Then, to get a maximum value of 3, we can apply a vertical translation of $+2$, making $d = 2$.

12. The midline is $y = -1$, so $d = 1$. The period is $\frac{7\pi}{6} - \frac{\pi}{6} = \pi$, so we can set $b = \frac{2\pi}{\pi} = 2$. The function is at its midline when $x = \frac{\pi}{6}$, and the parent $\sin x$ is at its midline when $x = 0$, so we can apply a phase shift of $\frac{\pi}{6}$, making $c = -\frac{\pi}{6}$. Finally, the amplitude is 3, but the function is immediately decreasing at the beginning of the period starting at $x = \frac{\pi}{6}$, so we must make $a < 0$. Therefore, $a = -3$.

13. The period is 8. We observe that a maximum occurs at $x = 6$. Without a phase shift, the maximum would occur one quarter of the way through a period, or at $x = 2$. In order to get the maximum to occur at 8, we must have a phase shift of 4, making $k = -4$ one possible answer.

320 TOPIC 3.6 ~ SINUSOIDAL TRANSFORMATIONS UNIT 3

> **Notes**
>
> To graph a sinusoidal function, simply utilize the five key points from one period of that function and keep track of the transformations. A few important notes:
>
> - Do not forget the absolute value in dilations of any kind: the sign only determines whether a reflection occurs or not.
>
> - Avoid using "compress" or "stretch." Rather, stick with "dilate by a factor of ...".
>
> - All sinusoidal functions can be written in infinitely many different equivalent forms because of their periodic nature.
>
> Given the graph or a description of a sinusoidal function $f(x) = a\sin(b(x+c)) + d$:
>
> - $y = d$ is the midline.
>
> - $|a|$ is the difference between the maximum and the midline: pay attention to whether the function is increasing or decreasing at the beginning of a period to determine the sign of a.
>
> - The period is $p = \left|\frac{1}{b}\right| \cdot 2\pi$, which can be rearranged to give $|b| = \frac{2\pi}{p}$. This is very handy.

Homework

1. $|a| = \frac{1}{2}$. The period is 4, so $|b| = \frac{2\pi}{4} = \frac{\pi}{2}$. Since $a > 0$ and $b > 0$, we have $a = \frac{1}{2}$ and $b = \frac{\pi}{2}$.

2.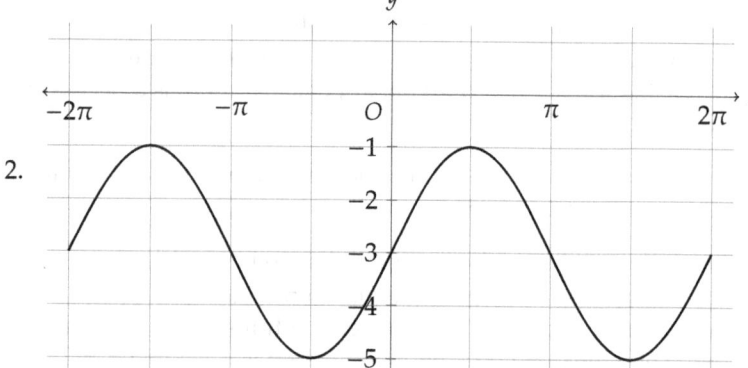

3. (a) $\frac{2\pi}{2\pi} = 1$ (b) $\frac{2\pi}{4} = \frac{\pi}{2}$ (c) $\frac{2\pi}{\frac{1}{3}} = 6\pi$ (d) $\frac{2\pi}{\frac{45}{2\pi}} = 45$ (e) $\frac{2\pi}{\left|-\frac{3}{4}\right|} = \frac{8\pi}{3}$ (f) $\frac{2\pi}{\frac{\pi}{3}} = 6$

4. Since the maximum value occurs when $x = 0$, the cosine model is appropriate. If the minimum is 2 and the maximum is 8, the amplitude is $\frac{8-2}{2} = 3$. The maximum next occurs when $x = 2$, so the period is 2. Therefore, $b = \frac{2\pi}{2} = \pi$. Finally, the midline is $y = \frac{8+2}{2} = 5$, so $c = 5$. An appropriate model is thus $g(x) = 3\cos(\pi x) + 5$.

5. (a) $24 - 6 = 18$ (remember: "best" is the fewest amount of throws!)
   (b) 24
   (c) $\frac{2\pi}{0.17592} \approx 35.7$, or about 5 weeks
   (d) Getting worse ($\cos x$ is decreasing for the first half of its period)

Reflection

Suggested Class Time. 60 minutes

Prerequisites. Students should have familiarity with the graphs of (and characteristics of the graphs of) sine and cosine functions.

Instructional Strategies. Things went smoothly, though they started, as they often do, somewhat slowly. Students made mistakes with phase shifts in Problem 3 (subtracting instead of adding). Some took a really long time to finish that first table in Problem 3. To help speed things along, I went up to the board after it seemed like most students were done with or close to done with Problem 3 and filled in the table, as well as the graph. Then, students got moving quickly.

There weren't really any other snags. After students finished Problem 8, I did go to the board and check with the class about what they got. Everyone seemed correct, only most students did forget the absolute value on all three of the amplitude, period, and frequency. At this point, we also talked about Problem 6, and surprisingly, literally no one realized that the graph of $h(x) = 2\cos(2x)$ was identical to the graph from Problem 5. I again got to reiterate that $\cos x$ is an even function, and we briefly talked about why with relation to the unit circle.

In my original lesson, Problem 11 did not include a part (b) and Problems 12-13 did not exist at all. I was upset with myself for not including more questions with phase shifts, though, so I printed out the questions that became Problems 11b, 12, and 13 and we worked them one at a time. The explanations I gave to students (after they attempted them) were identical to those provided in the solutions to the activity.

At first, I thought the lesson didn't go as well as I'd hoped, but I gave students 4 Topic Questions from AP Classroom at the end of the lesson, and the class average was 72.5%, so I ended up relatively pleased.

Technology. None was required.

Problems.

1. The height of the tide on one coastline can be modeled by the function

$$h(t) = 1.25 \sin\left(\frac{2\pi}{12.25}(t - 1.5)\right) + 1.5,$$

    where h is height in feet and t is time in hours. What is the maximum height of the tide on this coastline?

    (A) 1.5 feet   **(B)** 2.5 feet   **(C)** 2.75 feet   **(D)** 3 feet

2. The function f expressed as $f(x) = \sin(b(x + c))$ is graphed below.

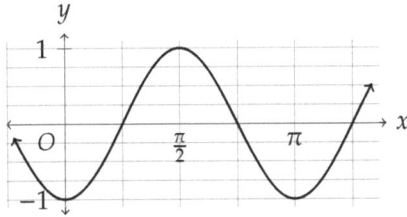

    Which of the following could be the values of b and c?

    (A) $b = 2$ and $c = \dfrac{\pi}{4}$ **(B)** $b = 2$ and $c = -\dfrac{\pi}{4}$ **(C)** $b = \dfrac{1}{2}$ and $c = \dfrac{\pi}{4}$ **(D)** $b = \dfrac{1}{2}$ and $c = -\dfrac{\pi}{4}$

3. The function f is defined by $f(x) = \cos(x)$. In the xy-plane, the graph of the function g is the image of the graph of f under the transformations of a horizontal dilation by a factor of 2 and a phase shift of $+\frac{\pi}{3}$. Which of the following could define g?

   (A) $g(x) = \cos\left(2\left(x - \frac{\pi}{3}\right)\right)$
   (B) $g(x) = \cos\left(2\left(x + \frac{\pi}{3}\right)\right)$
   (C) $g(x) = \cos\left(\frac{1}{2}\left(x - \frac{\pi}{3}\right)\right)$
   (D) $g(x) = \cos\left(\frac{1}{2}\left(x + \frac{\pi}{3}\right)\right)$

4. Consider the sinusoidal function graphed below.

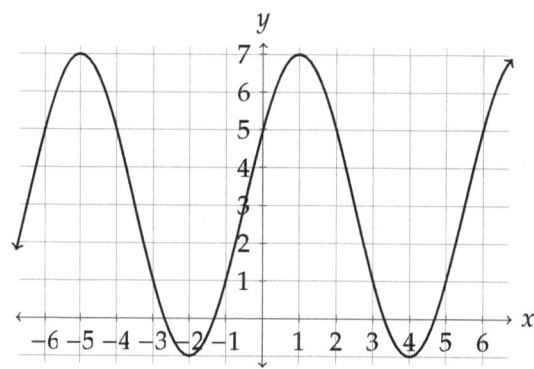

   What are the period and the amplitude of this sinusoidal function?

   (A) The period is 12 and the amplitude is 8.
   (B) The period is 6 and the amplitude is 8.
   (C) The period is 12 and the amplitude is 4.
   (D) The period is 6 and the amplitude is 4.

5. A sinusoidal function f has an amplitude of 5, a period of 8, and a phase shift of $+3$. Which of the following could be an expression for f?

   (A) $f(x) = 5\sin(8(x - 3))$
   (B) $f(x) = 5\sin\left(\frac{\pi}{4}(x - 3)\right)$
   (C) $f(x) = 5\sin(8(x + 3))$
   (D) $f(x) = 5\sin\left(\frac{\pi}{4}(x + 3)\right)$

Solutions.

1. The midline is $y = 1.5$ and the amplitude is 1.25, so the maximum is $1.5 + 1.25 = 2.75$. The correct answer is therefore **(C)**.

2. The phase shift is $+\frac{\pi}{4}$ so $\frac{\pi}{4}$ is subtracted from x, implying $c = -\frac{\pi}{4}$. The period is π so solving $\frac{2\pi}{b} = \pi$ gives $b = 2$. The correct answer is therefore **(B)**.

3. The horizontal dilation by a factor of 2 multiplies the input x by $\frac{1}{2}$. Then, the phase shift of $+\frac{\pi}{3}$ is achieved by subtracting $\frac{\pi}{3}$ from the input x. Therefore x becomes $\frac{1}{2}(x - \frac{\pi}{3})$. The correct answer is therefore **(C)**.

4. The vertical distance between the maximum and minimum is $7 - (-1) = 8$ so the amplitude is 4. The horizontal distance between consecutive minima is $4 - (-2) = 6$, so the period is 6. The correct answer is therefore **(D)**.

5. The period is 8, so solving $\frac{2\pi}{b} = 8$ yields $b = \frac{\pi}{4}$. The phase shift is 3 so 3 is subtracted from x. Hence, the argument of the sine function is $\frac{\pi}{4}(x - 3)$. The correct answer is therefore **(B)**.

Topic 3.7 ~ Modeling with Sinusoidal Functions

Learning Objectives
1. (3.7.A) Construct sinusoidal function models of periodic phenomena.

Success Criteria
1. I can understand the relationship between the differences in inputs of minima and maxima and the period of a periodic function.
2. I can use the period of a function to determine an appropriate horizontal dilation.
3. I can use context to determine the sign of the leading coefficient of a sinusoidal function model.
4. I can construct functions to model periodic phenomena.

In this lesson, you're going to practice constructing sinusoidal function models given data described graphically, tabularly, analytically, or verbally. Before you move into creating models, let's take a few minutes to analyze some characteristics of sinusoidal functions.

Sinusoidal Modeling Toolkit
• For a sinusoidal function $f(x) = a\sin(b(x+c)) + d$, the values a, b, d can be estimated by using $$d = \qquad b = \qquad |a| =$$
• $\sin x$ begins at its _____, while $\cos x$ begins at its _____.   Diagram:
• $\sin x$ is _____ immediately after its initial value, while $\cos x$ is _____ immediately after its initial value. This can be used to determine the desired sign of a.   Diagram:
• The difference between inputs of consecutive maxima or consecutive minima is equal to the _____ of a sinusoidal function.   Diagram:
• The distance between inputs of a maximum and a consecutive minimum (or vice-versa) is _____ of a sinusoidal function.
• The maximum of $\sin x$ is _____ of the way through a period, and the minimum is _____ of the way through a period. This can be used to determine _____.   Diagram:

1. Rutgers University, in conjunction with the US National Ice Center, tracks how much of the Northern Hemisphere is covered in snow at any given point in millions of square kilometers. This can be summarized by the annual snow cover extent (SCE), which has been recorded for the last 55 years. The table below shows the mean SCE for the Northern Hemisphere for the 55 years from 1966 to 2020. (Source: https://climate.rutgers.edu/snowcover/files/Robinson_snowdata2021.pdf)

Month	Mean SCE
January	47.1
February	46.0
March	40.4
April	30.5
May	19.1
June	9.4
July	3.9
August	3.0
September	5.4
October	18.6
November	34.3
December	43.7

As scientists learn more about climate change, the SCE data may provide evidence as to the speeding up or slowing down of certain climate trends.

(a) Use your calculator to input the data and create a scatterplot.

(b) Construct a sinusoidal function of the form $S(x) = a \sin(b(x + c)) + d$ or $S(x) = a \cos(b(x + c)) + d$ that could be used to model the mean SCE in month x, where $x = 1$ represents January and $x = 12$ represents December.

(c) Test out your function by graphing it in Y_1. How'd you do?

(d) Do you think that your function is the *best possible* approximation of the data?

You likely won't be surprised that technology can be used to perform SINUSOIDAL REGRESSION.

(e) Press stat, then CALC, then scroll down and select C:SinReg. Input the following settings:

$$\text{Iterations:3}$$
$$\text{Xlist:L1}$$
$$\text{Ylist:L2}$$
$$\text{Store RegEq:Y2}$$

For Period, we have the option of inputting what we *think* the period should be. What should you put here, if anything?

Note: *Iterations* means the number of times the calculator will complete a minimization algorithm. The default number is 3, which is perfectly acceptable; you can go all the way up to 16, though the calculator may take as much as a minute to perform the regression!

(f) Press Calculate. Record the values of a, b, c, d below, as well as the written-out function $S(x)$.

$a =$ $\qquad$ $b =$ $\qquad$ $c =$ $\qquad$ $d =$

$$S(x) =$$

(g) The calculator only produces sine functions. Does this make your model wrong?

(h) Press graph. The actual regression equation will be graphed alongside your function. How close was your function to the regression model?

2. In a blood pressure study, a participant's blood pressure was measured and automatically recorded multiple times over the course of 1 second. The table below displays the blood pressure, in millimeters of mercury (mmHg) recorded at each time.

Time (s)	0.2	0.4	0.6	0.8	1
Blood pressure (mmHg)	85	50	86	120	84

The minimum blood pressure was 50, the maximum blood pressure was 120, and there were no peaks in blood pressure between 0.4 and 0.8 second.

(a) Explain why a sine function would fit this data well.

(b) Construct a function $B(t) = a \sin(b(t + c)) + d$ to model the blood pressure B after t seconds.

Topic 3.7 Homework

1. Each of the following describes a relationship between two variables for which a sinusoidal model would be appropriate. Create a sinusoidal function $f(x) = a\sin(b(x+c))+d$ or $f(x) = a\cos(b(x+c))+d$ for each set.

   (a) On a Ferris wheel, the height f, in feet, of a car can be modeled based on the time x, in minutes, after the wheel started. The height of a car is at its peak of 65 feet after 7.75 minutes and is next at its lowest of 5 feet after 9.25 minutes.

   (b) The pressure f of gas in a chamber over an extended period of time has an average of 80 psi, but it fluctuates periodically every 5 minutes. The difference between the maximum pressure and the minimum pressure is 4 psi. The gas is at its highest pressure at $x = 0$, or 0 minutes.

   (c) The ball at the end of a pendulum begins at its peak height of 4 feet. The height f, in feet, of the ball after x seconds varies periodically (within reason, accounting for air resistance and gravity). The minimum height of the ball is 0.5 feet, which the ball reaches after 0.4 second.

2. The function $S(t)$ can be used to model the number of hours of peak sunlight s that a solar panel receives t days after being installed. The table provides values of t.

t	$S(t)$
7	4.74
14	4.82
21	4.97
28	5.09
35	5.12
42	5.16
49	5.14
56	5.03
63	5.00
70	4.88
77	4.74

   A homeowner is convinced that $S(t)$ is sinusoidal and is going to write $S(t) = a\cos(b(t+c)) + d$ or $S(t) = a\sin(b(t+c)) + d$.

   (a) Initially, the homeowner has no other data other than that in the table. Find values of $a, b, c,$ and d that could be appropriate.

   (b) The homeowner finds out that 4.74 is actually the *average* number of hours of peak sunlight that his solar panel receives. Without any other data in the table, how can the model from (a) be modified to reflect this new information?

3. The function $g(x) = 3\sin\left(\dfrac{2\pi}{k}(x+c)\right) + 1$ is graphed below.

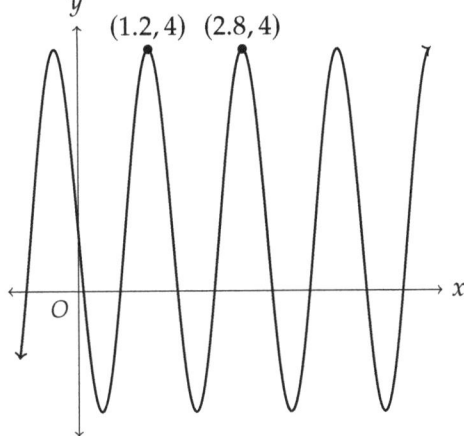

Find possible values of k and c.

Topic 3.7 Solutions/Notes

Activity

Sinusoidal Modeling Toolkit

- For a sinusoidal function $f(x) = a\sin(b(x+c)) + d$, the values a, b, d can be estimated by using

$$d = \frac{\max + \min}{2} \qquad b = \frac{\text{Period}}{2\pi} \qquad |a| = \max - d$$

- $\sin x$ begins at its <u>average/midline</u>, while $\cos x$ begins at its <u>maximum</u>.

Diagram: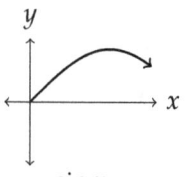
sin x   cos x

- $\sin x$ is <u>increasing</u> immediately after its initial value, while $\cos x$ is <u>decreasing</u> immediately after its initial value. This can be used to determine the desired sign of a.

Diagram: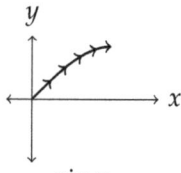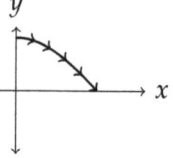
sin x   cos x

- The difference between inputs of consecutive maxima or consecutive minima is equal to the <u>period</u> of a sinusoidal function.

Diagram: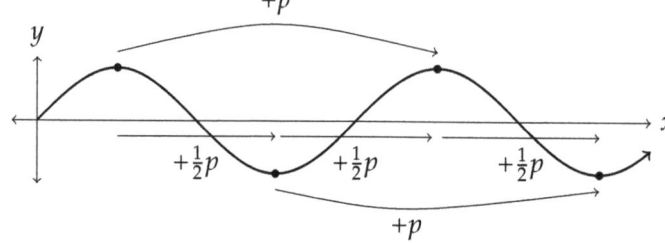

- The distance between inputs of a maximum and a consecutive minimum (or vice-versa) is <u>half the period</u> of a sinusoidal function.

- The maximum of $\sin x$ is <u>one quarter</u> of the way through a period, and the minimum is <u>three quarters</u> of the way through a period. This can be used to determine <u>phase shift</u>.

Diagram: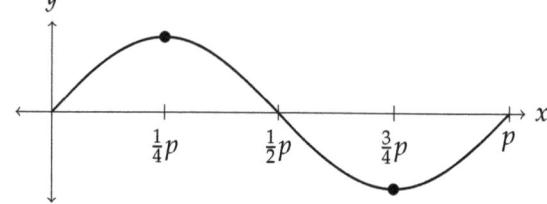

1. (a)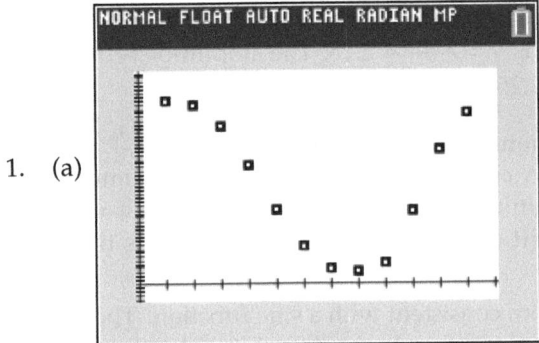

(b) The function immediately begins decreasing from a maximum, so a cosine function is a good fit. There are many sample models, but one uses $d = \frac{47.1+3}{2} = 25.05$, $|a| = 47.1 - 25.05 = 22.05$, and $b = \frac{2\pi}{12}$. Because the SCE is decreasing for small x, just like the parent cosine function, we use positive $a = 22.05$. The maximum of cosine typically occurs at the very beginning of one period, or at $x = 0$, but the maximum of these data occurs when $x = 1$. Therefore, a phase shift of $+1$ needs to occur, so $c = -1$. The model is thus $S(x) = 22.05 \cos\left(\frac{2\pi}{12}(x - 1)\right) + 25.05$.

(c)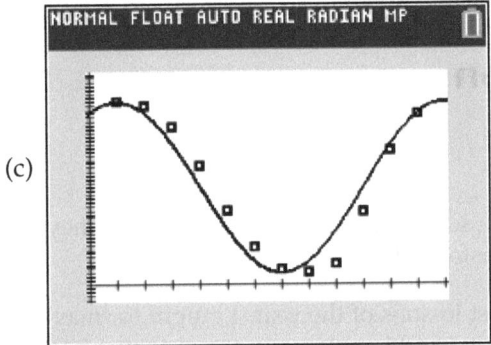

(d) Probably not!

(e) The period is likely 12.

(f) $a = 23.108, b = 0.534, c = 0.726, d = 24.715$, giving $S(x) = 23.108 \sin(0.534(x + 0.726)) + 24.715$.

(g) No - cosine functions are sinusoidal. Any cosine function can be expressed as a sine function.

(h) The regression function did a bit better (closer to the actual SCE values for the most part).

2. (a) The blood pressure starts near its average.

(b) $d = \frac{120+50}{2} = 85, |a| = 120 - 85 = 35$. Because the blood pressure is initially decreasing, we'll choose $a = -35$. The period appears to be $2(0.4) = 0.8$, so $b = \frac{2\pi}{0.8}$. The blood pressure is at its midline at $t = 0.2$, so we need a phase shift of 0.2. This gives $c = -0.2$. Therefore, $B(t) = -35 \sin\left(\frac{2\pi}{0.8}(t - 0.2)\right) + 85$.

Homework

1. (a) The period is $2(1.5) = 3$ minutes. The difference between the maximum and minimum is 60, so the amplitude is 30. The midline is $y = 65 - 30 = 35$. Neither a sine nor a cosine function makes any more sense than the other, so we'll just use sine. This gives $f(x) = 30 \sin\left(\frac{2\pi}{3}(x + c)\right) + 35$. The car will be at its average height and decreasing at $t = 8.5$ (can you see why?). On the parent function $f(x) = 30 \sin\left(\frac{2\pi}{3}(x)\right) + 35$, the height will be at its average and decreasing halfway along its period, or at $t = 1.5$. Therefore, a horizontal translation of $8.5 - 1.5 = 7$, or $+7$, will result in the correct function. This gives $f(x) = 30 \sin\left(\frac{2\pi}{3}(x - 7)\right) + 35$.

(b) The period is 5, the midline is $y = 80$, and the amplitude is 2. Because the pressure begins at its maximum, a cosine model is simplest. This gives $f(x) = 2 \cos\left(\frac{2\pi}{5}x\right) + 80$.

(c) Again, a cosine model is appropriate, as the height of the ball begins at its peak and then decreases. the period is 2(0.4) = 0.8 seconds, and the midline is $y = \frac{4+0.5}{2} = 2.25$. The amplitude is therefore $4 - 2.25 = 1.75$. This gives $f(x) = 1.75\cos\left(\frac{2\pi}{0.8}x\right) + 2.25$.

2. (a) The sunlight appears to start and end at its minimum. The midline is $d = \frac{5.16+4.74}{2} = 4.95$, giving an amplitude of $|a| = 5.16 - 4.95 = 0.21$. A cosine function begins at its maximum and then decreases, while the sunlight begins at its minimum and increases. This gives $a = -0.21$. Finally, the period is $77 - 7 = 70$, and a phase shift of $+7$ is needed. The function is therefore $S(t) = -0.21\cos\left(\frac{2\pi}{70}(t-7)\right) + 4.95$.

   (b) Now, the sunlight starts at its midline, which is more consistent with a sine function. The period is now not 70, but double that, or 140. The amplitude is also doubled, and it no longer needs to be negative, as the sunlight begins increasing, as is consistent with a sine function. Therefore, $S(t) = 0.42\sin\left(\frac{2\pi}{140}(t-7)\right) + 4.74$.

3. The difference between the inputs of two consecutive maxima is equal to the period, so $k = 2.8 - 1.2 = 1.6$. However, the function starts at its midline and begins decreasing, where the function g without the phase shift would start at its midline and begin increasing. A positive phase shift of half the period (or a negative phase shift of half the period) would adjust the graph accordingly, so $c = \pm 0.8$.

Reflection

Suggested Class Time. 50 minutes

Prerequisites. Students should be comfortable with making scatterplots in the calculator, and they should at this point be reasonably comfortable with graphing transformations of sine and cosine.

Instructional Strategies. This may have been one of the best lessons of the year. I taught for maybe 10 or 15 minutes, going through the "Sinusoidal Modeling Toolkit," and I made sure to use plenty of diagrams. One thing that I think helps students understand the quarters of a sine wave are going back to the unit circle: each quarter is related to a quadrant.

For identifying phase shift, I taught students the following.

- Determine valid $a, b,$ and d.

- Sketch $a\sin(bx) + d$ (or $a\cos(bx) + d$) - in other words, pretend there is no phase shift. Make sure you know what the period is.

- Identify the location of an easily observable point like a minimum or maximum on the un-shifted graph (using quarters of the period) and then in the actual data.

- Determine how much of a phase shift is required.

This seemed to work well for students: some took multiple practice questions (even stretching into the next lesson), including going over the homework the following class period, but others got it immediately. In essence, this "method" just focuses on the parent function.

For Problem 2, students did well except for one thing: many called the period 1. I've noticed this even in non-AP Precalculus classes: there is a tendency to assume the period is just "the end of the period," rather than the shortest distance needed to repeat. We talked about it further in detail.

Technology. Students needed a TI-84 or other calculator that can perform regression.

UNIT 3   TOPIC 3.7 ~ MODELING WITH SINUSOIDAL FUNCTIONS   331

Problems.

1. (▣) An antique seller has a piece of memorabilia they are looking to sell. They have found that, over the last few months, the price of similar memorabilia has varied seemingly periodically on a website where memorabilia is sold. The table below provides values of the price, in dollars, of similar memorabilia on the website.

Month t	2	3	5	7	11
Price p	$45	$53	$50	$30	$45

   The seller is unsure of what exactly the period would be, so they use technology to create a regression model $p(t) = a\sin(bt + c) + d$ using all five data points to model the price p based on the month t. Based on the regression model, which of the following is the best estimate for the difference between the highest price and the lowest price that similar memorabilia has sold for?

   (A) $15   **(B)** $31   **(C)** $40   **(D)** $55

2. The table displays the daily average temperature D, in degrees Fahrenheit, for each month in a certain city.

Month t	1	2	3	4	5	6	7	8	9	10	11	12
D ($F°$)	45	43	46	54	62	70	77	79	78	72	62	53

   A sinusoidal function $D(t) = a\sin(b(t + c)) + d$ is going to be used to model the daily average temperature D based on the month t. The function will use the assumption that the maximum of D is 79, the minimum of D is 43, and the period is 12. Which of the following is the best choice for the value of the constant c?

   (A) -8 **(B)** -5 **(C)** -3 **(D)** -2

3. An architect is designing a skate park, and one obstacle in the park will be a concrete wave that skateboarders can ride on. The architect is going to use a sinusoidal function to design the wave. The inputs of the function will be x, the distance along the ground, so that the x-axis will represent the ground, and the outputs will be h, the height above the ground. The wave will start, when $x = 0$, at its peak 3 feet off the ground. The wave will then reach the ground 8 feet away, when $x = 8$. The wave will continue up to its next peak, after which it will continue to decrease and increase periodically. Which of the following sinusoidal functions $h(x)$ will satisfy the architect's design?

   (A) $h(x) = 3\cos\left(\frac{2\pi}{8} \cdot x\right)$
   (B) $h(x) = 3\cos\left(\frac{2\pi}{16} \cdot x\right)$
   (C) $h(x) = 1.5\cos\left(\frac{2\pi}{8} \cdot x\right) + 1.5$
   (D) $h(x) = 1.5\cos\left(\frac{2\pi}{16} \cdot x\right) + 1.5$

4. (▣) The following data were collected in an experiment.

input values	5	8	11	14	17	20	23	26	29	32	35
output values	16	15	4	-9	-16	-15	-4	9	16	15	4

   A sinusoidal regression was performed on the data so that the function $f(x) = a\sin(bx + c) + d$ models the output values $f(x)$ as a function of the input values of x. What is the value of $f(100)$ as predicted by the model?

   (A) 0.024   **(B)** 0.262   **(C)** 15.024   **(D)** 16.978

5. (🖩) The function g is given by $g(x) = 3\cos(3x) + 2\sin(5x)$. Using the period of g, which of the following is the number of complete cycles of g over the interval $0 \leq x \leq 2025$?

   (A) 80   (B) 135   (C) 253   (D) 322

Solutions.

1. The calculator's sinusoidal regression gives $y = 15.3616\sin(0.696x - 1.059) + 39.875$. The amplitude is 15.3616 so the difference between the maximum and minimum is $2 \cdot 15.3616 \approx 31$. The correct answer is therefore **(B)**.

2. The maximum of a sine function without a horizontal translation occurs $\frac{1}{4}$ of the way along its period; this implies the maximum should be at $t = \frac{1}{4}(12) = 3$. But the maximum is given at $t = 8$. Hence, there must be a horizontal translation of $+5$, which makes $c = -5$. The correct answer is therefore **(B)**.

3. The period is 16 so $\frac{2\pi}{b} = 16$, which implies $b = \frac{2\pi}{16}$. The midline is $\frac{0+3}{2} = 1.5$. The correct answer is therefore **(D)**.

4. The calculator's sinusoidal regression gives $y = 16.987\sin(0.262x + 0.006) - 0.0236$. The value at $x = 100$ is 15.024. The correct answer is therefore **(C)**.

5. Using the calculator, we find that the period is 2π. Then there are $\frac{2025}{2\pi} \approx 322$ complete cycles. The correct answer is therefore **(D)**.

Topic 3.8 ~ The Tangent Function

Learning Objectives

1. (3.8.A) Construct representations of the tangent function using the unit circle.
2. (3.8.B) Describe key characteristics of the tangent function.
3. (3.8.C) Describe additive and multiplicative transformations involving the tangent function.

Success Criteria

1. I can describe the tangent of an angle as a slope of a ray or the ratio of the sine and cosine values.
2. I can identify characteristics of the graphs of tangent functions, with or without transformations.

Up to now, we have studied primarily the sine and cosine functions. However, in Topic 3.2, we also introduced the TANGENT function: given an angle θ in standard position in the coordinate plane with a terminal ray intersecting the unit circle at point P, like that shown below...

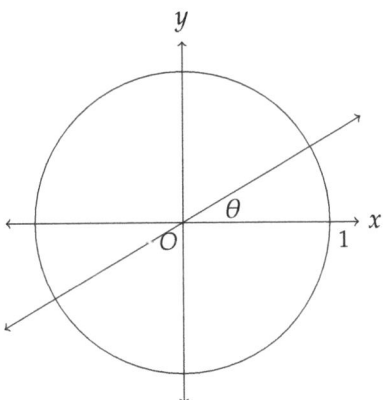

...the tangent function was defined as

- the _____ of the terminal ray, and as
- the ratio of _____ to _____, or the _____ displacement to the _____ displacement

These two representations can be utilized to study different characteristics of the tangent function. Using the slope interpretation leads to understanding the graph of $f(\theta) = \tan \theta$ and computing quadrantal values. Answer the following focusing on the *slope* interpretation of $\tan \theta$.

1. What is $\tan(0)$?
2. Find an angle θ, $0 < \theta < \pi$, for which $\tan \theta = 1$.
3. Find another angle θ, $\pi < \theta < 2\pi$, for which $\tan \theta = 1$.
4. The period of each of $\sin \theta$ and $\cos \theta$ is 2π, as it takes a full circle for the vertical and horizontal displacements to repeat, respectively. Does it take a full circle for the *slope* of a ray to begin to repeat?
5. What is the period of $\tan \theta$?
6. Draw a diagram with the terminal ray for an arbitrary angle θ and also a terminal ray for the angle $-\theta$. How are the slopes of the rays related?

7. Based on Problem 6, $\tan(-\theta) =$ _____, which makes $f(\theta) = \tan \theta$ an _____ function.

8. For what angles θ will the slope of the terminal ray of θ be undefined?

For the last question, you may have listed only a couple of angles. In fact, there are infinitely many such angles. One common way to express this is using a variable like k. For instance, suppose we wanted to write all of the angles for which $\tan \theta = -1$. This occurs at $\theta = \frac{3\pi}{4}$, but it also occurs after any additional period of $\tan \theta$. Since the period of $\tan \theta$ is π, we could write the solution set as

$$\theta = \frac{3\pi}{4} + k\pi, \text{ where } k \text{ is an integer.}$$

9. Rewrite the solution to Problem 8 similarly as above: $\theta =$ _____.

We can use the characteristics from Problems 1-9 to sketch a nice graph of $\tan \theta$. It has a few notable characteristics:

- $\tan 0 =$ _____
- $\tan \frac{\pi}{4} =$ _____
- $\tan\left(-\frac{\pi}{4}\right) =$ _____
- $\tan \theta$ has a period of _____
- $\tan \theta$ is undefined for $\frac{\pi}{2} + \pi k$, which corresponds graphically to _____

Use the characteristics above to sketch a graph of $f(\theta) = \tan \theta$.

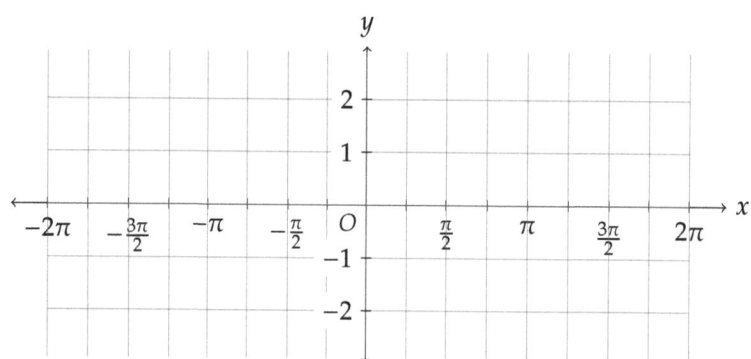

Given the graph of the function $f(\theta) = \tan(\theta)$, we can use our knowledge of transformations to identify characteristics of graphs of transformed tangent functions as well. Try some simple questions.

10. What is the period of $g(\theta) = 4\tan(2\pi\theta) - 1$?

11. Write an expression for all angles θ such that $h(x) = \tan\left(x + \frac{\pi}{3}\right)$ is undefined.

$$\theta =$$

12. A trigonometric function $f(\theta) = a\tan(b\theta) + 1$ is graphed below. What are the values of a and b?

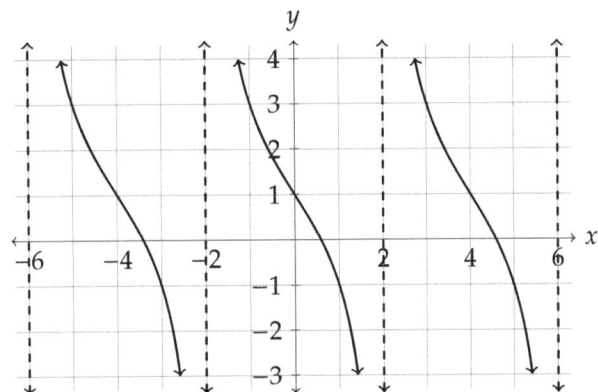

Now, to actually *compute* the slopes of these terminal rays, or to simply compute $\tan\theta$, we will use the other representation of $\tan\theta$ as $\frac{\sin\theta}{\cos\theta}$, along with knowledge of the unit circle and displacements for multiples of $\frac{\pi}{6}$ and $\frac{\pi}{4}$.

As a quick example: say we want to compute $\tan\frac{\pi}{6}$. We can use $\tan\theta = \frac{\sin\theta}{\cos\theta}$ to compute $\tan\frac{\pi}{6}$ as follows:

$$\tan\frac{\pi}{6} = \frac{\sin\frac{\pi}{6}}{\cos\frac{\pi}{6}}$$

$$= \frac{\frac{1}{2}}{\frac{\sqrt{3}}{2}} \qquad \text{(evaluate trig functions)}$$

$$= \frac{1}{2} \cdot \frac{2}{\sqrt{3}} \qquad \text{(multiply by reciprocal of denominator)}$$

$$= \frac{1}{\sqrt{3}} \qquad \text{(divide out 2/2)}$$

$$= \frac{1}{\sqrt{3}} \cdot \frac{\sqrt{3}}{\sqrt{3}} \qquad \text{(rationalize denominator)}$$

$$= \frac{\sqrt{3}}{3} \qquad \text{(simplify)}$$

Luckily, this process is much easier for odd multiples of $\frac{\pi}{4}$, as the tangent of any of these such angles will equal 1 or -1.

Try the following.

13. Compute $\tan\frac{\pi}{3}$.

14. Compute $\tan\left(-\frac{7\pi}{6}\right)$.

15. Compute $\tan\left(\frac{7\pi}{4}\right)$.

16. Find the slope of the ray shown below.

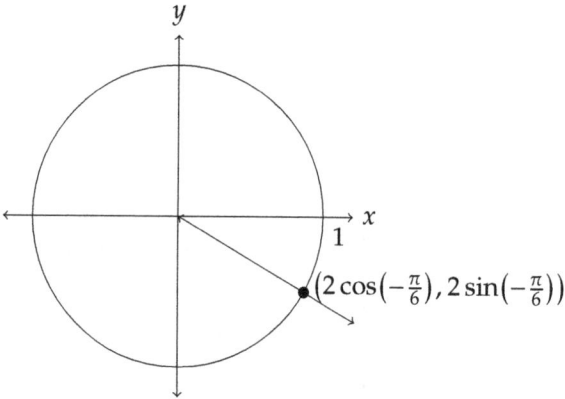

17. A ray starting at the origin passes through the point $(1, \sqrt{3})$. The angle the ray makes with the positive x-axis is θ. Find $\tan(\theta)$.

Topic 3.8 Homework

1. Identify the period of each of the following functions.

   (a) $f(\theta) = -3\tan\left(\dfrac{\theta}{2}\right)$

   (b) $g(x) = \tan(\pi(x-3)) + 4$

   (c) $f(x) = \tan(6x)$

   (d) $h(\theta) = 1 - \tan\left(\dfrac{\pi}{4}\cdot\theta\right)$

   (e) $g(\theta) = 4\tan(3\pi\theta) + 2$

   (f) $f(x) = \tan x + \sin x$

2. Compute the following.

   (a) $\tan \pi$

   (b) $\tan \dfrac{5\pi}{3}$

   (c) $\tan\left(-\dfrac{\pi}{6}\right)$

   (d) $\tan \dfrac{5\pi}{4}$

3. What is the slope of the ray shown below?

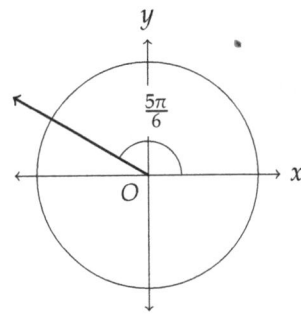

4. The line $y = kx$ passes through the point $\left(\cos\frac{3\pi}{4}, \sin\frac{3\pi}{4}\right)$. What is the value of k?

5. Find all vertical asymptotes of $f(x) = 3\tan\left(\frac{x}{2}\right) + 1$.

6. The function $g(\theta) = a\tan(b\theta) + d$ is graphed below. Find the values of a, b, and d.

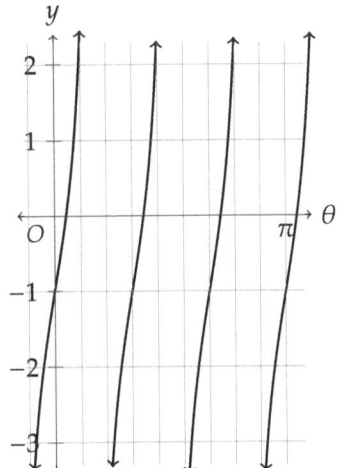

7. Sketch a graph of $f(\theta) = \tan(\pi(\theta - 1))$ for $-2 \leq x \leq 2$.

8. Determine whether each of the following is true for all θ or is not true for all θ.

   (a) $\tan(-\theta) = -\tan(\theta)$

   (b) $\tan(\theta + k\pi) = \tan(\theta)$, for any integer k

   (c) $\tan(\theta) = \tan(\pi - \theta)$

Topic 3.8 Solutions/Notes

Activity

... the tangent function was defined as

- the <u>slope</u> of the terminal ray, and as
- the ratio of <u>sin</u> θ to <u>cos</u> θ, or the <u>vertical</u> displacement to the <u>horizontal</u> displacement

1. The line/ray would be horizontal, or have a slope of 0. Therefore $\tan(0) = 0$.

2. $\dfrac{\pi}{4}$

3. $\dfrac{5\pi}{4}$

4. It only takes half of a circular rotation for a line to end back on top of itself.

5. The period of $\tan \theta$ is half of a circular rotation, or π.

6. Given any angle θ and its terminal ray, the terminal ray for the angle $-\theta$ is a reflection of the original ray over the x-axis. Therefore, its slope has an opposite sign.

7. $\tan(-\theta) = -\tan \theta$, which makes $f(\theta) = \tan \theta$ an <u>odd</u> function.

8. The slope of the ray will be undefined when the ray is vertical. This occurs when the terminal ray coincides with the y-axis, which occurs at $\frac{\pi}{2}, \frac{3\pi}{2}, \frac{5\pi}{2}$, etc.

9. $\theta = \dfrac{\pi}{2} + k\pi$

We can use the characteristics from Problems 1-9 to sketch a nice graph of $\tan \theta$. It has a few notable characteristics:

- $\tan 0 = \underline{0}$
- $\tan \dfrac{\pi}{4} = \underline{1}$
- $\tan\left(-\dfrac{\pi}{4}\right) = \underline{-1}$
- $\tan \theta$ has a period of $\underline{\pi}$
- $\tan \theta$ is undefined for $\dfrac{\pi}{2} + \pi k$, which corresponds graphically to <u>vertical asymptotes</u>

Graph of $f(\theta) = \tan \theta$:

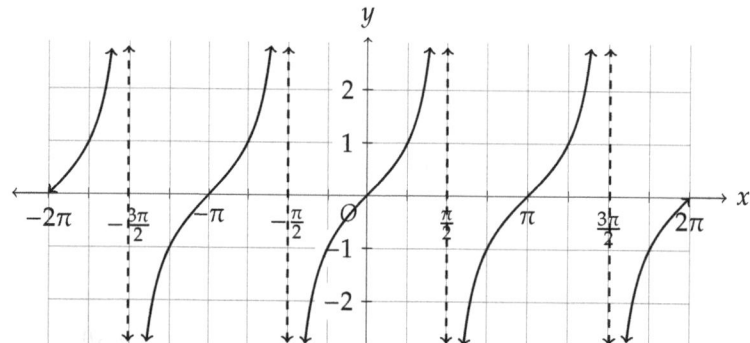

10. The period of g is $\dfrac{\pi}{2\pi} = \dfrac{1}{2}$.

11. The graph of $\tan x$ has been translated by $-\frac{\pi}{3}$, so the first positive asymptote will occur at $\theta = \frac{\pi}{2} - \frac{\pi}{3} = \frac{\pi}{6}$. The period remains π, however, so the asymptotes occur at $x = \theta = \frac{\pi}{6} + k\pi$ for k equal to any integer.

12. The period is 2, so we can solve $\frac{\pi}{b} = 2$ to get $b = \frac{\pi}{2}$. The value of a is negative, as the function is immediately decreasing for $\theta > 0$, and the fact that it decreases by 2 (from 1 at $\theta = 0$ to -1 at $\theta = 1$) indicates that $a = -2$.

13. $\tan \frac{\pi}{3} = \frac{\sin \frac{\pi}{3}}{\cos \frac{\pi}{3}} = \frac{\frac{\sqrt{3}}{2}}{\frac{1}{2}} = \sqrt{3}$

14. $\tan\left(-\frac{7\pi}{6}\right) = \tan\left(\frac{5\pi}{6}\right) = \frac{\sin \frac{5\pi}{6}}{\cos \frac{5\pi}{6}} = \frac{\frac{1}{2}}{-\frac{\sqrt{3}}{2}} = -\frac{\sqrt{3}}{3}$

15. -1

16. We simply need to compute $\tan\left(-\frac{\pi}{6}\right)$. We already computed that $\tan\left(\frac{\pi}{6}\right) = \frac{\sqrt{3}}{3}$, so using the fact that $\tan \theta$ is odd, we get $\tan\left(-\frac{\pi}{6}\right) = -\frac{\sqrt{3}}{3}$.

17. $\tan \theta$ is just the slope of the ray through this point, which is the ratio of the y-coordinate to the x-coordinate, which is precisely $\frac{\sqrt{3}}{1} = \sqrt{3}$. (The point $(1, \sqrt{3})$ is not on the unit circle, but each of 1 and $\sqrt{3}$ are double the values $\frac{1}{2}$ and $\frac{\sqrt{3}}{2}$, respectively. Therefore, $(1, \sqrt{3})$ lies on a circle of radius 2. Hence the coordinates can be written as $\left(2 \cos \frac{\pi}{3}, 2 \sin \frac{\pi}{3}\right)$. When we divide the coordinates, the 2's divide out and we get $\frac{\sin \frac{\pi}{3}}{\cos \frac{\pi}{3}}$, which is an expression for $\tan \frac{\pi}{3}$.)

Notes

Given an angle θ in standard position, the slope of the terminal ray is $\tan \theta$. This is equivalent to the ratio $\frac{\sin \theta}{\cos \theta}$.

The graph of $\tan \theta$ has a period of π and has vertical asymptotes at every odd multiple of $\frac{\pi}{2}$, which can be written as $\theta = \frac{\pi}{2} + k\pi$ for any integer k.

$\tan \theta$ is an odd function, so $\tan(-\theta) = -\tan(\theta)$.
The values of $\tan \theta$ are also very predictable for multiples of $\frac{\pi}{4}$ or $\frac{\pi}{3}$.

- $\tan(k\pi) = 0$ for any integer k
- $\tan\left(k \cdot \frac{\pi}{4}\right) = \pm 1$ for any odd integer k
- $\tan\left(k \cdot \frac{\pi}{6}\right)$ equals $\pm\sqrt{3}$ or $\pm\frac{1}{\sqrt{3}} = \pm\frac{\sqrt{3}}{3}$ for any integer k that isn't a multiple of 3

Homework

1. (a) $\frac{\pi}{1/2} = 2\pi$ (b) $\frac{\pi}{\pi} = 1$ (c) $\frac{\pi}{6}$ (d) $\frac{\pi}{\pi/4} = 4$ (e) $\frac{\pi}{3\pi} = \frac{1}{3}$

   (f) While $\tan x$ will repeat periodically after every increase of π in the inputs, $\sin x$ will only repeat periodically after every increase of 2π. Hence, $\tan x$ will also repeat after every increase of 2π in the inputs, so the period is 2π.

2. (a) $\frac{0}{1} = 0$ (b) $\frac{-\frac{\sqrt{3}}{2}}{\frac{1}{2}} = -\sqrt{3}$ (c) $\frac{-\frac{1}{2}}{\frac{\sqrt{3}}{2}} = -\frac{\sqrt{3}}{3}$ (d) 1

3. The slope of the ray is simply $\tan \frac{5\pi}{6}$, which is $\frac{\frac{1}{2}}{-\frac{\sqrt{3}}{2}} = -\frac{\sqrt{3}}{3}$.

4. k is the slope of the line, which is precisely the slope of the ray through the given point since the line passes through the origin. Therefore, $k = \tan \frac{3\pi}{4} = -1$.

5. The graph of $y = \tan x$ is being horizontally dilated by a factor of 2, and there is no phase shift; therefore, the asymptotes are $x = 2(\frac{\pi}{2} + n\pi) = \pi + 2\pi \cdot n$ for any integer n. (We use n here to illustrate that the variable can be any letter!)

6. The period is $\frac{\pi}{3}$, so $b = 3$. The y-intercept is $(0, -1)$, so $d = -1$. An increase from $\theta = 0$ to $\theta = 1$ in $y = \tan \theta$ leads to an increase in outputs by 1, so an increase from $\theta = 0$ to $\theta = \frac{1}{3}$ in $g(\theta)$ will lead to an increase in outputs by a. The graph increases from $y = -1$ to $y = 2$, and the orientation has not changed, so $a = 3$.

7. The period is $\pi/\pi = 1$, and there is a phase shift of $+1$ (interestingly, this phase shift does nothing, as the shift is by precisely one period!).

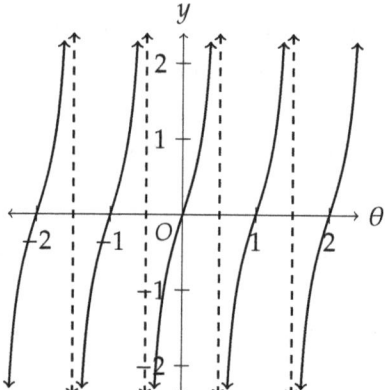

8. (a) True: this is the fact that $f(x) = \tan x$ is odd
   (b) True: this reflects the period of $f(x) = \tan x$ being π
   (c) False: for an angle θ in standard position with its terminal ray, the angle $\pi - \theta$ will have a terminal ray that is reflected across the y-axis. The horizontal displacement will change signs while the vertical displacement will not, so $\tan(\theta) = -\tan(\pi - \theta)$.

Reflection

Suggested Class Time. 50-60 minutes

Prerequisites. Students should be familiar with the tangent function and its output as the slope of a ray or the ratio of sine to cosine.

Instructional Strategies. This lesson went surprisingly smoothly: I thought students may struggle a bit with tangent, but by and large there were no issues except for trying to sketch the graph.

I gave students 55 minutes to complete the activity, and some were done with 10-15 minutes to spare. Over the course of the period, I would go to the board and fill in some of the answers to the activity: I filled in Problems 1-6 at one point, Problem 7 at another, Problem 9 at another, and Problems 10-12 at another. Students did have trouble drawing the graph of tangent from the given information, but I used an applet to help students see precisely why the graph of tangent looks the way it does (link: https://www.geogebra.org/m/jhmwk4rz).

Before filling in the notes, we went over Problems 13-17 together, and students were very confident. Having already done some exact values of tangent in Unit 3A seemed to really help.

Technology. None was required, though a few students did use their calculators to check their graphs of $\tan x$.

Problems.

1. The figure shows point P in the coordinate plane at the intersection of the terminal ray of an angle measuring θ radians and a circle with radius 3. It is known that $\tan \theta = -\frac{2}{5}$. What is the slope of the terminal ray?

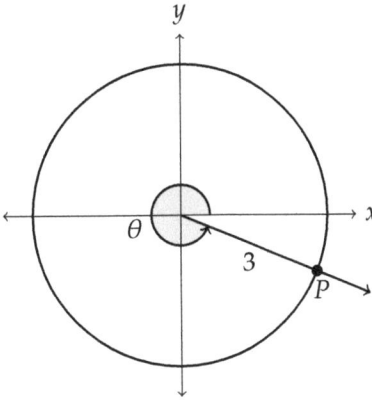

   (A) $-\dfrac{6}{5}$ (B) $-\dfrac{2}{15}$ (C) $-\dfrac{2}{5}$ (D) $-\dfrac{5}{2}$

2. The function f is given by $f(x) = 2\tan\left(x + \frac{\pi}{3}\right)$. Which of the following gives a general expression for the inputs x corresponding to vertical asymptotes of the graph of f?

   (A) $x = \dfrac{\pi}{6} + \dfrac{\pi}{2}k$, where k is an integer

   (B) $x = \dfrac{5\pi}{6} + \dfrac{\pi}{2}k$, where k is an integer

   (C) $x = \dfrac{\pi}{6} + \pi k$, where k is an integer

   (D) $x = \dfrac{5\pi}{6} + \pi k$, where k is an integer

3. The function f is given by $f(x) = \tan(x + a)$ for some real number a. The vertical asymptotes of f occur at the inputs $x = \frac{3\pi}{4} + \pi k$, where k is an integer. What is the value of a?

   (A) $-\dfrac{3\pi}{4}$ (B) $-\dfrac{\pi}{4}$ (C) $\dfrac{\pi}{4}$ (D) $\dfrac{3\pi}{4}$

4. Which of the following is the period of the function $f(\theta) = 4\tan\left(\frac{\pi}{3}(\theta - 2)\right) - 1$?

   (A) 1 (B) $\dfrac{\pi}{3}$ (C) 3 (D) 5

5. The lines $y = x$ and $y = \frac{\sqrt{3}}{3}x$ intersect at the origin and form an angle θ in the first quadrant. What is the measure of θ?

   (A) $\dfrac{\pi}{12}$ (B) $\dfrac{\pi}{8}$ (C) $\dfrac{\pi}{6}$ (D) $\dfrac{\pi}{4}$

Solutions.

1. The tangent of the angle is also the slope of the ray. Thus, the slope is $-2/5$. The correct answer is therefore **(C)**.

2. The graph of $\tan x$ has been translated by $-\frac{\pi}{3}$, so the asymptotes will occur at $x + \frac{\pi}{3} = \frac{\pi}{2} + \pi k$ where k is an integer. Thus $x = \left(\frac{\pi}{2} - \frac{\pi}{3}\right) + \pi k = \frac{\pi}{6} + \pi k$. The correct answer is therefore **(C)**.

3. The graph of $\tan x$ has been translated by $-a$, and the period remains π, so the asymptotes will occur at $x + a = \frac{\pi}{2} + \pi k$ where k is an integer Thus $x = \left(\frac{\pi}{2} - a\right) + \pi k$ and this must be equal to $\frac{3\pi}{4} + \pi k$. It follows that $\frac{\pi}{2} - a = \frac{3\pi}{4}$ so that $a = \frac{\pi}{2} - \frac{3\pi}{4} = -\frac{\pi}{4}$. The correct answer is therefore **(B)**.

4. The period is π divided by $\frac{\pi}{3}$, which is 3. The correct answer is therefore **(C)**.

5. The tangent of the angle α that $y = x$ makes with the x-axis is given by $\tan \alpha = 1$. Hence, $\alpha = \frac{\pi}{4}$. The tangent of the angle β that $y = \frac{\sqrt{3}}{3}x$ makes with the x-axis is given by $\tan \beta = \frac{\sqrt{3}}{3}$. Hence, $\beta = \frac{\pi}{6}$. Then $\theta = \alpha - \beta = \frac{\pi}{4} - \frac{\pi}{6} = \frac{\pi}{12}$. The correct answer is therefore **(A)**.

Topic 3.9 ~ Inverse Trigonometric Functions

Learning Objectives

1. (3.9.A) Construct analytical and graphical representations of the inverse of the sine, cosine, and tangent functions over a restricted domain.

Success Criteria

1. I can evaluate inverse trigonometric expressions, keeping in mind domain restrictions.
2. I can sketch a graph of $\arcsin x$, $\arccos x$, and $\arctan x$.
3. I can identify the domain and range of inverse trigonometric functions.

The functions $\sin x, \cos x$, and $\tan x$ each take angles as inputs and output some measure (vertical displacement, horizontal displacement, and slope, respectively). Given a particular one of these measures, could we work *backwards* to compute an angle?

In short: do the trigonometric functions have *inverses*?

For a function $y = f(x)$ graphed in the plane, recall that the graph of $y = f^{-1}(x)$, if it exists, is a transformation of the graph of $f(x)$.

1. What transformation is it?

Recall that a function only has an inverse (is _____) if its reflection over $y = x$ is a function. This means that each input can only have _____. If the inputs and outputs reverse to create an inverse, though, this means that the original function must have _____ for each unique _____.

2. Do $\sin x, \cos x$, or $\tan x$ have a single input for each unique output?

In order for $\sin x, \cos x$, and $\tan x$ to have inverse functions, they must have their domains *restricted*. We can visualize this with the applet at https://www.geogebra.org/m/ycs7zexs.

3. Use the applet to find the largest possible domain $[a, b]$ ($a \leq x \leq b$) such that $f(x) = \sin x$ would have an inverse function on $[a, b]$. What are a and b?

4. There is a convenient rule that can be used for finding domain restrictions.

   If each output of a function has a unique input, then a _____ line must only pass through the function ____ time. Checking for this is called the _____. Put another way, a function will only pass this test on a given domain restriction if the function is strictly _____ or strictly _____ on this interval.

5. The inverse function of $f(x) = \sin x$ is called *arcsine* and is written as $\arcsin x$ or $\sin^{-1} x$.

   (a) Graph $y = \arcsin x$ in your graphing calculator to confirm the graph you found in the applet.
   (b) For the function $\sin x$, the inputs x are angles in radians and the outputs are vertical displacements. What are the inputs and outputs of $\arcsin x$?

(c) What are the domain and range of sin x?

(d) What are the domain and range of arcsin x?

We can repeat the process with cos x and tan x to create the inverse functions arccos x and arctan x.

Function

	arcsin x	arccos x	arctan x
Domain restriction			
Picture of the domain restriction			
Domain			
Range			
Graph			

We can use the inverse trigonometric functions to find inverses of transformed trigonometric functions using the procedure we covered in Topic 2.8.

6. Find the inverse of $f(x) = 2\sin(3x) - 1$.

7. Find the inverse of $g(x) = \tan(\pi(x-3))$.

When actually evaluating inverse trigonometric expressions, it is incredibly important to remember the domain restrictions. We can use the following process.

- Given $f^{-1}(y) = \theta$, rewrite as $y = f(\theta)$.
- Find the value of θ satisfying $y = f(\theta)$ _____.

8. Evaluate $\sin^{-1}(1)$.

9. Evaluate $\tan^{-1}(-1)$.

10. Evaluate $\sin^{-1}\left(-\dfrac{\sqrt{3}}{2}\right)$.

11. Evaluate $\cos^{-1}\left(-\dfrac{1}{2}\right)$.

12. Confirm the answers from Problems 8, 9, 10, and 11 using your graphing calculator. (Can you find the inverse trig functions?)

For the inverse trigonometric functions, we'll also want to keep in mind our knowledge of function transformations.

13. Find the domain and range of $f(x) = \sin^{-1}(2x) + 1$.

14. Shown below is the graph of the function $g(x) = a \arctan(x - b)$ in the xy-plane. What are a and b?

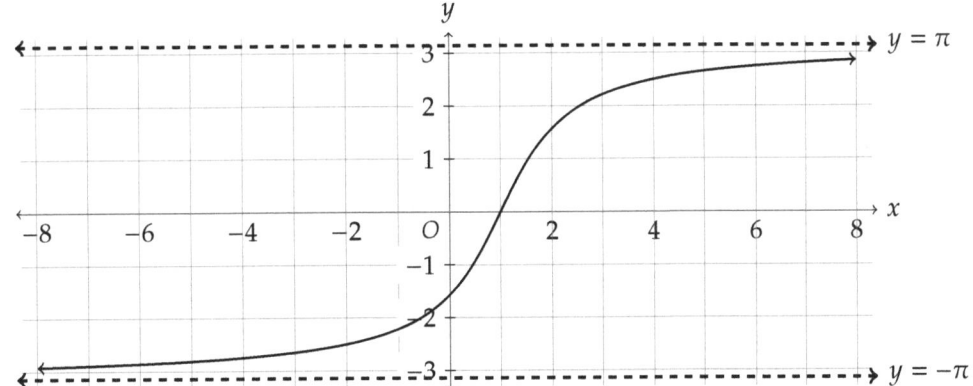

Topic 3.9 Homework

1. Compute the following.

    (a) $\sin^{-1}(-1)$

    (b) $\cos^{-1}(-1)$

    (c) $\tan^{-1}\sqrt{3}$

    (d) $\sin^{-1}\left(-\dfrac{1}{2}\right)$

    (e) $\arccos 0$

    (f) $\tan^{-1}\left(-\dfrac{\sqrt{3}}{3}\right)$

    (g) $\sin^{-1}\left(\dfrac{\sqrt{2}}{2}\right)$

    (h) $\arctan 0$

2. A calculator has output the following: $\cos^{-1}(-0.7)=2.346$. Interpret what this means both verbally and by drawing something on the diagram below.

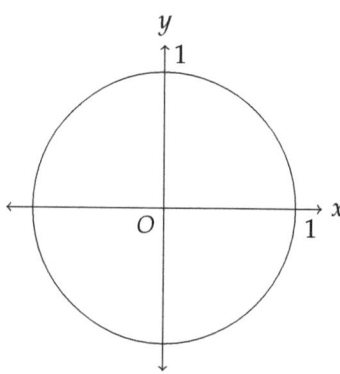

3. Find the domain and range of each of the following functions.

    (a) $f(x) = 2\sin^{-1}(x) - 3$

    (b) $g(x) = \dfrac{2}{\pi}\arctan(x+2)$

    (c) $h(x) = 4\cos^{-1}\left(\dfrac{1}{3}x\right)$

4. Recall the definitions you learned in a previous year for the sine, cosine, and tangent of acute angles.

$$\sin\theta = \dfrac{\text{opposite}}{\text{hypotenuse}} \qquad \cos\theta = \dfrac{\text{adjacent}}{\text{hypotenuse}} \qquad \tan\theta = \dfrac{\text{opposite}}{\text{adjacent}}$$

   (a) A wheelchair ramp is to be built that will be 15 feet long. It will go from the ground to a doorstep that is 2 feet off the ground. What angle will the ramp elevate at from the ground?

   (b) Consider the triangle below.

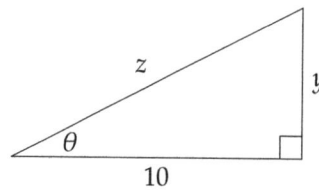

   i. If the function f gives values of θ as a function of y, find an expression for $f(y)$.

   ii. If the function f gives values of θ as a function of z, find an expression for $f(z)$.

Topic 3.9 Solutions/Notes

Activity

1. Reflection across the line $y = x$

Recall that a function only has an inverse (is <u>invertible</u>) if its reflection over $y = x$ <u>is a function</u>. This means that each input can only have <u>one output</u>. If the inputs and outputs reverse to create an inverse, though, this means that the original function must have <u>one input</u> for each unique <u>output</u>.

2. No. These functions are periodic, so each output corresponds to infinitely many inputs.

3. There are numerous correct answers, but we want to steer students to $a = -\frac{\pi}{2}$ and $b = \frac{\pi}{2}$.

4. If each output of a function has a unique input, then a <u>horizontal</u> line must only pass through the function <u>one</u> time. Checking for this is called the <u>Horizontal Line Test</u>. Put another way, a function will only pass this test on a given domain restriction if the function is strictly <u>increasing</u> or strictly <u>decreasing</u> on this interval.

5. (b) The inputs are vertical displacements, and the outputs are angles in radians.

   (c) Domain: $\mathbb{R}$, range: $[-1, 1]$

   (d) Domain: $[-1, 1]$, range: $\left[-\frac{\pi}{2}, \frac{\pi}{2}\right]$ (the domain restriction)

	Function		
	arcsin x	**arccos** x	**arctan** x
Domain restriction	$\left[-\frac{\pi}{2}, \frac{\pi}{2}\right]$	$[0, \pi]$	$\left(-\frac{\pi}{2}, \frac{\pi}{2}\right)$
Picture of the domain restriction	$\sin x \geq 0$ / $\sin x \leq 0$	$\cos x \leq 0$ / $\cos x \geq 0$	$\tan x \geq 0$ / $\tan x \leq 0$
Domain	$[-1, 1]$	$[-1, 1]$	$\mathbb{R}$
Range	$\left[-\frac{\pi}{2}, \frac{\pi}{2}\right]$	$[0, \pi]$	$\left(-\frac{\pi}{2}, \frac{\pi}{2}\right)$
Graph	endpoints $\left(1, \frac{\pi}{2}\right)$ and $\left(-1, -\frac{\pi}{2}\right)$	endpoints $(-1, \pi)$ and $(1, 0)$	horizontal asymptotes $y = \frac{\pi}{2}$ and $y = -\frac{\pi}{2}$

6. Let $y = 2\sin(3x) - 1$. Reversing x and y gives us $x = 2\sin(3y) - 1$. From here:

$$x = 2\sin(3y) - 1$$
$$x + 1 = 2\sin(3y)$$
$$\frac{x+1}{2} = \sin(3y)$$
$$\sin^{-1}\left(\frac{x+1}{2}\right) = 3y$$
$$\frac{1}{3}\sin^{-1}\left(\frac{x+1}{2}\right) = y$$

The inverse is $f^{-1}(x) = \frac{1}{3}\sin^{-1}\left(\frac{x+1}{2}\right)$.

7. Following the same process as in Problem 6:

$$x = \tan(\pi(y - 3))$$
$$\tan^{-1}(x) = \pi(y - 3)$$
$$\frac{1}{\pi}\tan^{-1}(x) = y - 3$$
$$\frac{1}{\pi}\tan^{-1}(x) + 3 = y$$

The inverse is $g^{-1}(x) = \frac{1}{\pi}\tan^{-1}(x) + 3$.

- Given $f^{-1}(y) = \theta$, rewrite as $y = f(\theta)$.
- Find the value of θ satisfying $y = f(\theta)$ within the domain restriction of f.

8. We want θ such that $\sin\theta = 1$. The only such θ in $\left[-\frac{\pi}{2}, \frac{\pi}{2}\right]$ is $\theta = \frac{\pi}{2}$, so $\sin^{-1}(1) = \frac{\pi}{2}$.

9. We want θ such that $\tan\theta = -1$. The only such θ in $\left(-\frac{\pi}{2}, \frac{\pi}{2}\right)$ is $\theta = -\frac{\pi}{4}$, so $\tan^{-1}(-1) = -\frac{\pi}{4}$.

10. We want θ such that $\sin\theta = -\frac{\sqrt{3}}{2}$. The only such θ in $\left[-\frac{\pi}{2}, \frac{\pi}{2}\right]$ is $\theta = -\frac{\pi}{3}$, so $\sin^{-1}\left(-\frac{\sqrt{3}}{2}\right) = -\frac{\pi}{3}$.

11. We want θ such that $\cos\theta = -\frac{1}{2}$. The only such θ in $[0, \pi]$ is $\theta = \frac{2\pi}{3}$, so $\cos^{-1}\left(-\frac{1}{2}\right) = \frac{2\pi}{3}$.

12. Students will find the TI-84 is not terribly receptive to outputting fractions of π, so they'll have to confirm by evaluating their answers against the decimals the calculator provides.

13. The domain and range of $y = \sin^{-1}(x)$ are $[-1, 1]$ and $\left[-\frac{\pi}{2}, \frac{\pi}{2}\right]$. The function f is a horizontal dilation of $y = \sin^{-1}(x)$ by $\frac{1}{2}$ and a vertical translation of $+1$, so the domain of f is $\left[-\frac{1}{2}, \frac{1}{2}\right]$ and the range of f is $\left[1 - \frac{\pi}{2}, 1 + \frac{\pi}{2}\right]$.

14. There has been a horizontal translation $+1$ to the graph of $y = \arctan x$, so $b = 1$. There has also been a vertical dilation of $y = \arctan x$ by a factor of 2, so $a = 2$.

UNIT 3   TOPIC 3.9 ~ INVERSE TRIGONOMETRIC FUNCTIONS   349

Homework

1. (a) $-\dfrac{\pi}{2}$ (c) $\dfrac{\pi}{3}$ (e) $\dfrac{\pi}{2}$ (g) $\dfrac{\pi}{4}$

    (b) π (d) $-\dfrac{\pi}{6}$ (f) $-\dfrac{\pi}{6}$ (h) 0

2. For horizontal displacement of -0.7, the angle θ within $0 \le \theta \le \pi$ is 2.346 radians.

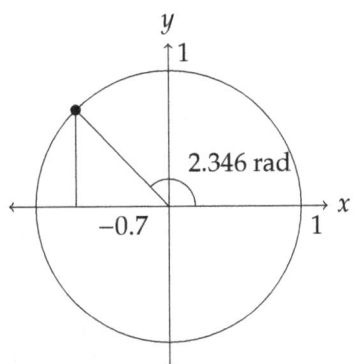

3. (a) Domain: $[-1,1]$, range: $\left[2 \cdot (-\dfrac{\pi}{2}) - 3, 2 \cdot (\dfrac{\pi}{2}) - 3\right] \to [-\pi - 3, \pi - 3]$

    (b) Domain: $\mathbb{R}$, range: $\left(\dfrac{2}{\pi} \cdot \left(-\dfrac{\pi}{2}\right), \dfrac{2}{\pi} \cdot \dfrac{\pi}{2}\right) \to (-1, 1)$

    (c) Domain: $[-1 \cdot 3, 1 \cdot 3] \to [-3, 3]$, range: $[0 \cdot 4, \pi \cdot 4] \to [0, 4\pi]$

4. (a) We can draw a diagram of the ramp.

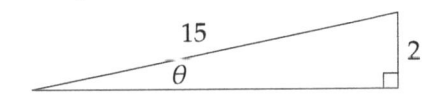

Therefore, $\sin\theta = \dfrac{2}{15}$, so $\theta = \sin^{-1}\left(\dfrac{2}{15}\right) \approx 0.133$ rad $= 7.662°$.

(b) i. $\tan\theta = \dfrac{y}{10}$, so $\tan^{-1}\left(\dfrac{y}{10}\right) = \theta$. therefore, $f(y) = \tan^{-1}\left(\dfrac{y}{10}\right)$.

     ii. $\cos\theta = \dfrac{10}{z}$, so $\cos^{-1}\left(\dfrac{10}{z}\right) = \theta$. therefore, $f(z) = \cos^{-1}\left(\dfrac{10}{z}\right)$.

Reflection

Suggested Class Time. 60 minutes

Prerequisites. Beyond the obvious trigonometric content needed, students need to be comfortable with Topic 2.8 (inverse functions, domain restrictions, reversing inputs and outputs, etc.).

Instructional Strategies. I started by giving a 5-minute introduction to the topic by revisiting inverse functions. We discussed how a function must be one-to-one to have an inverse, and how a function that isn't one-to-one must have part of its domain "chopped off" in order for its inverse to exist. We talked in particular about the function $f(x) = x^2$ and how one can determine both graphically and analytically that f does not have an inverse.

Students said the applet was quite helpful, but like other activities, some of my weaker groups of students needed a bit of a push. After probably 15-20 minutes, I decided to go over Problems 2-5 (some groups were well past this at this point). Once I did that, all of the weaker groups moved much more quickly. I then waited another 15-20 minutes and filled in the table, as well as put up solutions to Problems 7-8. After another few minutes, we summarized the activity and went over everyone's answers to Problems 8-14.

Technology. Students used laptops to open the applet and also used their graphing calculators.

Problems.

1. The function f is defined by $f(x) = \cos(3x) + 1$. The function g is defined as $g(x) = f^{-1}(x)$. Which of the following are the domain and range of g?

    (A) The domain is $0 \leq x \leq 2$ and the range is $0 \leq g(x) \leq \dfrac{\pi}{3}$.

    (B) The domain is $-2 \leq x \leq 0$ and the range is $0 \leq g(x) \leq \dfrac{\pi}{3}$.

    (C) The domain is $0 \leq x \leq 2$ and the range is $0 \leq g(x) \leq 3\pi$.

    (D) The domain is $-2 \leq x \leq 0$ and the range is $0 \leq g(x) \leq 3\pi$.

2. The function f is an inverse trigonometric function. The figure shows the graph of the functions f and $g(x) = b$. The point A lies at the intersection of the graphs of f and g.

    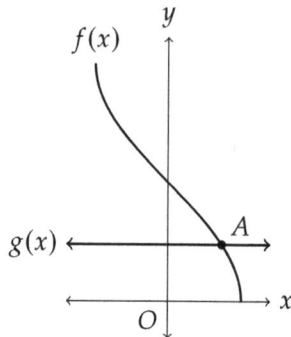

    If the coordinates of A are (a, b), which of the following must be true?

    (A) $\sin a = b$ (B) $\sin b = a$ (C) $\cos a = b$ (D) $\cos b = a$

3. (▣) The function f, whose period is 2π, is given by $f(x) = \sin x + \cos x$. In order to define the inverse function of f, which of the following specifies a restricted domain for f and provides a rationale for why f is invertible on that domain?

    (A) $0 \leq x \leq \pi$, because all possible values of f occur without repeating on this interval.

    (B) $\dfrac{\pi}{4} \leq x \leq \dfrac{3\pi}{4}$, because all possible values of f occur without repeating on this interval.

    (C) $0 \leq x \leq \pi$, because the length of this interval is half the period.

    (D) $\dfrac{\pi}{4} \leq x \leq \dfrac{3\pi}{4}$, because the length of this interval is half the period.

4. (■) Tasha is a building inspector charged with ensuring businesses are compliant with accessibility regulations. Regulations state that all wheelchair ramps for businesses should have an angle of elevation of $\frac{\pi}{36}$ radians or less. Tasha measures a wheelchair ramp and its height, and then draws a diagram like the one pictured below, where θ is the angle of elevation.

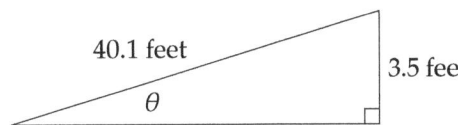

Which of the following is the angle of elevation of the ramp, and which is an appropriate recommendation for Tasha to make to the business?

(A) The angle of elevation is 0.0874 radians and the ramp should be extended to at least 40.16 feet.
(B) The angle of elevation is 0.0874 radians and the ramp should be shortened to at most 40.01 feet.
(C) The angle of elevation is 0.0871 radians and the ramp should be extended to at least 40.16 feet.
(D) The angle of elevation is 0.0871 radians and no changes to the ramp are required.

5. Which of the following is the inverse of the function $g(x) = 4\tan(5x - 2) + 3$?

(A) $g^{-1}(x) = \frac{1}{5}\tan^{-1}\left(\frac{x}{4} - 3\right) + 2$

(B) $g^{-1}(x) = \frac{1}{5}\left(\tan^{-1}\left(\frac{x}{4} - 3\right) + 2\right)$

(C) $g^{-1}(x) = \frac{1}{5}\tan^{-1}\left(\frac{x-3}{4}\right) + 2$

(D) $g^{-1}(x) = \frac{1}{5}\left(\tan^{-1}\left(\frac{x-3}{4}\right) + 2\right)$

Solutions.

1. *First Solution.* The range of f is $[-1+1, 1+1] = [0, 2]$, so the domain of $g = f^{-1}$ is $0 \le x \le 2$. The period of f is $2\pi/3$ with one complete period on the interval $[0, 2\pi/3]$, so the range of f^{-1} is the first half of this period, or $0 \le g(x) \le \pi/3$. The correct answer is therefore **(A)**.

    Second Solution. The function $g(x) = f^{-1}(x)$ is given by $g(x) = \frac{1}{3}\arccos(x-1)$. The domain is therefore $[-1+1, 1+1] = [0, 2]$, and because the domain restriction of $y = \cos x$ is $[0, \pi]$, the range of g is $\frac{1}{3}[0, \pi] = [0, \frac{\pi}{3}]$. The correct answer is therefore **(A)**.

2. The graph shown is that of the inverse cosine function. Hence, (a, b) implies $\cos^{-1} a = b$ so that $a = \cos b$. The correct answer is therefore **(D)**.

3. To define the inverse function, we need the interval to be half the period and to include all possible values of f without repeating any of them. Note that $f(0) = 1$, $f(\pi) = -1$, $f(\frac{\pi}{4}) = \sqrt{2}$, and $f(\frac{3\pi}{4}) = -\sqrt{2}$. Then since $f(\frac{\pi}{4}) > f(0)$ and $f(\frac{3\pi}{4}) < f(\pi)$, it is the interval $\frac{\pi}{4} \le x \le \frac{3\pi}{4}$ that will include all possible values with none repeated. The correct answer is therefore **(B)**.

4. The angle of elevation is
$$\sin^{-1}\left(\frac{3.5}{40.1}\right) = 0.0874.$$
Since this is larger than $\frac{\pi}{36} \approx 0.0873$, we need to calculate a new length of the the ramp. Let R be the length the ramp should be. Since the regulation is $\frac{\pi}{36}$, we must have $\sin\frac{\pi}{36} = 3.5/R$, or
$$R = \frac{3.5}{\sin\frac{\pi}{36}} = 40.156 \text{ feet}.$$

The correct answer is therefore **(A)**.

5. Let $y = 4\tan(5x - 2) + 3$. Reversing x and y gives $x = 4\tan(5y - 2) + 3$. Then

$$x = 4\tan(5y - 2) + 3$$
$$x - 3 = 4\tan(5y - 2)$$
$$\frac{x-3}{4} = \tan(5y - 2)$$
$$\tan^{-1}\left(\frac{x-3}{4}\right) = 5y - 2$$
$$\tan^{-1}\left(\frac{x-3}{4}\right) + 2 = 5y$$
$$\frac{1}{5}\left(\tan^{-1}\left(\frac{x-3}{4}\right) + 2\right) = y.$$

The inverse is $g^{-1}(x) = \frac{1}{5}\left(\tan^{-1}\left(\frac{x-3}{4}\right) + 2\right)$. The correct answer is therefore **(D)**.

Topic 3.10 ~ Trigonometric Equations and Inequalities (Day 1)

Learning Objectives
1. (3.10.A) Solve equations and inequalities involving trigonometric functions.

Success Criteria
1. I can use analytical techniques, knowledge of the unit circle, technology, and graphs to solve equations and inequalities involving trigonometric functions.

In this lesson and the next, we'll discuss solving equations involving trigonometric functions.

While equations involving polynomial functions of degree n will result in n solutions, trigonometric functions are periodic, which means they will typically have _____ solutions.

Example 1: Let f be the function given by $f(x) = 2\sin x$.

(a) Find all x in the interval $0 \leq x \leq 2\pi$ such that $f(x) = 1$.

(b) Find all x such that $f(x) = 1$.

Solution:

The solution in part (a) required that our angle x be in a particular interval. In (b), we wanted *all* solutions. The solutions provided in (b) are often referred to as the _____ of the equation.

When identifying the general solution, it will be important to remember the _____ of the trigonometric function at hand.

Example 2: Let g be the function given by $g(\theta) = 1 - \tan^2(2\theta)$. Find the general solution of $g(\theta) = 0$.

Solution:

When the trigonometric outputs aren't convenient values like $\pm\frac{1}{2}, \pm 1, \pm\frac{\sqrt{3}}{2}$, etc., then we will leave our answer in terms of inverse trigonometric functions.

Example 3: If g is the function $g(x) = 4 - 3\cos(x)$, find the solutions of $g(x) = 5$.

Solution:

Other times, we'll need our calculators.

Example 4: (📱) The amount of daily rainfall in one region is roughly periodic as time goes on. The rainfall R, in inches, is given by

$$R(t) = 1.2\cos\left(\frac{2\pi}{365}(t - 170)\right) + 1.4,$$

for day t. The value $t = 0$ corresponds to January 1 of a certain year.

(a) For approximately how many days in one year will the rainfall be greater than 2 inches?

(b) On which days of the year t will the rainfall be 1.5 inches?

(c) One period of the graph of R is displayed.

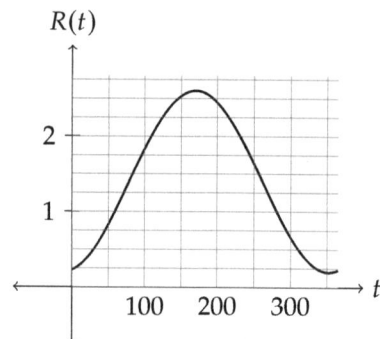

For approximately how many days in this year will the rainfall

  i. be less than 0.75 inch?

  ii. be between 1 and 2 inches?

Topic 3.10 (Day 1) Homework

1. Let f be the function given by $f(x) = 2\cos x + 1$. Find the zeros of f on the interval $0 \leq x < 2\pi$.

2. Find all of f from Problem 1.

3. Let g and h be the functions expressed as $g(x) = 2\tan x + 1$ and $h(x) = \tan x + 5$. Find the general solution of the equation $g(x) = h(x)$.

4. If f is the function $f(\theta) = 4\cos^2 \theta - 3$, find all inputs θ that are zeros of $f(\theta)$ on the interval $[-\pi, \pi]$.

5. The table below shows input-output pairs for a function $f(x)$.

x	-1	$-\dfrac{\sqrt{3}}{2}$	$\dfrac{1}{2}$	$\dfrac{\sqrt{3}}{2}$
$f(x)$	π	$\dfrac{5\pi}{6}$	$\dfrac{\pi}{3}$	$\dfrac{\pi}{6}$

   Find an expression for the function $f(x)$.

6. The function f is given by $f(x) = \cos(8x)$. For how many values of x in the interval $0 \leq x \leq 2\pi$ will $f(x) = \frac{2}{3}$?

7. (🔳) Two runners, Adam and Sydnee, are running a 10-kilometer race. Suppose that Adam's velocity is given by the function
$$v_A(t) = 120\sin\left(\frac{\pi}{15}(t-1)\right) + 496$$
where v is in feet per minute and t is in minutes. Sydnee, on the other hand, runs at a constant rate of $v_S(t) = 550$ feet per minute for the entire race. Both runners start at the same time.

   (a) At approximately what times t in the first hour will Adam's velocity be equal to Sydnee's?

   (b) In the first 30 minutes of the race, for how many minutes will Sydnee be running faster than Adam?

Topic 3.10 (Day 1) Solutions/Notes

Activity
While equations involving polynomial functions of degree n will result in n solutions, trigonometric functions are periodic, which means they will typically have <u>infinitely many</u> solutions.

Solution 1:

(a) $2\sin x = 1$ means that $\sin x = 1$. From the unit circle, this can occur in Quadrant I, when $x = \frac{\pi}{6}$, or in Quadrant II, when $x = \frac{5\pi}{6}$.

(b) For either of the inputs x in (a), the output of $\sin x$ will still equal $\frac{1}{2}$ if any number of periods is added to the input. In short, $\sin(x + 2\pi k) = \frac{1}{2}$, where k is any integer, for either of the x in (a). Therefore, the solutions are

$$x = \frac{\pi}{6} + 2\pi k, \text{where } k \text{ is any integer}$$

$$x = \frac{5\pi}{6} + 2\pi k, \text{where } k \text{ is any integer}$$

The solutions provided in (b) are often referred to as the <u>general solution</u> of the equation.

When identifying the general solution, it will be important to remember the <u>period</u> of the trigonometric function at hand.

Solution 2: We can factor this expression into $g(\theta) = (1 + \tan(2x))(1 - \tan(2x))$. Setting each factor to 0, we get $\tan(2x) = -1$ and $\tan(2x) = 1$. From here, we solve each equation.

$$\tan(2x) = -1 \qquad\qquad \tan(2x) = 1$$
$$\arctan(\tan(2x)) = \arctan -1 \qquad\qquad \arctan(\tan(2x)) = \arctan 1$$
$$2x = -\frac{\pi}{4} \qquad\qquad 2x = \frac{\pi}{4}$$

At this point, we want to divide by 2, but we have to first consider the general solution. $\tan x$ has a period of π, so our general solutions of $\tan(2x) = -1$ and $\tan(2x) = 1$ are

$$2x = -\frac{\pi}{4} + \pi k, \text{where } k \text{ is any integer}$$
$$2x = \frac{\pi}{4} + \pi k, \text{where } k \text{ is any integer}$$

Now we can divide by 2.

$$x = \frac{1}{2}\left(-\frac{\pi}{4} + \pi k\right) = -\frac{\pi}{8} + \frac{\pi}{2} \cdot k, \text{where } k \text{ is any integer}$$
$$2x = \frac{1}{2}\left(\frac{\pi}{4} + \pi k\right) = \frac{\pi}{8} + \frac{\pi}{2} \cdot k, \text{where } k \text{ is any integer}$$

Solution 3: We solve for x in a traditional manner.

$$4 - 3\cos(x) = 5$$
$$-3\cos(x) = 1$$
$$\cos(x) = -\frac{1}{3}$$
$$\arccos(\cos(x)) = \arccos\left(-\frac{1}{3}\right)$$
$$x = \arccos\left(-\frac{1}{3}\right)$$

Finally, we get the general solution by considering adding integer multiples of the period to the input x. Our general solution is $x = \arccos\left(-\frac{1}{3}\right) + 2\pi k$ for any integer k.

Solution 4:

(a) We graph R in Y_1 and the line $y = 2$ in Y_2. The period of R is 365, so we change our window to have Xmin=0 and Xmax=365 and adjust our Ymin and Ymax to include the entire graph.

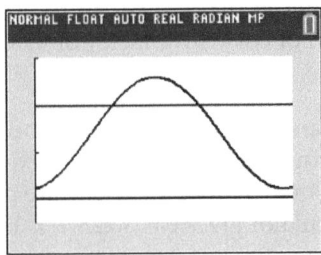

The intersections occur at $x = 109.167$ and $x = 230.833$. The rainfall will be greater than 2 inches between these days, so the rainfall will be greater than 2 inches for approximately $230.833 - 109.167 = 121.666$, or 121 (or 122) days.

(b) We change Y_2 to Y2=1.5 and find the intersections. The inputs at which the intersections occur are $x = 83.597$ and $x = 256.403$, so the rainfall will be 1.5 on approximately the 84th and 256th day of the year.

(c) We can look at the graph and draw the line $y = 0.75$, $y = 1$, and $y = 2$.

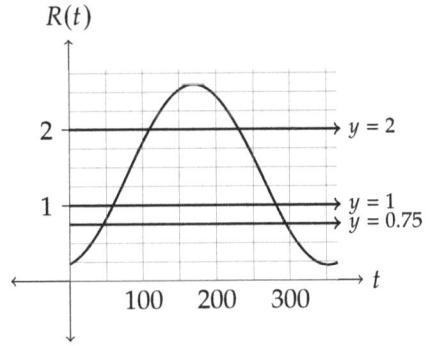

  i. The rainfall will equal 0.75 inch on approximately days 40 and 290. The rainfall will be less than 1.5 inch for days before day 40 and after day 290, so there are approximately $39+(365-290) = 114$ days where the rainfall will be less than 0.75 inch.

  ii. The intersections occur around the input $t = 60$, $t = 100$, $t = 230$, and $t = 280$. Therefore, the rainfall will be between 1 and 2 inches for approximately $(100 - 60) + (280 - 230) = 90$ days.

Homework

1. $2\cos(x) + 1 = 0 \Rightarrow \cos(x) = -\frac{1}{2} \Rightarrow \arccos(\cos(x)) = \arccos\left(-\frac{1}{2}\right) \Rightarrow x = \frac{2\pi}{3}$ or $x = \frac{4\pi}{3}$

2. $x = \frac{2\pi}{3} + 2\pi m$ or $x = \frac{4\pi}{3} + 2\pi m$ for any integer m

3. $2\tan x + 1 = \tan x + 5 \Rightarrow \tan x = 4 \Rightarrow \arctan(\tan x) = \arctan 4 \Rightarrow x = \arctan 4$. The general solution is $x = \arctan 4 + \pi k$ for any integer k.

4. $4\cos^2\theta - 3 = 0 \Rightarrow 4\cos^2\theta = 3 \Rightarrow \cos^2\theta = \dfrac{3}{4} \Rightarrow \cos\theta = \pm\sqrt{\dfrac{3}{4}} = \dfrac{\sqrt{3}}{2}$. On the interval $[-\pi, \pi]$, $\cos\theta = \pm\dfrac{\sqrt{3}}{2}$ at $\theta = \dfrac{\pi}{6}, \theta = \dfrac{5\pi}{6}, \theta = -\dfrac{\pi}{6}$, and $\theta = -\dfrac{5\pi}{6}$.

5. If the inputs and outputs are reversed, these input-output pairs match that of the function $g(x) = \cos x$. Therefore, $f(x) = g^{-1}(x) = \cos^{-1} x$.

6. For any given output y, there will be two inputs for which the function $\cos x$ will output y. The function f is a horizontal dilation of $\cos x$ by a factor of 8, so this will result in there being $8(2) = 16$ solutions of $\cos(8x) = \dfrac{2}{3}$.

7. (a) We graph v_A and v_S in Y₁ and Y₂ and set Xmin=0 and Xmax=60 to restrict to the first hour (60 minutes). We adjust the Ymin and Ymax to fit the maximum and minimum of v_A. The times at which Adam's velocity will equal Sydnee's velocity are precisely the inputs at which the graphs of v_A and v_S intersect. The calculator gives $t = 3.228, t = 13.771, t = 33.228$, and $t = 43.771$ minutes.

   (b) We keep the same graph but change Xmax=30 to reflect the first 30 minutes. The intersections of v_A and v_S occura t $t = 3.228$ and $t = 13.771$. The outputs of v_S are higher than the outputs of v_A for $t < 3.228$ and $13.771 < t < 30$, so Sydnee's velocity will be higher than Adam's velocity for approximately $3.228 + (30 - 13.771) = 19.457$ minutes in the first 30 minutes.

Reflection

Suggested Class Time. 50-60 minutes

Prerequisites. Students should of course be very familiar with unit circle values, the inverse trigonometric functions and their domain restrictions, and have a good feel for utilizing the symmetry of the unit circle.

Instructional Strategies. One main thing jumps out to me. Going back to Algebra I, when students would solve equations like $x^2 = 25$, it would be common for students to forget their ±. In other words, students would only remember the domain restriction of $x > 0$ for $f(x) = x^2$. This exact same thing can happen with trigonometric functions, only it's worse, because we (I'm guilty, too) often place such an emphasis on the domain restrictions required for inverses.

Therefore, in Example 1 of this lesson, when we got to the equation $\sin x = \dfrac{1}{2}$, and a student said, "now we take the arcsine," I actually said, "We could, but do we need to?" I immediately transitioned everyone to thinking about the unit circle. Rather than asking "What's the arcsine of $\dfrac{1}{2}$?", we asked "For what angle/s does $\sin x = \dfrac{1}{2}$?" The difference is that the former includes a domain restriction, while the second – and the question at hand – doesn't. We went on our merry way with finding solutions.

In Example 2, we ran into the same thing. This time, I did include writing a ± for $\sqrt{1}$, but I again emphasized that we could solve $\tan^2(2\theta) = 1$ by simply thinking about what values, when squared, equal 1.

For Example 2, we had the added complexity of the argument. I taught my students this term (argument), and told them to, in general, solving as if the argument were x (or θ) to begin with. Then, we simply got, for instance, argument $= \dfrac{\pi}{4} + \pi k$. Then, we erased "argument" and input 2θ, afterward isolating θ. I think putting this word there helped students to compartmentalize and not get overwhelmed (or make silly mistakes).

For Example 3, I want to make note of something I've found very helpful for students. In my own classroom, I cannot count the number of times that students have said the strategy to solve something like $\cos(x) = -\dfrac{1}{3}$ is to "divide by cosine." While this is much less of a problem this year, I still have students who struggle or are slow to identify that we should take the arccosine of both sides. What I found so helpful is the following verbiage:

We have an input of x that gives an output of $-\frac{1}{3}$. If we have the output and want the input, then we…

One of my students in particular lit up immediately when I said this (which happened twice) and blurted out "We take the inverse!" The emphasis that this class places on inputs and outputs truly works!

For Example 4, we only worked (a) as a class; I then had students work (b) and (c) in their groups, after which we went over it.

After the lesson, I actually gave students the first three questions provided at the end of the Reflection for Day 2. They were meant to be challenging versions of the ones we worked in class. As such, I let students work on them in groups, after which we discussed.

Technology. Students only need a graphing calculator for Example 4.

Topic 3.10 ~ Trigonometric Equations and Inequalities (Day 2)

Learning Objectives
1. (3.10.A) Solve equations and inequalities involving trigonometric functions.

Success Criteria
1. I can use analytical techniques (including factoring), knowledge of the unit circle, technology, and graphs to solve equations and inequalities involving trigonometric functions.

Today, you will continue with solving equations and inequalities involving trigonometric functions. Now, we'll incorporate slightly more advanced techniques and knowledge of trigonometric functions.

1. Here is an example involving factoring. Read through the example, making sure you understand each step and the final solution.

   > **Example 1:** Let g be the function given by $g(\theta) = 2\sin^2\theta + \sin\theta - 1$. Find all θ in the interval $[0, 2\pi)$ satisfying $g(\theta) = 0$.
   >
   > **Solution:** g has the same structure as $g(x) = 2x^2 + x - 1$, where $x = \sin\theta$. Therefore, $g(\theta)$ can be factored. Since $g(x)$ factors into $(2x-1)(x+1)$, $g(\theta)$ can factor into $g(\theta) = (2\sin\theta - 1)(\sin\theta + 1)$. Now, we can set the factors equal to 0 and solve.
   >
   > $$2\sin\theta - 1 = 0 \Rightarrow \sin\theta = \frac{1}{2} \Rightarrow \theta = \frac{\pi}{6}, \frac{5\pi}{6}$$
   > $$\sin\theta + 1 = 0 \Rightarrow \sin\theta = -1 \Rightarrow \theta = \frac{3\pi}{2}$$
   >
   > The solutions of $g(\theta) = 0$ in the interval $0 \leq \theta < 2\pi$ are therefore $\theta = \frac{\pi}{6}, \theta = \frac{5\pi}{6}$, and $\theta = \frac{3\pi}{2}$.

2. Try a similar example: if f is the function $f(x) = 2\cos^2 x + 7\cos x$, find the input values x in the interval $0 \leq x < 2\pi$ such that $f(x) = -3$.

3. Let h be the function given by $h(\theta) = \sin\theta\tan^2\theta - 3\sin\theta$. Find the zeros of h in the interval $[0, 2\pi]$.

4. Sometimes, domain restrictions will influence our solutions.

Example 2: Let f be the function given by $f(x) = 5 + 4\sin x$. Find the solution/s of $f(x) = 2$ on the restricted domain $\pi < x < 2\pi$.

Solution: We solve for x.

$$5 + 4\sin x = 2$$
$$\sin x = -\frac{3}{4}$$
$$x = \arcsin\left(-\frac{3}{4}\right)$$

Because $y = \sin x$ has a domain restriction of $-\frac{\pi}{2} \leq x \leq \frac{\pi}{2}$, the value of $\arcsin\left(-\frac{3}{4}\right)$ is in Quadrant IV, meaning $-\frac{\pi}{2} < \arcsin\left(-\frac{3}{4}\right) < 0$. To get value/s of x between π and 2π, we must use the period of $\sin x$ a to find the solutions. If we add one full period of 2π to $\arcsin\left(-\frac{3}{4}\right)$, we will get another angle in Quadrant IV, but this time between $\frac{3\pi}{2}$ and 2π, which satisfies our desired interval of $\pi < x < 2\pi$. Therefore, our solution is $x = \arcsin\left(-\frac{3}{4}\right) + 2\pi$. See the diagram below.

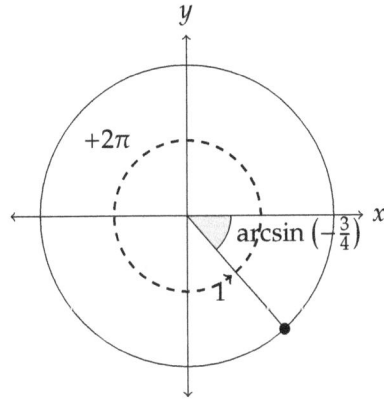

5. Let g be the function given by $g(\theta) = 7 - 2\tan\theta$. Find the value of θ with $0 < \theta < \pi$ such that $g(\theta) = 8$.

6. Now for an example involving an inequality. To solve trigonometric inequalities, you will find (a) solutions and then (b) use the behavior of the graphs of the trigonometric functions. Sketching always helps!

> **Example 3:** Let f be the function given by $f(x) = 4\sin x + 3$. For $0 \leq x < 2\pi$, find all x such that $f(x) < 1$.
>
> **Solution:** First, we solve $f(x) = 1$.
> $$4\sin x + 3 = 1$$
> $$\sin x = -\frac{1}{2}$$
>
> The inputs x in $0 \leq x < 2\pi$ with $\sin x = -\frac{1}{2}$ are $x = \frac{7\pi}{6}$ and $x = \frac{11\pi}{6}$. Now, we consider the graph of $\sin x$ for $0 \leq x < 2\pi$.
>
>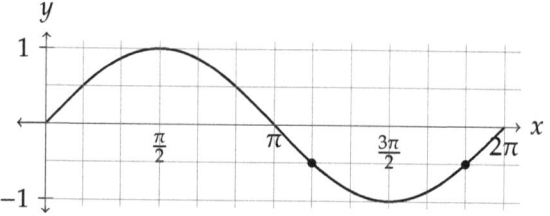
>
> For inputs between $\frac{7\pi}{6}$ and $\frac{11\pi}{6}$, $\sin x$ decreases from $-\frac{1}{2}$ to -1 and then increases to $-\frac{1}{2}$ again. Therefore, $\sin x < -\frac{1}{2}$ on the interval $\frac{7\pi}{6} < x < \frac{11\pi}{6}$, and hence this is the interval on which $f(x) < 1$.

7. Let f be the function given by $f(\theta) = 1 - \tan \theta$. Find the values of θ in the interval $0 \leq \theta < \pi$ such that $f(\theta) < 0$.

Topic 3.10 (Day 2) Homework

1. Solve $\sin^2 x - \sin x = 0$ on the interval $0 \leq x < 2\pi$.

2. Let f and g be the functions be given by $f(\theta) = 3\cos^2\theta - 1$ and $g(\theta) = 2\cos\theta$. If h is the function given by $h(\theta) = f(\theta) + g(\theta)$, find the zeros of h on the interval $[0, \pi]$.

3. The function s is given by $s(\theta) = \sqrt{3}\cos\theta - 2\sin\theta\cos\theta$. Find the solutions of $s(\theta) = 0$ such that $-\pi < \theta < \pi$.

4. The function g can be expressed as $g(x) = \tan x$. Find all values of x with $0 < x < \pi$ such that $g(x) < 2$.

5. Refer back to the function from Problem 4. Find x with $\frac{\pi}{2} < x < \frac{3\pi}{2}$ such that $g(x) = \frac{4}{3}$.

6. (Multiple Choice) Let f be the function given by $f(x) = 5\cos x$. Which of the following gives all solutions x with $0 < x < 2\pi$ of $f(x) = 1$?

    (A) $x = \arccos\frac{1}{5}$ only
    (B) $x = \arccos\frac{1}{5}$ and $x = -\arccos\frac{1}{5}$
    (C) $x = \arccos\frac{1}{5}$ and $x = \arccos\frac{1}{5} + 2\pi$
    (D) $x = \arccos\frac{1}{5}$ and $x = 2\pi - \arccos\frac{1}{5}$

Topic 3.10 (Day 2) Solutions/Notes

Activity

2. $2\cos^2 x + 7\cos x = -3 \Rightarrow 2\cos^2 x + 7x + 3 = 0 \Rightarrow (2\cos x + 1)(\cos x + 3)$. Setting each factor to 0 and solving gives $\cos x = -\frac{1}{2}$ and $\cos x = -3$. The second equation has no solutions, and the first equation has solutions of $x = \frac{2\pi}{3}$ and $x = \frac{4\pi}{3}$. Therefore, $f(x) = -3$ for $x = \frac{2\pi}{3}$ and $x = \frac{4\pi}{3}$.

3. We set $h(\theta) = 0$ and factor out $\sin\theta$ to get $\sin\theta \left(\tan^2\theta - 3\right) = 0$. Setting $\sin\theta = 0$, we get $\theta = 0, \pi, 2\pi$ in the given domain. Solving $\tan^2\theta - 3 = 0$ gives $\tan\theta = \pm\sqrt{3}$, which has solutions of $\theta = \frac{\pi}{3}, \theta = \frac{2\pi}{3}, \theta = \frac{4\pi}{3}$, and $\theta = \frac{5\pi}{3}$. Our complete solution set is therefore $\left\{0, \frac{\pi}{3}, \pi, \frac{2\pi}{3}, \frac{4\pi}{3}, \frac{5\pi}{3}, 2\pi\right\}$.

5. $7 - 2\tan x = 8 \Rightarrow \tan x = -\frac{1}{2} \Rightarrow x = \arctan\left(-\frac{1}{2}\right)$. Given the domain restriction of $-\frac{\pi}{2} < x < \frac{\pi}{2}$ for $\tan x$ to have an inverse, we must add one period of $\tan x$ to get an angle in the desired interval. This results in $x = \arctan\left(-\frac{1}{2}\right) + \pi$.

7. $1 - \tan\theta < 0$ gives $\tan\theta > 1$. Solving $\tan\theta = 1$ for $0 \leq \theta < \pi$ gives $\theta = \frac{\pi}{4}$. The graph of $\tan\theta$ is increasing for $0 < \theta < \frac{\pi}{2}$ and negative for $\frac{\pi}{2} < \theta < \pi$, so $\tan\theta$ is greater than 1 only for $\frac{\pi}{4} < \theta < \frac{\pi}{2}$ on the given interval.

Homework

1. $\sin^2 x - \sin x = 0 \Rightarrow \sin x(\sin x - 1) = 0$, so setting each factor equal to 0 results in $\sin x = 0$ or $\sin x = 1$. The first equation gives $x = 0$ and the second equation gives $x = \frac{\pi}{2}$, so the solutions are $x = 0$ and $x = \frac{\pi}{2}$.

2. $h(\theta) = 0 \Rightarrow 3\cos^2\theta + 2\cos\theta - 1 = 0 \Rightarrow (3\cos\theta - 1)(\cos\theta + 1) = 0$. Setting each factor equal to 0 gives $\cos\theta = \frac{1}{3}$ or $\theta = -1$. The solutions of these equations on the interval $[0, \pi]$ are $\theta = \arccos\frac{1}{3}$ and $\theta = \pi$.

3. We set $s(\theta) = 0$ and factor out $\cos\theta$ to get $\cos\theta(\sqrt{3} - 2\sin\theta)$. Setting each factor equal to 0 results in $\cos\theta = 0$ and $\sin\theta = -\frac{\sqrt{3}}{2}$. The first equation yields $\theta = \frac{\pi}{2}$ and $\theta = \frac{3\pi}{2}$, while the second equation yields $\theta = \frac{4\pi}{3}$ and $\theta = \frac{5\pi}{3}$. However, we need our angles to be in the interval $-\pi < \theta < \pi$, so we adjust them to get $\theta = \frac{\pi}{2}, \theta = -\frac{\pi}{2}, \theta = -\frac{2\pi}{3}$, and $\theta = -\frac{\pi}{3}$, respectively.

4. For $0 < x < \pi$, $\tan x$ is positive only for $0 < x < \frac{\pi}{2}$ and is increasing. Therefore, $\tan x < 2$ only if $0 < x < \arctan 2$.

5. $\tan x = \frac{4}{3} \Rightarrow x = \arctan\frac{4}{3}$. However, given the domain restriction of $-\frac{\pi}{2} < x < \frac{\pi}{2}$, the value $\arctan\frac{4}{3}$ is between 0 and $\frac{\pi}{2}$. To get an input value x that will also yield $\tan x = \frac{4}{3}$ in the interval $\frac{\pi}{2} < x < \frac{3\pi}{2}$, we must add one period to our solution, giving $x = \arctan\frac{4}{3} + \pi$.

6. One solution is obviously $x = \arccos\frac{1}{5}$. However, because $\cos x$ is even, $\cos x$ can equal $\frac{1}{5}$ for an angle in Quadrant IV. This would give $x = -\arccos\frac{1}{5}$, but this is not in the interval $0 < x < 2\pi$. if we add one period of $\cos x$, though, we will get a value in the desired interval. This gives $x = -\arccos\frac{1}{5} + 2\pi = 2\pi - \arccos\frac{1}{5}$. The answer is (D).

Reflection

Suggested Class Time. 45-60 minutes

Prerequisites. Students must be comfortable with factoring techniques.

Instructional Strategies. This lesson was quite straightforward. We actually spent the first 30 minutes of this class going over homework and then talking about the AP exam structure. From there, students worked diligently on the practice questions. I'd put up solutions for each question after 5-10 minutes and field questions.

The main things I noticed students struggling with were the domain restrictions, or the issue of the students *not reading them*. On Problem 3 in particular, a female student asked, "Why didn't you include $-\frac{\pi}{3}$?" I then had the student ask me to identify every sophomore male in the class (I have a class with both sophomores and juniors). I went up to each sophomore male and asked the female student, "Do they count?", to which she would say yes. Then, I went up to a junior, and my conversation with the female student went as follows.

> Me: "What about him?"
> Student: "No, he's not a sophomore."
> Me: "Yeah, but he's a male!"
> Student: "... but I asked about sophomores."
> Me: "I know! But he's male!"
> Student: "But he's not a sophomore!!"

I played this up a bit for effect, after which I said, "EXACTLY! Yes, he's a male, but you specifically asked for *sophomores*. It wasn't enough to just be male." Looking back at the domain restriction on Problem 3, the student quickly said "Oh, I get it."

It was clear after the lesson that no one concept of this topic is anywhere near difficult for my students: it's just the level of precision and attention to detail required that will trip them up. As such, I decided to assign a rather lengthy Delta Math assignment on top of the homework, which I made optional for students (but will give a nice fat class grade for if they complete it).

Technology. None was required.

Problems.

1. Let f be the function given by $f(x) = 4\sin^2 x$. Which of the following gives the solutions of $f(x) = 3$ on the interval $0 \leq x \leq 2\pi$?

   (A) $x = \frac{\pi}{6}, x = \frac{5\pi}{6}$
   (B) $x = \frac{\pi}{3}, x = \frac{2\pi}{3}$
   (C) $x = \frac{\pi}{6}, x = \frac{5\pi}{6}, x = \frac{7\pi}{6}, x = \frac{11\pi}{6}$
   (D) $x = \frac{\pi}{3}, x = \frac{2\pi}{3}, x = \frac{4\pi}{3}, x = \frac{5\pi}{3}$

2. Let $\theta = \arcsin\left(-\frac{1}{4}\right)$. The function f is defined by $f(x) = 4\sin(x) - 1$. Which of the following is a solution to $f(x) = 0$?

   (A) θ (B) $\pi - \theta$ (C) $\pi + \theta$ (D) $2\pi + \theta$

3. The function g is defined by $g(\theta) = \tan(2\theta) + 1$. How many solutions does $g(x) = 4$ have on the interval $0 \leq \theta \leq \pi$?

   (A) 1 (B) 2 (C) 4 (D) Infinitely many

4. The function h is defined by
$$h(\theta) = 3\tan\left(\frac{\pi\theta}{4}\right) + 5.$$
The solutions of the equation $h(\theta) = -1$ can be expressed as $\theta = a + bk$, where b is an integer. What is the value of b?

   (A) $b = 1$ (B) $b = 2$ (C) $b = 4$ (D) $b = 8$

5. (■) The height h, in feet, of a pirate ship ride at a carnival t seconds after beginning can be modeled by the function
$$h(t) = 13\sin\left(\frac{2\pi}{25}(t-6.25)\right) + 15.$$
If the ride lasts 50 seconds, then for how many seconds will the pirate ship be at least 20 feet high?

(A) 9.358 seconds   (B) 18.717 seconds   (C) 34.358 seconds   (D) 42.179 seconds

6. The function g is defined by $g(x) = 3\cos(x) + 1$. For what value of x, with $\pi < x < 2\pi$, will $g(x) = 2$?

(A) $\arccos\left(\frac{1}{3}\right)$ (B) $\arccos\left(\frac{1}{3}\right) + \pi$ (C) $2\pi - \arccos\left(\frac{1}{3}\right)$ (D) $\arccos\left(\frac{1}{3}\right) + 2\pi$

Solutions.

1. Setting $f(x) = 3$, dividing by 4, taking the square root of both sides yields $\sin x = \pm\frac{\sqrt{3}}{2}$. The vertical displacement of an angle x is equal to $\pm\frac{\sqrt{3}}{2}$ at $x = \frac{\pi}{3}, x = \frac{2\pi}{3}, x = \frac{4\pi}{3}$, and $x = \frac{5\pi}{3}$. The correct answer is therefore **(D)**.

2. Setting $f(x) = 0$ and solving for x yields $\sin x = \frac{1}{4}$. From the definition of θ, we know that $\sin\theta = -\frac{1}{4}$. Because the range of $g(x) = \arcsin x$ is $[-\frac{\pi}{2}, \frac{\pi}{2}]$, we know that $-\frac{\pi}{2} < \theta < 0$. To transform θ into an angle with a positive vertical displacement of $\frac{1}{4}$, we must rotate back into Quadrant II or Quadrant III. The angle $\pi + \theta$ will rotate θ to be symmetric over the x-axis is Quadrant II and therefore have a vertical displacement of $\frac{1}{4}$. The correct answer is therefore **(C)**.

3. Let $f(\theta) = \tan\theta$. On the interval $[0, \pi]$, $f(\theta)$ will equal 4 exactly one time - for an angle in Quadrant I. The function $g(\theta)$ is simply a horizontal dilation of $f(\theta)$ by a factor of $\frac{1}{2}$, along with a vertical translation which will not affect the number of solutions. This horizontal dilation will create a second solution of $g(\theta) = 4$. The correct answer is therefore **(B)**.

4. We solve for θ.
$$3\tan\left(\frac{\pi}{4}\cdot\theta\right) + 5 = -1$$
$$\tan\left(\frac{\pi}{4}\cdot\theta\right) = -2$$

The general solution of $\tan x = -2$ is $x = p + \pi k$ for some value of p. Setting $\frac{\pi}{4}\cdot\theta$ as x, we get
$$\frac{\pi}{4}\cdot\theta = p + \pi k$$
$$\theta = \frac{4}{\pi}p + \frac{4}{\pi}\cdot\pi k$$
$$\theta = \frac{4}{\pi}p + 4k$$

This has the form $a + bk$, so the value of b is 4. The correct answer is therefore **(C)**.

5. Graphing h and the line $y = 20$ with a window of $0 \le x \le 50$ and $0 \le y \le 30$ yields two portions of the graph above the line $y = 20$. The first two intersections occur at $t = 7.821$ and $t = 17.179$ for a total time of $17.179 - 7.821 = 9.358$ seconds. Because h is periodic, the second interval of time that the function is above 20 is also 9.358 seconds, giving a total of $2(9.358) = 18.716$ seconds. The correct answer is therefore **(B)**.

6. Solving for x yields $x = \arccos\frac{1}{3}$, where $0 < x < \frac{\pi}{2}$. To get a value of x in the interval $\pi < x < 2\pi$ where $\cos x = \frac{1}{3}$, our angle must be in Quadrant IV. The desired angle is $x = 2\pi - \arccos\frac{1}{3}$. The correct answer is therefore **(C)**.

Topic 3.11 ~ The Secant, Cosecant, and Cotangent Functions

Learning Objectives
1. (3.11.A) Identify key characteristics of functions that involve quotients of the sine and cosine functions.

Success Criteria
1. I can identify graphs of functions involving quotients of sine and cosine functions.
2. I can identify the asymptotes of functions involving quotients of sine and cosine functions.
3. I can identify the domain and range of functions involving quotients of sine and cosine functions.

In the previous 10 topics, we have concerned ourselves with certain measurements related to circular motion: vertical displacement, horizontal displacement, and slope. It turns out that the *reciprocals* of these measurements are themselves useful measurements!

Reciprocal Trigonometric Functions
The <u>COSECANT</u> function is the reciprocal of the sine function: $\csc x = \dfrac{1}{\sin x}$.
The <u>SECANT</u> function is the reciprocal of the cosine function: $\sec x = \dfrac{1}{\cos x}$.
The <u>COTANGENT</u> function is the reciprocal of the tangent function: $\cot x = \dfrac{1}{\tan x} = \dfrac{\cos x}{\sin x}$.

Let's look at some simple examples.

1. What is $\sin \dfrac{\pi}{6}$? Based on this, what is $\csc \dfrac{\pi}{6}$?

2. What is $\cos \dfrac{\pi}{6}$? Based on this, what is $\sec \dfrac{\pi}{6}$?

3. Find the following.

   (a) $\csc \dfrac{\pi}{2}$ \qquad (b) $\cot \dfrac{\pi}{6}$ \qquad (c) $\sec \dfrac{2\pi}{3}$

The outputs of $\csc x, \csc x$, and $\cot x$ are literally reciprocals of the outputs $\sin x, \cos x$, and $\tan x$, which means their graphs are reasonably predictable. Use number sense and the definition of a reciprocal to fill in the following about just one of the functions, $\csc x$.

4. As $\sin x$ increases to 1, $\csc x$ will _____ to _____.

5. When $\sin x = 1$, $\csc x$ will equal _____.

6. As $\sin x$ decreases from 1 to arbitrarily small (0), $\csc x$ will _____ from _____ to arbitrarily _____ (_____).

7. As $\sin x$ decreases from 0 to −1, $\csc x$ will _____ from _____ to _____.

8. Finally, as sin x increases from −1 to becoming arbitrarily small and negative, csc x will _____ to becoming arbitrarily _____ and negative.

9. In general, when sin x is at a local maximum, csc x will be at a local _____. When sin x is at a local minimum, csc x will be at a local _____.

10. Finally, when sin x equals 0, csc x will be _____.

11. Now, we can sketch a graph of csc x using this reciprocal relationship. On the same plane with the graph of sin x shown below, sketch a graph of $y = \csc x$.

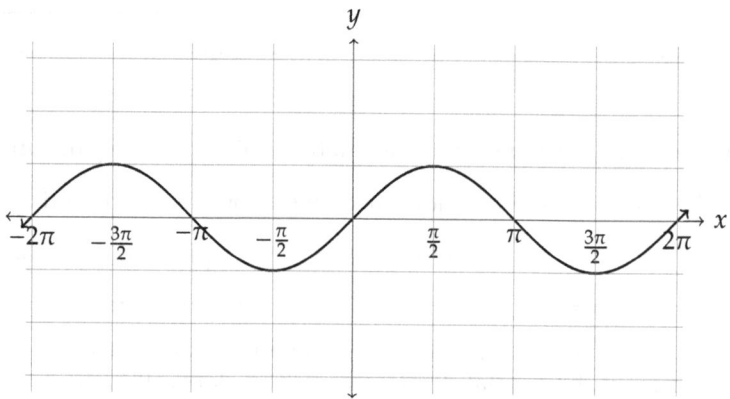

12. Now, try sketching a graph of $y = \sec x$ in the sam grid as the graph of cos x.

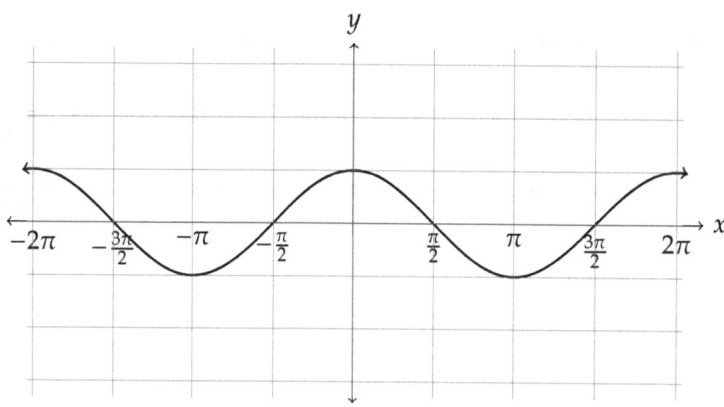

13. Sketch $y = \cot x$ in the same grid as the graph of tan x.

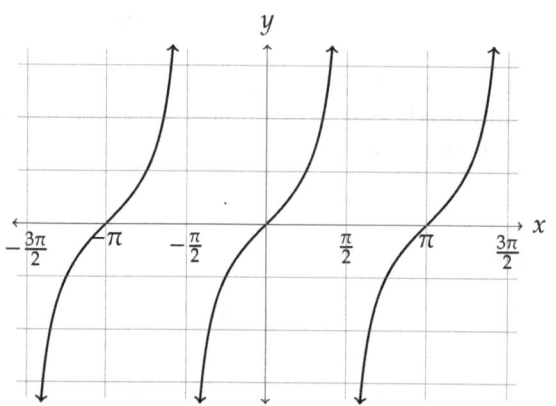

Unlike the graphs of sin x and cos x, the graphs of csc x and cot x will have VERTICAL ASYMPTOTES. The graph of cot x will have asymptotes similar to tan x, but at different input values. Complete the following.

Vertical Asymptotes of Reciprocal Trigonometric Functions

The graphs of csc θ and cot θ will have a vertical asymptote when _____ $\theta =$ _____. This occurs when $\theta =$ _____ for any integer k.

The graph of sec θ will have a vertical asymptote when _____ $\theta =$ _____. This occurs when $\theta =$ _____ for any integer k.

Now for some problems! One note, however: phase shifts are tricky, so here's a useful way of determining asymptotes (inputs where a function is undefined) when a phase shift is involved.

Step 1: Identify the general solution of the asymptote you know for the *parent function*.

Step 2: Set the argument of the function (what's inside the parentheses of $f(\ )$) equal to the general solution.

Step 3: Solve for x (or whatever the input variable).

If you only want a single asymptote, you can ignore the "general solution" part and just find a single known asymptote before solving.

Example: Find the domain of $f(x) = 3\sec\left(\frac{1}{3}\left(x + \frac{\pi}{4}\right)\right)$.

Solution:

Step 1: sec x is undefined when cos $x = 0$. This occurs for $x = \frac{\pi}{2} + \pi k$ for any integer k.

Step 2: We set $\frac{1}{3}\left(x + \frac{\pi}{4}\right) = \frac{\pi}{2} + \pi k$.

Step 3: We solve for x.

$$\frac{1}{3}\left(x + \frac{\pi}{4}\right) = \frac{\pi}{2} + \pi k$$

$$x + \frac{\pi}{4} = \frac{3\pi}{2} + 3\pi k \quad \text{(multiply both sides by 3)}$$

$$x = \frac{3\pi}{2} - \frac{\pi}{4} + 3\pi k \quad \text{(subtract } \frac{\pi}{4}\text{)}$$

$$x = \frac{5\pi}{4} + 3\pi k \quad \text{(simplify)}$$

The domain is all real numbers except inputs of the form $x = \frac{5\pi}{4} + 3\pi k$ for any integer k.

---

14. Find the first 2 positive inputs for which $f(x) = 2\sec\left(x + \frac{\pi}{3}\right)$ will be undefined.

15. Find a value θ for which $g(x) = \csc\left(2x - \frac{\pi}{4}\right)$ will have a vertical asymptote at $x = \theta$.

16. The function g is defined by $g(x) = \cot(2(x - \pi))$. What is the domain of g?

Topic 3.11 Homework

1. Find the range of each of $f(x) = \csc x$, $g(x) = \sec x$, and $h(x) = \cot x$.

2. Find the domain of each of the following functions.

    (a) $f(x) = 2\sec\left(x + \dfrac{\pi}{6}\right) - 1$
    (b) $g(x) = \cot\left(4\left(x - \dfrac{\pi}{3}\right)\right)$
    (c) $f(\theta) = -3\csc\left(\dfrac{1}{2}\theta - \dfrac{\pi}{2}\right) + 2$

3. For each function from Problem 2, find the input value where each function will have its first positive asymptote.

4. Use transformations of functions to determine the range of each of the functions from Problem 2.

5. Each of the functions graphed below can be written in the form $g(x) = af(bx) + c$, where f is either $f(x) = \sec x$, $f(x) = \csc x$, or $f(x) = \cot x$.

    Write an expression for each function.

    (a)
    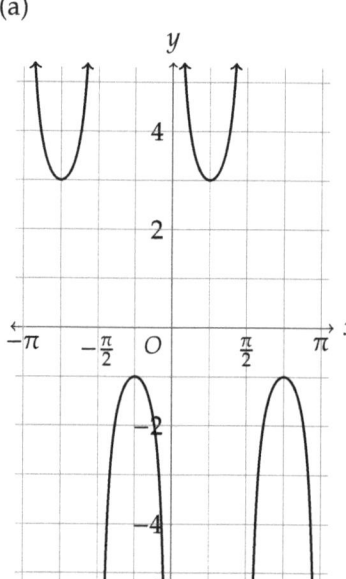

    (b)
    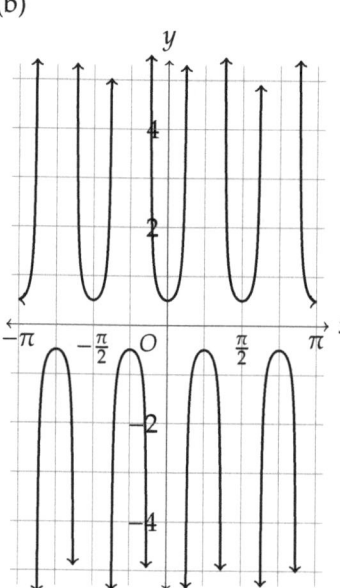

    (c)
    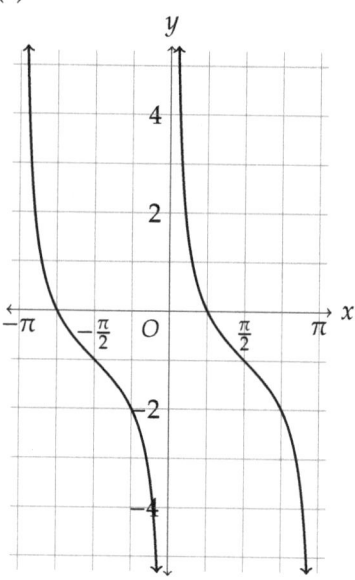

6. Let g be the function defined by
$$g(x) = \dfrac{1}{(2\sin x - 1)(\cos x + 1)}.$$
For what values of x in the interval $0 \le x \le 2\pi$ will $g(x)$ be undefined?

Topic 3.11 Solutions/Notes

Activity

1. $\sin \frac{\pi}{6} = \frac{1}{2}$, so $\csc \frac{\pi}{6} = \frac{1}{\frac{1}{2}} = 2$

2. $\cos \frac{\pi}{6} = \frac{\sqrt{3}}{2}$, so $\sec \frac{\pi}{6} = \frac{2}{\sqrt{3}} = \frac{2\sqrt{3}}{3}$

3. (a) $\dfrac{1}{\sin \frac{\pi}{2}} = \dfrac{1}{1} = 1$ (b) $\dfrac{\cos \frac{\pi}{6}}{\sin \frac{\pi}{6}} = \dfrac{\frac{\sqrt{3}}{2}}{\frac{1}{2}} = \sqrt{3}$ (c) $\dfrac{1}{\cos \frac{2\pi}{3}} = \dfrac{1}{-\frac{1}{2}} = -2$

4. As $\sin x$ increases to 1, $\csc x$ will <u>decrease</u> to <u>1</u>.

5. When $\sin x = 1$, $\csc x$ will equal <u>1</u>.

6. As $\sin x$ decreases from 1 to arbitrarily small (0), $\csc x$ will <u>increase</u> from <u>1</u> to arbitrarily <u>large</u> (∞).

7. As $\sin x$ decreases from 0 to -1, $\csc x$ will <u>increase</u> from <u>$-\infty$</u> to <u>-1</u>.

8. Finally, as $\sin x$ increases from -1 to becoming arbitrarily small and negative, $\csc x$ will <u>decrease</u> to becoming arbitrarily <u>large</u> and negative.

9. In general, when $\sin x$ is at a local maximum, $\csc x$ will be at a local <u>minimum</u>. When $\sin x$ is at a local minimum, $\csc x$ will be at a local <u>maximum</u>.

10. Finally, when $\sin x$ equals 0, $\csc x$ will be <u>undefined</u>.

11.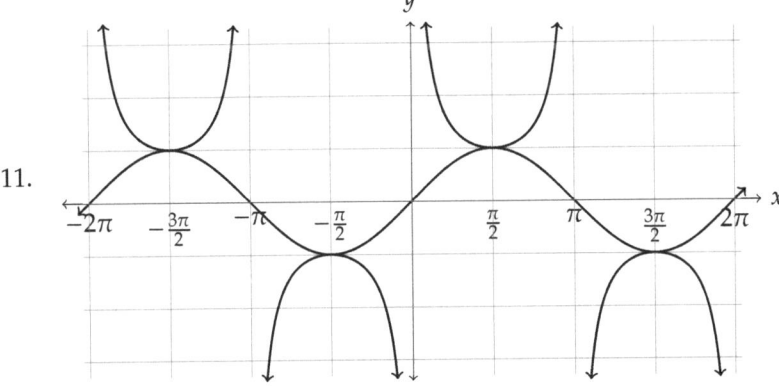

12.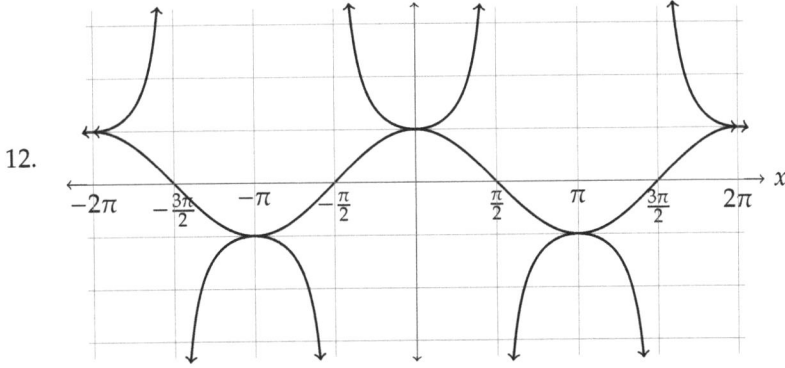

13.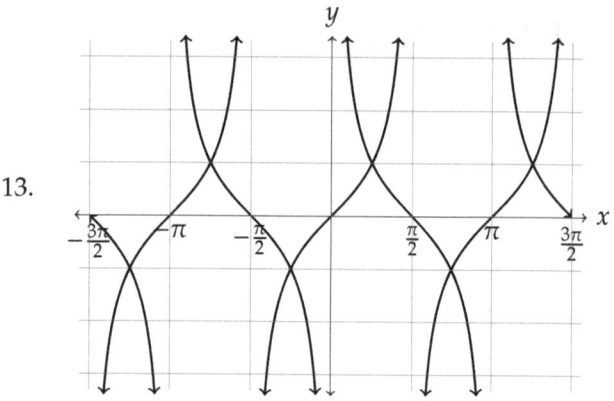

Vertical Asymptotes of Reciprocal Trigonometric Functions

The graphs of csc θ and cot θ will have a vertical asymptote when $\underline{\sin\theta = 0}$. This occurs when $\theta = \underline{k\pi}$ for any integer k.

The graph of sec θ will have a vertical asymptote when $\underline{\cos\theta = 0}$. This occurs when $\theta = \underline{\frac{\pi}{2} + \pi k}$ for any integer k.

14. sec x is undefined when cos x equals 0, which occurs for $x = \frac{\pi}{2} + \pi k$. We set $\frac{\pi}{2} + \pi k = x + \frac{\pi}{3}$ and get $x = \left(\frac{\pi}{2} - \frac{\pi}{3}\right) + \pi k = \frac{\pi}{6} + \pi k$. For $k = 0$, we get $x = \frac{\pi}{6}$. For $k = 1$, we get $x = \frac{\pi}{6} + \pi = \frac{7\pi}{6}$. The first two inputs for which the function is undefined are $x = \frac{\pi}{6}$ and $x = \frac{7\pi}{6}$.

15. csc x is undefined when sin $x = 0$, which occurs for $x = \pi k$. The simplest of these inputs for which csc x is undefined is $x = 0$. Therefore, we can set $2x - \frac{\pi}{4} = 0$ and solve for x: $2x - \frac{\pi}{4} = 0 \Rightarrow 2x = \frac{\pi}{4} \Rightarrow x = \frac{\pi}{8}$. The function g will have a vertical asymptote at $\theta = \frac{\pi}{8}$.

16. The function cot x is defined for all real numbers except inputs of the form $x = \pi k$. Therefore, g will be undefined if $2(x - \pi) = \pi k$. We solve this for x:

$$2(x - \pi) = \pi k$$
$$x - \pi = \frac{\pi}{2}k \qquad \text{(multiply by } \tfrac{1}{2}\text{)}$$
$$x = \pi + \frac{\pi}{2}k \qquad \text{(add } \pi\text{)}$$

The domain of g is all real numbers except those of the form $x = \pi + \frac{\pi}{2}k$. (Note: This can be rewritten in the form $x = \frac{\pi}{2}k$ as well, since $x = \pi + \frac{\pi}{2}(-1) = \frac{\pi}{2}$.)

Notes

The functions sec x, csc x, and cot x are reciprocals of cos x, sin x, and tan x, respectively. The reciprocal trigonometric functions are undefined, and hence their graphs will have vertical asymptotes, when cos x, sin x, and tan x (or sin x) equal 0, respectively.

Because the ranges of both sin x and cos x are $[-1, 1]$, the ranges of the reciprocal functions csc x and sec x are all numbers *greater than* or equal to 1 or *less than* or equal to -1, which can be expressed as $(-\infty, -1] \cup [1, \infty)$.

Homework

1. The range of both f and g are $(-\infty, -1] \cup [1, \infty)$, while the range of h is $(-\infty, \infty)$.

2. (a) $\sec x$ is undefined when $x = \frac{\pi}{2} + \pi k$ for any integer k, so we set $x + \frac{\pi}{6}$ equal to this and solve for x where f will be undefined.

$$x + \frac{\pi}{6} = \frac{\pi}{2} + \pi k \Rightarrow x = \left(\frac{\pi}{2} - \frac{\pi}{6}\right) + \pi k \Rightarrow x = \frac{\pi}{3} + \pi k$$

The domain of f is all real numbers except those of the form $x = \frac{\pi}{3} + \pi k$ for any integer k.

(b) $\cot x$ is undefined when $x = \pi k$ for any integer k, so we set $4\left(x - \frac{\pi}{3}\right) = \pi k$ and solve for x where g will be undefined.

$$4\left(x - \frac{\pi}{3}\right) = \pi k \Rightarrow x - \frac{\pi}{3} = \frac{\pi}{4}k \Rightarrow x = \frac{\pi}{3} + \frac{\pi}{4}k$$

The domain of g is all real numbers except those of the form $x = \frac{\pi}{3} + \frac{\pi}{4}k$ for any integer k.

(c) $\csc x$ is undefined when $\theta = \pi k$ for any integer k, so we set $\frac{1}{2}\theta - \frac{\pi}{2} = \pi k$ and solve for θ where g will be undefined.

$$\frac{1}{2}\theta - \frac{\pi}{2} = \pi k \Rightarrow \frac{1}{2}\theta = \frac{\pi}{2} + \pi k \Rightarrow \theta = \pi + 2\pi k$$

The domain of f is all real numbers except those of the form $\theta = \pi + 2\pi k$.

3. (a) $x = \frac{\pi}{3}$ (b) $x = \frac{\pi}{3}$ (c) $x = \pi$

4. (a) We have a vertical dilation by a factor of 2 and a vertical translation of -1:

$$(-\infty, -1] \cup [1, \infty) \to (-\infty, -2] \cup [2, \infty) \to (-\infty, -3] \cup [1, \infty)$$

The range is $(-\infty, -3] \cup [1, \infty)$.

(b) The range is all real numbers.

(c) We have a vertical dilation by a factor of 3, a reflection over the x-axis (which is irrelevant for range because of the symmetry of the range of $\csc x$ over the x-axis), and a vertical translation of 2.

$$(-\infty, -1] \cup [1, \infty) \to (-\infty, -3] \cup [3, \infty) \to (-\infty, -1] \cup [5, \infty)$$

5. (a) The midline is $y = 1$, and the local minima and maxima are each 2 units away from the midline. Therefore, $a = 2$ and $c = 1$. The function has a vertical asymptote at $x = 0$, which rules out $\sec x$, and is not decreasing (or increasing) for all x, which rules out $\cot x$. This means it must be $\csc x$. Finally, the asymptotes occur over intervals of length $\frac{\pi}{2}$, which is twice as frequently as the graph of $\csc x$ (with asymptotes occurring over intervals of length π). Therefore, $b = 2$, so $g(x) = 2\csc(2x) + 1$.

(b) The function is defined for $x = 0$, which rules out both $\csc x$ and $\cot x$, leaving us with $\sec x$. The midline is $y = 0$, so $c = 0$, but the range is $(-\infty, -\frac{1}{2}] \cup [\frac{1}{2}, \infty)$, making $a = \frac{1}{2}$. The period is only $\frac{\pi}{2}$, so $\frac{\pi}{2} = \frac{2\pi}{b}$ gives $b = 4$. Therefore, $g(x) = \frac{1}{2}\sec(4x)$.

(c) This is just the graph of $\cot x$ translated down 1 unit, so $g(x) = \cot x - 1$.

6. g will be undefined if its denominator equals 0. Setting each factor of the denominator equal to 0 yields $2\sin x - 1 = 0$ and $\cos x + 1 = 0$. These give $\sin x = \frac{1}{2}$, which has solutions of $x = \frac{\pi}{6}$ and $x = \frac{5\pi}{6}$, and $\cos x = -1$, which has a solution of $x = \pi$. Therefore, g is undefined for $x = \frac{\pi}{6}$, $x = \frac{5\pi}{6}$, and $x = \pi$ on the interval $0 \le x \le 2\pi$.

Reflection

Suggested Class Time. 60 minutes

Prerequisites. A comfort level with asymptotes, among others.

Instructional Strategies. Full disclosure: I was absent this day, and I actually instructed my students to work through this in my absence. I posted the solutions to the activity at the front of the room and told students to e-mail me any questions: I received exactly one question. When I returned, students said everything made sense, and I actually had fewer questions on the homework than usual (they were on Problem 2(b) and Problem 5, which were reasonably awkward for students). As we moved into Topic 3.12, I didn't notice any seeming deficiencies with regards to this topic.

Technology. None was intended, but in the Notes section, I would have made a point to show students how to graph these in the TI-84 ("there's no secant button," etc.).

Problems.

1. The graph of the function $y = f(x)$ is shown below.

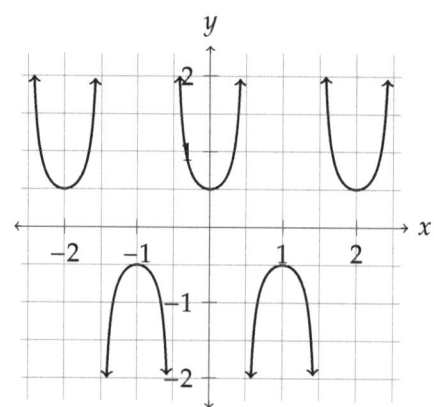

   Which of the following could be an expression for $f(x)$?

   (A) $f(x) = \dfrac{1}{2}\sec(\pi x)$ **(B)** $f(x) = \dfrac{1}{2}\sec(2\pi x)$ **(C)** $f(x) = \dfrac{1}{2}\csc(\pi x)$ **(D)** $f(x) = \dfrac{1}{2}\csc(2\pi x)$

2. The functions f and g are defined by $f(x) = \sec^2 x$ and $g(x) = \tan x$. The function h is defined by $h(x) = f(x) - g(x)$. Which of the following is the value of $h(-\frac{\pi}{4})$?

   (A) $-\dfrac{3}{2}$ **(B)** 1   **(C)** $\dfrac{3}{2}$ **(D)** 3

3. The function g is given by $g(x) = 2\sec(x + \frac{\pi}{6})$. Which of the following gives a general expression for the inputs x corresponding to the vertical asymptotes of the graph of g?

   (A) $x = -\dfrac{\pi}{6} + \pi k$ where k is an integer

   (B) $x = \dfrac{\pi}{3} + \pi k$ where k is an integer

   (C) $x = -\dfrac{\pi}{6} + \dfrac{\pi}{2}k$ where k is an integer

   (D) $x = \dfrac{\pi}{3} + \dfrac{\pi}{2}k$ where k is an integer

4. Let f be the function defined by $f(x) = 3\csc(4x - \pi) + 1$. Which of the following cannot be an output value of the function f?

   (A) -4 **(B)** -2 **(C)** 2   **(D)** 4

5. The input value $x = \frac{3\pi}{4}$ makes which of the following functions undefined?

   (A) $f(x) = 3\cot(3x - \pi)$ **(B)** $g(x) = -2\csc(4x - \pi)$ **(C)** $h(x) = -\sin(8x - 2\pi)$ **(D)** $k(x) = 4\sec(4x - \pi)$

Solutions.

1. The function f has an output value when $x = 0$ is the input, so the function is the secant function. The period of the secant pictured is 2, and $\sec(\pi x)$ has period $2\pi/\pi = 2$. The correct answer is therefore **(A)**.

2. We have
$$h\left(-\frac{\pi}{4}\right) = f\left(-\frac{\pi}{4}\right) - g\left(-\frac{\pi}{4}\right) = \sec^2\left(-\frac{\pi}{4}\right) - \tan\left(-\frac{\pi}{4}\right) = \frac{1}{\cos^2(-\pi/4)} - (-1)$$
$$= \frac{1}{(\sqrt{2}/2)^2} + 1 = 2 + 1 = 3.$$

    The correct answer is therefore **(D)**.

3. The graph of $\sec \theta$ has a vertical asymptote when $\cos \theta = 0$. This happens when $\theta = \frac{\pi}{2} + \pi k$ for any integer k. For the function g, this implies vertical asymptotes occur when $x + \frac{\pi}{6} = \frac{\pi}{2} + \pi k$. Then $x = \frac{\pi}{2} - \frac{\pi}{6} + \pi k = \frac{\pi}{3} + \pi k$. The correct answer is therefore **(B)**.

4. We have a vertical dilation by a factor of 3 and a vertical translation of 1. The range $(-\infty, -1] \cup [1, \infty)$ therefore becomes $(-\infty, -1 \cdot 3 + 1] \cup [1 \cdot 3 + 1, \infty)$, which is $(-\infty, -2] \cup [4, \infty)$. The only value of those listed that cannot be an output value is 2. The correct answer is therefore **(C)**.

5. We see that $g(\frac{3\pi}{4}) = -2\csc(4 \cdot \frac{3\pi}{4} - \pi) = -2\csc(2\pi)$, which is undefined. The correct answer is therefore **(B)**.

Topic 3.12 ~ Equivalent Representations of Trigonometric Functions (Day 1)

Learning Objectives

1. (3.12.A) Rewrite trigonometric expressions in equivalent forms with the Pythagorean identity.
2. (3.12.B) Rewrite trigonometric expressions in equivalent forms with sine and cosine sum identities.
3. (3.12.C) Solve equations using equivalent representations of trigonometric functions.

Success Criteria

1. I can rewrite trigonometric equations in terms of a single trigonometric function and use this to solve trigonometric equations.

In order to solve certain trigonometric equations and inequalities, we will make use of clever rewritings called *identities*. An IDENTITY is an expression that is known to be true. Technically, any equation, even $2 + 2 = 4$, is an identity, but we usually reserve the term *identity* for an equation that may not appear immediately obvious.

One set of identities that may not feel like identities are the definitions of $\tan \theta$, $\csc \theta$, $\sec \theta$, and $\tan \theta$.

$\tan \theta =$ $\qquad$ $\csc \theta =$ $\qquad$ $\sec \theta =$ $\qquad$ $\cot \theta =$

There are *many* other trigonometric identities, but the purview of this course is limited to a select few as shown in the table on the next page.

AP Precalculus Trigonometric Identities

Name	Identify	Alt. Version/s	Common Uses
Pythagorean	_____ = 1 _____ = __ _____ = __	$\sin^2 x =$ _____ $\cos^2 x =$ _____ $\tan^2 x =$ _____ $\cot^2 x =$ _____	Rewriting $\sin x$ or $\cos x$ in terms of each other Rewriting $\sec x$ or $\tan x$ in terms of each other Rewriting $\csc x$ or $\cot x$ in terms of each other
Sum/Difference	$\sin(\alpha + \beta) =$ _____ $\cos(\alpha + \beta) =$ _____ $\sin(\alpha - \beta) =$ _____ $\cos(\alpha - \beta) =$ _____		Computing $\sin x$ and $\cos x$ for angles that aren't multiples of $\frac{\pi}{6}$ or $\frac{\pi}{4}$
Double Angle	$\sin(2\theta) =$ _____ $\cos(2\theta) =$ _____ _____ = _____ _____ = _____		Converting $\sin\theta\cos\theta$ into one single trig function More flexibility in rewriting $\sin x$ and $\cos x$ in terms of each other (or removing *squared* terms)

The goal with most identities is not mindless rewriting, but making an expression easier to interpret or an equation easier to solve. This often amounts to trying to _____ the number of different trigonometric functions in a given function.

Let's see some of these in action!

Examples

1. Rewrite each of the following functions in terms of a single trigonometric function.

   (a) $g(\theta) = \sin\theta \cos\theta \cot\theta$

   (b) $f(x) = \dfrac{1 - \sin^2 x}{\sec x}$

   (c) $s(\theta) = \sin\theta(1 + \cot^2\theta)$

   (d) $h(x) = \cos x - \sin^2 x + 2$

   (e) $f(\theta) = \sin\theta \cos\theta$

   (f) $f(x) = \dfrac{\sec^2 x - \tan^2 x}{\sin x}$

   (g) $g(\theta) = \cos(2\theta) + \sin\theta$

   (h) $h(x) = \cos x + \sqrt{1 - \sin^2 x}$, where $x \in \left[0, \frac{\pi}{2}\right]$

2. Compute each of the following without a calculator.

   (a) $\sin\left(\dfrac{\pi}{3} + \dfrac{\pi}{4}\right)$

   (b) $\cos\left(\dfrac{2\pi}{3} - \dfrac{\pi}{4}\right)$

3. Let f be the function given by $f(x) = 3\cos x - 2\sin^2 x$. Find all inputs x such that $f(x) = 0$.

Topic 3.12 (Day 1) Homework

1. Try to complete each identity without your notes. For $\cos(2\theta)$, write all 3 forms.

    (a) $\sin^2\theta + \cos^2\theta =$ _____

    (b) $\tan^2\theta + 1 =$ _____

    (c) $1+$ _____ $= \csc^2 x$

    (d) $\sin(\alpha + \beta) =$ _____

    (e) $\cos(x - y) =$ _____

    (f) $\sin(\alpha - \beta) =$ _____

    (g) $\cos(\theta + \gamma) =$ _____ (Note: γ is the Greek letter *gamma*.)

    (h) $\sin(2\theta) =$ _____

    (i) $\cos(2\theta) =$ _____ $=$ _____ $=$ _____

2. The function f is defined by $f(x) = \sin^2 x + \cos x$. Rewrite f in terms of only $\cos x$.

3. The function g is defined by $g(\theta) = \dfrac{\sec^2\theta - 1}{\tan\theta}$. Rewrite g as a single trigonometric function.

4. Let $\theta = \dfrac{2\pi}{3} - \dfrac{3\pi}{4}$. Compute $\sin\theta$ and $\cos\theta$.

5. Shown below are expressions for the five functions $f, g, h, j,$ and k.

    $f(x) = \sec^2 x - 1$ $g(x) = \sin x \sec x$ $h(x) = \dfrac{1}{\cot x}$ $j(x) = \dfrac{\sin x + \cos x}{\cos^2 x}$ $k(x) = \dfrac{\cot x}{\csc^2 x - 1}$

    Which of the five functions are equivalent to $t(x) = \tan x$ for all x where the functions are defined?

6. Remember that identities don't *have* to be used to solve a trigonometric equation. In general, identities are used to rewrite a given function in terms of only *one* trigonometric function - this can enable you to use algebraic techniques like factoring.

    (a) Let f be the function given by $f(\theta) = \sin^2\theta + \sin\theta$. Find the zeros of f on the interval $0 \leq \theta \leq 2\pi$.

    (b) The function g is expressed as $g(x) = \sin x - \cos^2 x$. Suppose we wanted to find the values of x on the interval $0 \leq x \leq 2\pi$ such that $g(x) = -1$. Show that solving $g(x) = 1$ can be reduced to the exact problem given in (a).

    (c) The function f is defined by $f(x) = 2\sin^2 x - 5\cos x$. Find all values of x on the interval $-\pi \leq x \leq \pi$ such that $f(x) = 4$.

Topic 3.12 (Day 1) Solutions/Notes

Activity

AP Precalculus Trigonometric Identities

Name	Identify	Alt. Version/s	Common Uses
Pythagorean	$\sin^2 x + \cos^2 x = 1$	$\sin^2 x = 1 - \cos^2 x$ $\cos^2 x = 1 - \sin^2 x$	Rewriting $\sin x$ or $\cos x$ in terms of each other
	$\tan^2 x + 1 = \sec^2 x$	$\tan^2 x = \sec^2 x - 1$	Rewriting $\sec x$ or $\tan x$ in terms of each other
	$1 + \cot^2 x = \csc^2 x$	$\cot^2 x = \csc^2 x - 1$	Rewriting $\csc x$ or $\cot x$ in terms of each other
Sum/Difference	$\sin(\alpha + \beta) = \sin\alpha\cos\beta + \sin\beta\cos\alpha$ $\cos(\alpha + \beta) = \cos\alpha\cos\beta - \sin\alpha\sin\beta$ $\sin(\alpha - \beta) = \sin\alpha\cos\beta - \sin\beta\cos\alpha$ $\cos(\alpha - \beta) = \cos\alpha\cos\beta + \sin\alpha\sin\beta$		Computing $\sin x$ and $\cos x$ for angles that aren't multiples of $\frac{\pi}{6}$ or $\frac{\pi}{4}$
Double Angle	$\sin(2\theta) = \sin\theta\cos\theta + \sin\theta\cos\theta = 2\sin\theta\cos\theta$ $\cos(2\theta) = \cos\theta\cos\theta - \sin\theta\sin\theta = \cos^2\theta - \sin^2\theta$ $ = 1 - 2\sin^2\theta$ $ = 2\cos^2\theta - 1$		Converting $\sin\theta\cos\theta$ into one single trig function More flexibility in rewriting $\sin x$ and $\cos x$ in terms of each other (or removing *squared* terms)

This often amounts to trying to <u>reduce</u> the number of different trigonometric functions in a given function.

Examples

1. Rewrite each of the following functions in terms of a single trigonometric function.

   (a) $g(\theta) = \sin\theta\cos\theta\cot\theta = \sin\theta\cos\theta \cdot \dfrac{\cos\theta}{\sin\theta} = \cos^2\theta$

   (b) $f(x) = \dfrac{1 - \sin^2 x}{\sec x} = \dfrac{\cos^2 x}{\frac{1}{\cos x}} = \cos^2 x \cdot \cos x = \cos^3 x$

   (c) $s(\theta) = \sin\theta(1 + \cot^2\theta) = \sin\theta(\csc^2\theta) = \sin\theta \cdot \dfrac{1}{\sin^2\theta} = \dfrac{1}{\sin\theta} = \csc\theta$

   (d) $h(x) = \cos x - \sin^2 x + 2 = \cos x - (1 - \cos^2 x) + 2 = \cos^2 x + \cos x + 1$

   (e) $f(\theta) = \sin\theta\cos\theta = \frac{1}{2} \cdot (2\sin\theta\cos\theta) = \frac{1}{2}\sin(2\theta)$

   (f) $f(x) = \dfrac{\sec^2 x - \tan^2 x}{\sin x} = \dfrac{(1 + \tan^2 x) - \tan^2 x}{\sin x} = \dfrac{1}{\sin x} = \csc x$

   (g) $g(\theta) = \cos(2\theta) + \sin\theta = (1 - 2\sin^2\theta) + \sin\theta = -2\sin^2\theta + \sin\theta + 1 = -(2\sin\theta + 1)(\sin\theta - 1)$

   (h) $h(x) = \cos x + \sqrt{1 - \sin^2 x} = \cos x + \sqrt{\cos^2 x} = \cos x + \cos x = 2\cos x$, where $x \in \left[0, \frac{\pi}{2}\right]$

2. Compute each of the following without a calculator.

    (a) $\sin\left(\dfrac{\pi}{3}+\dfrac{\pi}{4}\right) = \sin\dfrac{\pi}{3}\cos\dfrac{\pi}{4}+\sin\dfrac{\pi}{4}\cos\dfrac{\pi}{3} = \dfrac{\sqrt{3}}{2}\cdot\dfrac{\sqrt{2}}{2}+\dfrac{\sqrt{2}}{2}\cdot\dfrac{1}{2} = \dfrac{\sqrt{6}+\sqrt{2}}{4}$

    (b) $\cos\left(\dfrac{2\pi}{3}-\dfrac{\pi}{4}\right) = \cos\dfrac{2\pi}{3}\cos\dfrac{\pi}{4}+\sin\dfrac{2\pi}{3}\sin\dfrac{\pi}{4} = -\dfrac{1}{2}\cdot\dfrac{\sqrt{2}}{2}+\dfrac{\sqrt{3}}{2}\cdot\dfrac{\sqrt{2}}{2} = \dfrac{\sqrt{6}-\sqrt{2}}{4}$

3. We can replace $\sin^2 x$ with $1-\cos^2 x$, which gives us

$$f(x) = 3\cos x - 2(1-\cos^2 x) = 3\cos x - 2 + 2\cos^2 x = 2\cos^2 x + 3\cos x - 2$$

Now, we can set $f(x) = 0$ and factor.

$$2\cos^2 x + 3\cos x - 2 = 0$$
$$(2\cos x - 1)(\cos x + 2) = 0$$

This means that either $2\cos x - 1 = 0$ or $\cos x + 2 = 0$. The second equation yields no solutions, and the first equation yields $\cos x = \frac{1}{2}$. This occurs for $x = \frac{\pi}{3}$ and $x = \frac{5\pi}{3}$ in the first period of $\cos x$, so our general solutions are $x = \frac{\pi}{3} + 2\pi k$ and $x = \frac{5\pi}{3} + 2\pi k$ for any integer k.

Homework

1. (a) $\sin^2\theta + \cos^2\theta = 1$

   (b) $\tan^2\theta + 1 = \sec^2\theta$

   (c) $1 + \cot^2 x = \csc^2 x$

   (d) $\sin(\alpha+\beta) = \sin\alpha\cos\beta + \sin\beta\cos\alpha$

   (e) $\cos(x-y) = \cos x\cos y + \sin x\sin y$

   (f) $\sin(\alpha-\beta) = \sin\alpha\cos\beta - \sin\beta\cos\alpha$

   (g) $\cos(\theta+\gamma) = \cos\theta\cos\gamma - \sin\theta\sin\gamma$

   (h) $\sin(2\theta) = 2\sin\theta\cos\theta$

   (i) $\cos(2\theta) = \cos^2\theta - \sin^2\theta = 2\cos^2\theta - 1 = 1 - 2\sin^2\theta$

2. $f(x) = (1-\cos^2 x) + \cos x = -\cos^2 x + \cos x + 1$

3. $g(\theta) = \dfrac{(1+\tan^2\theta)-1}{\tan\theta} = \dfrac{\tan^2\theta}{\tan\theta} = \tan\theta$

4. $\sin\theta = \sin\left(\dfrac{2\pi}{3}-\dfrac{3\pi}{4}\right) = \sin\left(\dfrac{2\pi}{3}\right)\cos\left(\dfrac{3\pi}{4}\right) - \sin\left(\dfrac{3\pi}{4}\right)\cos\left(\dfrac{2\pi}{3}\right) = \dfrac{\sqrt{3}}{2}\left(-\dfrac{\sqrt{2}}{2}\right) - \dfrac{\sqrt{2}}{2}\left(-\dfrac{1}{2}\right) = \dfrac{-\sqrt{6}+\sqrt{2}}{4}$

   $\cos\theta = \cos\left(\dfrac{2\pi}{3}-\dfrac{3\pi}{4}\right) = \cos\left(\dfrac{2\pi}{3}\right)\cos\left(\dfrac{3\pi}{4}\right) + \sin\left(\dfrac{3\pi}{4}\right)\sin\left(\dfrac{2\pi}{3}\right) = -\dfrac{1}{2}\left(-\dfrac{\sqrt{2}}{2}\right) + \dfrac{\sqrt{2}}{2}\cdot\dfrac{\sqrt{3}}{2} = \dfrac{\sqrt{2}+\sqrt{6}}{4}$

5. $f(x) = \tan^2 x$, not $t(x) = \tan x$. $g(x) = \sin x \cdot \dfrac{1}{\cos x} = \dfrac{\sin x}{\cos x} = \tan x$. $h(x) = \dfrac{1}{\frac{1}{\tan x}} = \tan x$

   j does not simplify quickly, but most certainly does not equal $t(x) = \tan x$. The "cleanest" equivalent expression is

$$j(x) = \dfrac{\sin x}{\cos^2 x} + \dfrac{\cos x}{\cos^2 x} = \tan x\sec x + \sec x.$$

$k(x) = \dfrac{\cot x}{(\cot^2 x + 1) - 1} = \dfrac{\cot x}{\cot^2 x} = \dfrac{1}{\cot x} = \tan x.$

We conclude that the functions g, h, and k are equivalent to the function t.

6. (a) $f(\theta) = 0 \Rightarrow \sin\theta(\sin\theta + 1) = 0$, which gives $\sin\theta = 0$ and $\sin\theta = -1$. On the interval $0 \leq \theta \leq 2\pi$, the solutions of the first equation are $\theta = 0$ and $\theta = \pi$ and the solution of the second equation is $\theta = \frac{3\pi}{2}$. The solutions are therefore $\theta = 0$, $\theta = \pi$, and $\theta = \frac{3\pi}{2}$.

(b) $g(x) = 1 \Rightarrow \sin x - (1 - \sin^2 x) = 1 \Rightarrow \sin^2 x + \sin x - 1 = -1 \Rightarrow \sin^2 x + \sin x = 0$

(c) $f(x) = 4 \Rightarrow 2(1 - \cos^2 x) - 5\cos x = 4 \Rightarrow -2\cos^2 x - 5\cos x - 2 = 0 \Rightarrow 2\cos^2 x + 5\cos x + 2 = 0$

Factoring gives $(2\cos x + 1)(\cos x + 2) = 0$. Setting each factor equal to 0 and solving gives $\cos x = -2$, which has no solutions, and $\cos x = -\frac{1}{2}$, which has solutions of $x = \frac{2\pi}{3}$ and $x = -\frac{2\pi}{3}$ on the interval $-\pi \leq x \leq \pi$. The only solutions are therefore $x = \frac{2\pi}{3}$ and $x = -\frac{2\pi}{3}$.

Reflection

Suggested Class Time. 60-70 minutes

Prerequisites. Students should be familiar with the reciprocal identities/definitions.

Instructional Strategies. Full disclosure: I had to blitz through this lesson. My students had to take a district-mandated assessment when I was scheduled to do this lesson, so we were only able to dedicate about 40-45 minutes to it.

That said, I did make the choice with this lesson to just rip the Band-aid off and introduce all of the identities at once. While in my other precalculus classes I like to spend more time developing proofs, I figured it was more important that students be fluent with using the identities rather than being able to derive them. I did show a diagram with the historical definitions of the six trigonometric functions (shown below)

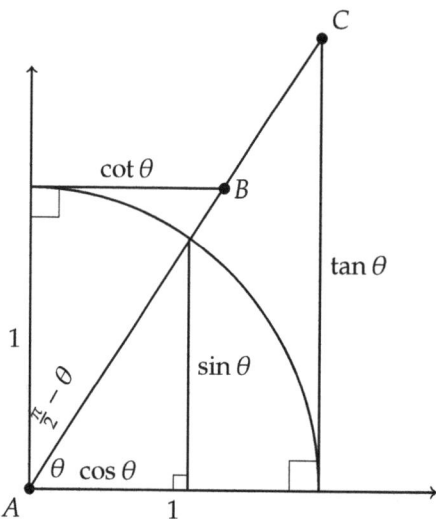

Note: $AC = \sec\theta$ and $AB = \csc\theta$.

I developed them by having students recall the circle definitions they learned in geometry: radius, diameter, chord, tangent, and secant. We then briefly discussed how the diagram results in $\tan\theta = \frac{\sin\theta}{\cos\theta}$, as well as what the "co" represents: complementary! In the diagram, AB is simply the secant of the angle complementary to θ, or the <u>complementary secant of θ</u>, i.e. $\csc\theta$. From there, we quickly wrote all of the Pythagorean identities.

For the sum and difference identities, I pointed out the structure, but that was about it. For the double angle, we did derive them using $\alpha = \theta$ and $\beta = \theta$.

We then were able to complete all of Problems 1 and 2(a), and nearly finish Problem 3. By and large, I was telling my students they need to be flexible. Over and over, I told them to be open-minded and to know multiple equivalent representations of just about everything. The sheer shock of doing 8 of these problems (those in Problem 1) in the span of about 10 minutes seemed to work wonders for normalizing the process for students.

Note: The use of x versus θ in the notes was intentional. They need to be comfortable with both.

Technology. None

Topic 3.12 ~ Equivalent Representations of Trigonometric Functions (Day 2)

Learning Objectives

1. (3.12.A) Rewrite trigonometric expressions in equivalent forms with the Pythagorean identity.
2. (3.12.B) Rewrite trigonometric expressions in equivalent forms with sine and cosine sum identities.
3. (3.12.C) Solve equations using equivalent representations of trigonometric functions.

Success Criteria

1. I can rewrite trigonometric equations in terms of a single trigonometric function and use this to solve trigonometric equations.

This lesson is all about practice!

Before doing so, it's worth reviewing somewhat of an "identity toolkit." You've got lots of identities, and many trigonometric equations can be rewritten or solved in numerous ways. However, there are certain patterns to look for and common rewritings.

Identities Toolkit

- Any squared trigonometric function can be rewritten in terms of a different squared trigonometric function using a Pythagorean Identity.

- For $\sin^2 \theta$ and $\cos^2 \theta$, if a Pythagorean Identity won't get the job done, don't forget the *three* different representations of $\cos(2\theta)$.

- It often helps to rewrite expressions in terms of only $\sin \theta$ and $\cos \theta$.

- The sum and difference identities of sine and cosine follow certain patterns.

$$\sin(\alpha + \beta) = \sin(\alpha)\cos(\beta) + \sin(\beta)\cos(\alpha)$$
$$\sin(\alpha - \beta) = \sin(\alpha)\cos(\beta) - \sin(\beta)\cos(\alpha)$$

Observe how (1) the operations are *same* on both sides of both equations and (2) the terms each have the form $\sin(\text{angle})\cos(\text{other angle})$.

$$\cos(\alpha + \beta) = \cos(\alpha)\cos(\beta) - \sin(\beta)\sin(\alpha)$$
$$\cos(\alpha - \beta) = \cos(\alpha)\cos(\beta) + \sin(\beta)\sin(\alpha)$$

Observe how (1) the operations are *inverses* on both sides of both equations and (2) the terms go $\cos(\text{angle})\cos(\text{other angle})$ and $\sin(\text{angle})\sin(\text{other angle})$ *in order*.

- Any expression with a $\sin \theta \cos \theta$ term can be rewritten using the identity for $\sin(2\theta)$.

For this set of practice questions, they'll be broken into skills and difficulty levels.

Topic 3.12 ~ Equivalent Representations of Trigonometric Functions (Day 2)

Skill 1: Rewrite trigonometric functions in equivalent forms.

LEVEL ONE: Rewrite $f(x) = 1 + \cot^2 x$ in terms of a single trigonometric function with no constant term.

LEVEL TWO: The function $f(\theta) = \sec \theta \cot \theta$ can be rewritten as $g(\theta)$, where g is a trigonometric function. What is the function g?

LEVEL THREE: Let g and h be the functions defined by $g(x) = \cos^2 x$ and $h(x) = \sin x + \sin^2 x$. If k is the function defined by $k(x) = h(x) - g(x)$, find an expression for k in terms of only $\sin x$.

LEVEL FOUR: Let the function f be given by

$$f(\theta) = \frac{1}{\sec \theta - \sin \theta \tan \theta}.$$

The function f can be rewritten in the form $f(x) = t(x)$, where t is a trigonometric function. What is the function $t(x)$?

LEVEL FIVE: The function f is given by

$$f(x) = \frac{6 + 3\sin(2x)}{1 + (\sin x + \cos x)^2}.$$

It can be shown that $f(x) = k$ for some constant k. What is the value of k?

Skill 2: Utilize the sum and difference identities for sine and cosine.

LEVEL ONE: Compute $\cos\left(\dfrac{5\pi}{4} - \dfrac{2\pi}{3}\right)$.

LEVEL TWO: Consider the angle θ in the triangle.

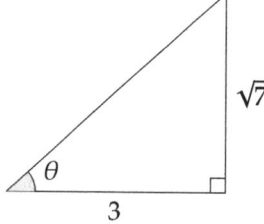

Compute $\cos(2\theta)$.

LEVEL THREE: For angles α and β with $0 < \alpha < \beta < \dfrac{\pi}{2}$, it is known that $\sin\alpha = \dfrac{5}{13}$ and $\sin\beta = \dfrac{3}{5}$. What is $\sin(\alpha + \beta)$?

LEVEL FOUR: Compute $\tan\left(\dfrac{\pi}{3} + \dfrac{\pi}{4}\right)$.

LEVEL FIVE: Find an identity for $\tan(\alpha + \beta)$ in terms of $\tan\alpha$ and $\tan\beta$.

Skill 3: Solve trigonometric equations using identities.

LEVEL ONE: Let h be the function given by $h(\theta) = \sin\theta \cos\theta \sec^2\theta$. For what value/s of θ in the interval $0 \leq \theta \leq 2\pi$ will $h(\theta) = 1$?

LEVEL TWO: Let f and g be the functions defined by $f(\theta) = \sin(2\theta)$ and $g(\theta) = \sin(\theta)$. Find all θ in the interval $0 \leq \theta < 2\pi$ such that $f(\theta) = g(\theta)$.

LEVEL THREE: Let g be the function defined by $g(\theta) = \sec^2 x - \tan x$. Find all x such that $g(x) = 1$.

LEVEL FOUR: The function f is defined by $f(\theta) = \tan^2\theta - 2\sec\theta$. Find all θ in the interval $0 \leq \theta \leq 2\pi$ such that $f(\theta) = -1$.

LEVEL FIVE: Find the zeros of $f(x) = 16\sin^2 x \cos^2 x - 3$ on the interval $0 \leq x \leq \frac{\pi}{2}$.

Skill 4: Apply identities in novel settings.

LEVEL ONE: The functions f and g are given by $f(x) = \sec x$ and $g(x) = \tan x$. The function h is defined by
$$h(x) = \frac{1}{(f(x))^2 - (g(x))^2}.$$
What is $h\left(\frac{\pi}{6}\right)$?

LEVEL TWO: Use the identity for $\sin(\alpha + \beta)$ to prove the identity $\sin(2\theta) = 2\sin(\theta)\cos(\theta)$.

LEVEL THREE: For certain angles α and θ, it is known that $\cos(\alpha + \theta) = \dfrac{3}{8}\cos\alpha - \dfrac{\sqrt{55}}{8}\sin\alpha$. What is $\tan\theta$?

LEVEL FOUR: The point A in the coordinate plane below lies at the intersection of the terminal ray of an angle of measure θ and a circle with a radius of 1. The point B lies at the intersection of another ray and the unit circle.

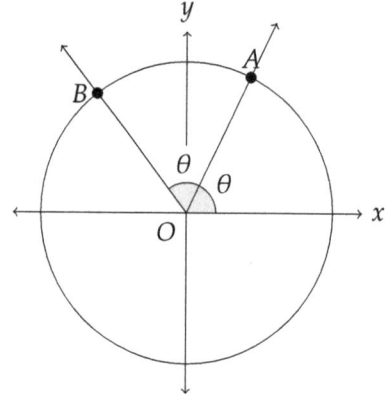

The coordinates of A are (x, y). If the acute angle between the rays OA and OB is θ, find expressions for the coordinates of B.

LEVEL FIVE: Consider the expression $\arcsin x = \arccos \theta$, where $0 < \theta < \frac{\pi}{2}$ and $0 < x < \frac{\pi}{2}$. Find an expression for θ in terms of x with no trigonometric functions.

Topic 3.12 (Day 2) Homework

1. Rewrite each function below in terms of a single trigonometric function.

    (a) $f(x) = \sin x \cot x$

    (b) $r(\theta) = \sec\theta \left(1 + \tan^2\theta\right)$

    (c) $f(\theta) = \dfrac{\cos^2\theta + \sin^2\theta}{\tan\theta}$

    (d) $g(\theta) = 2\cos^2\theta - \sin^2\theta$

    (e) $f(x) = \sec^2 x + \tan^2 x$

    (f) $h(x) = \dfrac{\csc^2 x - 1}{\cot x}$

    (g) $f(\theta) = 4\sin\theta\cos\theta + \cos^2\theta$

    (h) $g(\theta) = (\sin\theta + \cos\theta)^2$

2. It can be shown that $\dfrac{\pi}{12} = \dfrac{\pi}{3} - \dfrac{\pi}{4}$. Compute $\sin\dfrac{\pi}{12}$ and $\cos\dfrac{\pi}{12}$.

3. For angles θ and γ, the following values are known.

Angle	Value of $\sin x$	Value of $\cos x$
$x = \theta$	$\dfrac{3}{5}$	$-\dfrac{4}{5}$
$x = \gamma$	$-\dfrac{1}{4}$	$\dfrac{\sqrt{15}}{4}$

    Compute each of the following.

    (a) $\sin(2\gamma)$

    (b) $\cos(\theta + \gamma)$

    (c) $\cos(2\gamma)$

    (d) $\sin^2\gamma + \cos^2\gamma$

4. Let g be the function given by $g(x) = \cos x - \sin^2 x$. Find all solutions of $g(x) = -1$ on the interval $0 \le x \le 2\pi$.

5. Let f be the function given by $f(\theta) = 2(\cos^2\theta - \sin^2\theta)$. Find all solutions of $f(\theta) = 1$ on the interval $0 \le \theta \le 2\pi$.

Topic 3.12 (Day 2) Solutions/Notes

Activity

Skill 1:

LEVEL ONE: $f(x) = \csc^2 x$

LEVEL TWO: $f(\theta) = \dfrac{1}{\cos\theta} \cdot \dfrac{\cos\theta}{\sin\theta} = \dfrac{1}{\sin\theta} = \csc\theta$

LEVEL THREE: $k(x) = \sin x + \sin^2 x - \cos^2 x = \sin x + \sin^2 x - (1 - \sin^2 x) = 2\sin^2 x + \sin x - 1$

LEVEL FOUR: $f(\theta) = \dfrac{1}{\frac{1}{\cos\theta} - \sin\theta \cdot \frac{\sin\theta}{\cos\theta}} = \dfrac{1}{\frac{1-\sin^2\theta}{\cos\theta}} = \dfrac{1}{\frac{\cos^2\theta}{\cos\theta}} = \dfrac{1}{\cos\theta} = \sec\theta$

LEVEL FIVE: $f(x) = \dfrac{3(2+\sin(2x))}{1+\sin^2 x + \cos^2 x + 2\sin x \cos x} = \dfrac{3(2+\sin(2x))}{1+1+\sin(2x)} = \dfrac{3(2+\sin(2x))}{2+\sin(2x)} = 3$

Skill 2:

LEVEL ONE: $\cos\left(\dfrac{5\pi}{4}\right)\cos\left(\dfrac{2\pi}{3}\right) + \sin\left(\dfrac{5\pi}{4}\right)\sin\left(\dfrac{2\pi}{3}\right) = -\dfrac{\sqrt{2}}{2}\left(-\dfrac{1}{2}\right) + \left(-\dfrac{\sqrt{2}}{2}\right)\cdot\dfrac{\sqrt{3}}{2} = \dfrac{\sqrt{2}-\sqrt{6}}{4}$

LEVEL TWO: The hypotenuse has length $\sqrt{3^2 + (\sqrt{7})^2} = 4$, so $\cos(\theta) = \dfrac{3}{4}$ and $\sin(\theta) = \dfrac{\sqrt{7}}{4}$. Therefore,

$$\cos(2\theta) = \cos^2\theta - \sin^2\theta = \left(\dfrac{3}{4}\right)^2 - \left(\dfrac{\sqrt{7}}{4}\right)^2 = \dfrac{2}{16} = \dfrac{1}{8}.$$

LEVEL THREE: We need $\cos\alpha$ and $\cos\beta$ and can compute

$$\cos\alpha = \sqrt{1 - \sin^2\alpha} = \sqrt{1 - \left(\dfrac{5}{13}\right)^2} = \dfrac{12}{13} \quad \text{and} \quad \cos\beta = \sqrt{1 - \sin^2\beta} = \sqrt{1 - \left(\dfrac{3}{5}\right)^2} = \dfrac{4}{5}$$

(we know that $\cos\alpha > 0$ and $\cos\beta > 0$ because $0 < \alpha < \beta < \frac{\pi}{2}$, and $\cos\theta > 0$ for $0 \leq \theta < \frac{\pi}{2}$). Therefore,

$$\sin(\alpha + \beta) = \dfrac{5}{13}\cdot\dfrac{4}{5} + \dfrac{3}{5}\cdot\dfrac{12}{13} = \dfrac{56}{65}$$

LEVEL FOUR: We can write

$$\tan\left(\dfrac{\pi}{3} + \dfrac{\pi}{4}\right) = \dfrac{\sin\left(\frac{\pi}{3} + \frac{\pi}{4}\right)}{\cos\left(\frac{\pi}{3} + \frac{\pi}{4}\right)}.$$

The numerator is $\dfrac{\sqrt{6}+\sqrt{2}}{4}$, and the denominator is $\dfrac{\sqrt{6}-\sqrt{2}}{4}$, so

$$\tan\left(\dfrac{\pi}{3} + \dfrac{\pi}{4}\right) = \dfrac{\sqrt{6}+\sqrt{2}}{\sqrt{6}-\sqrt{2}}.$$

LEVEL FIVE: We have

$$\tan(\alpha + \beta) = \dfrac{\sin(\alpha+\beta)}{\cos(\alpha+\beta)}$$
$$= \dfrac{\sin\alpha\cos\beta + \sin\beta\cos\alpha}{\cos\alpha\cos\beta - \sin\alpha\sin\beta}$$

If we multiply the numerator and denominator by $\dfrac{1}{\cos\alpha\cos\beta}$, we get

$$\tan(\alpha+\beta) = \dfrac{\frac{1}{\cos\alpha\cos\beta}(\sin\alpha\cos\beta + \sin\beta\cos\alpha)}{\frac{1}{\cos\alpha\cos\beta}(\cos\alpha\cos\beta - \sin\alpha\sin\beta)}$$

$$= \dfrac{\frac{\sin\alpha}{\cos\alpha} + \frac{\sin\beta}{\cos\beta}}{1 - \frac{\sin\alpha}{\cos\alpha}\cdot\frac{\sin\beta}{\cos\beta}}$$

$$= \dfrac{\tan\alpha + \tan\beta}{1 - \tan\alpha\tan\beta}$$

Skill 3:

LEVEL ONE: Note that

$$h(\theta) = \sin\theta\cos\theta \cdot \dfrac{1}{\cos^2\theta} = \dfrac{\sin\theta}{\cos\theta} = \tan\theta.$$

Therefore, $h(\theta) = 1$ when $\tan\theta = 1$, which occurs at $\theta = \frac{\pi}{4}$ and $\theta = \frac{5\pi}{4}$ on the interval $0 \le \theta \le 2\pi$.

LEVEL TWO: Using the double angle identity, we can rewrite $\sin(2\theta)$ as $2\sin\theta\cos\theta$. Then, we get $2\sin\theta\cos\theta = \theta$. Subtracting $\sin\theta$ from both sides and factoring gives $\sin\theta(2\cos\theta - 1) = 0$. Now, setting the first factor equal to 0 yields $\theta = 0$ and $\theta = \pi$, while the second yields $\cos\theta = \frac{1}{2}$, which has solutions of $\theta = \frac{\pi}{3}$ and $\theta = \frac{5\pi}{3}$ on the interval $0 \le x < 2\pi$. Therefore, the solutions of $f(\theta) = g(\theta)$ are $\theta = 0, \theta = \frac{\pi}{3}, \theta = \pi,$, and $\theta = \frac{5\pi}{3}$.

LEVEL THREE: $g(\theta) = 1 \Rightarrow (\tan^2 x + 1) - \tan x = 1 \Rightarrow \tan^2 x - \tan x = 0 \Rightarrow \tan x(\tan x - 1) = 0$. Setting each factor to 0 gives us $\tan x = 0$, which occurs when $x = k\pi$ for any integer k, and $\tan x = 1$, which occurs when $x = \frac{\pi}{4} + \pi k$.

LEVEL FOUR: $f(\theta) = -1 \Rightarrow (\sec^2\theta - 1) - 2\sec\theta = -1 \Rightarrow \sec^2\theta - 2\sec\theta = 0 \Rightarrow \sec\theta(\sec\theta - 2) = 0$. Setting the first factor to 0 yields $\sec\theta = 0$, which has no solution. Setting the second factor to 0 gives $\sec\theta = 2$, which can be rewritten as $\cos\theta = \frac{1}{2}$. This has solutions of $\theta = \frac{\pi}{3}$ and $\theta = \frac{5\pi}{3}$.

LEVEL FIVE: We rewrite using the double angle identity for sine.

$$16\sin^2 x\cos^2 x - 3 = 0$$
$$(4\sin x\cos x)^2 - 3 = 0$$
$$(2(2\sin x\cos x))^2 - 3 = 0$$
$$(2\sin(2x))^2 - 3 = 0$$
$$4\cdot(\sin(2x))^2 - 3 = 0$$
$$(\sin(2x))^2 = \dfrac{3}{4}$$
$$\sin(2x) = \pm\sqrt{\dfrac{3}{4}} = \pm\dfrac{\sqrt{3}}{2}$$

Because our angle θ must be in the interval $[0, \frac{\pi}{2}]$, so $\sin(2x)$ must be positive and hence equal $\frac{\sqrt{3}}{2}$. Therefore, $2x = \frac{\pi}{3} + 2\pi k$ or $2x = \frac{2\pi}{3} + 2\pi k$ for any k, so $x = \frac{\pi}{6} + \pi k$ or $x = \frac{\pi}{3} + \pi k$. The only solutions in the interval $[0, \frac{\pi}{2}]$ is when $k = 0$, giving $x = \frac{\pi}{6}$ and $x = \frac{\pi}{3}$.

Skill 4:

LEVEL ONE: $h(x) = \dfrac{1}{\sec^2 x - \tan^2 x} = \dfrac{1}{(1 + \tan^2 x) - \tan^2 x} = \dfrac{1}{1} = 1$. h is a constant function equal to 1, so $h\left(\dfrac{\pi}{6}\right) = 1$.

LEVEL TWO: Let $\alpha = \theta$ and $\beta = \theta$. Then

$$\sin(\theta + \theta) = \sin\theta\cos\theta + \sin\theta\cos\theta = 2\sin\theta\cos\theta$$

LEVEL THREE: We can infer that $\cos\theta = \dfrac{3}{8}$ and $\sin\theta = \dfrac{\sqrt{55}}{8}$. Therefore,

$$\tan\theta = \dfrac{\sin\theta}{\cos\theta} = \dfrac{\sqrt{55}/8}{3/8} = \dfrac{\sqrt{55}}{3}.$$

LEVEL FOUR: The angle with a terminal ray through B measures 2θ, so the horizontal and vertical displacement of B, and hence the x- and y-coordinates, which we'll call x_B and y_B, are simply $x_B = \cos(2\theta)$ and $y_B = \sin(2\theta)$. Therefore, $x_B = \cos^2\theta - \sin^2\theta$ and $y_B = 2\sin\theta\cos\theta$. From the point A, we know that $\cos\theta = x$ and $\sin\theta = y$, so $x_B = x^2 - y^2$ and $y_B = 2xy$.

LEVEL FIVE: Firstly, we can take the cosine of both sides to get $\cos(\arcsin x) = \theta$. Now, for simplicity, let $\arcsin x = y$. We now want to rewrite $\cos(y)$. But if $\arcsin x = y$, then $\sin y = x$, and by the Pythagorean Identity, $\cos y = \pm\sqrt{1 - \sin^2 y} = \sqrt{1 - x^2}$ (ignoring the negative solution because both angles are in Quadrant I). This means $\theta = \cos y = \sqrt{1 - x^2}$.

Homework

1. (a) $f(x) = \sin x \cot x = \sin x \cdot \dfrac{\cos x}{\sin x} = \cos x$
    (b) $r(\theta) = \sec\theta(1 + \tan^2\theta) = \sec\theta \cdot \sec^2\theta = \sec^3\theta$
    (c) $f(\theta) = \dfrac{\cos^2\theta + \sin^2\theta}{\tan\theta} = \dfrac{1}{\tan\theta} = \cot\theta$
    (d) $g(\theta) = 2\cos^2\theta - \sin^2\theta = 2(1 - \sin^2\theta) - \sin^2\theta = 2 - 3\sin^2\theta$
    OR $g(\theta) = 2\cos^2\theta - (1 - \cos^2\theta) = 3\cos^2\theta - 1$
    (e) $f(x) = \sec^2 x + \tan^2 x = (1 + \tan^2 x) + \tan^2 x = 1 + 2\tan^2 x$
    OR $f(x) = \sec^2 x + (\sec^2 x - 1) = 2\sec^2 x - 1$
    (f) $h(x) = \dfrac{\csc^2 x - 1}{\cot x} = \dfrac{\cot^2 x}{\cot x} = \cot x$
    (g) $f(\theta) = 4\sin\theta\cos\theta + \cos^2\theta = 2(2\sin\theta\cos\theta) + (1 - \sin^2\theta) = 2\sin(2\theta) + 1 - \sin^2\theta$
    (h) $(\sin\theta + \cos\theta)^2 = (\sin\theta + \cos\theta)(\sin\theta + \cos\theta) = \sin^2\theta + \cos^2\theta + 2\sin\theta\cos\theta = 1 + \sin(2\theta)$

2. $\sin\left(\dfrac{\pi}{12}\right) = \sin\left(\dfrac{\pi}{3} - \dfrac{\pi}{4}\right) = \sin\dfrac{\pi}{3}\cos\dfrac{\pi}{4} - \sin\dfrac{\pi}{4}\cos\dfrac{\pi}{3} = \dfrac{\sqrt{3}}{2} \cdot \dfrac{\sqrt{2}}{2} - \dfrac{\sqrt{2}}{2} \cdot \dfrac{1}{2} = \dfrac{\sqrt{6} - \sqrt{2}}{4}$

    $\cos\left(\dfrac{\pi}{12}\right) = \cos\dfrac{\pi}{3}\cos\dfrac{\pi}{4} + \sin\dfrac{\pi}{3}\sin\dfrac{\pi}{4} = \dfrac{1}{2} \cdot \dfrac{\sqrt{2}}{2} + \dfrac{\sqrt{3}}{2} \cdot \dfrac{\sqrt{2}}{2} = \dfrac{\sqrt{2} + \sqrt{6}}{4}$

3. (a) $\sin(2\gamma) = 2\sin\gamma\cos\gamma = 2 \cdot -\dfrac{1}{4} \cdot \dfrac{\sqrt{15}}{4} = \dfrac{-2\sqrt{15}}{16} = -\dfrac{\sqrt{15}}{8}$
    (b) $\cos(\theta + \gamma) = \cos(\theta)\cos(\gamma) - \sin(\theta)\sin(\gamma) = -\dfrac{4}{5} \cdot \dfrac{\sqrt{15}}{4} - \dfrac{3}{5}\left(-\dfrac{1}{4}\right) = \dfrac{3 - 4\sqrt{15}}{20}$

(c) $\cos(2\gamma) = \cos^2\gamma - \sin^2\gamma = \dfrac{15}{16} - \dfrac{1}{16} = \dfrac{14}{16} = \dfrac{7}{8}$

(d) 1

4. We substitute $1 - \cos^2 x$ for $\sin x$.

$$g(x) = -1$$
$$\cos x - (1 - \cos^2 x) = -1$$
$$\cos x - 1 + \cos^2 x = -1$$
$$\cos^2 x + \cos x = 0$$
$$\cos x(\cos x + 1) = 0$$

This gives either $\cos x = 0$ or $\cos x + 1 = 0$. On the interval $0 \leq x \leq 2\pi$, the first equation has solutions of $x = \frac{\pi}{2}$ and $x = \frac{3\pi}{2}$ and the second equation has a solution of $x = \pi$. The solutions are therefore $x = \frac{\pi}{2}, x = \pi$, and $x = \frac{\pi}{2}$.

5. We recognize $\cos^2\theta - \sin^2\theta$ is an equivalent representation of $\cos(2\theta)$. Therefore:

$$f(\theta) = 1$$
$$2\cos(2\theta) = 1$$
$$\cos(2\theta) = \dfrac{1}{2}$$

The general solutions of $\cos x = \frac{1}{2}$ are $x = \frac{\pi}{3} + 2\pi k$ and $x = \frac{5\pi}{3} + 2\pi k$ for any integer k. We can set 2θ equal to each of these and solve for θ.

$$2\theta = \dfrac{\pi}{3} + 2\pi k \implies \theta = \dfrac{\pi}{6} + \pi k$$
$$2\theta = \dfrac{5\pi}{3} + 2\pi k \implies \theta = \dfrac{5\pi}{6} + \pi k$$

We only want solutions on the interval $0 \leq \theta \leq 2\pi$, though. Looking at $k = 0$ and $k = 1$, we get solutions of $\theta = \frac{\pi}{6}, \theta = \frac{5\pi}{6}, \theta = \frac{7\pi}{6}$, and $\theta = \frac{11\pi}{6}$.

Reflection

Suggested Class Time. 75+ minutes

Prerequisites. A lot of algebraic and arithmetic skill.

Instructional Strategies. This lesson was a beast, and I wanted it to not feel like just another worksheet, so here's how I structured it.

I gave students a handout consisting of the problems on pages 385-388. I told them that they were responsible for earning at least 20 "points," where the points would be calculated as the minimum of the number of problems gotten correct times 2 (so a minimum of 10) and the sum total of their "levels". This was to discourage students from simply working 4 or 5 high level problems and calling it a day (little did I realize that was not going to happen in a million years for my students!). I instructed students to check in with me after they did each problem so I could check them (really, to see mistakes, offer feedback, etc.), and then I collected the sheets at the end of class.

I quickly realized that my students were mostly fine with the trig, but absolutely atrocious with the arithmetic. Among the issues I saw multiple times, along with how I addressed them, are the following.

- Doing weird things like saying $\cos\left(\frac{2\pi}{3}\right) = \cos\left(-\frac{1}{2}\right)$

  This blew me away. I tried to emphasize the inputs and outputs to students: the inputs of trig functions are angles and the outputs are displacements/ratios/slopes, not more trig functions!

- $(\sin x + \cos x)^2 = 1$

  I was pretty harsh about this. I told my students they cannot take an AP math class and keep doing this (this being $(x+y)^2 = x^2 + y^2$). We had definitely done this numerous times before, and I've pulled out all the stops: showing the area model, encouraging students to expand it, even "proving" that $7^2 = 25$. With my particular group of students, I decided to try a different tact, and that was bluntness.

- Not knowing how to find sine or cosine from a triangle (or only assuming that cosine/sine is the vertical/horizontal side, thereby ignoring the radius/hypotenuse)

  This was reasonable, as all of Unit 3 has focused on displacement, not triangle trig. I just reminded students of SOH-CAH-TOA (which their previous teacher really harped).

- Committing order of operation felonies

  A number of students committed a version of $\frac{1}{x+y} = \frac{1}{x} + \frac{1}{y}$. I simply tell students, as precisely as I can, that this amounts to *changing what you're dividing by*, which you can't just... do. I also emphasize that order of operations must be obeyed, and that if, for instance, the denominator does not contain like terms and you "can't" add them before dividing, then that's just too bad, and you'll have to give up on any aspirations of carrying out the division.

  Another student wanted to "cancel" the sines in the expression $\frac{6+3\sin(2x)}{2+\sin(2x)}$. For this student, I said we were going to play a game show called "CAN... YOU... CANCEL... THAT?" I wrote the following five expressions.

  $$\frac{xy}{y} \qquad \frac{x+y}{x} \qquad \frac{x}{x+y} \qquad \frac{2x+y}{x+y} \qquad \frac{2(x+y)}{(x+y)}$$

  The student said yes correctly to the first and then unconvincingly said no to the next 3. I talked about the fact that addition makes "cancelling" complicated because of the hierarchy of addition and division in the order of operations, while "cancelling" is simple when only multiplication and division are involved. I then talked about how getting from a sum or difference to a product is precisely how we defined factoring earlier in the year!

After the lesson, I emphasized to my students that getting to this level of problem within just 1 class period of having learned identities was *tough*, and that they should be really proud of how well they were doing and to not get discouraged (they seemed fine and certainly got some problems right, but they also made a ton of mistakes). I collected their record sheets, fully intending to give them all the full credit for the day. I then encouraged all of them to come by office hours, where we could resume our practice.

Technology. None, much to students' dismay!

Problems.

1. The function h is defined by
$$h(\theta) = \frac{\tan^2\theta - \sec^2\theta}{1 - \cos^2\theta}.$$
Which of the following is an equivalent expression for h in terms of a single trigonometric function raised to a power?

    (A) $h(\theta) = -\sin^2\theta$ **(B)** $h(\theta) = -\csc^2\theta$ **(C)** $h(\theta) = \sin^2\theta$ **(D)** $h(\theta) = \csc^2\theta$

2. The function f is defined by $f(x) = \sin(x - \alpha)$, where α is an angle. It is known that $\sin \alpha = -\frac{3}{4}$ and $\cos \alpha = \frac{\sqrt{7}}{4}$. Which of the following is an equivalent representation of $f(x)$?

   (A) $f(x) = \frac{1}{4}(3\sin x - \sqrt{7}\cos x)$
   (B) $f(x) = \frac{1}{4}(3\sin x + \sqrt{7}\cos x)$
   (C) $f(x) = \frac{1}{4}(\sqrt{7}\sin x - 3\cos x)$
   (D) $f(x) = \frac{1}{4}(\sqrt{7}\sin x + 3\cos x)$

3. The function g is given by $g(\theta) = \tan\theta \sin\theta + \sin\theta$. For what values of θ, with $0 \leq \theta \leq 2\pi$, will $g(\theta) = 0$?

   (A) $\theta = 0, \theta = \frac{3\pi}{4}, \theta = \pi$
   (B) $\theta = \frac{\pi}{2}, \theta = \frac{3\pi}{4}$
   (C) $\theta = 0, \theta = \frac{3\pi}{4}, \theta = \pi, \theta = \frac{7\pi}{4}, \theta = 2\pi$
   (D) $\theta = \frac{\pi}{2}, \theta = \frac{3\pi}{4}, \theta = \frac{7\pi}{4}, \theta = \frac{3\pi}{2}$

4. The function g is defined by $g(x) = \cos x \cot x$. Which of the following expresses g as a fraction involving only powers of a single trigonometric function?

   (A) $g(x) = \dfrac{\cos^2 x}{1 - \cos^2 x}$ (B) $g(x) = \dfrac{\cos^2 x}{1 - \cos x}$ (C) $g(x) = \dfrac{\sin^2 x - 1}{\sin x}$ (D) $g(x) = \dfrac{1 - \sin^2 x}{\sin x}$

5. The function f is given by $f(x) = 8\cos(2x)$. Which of the following is an equivalent form for f?

   (A) $f(x) = 16\sin x \cos x$ (B) $f(x) = 8\cos^2 x - 8\sin^2 x$ (C) $f(x) = 16 - 16\sin^2 x$ (D) $f(x) = 8\sin x \cos x$

Solutions.

1. We have
$$h(\theta) = \frac{\tan^2\theta - \sec^2\theta}{1 - \cos^2\theta} = \frac{\sec^2\theta - 1 - \sec^2\theta}{\sin^2\theta} = \frac{-1}{\sin^2\theta} = -\csc^2\theta.$$
   The correct answer is therefore **(B)**.

2. Using the difference identity for sine, we get
$$\sin(x - \alpha) = \sin x \cos\alpha - \cos x \sin\alpha = \sin x \cdot \frac{\sqrt{7}}{4} - \cos x \cdot \left(-\frac{3}{4}\right) = \frac{1}{4}(\sqrt{7}\sin x + 3\cos x).$$
   The correct answer is therefore **(D)**.

3. We have
$$\tan\theta \sin\theta + \sin\theta = 0$$
$$\sin\theta(\tan\theta + 1) = 0$$
$$\sin\theta = 0 \text{ or } \tan\theta = -1.$$
   The sine will be zero on $0 \leq \theta \leq 2\pi$ for $\theta = 0$, $\theta = \pi$, and $\theta = 2\pi$. The tangent will be -1 on $0 \leq \theta \leq 2\pi$ for $\theta = 3\pi/4$ and $\theta = 7\pi/4$. The correct answer is therefore **(C)**.

4. We have
$$g(x) = \cos x \cot x = \cos x \cdot \frac{\cos x}{\sin x} = \frac{\cos^2 x}{\sin x} = \frac{1 - \sin^2 x}{\sin x}.$$
   The correct answer is therefore **(D)**.

5. Since $\cos(2x) = \cos^2 x - \sin^2 x$, $8\cos(2x) = 8\cos^2 x - 8\sin^2 x$. The correct answer is therefore **(B)**.

Topics 3.10-3.12 Circuit

Instructions: Start in Cell 1. Solve the problem, then find the cell with the corresponding answer. Mark this Cell 2. Proceed until you've finished the circuit. (Note: To save room, we're assuming that k is any integer in the answers with a k.)

Cell __1__
Problem: Find the zeros of $f(x) = 4\cos^2 x - 1$ for $0 \leq x \leq 2\pi$.
Answer: $a = \frac{\pi}{3}, b = \frac{2\pi}{3}$

Cell ____
Problem: The function f is defined by $f(x) = 3\sin^2 x + \cos(2x) - 2$. Find all x such that $f(x) = 0$.
Answer: $x = \frac{\pi}{6} + 2\pi k, x = \frac{5\pi}{6} + 2\pi k$

Cell ____
Problem: The function $f(\theta) = \sin(\theta)$ satisfies $f(\theta) \geq \frac{1}{2}$ on the interval $a \leq \theta \leq b$, where $0 < a < b < 2\pi$. Find a and b.
Answer: $f(\theta) = \frac{1}{1-\sin^2 \theta}$

Cell ____
Problem: The function $f(x) = 4\tan^2 x \cos^2 x - 3$ has two zeros a and b on the interval $[0, \pi]$. Find a and b.
Answer: $x = \frac{1}{6}$

Cell ____	Cell ____
Problem: The function $f(x) = \sin x + \cos x$ can be rewritten in the form $f(x) = a\sin(x+b)$, where $a > 0$ and $0 < b < 2\pi$. Find a and b. Answer: $f(\theta) = \cos^2 \theta$	Problem: Rewrite $f(\theta) = \frac{\sin(2\theta)}{2\tan\theta}$ as a power of a single trigonometric function. Answer: $x = -\frac{3}{2}$
Cell ____ Problem: Find an expression for $f(\theta) = \csc^2 \theta \tan^2 \theta$ that involves only powers of $\sin \theta$. Answer: $x = \frac{\pi}{3}, x = \frac{2\pi}{3}, x = \frac{4\pi}{3}, x = \frac{5\pi}{3}$	**Cell ____** Problem: Let $f(x) = 3\arctan\left(2\sqrt{3}x\right)$. Find the value of x in the domain of f such that $f(x) = \frac{\pi}{2}$. Answer: $a = \sqrt{2}, b = \frac{\pi}{4}$
Cell ____ Problem: The function f is defined by $f(x) = \frac{2\sec^2 x - 2\tan^2 x}{\csc x}$. Find all values of x such that $f(x) = 1$. Answer: $a = \frac{\pi}{6}, b = \frac{5\pi}{6}$	**Cell ____** Problem: Let $f(x) = 2\arccos(2x+3)$. Find all values of x in the domain of f such that $f(x) = \pi$. Answer: $x = \frac{\pi}{2} + 2\pi k, x = \frac{3\pi}{2} + 2\pi k$

Topics 3.10-3.12 Circuit Solutions

Cell 1
Problem: Find the zeros of $f(x) = 4\cos^2 x - 1$ for $0 \leq x \leq 2\pi$.
Answer: $a = \frac{\pi}{3}, b = \frac{2\pi}{3}$
Solution:
$$4\cos^2 x - 1 = 0$$
$$\cos^2 x = \frac{1}{4}$$
$$\cos x = \pm \frac{1}{2}$$
$$x = \frac{\pi}{3}, x = \frac{2\pi}{3}, x = \frac{4\pi}{3}, x = \frac{5\pi}{3}$$

Cell 5
Problem: The function f is defined by $f(x) = 3\sin^2 x + \cos(2x) - 2$. Find all x such that $f(x) = 0$.
Answer: $x = \frac{\pi}{6} + 2\pi k, x = \frac{5\pi}{6} + 2\pi k$
Solution:
$$3\sin^2 x + \cos(2x) - 2 = 0$$
$$3\sin^2 x + (1 - 2\sin^2 x) - 2 = 0$$
$$\sin^2 x - 1 = 0$$
$$(\sin x - 1)(\sin x + 1) = 0$$
$$\sin x = 1 \implies x = \frac{\pi}{2} + 2\pi k$$
$$\text{or} \quad \sin x = -1 \implies x = \frac{3\pi}{2} + 2\pi k$$

Cell 3
Problem: The function $f(\theta) = \sin(\theta)$ satisfies $f(\theta) \geq \frac{1}{2}$ on the interval $a \leq \theta \leq b$, where $0 < a < b < 2\pi$. Find a and b.
Answer: $f(\theta) = \frac{1}{1-\sin^2 \theta}$
Solution:
For $0 < \theta < 2\pi$, $f(\theta) = \frac{1}{2}$ only at $x = \frac{\pi}{6}$ and $x = \frac{5\pi}{6}$. The output $f(\theta)$ is greater than or equal to $\frac{1}{2}$ for all θ with $\frac{\pi}{6} \leq \theta \leq \frac{5\pi}{6}$, so $a = \frac{\pi}{6}$ and $b = \frac{5\pi}{6}$.

Cell 10
Problem: The function $f(x) = 4\tan^2 x \cos^2 x - 3$ has two zeros a and b on the interval $[0, \pi]$. Find a and b.
Answer: $x = \frac{1}{6}$
Solution:
$$4\tan^2 x \cos^2 x - 3 = 0$$
$$4 \cdot \frac{\sin^2 x}{\cos^2 x} \cdot \cos^2 x - 3 = 0$$
$$4\sin^2 x - 3 = 0$$
$$\sin^2 x = \frac{3}{4}$$
$$\sin x = \pm \frac{\sqrt{3}}{2}$$
$$x = \frac{\pi}{3} \quad \text{or} \quad x = \frac{2\pi}{3}$$

Cell __8__	Cell __7__
Problem: The function $f(x) = \sin x + \cos x$ can be rewritten in the form $f(x) = a\sin(x+b)$, where $a > 0$ and $0 < b < 2\pi$. Find a and b. Answer: $f(\theta) = \cos^2 \theta$ Solution: $$a\sin(x+b) = \sin x + \cos x$$ $$a(\sin x \cos b + \sin b \cos x) = \sin x + \cos x$$ $$a\sin x \cos b + a\sin b \cos x = \sin x + \cos x$$ $$(a\cos b)\sin x + (a\sin b)\cos x = \sin x + \cos x$$ This means $a\cos b = 1$ and $a\sin b = 1$. Therefore, $\frac{a\sin b}{a\cos b} = \frac{1}{1}$, or $\tan b = 1$, which gives $b = \frac{\pi}{4}$ (not $b = \frac{5\pi}{4}$ since $a > 0$). Therefore, $a\cos b = 1$ gives $a \cdot \frac{\sqrt{2}}{2} = 1$, or $a = \sqrt{2}$.	Problem: Rewrite $f(\theta) = \frac{\sin(2\theta)}{2\tan\theta}$ as a power of a single trigonometric function. Answer: $x = -\frac{3}{2}$ Solution: $$f(\theta) = \frac{2\sin\theta\cos\theta}{2\sin\theta/\cos\theta}$$ $$= 2\sin\theta\cos\theta \cdot \frac{\cos\theta}{2\sin\theta}$$ $$= \cos^2\theta$$
Cell __2__	Cell __9__
Problem: Find an expression for $f(\theta) = \csc^2\theta\tan^2\theta$ that involves only powers of $\sin\theta$. Answer: $x = \frac{\pi}{3}, x = \frac{2\pi}{3}, x = \frac{4\pi}{3}, x = \frac{5\pi}{3}$ Solution: $$f(\theta) = \frac{1}{\sin^2\theta} \cdot \frac{\sin^2\theta}{\cos^2\theta}$$ $$= \frac{1}{\cos^2\theta}$$ $$= \frac{1}{1-\sin^2\theta}$$	Problem: Let $f(x) = 3\arctan\left(2\sqrt{3}x\right)$. Find the value of x in the domain of f such that $f(x) = \frac{\pi}{2}$. Answer: $a = \sqrt{2}, b = \frac{\pi}{4}$ Solution: $$3\arctan\left(2\sqrt{3}x\right) = \frac{\pi}{2}$$ $$\arctan\left(2\sqrt{3}x\right) = \frac{\pi}{6}$$ $$\tan\left(\arctan\left(2\sqrt{3}x\right)\right) = \tan\left(\frac{\pi}{6}\right)$$ $$2\sqrt{3}x = \frac{\sqrt{3}}{3}$$ $$x = \frac{1}{6}$$
Cell __4__	Cell __6__
Problem: The function f is defined by $f(x) = \frac{2\sec^2 x - 2\tan^2 x}{\csc x}$. Find all values of x such that $f(x) = 1$. Answer: $a = \frac{\pi}{6}, b = \frac{5\pi}{6}$ Solution: $$\frac{2(\sec^2 x - \tan^2 x)}{\csc x} = 1$$ $$\frac{2(1+\tan^2 x - \tan^2 x)}{\csc x} = 1$$ $$\frac{2}{1/\sin x} = 1$$ $$2\sin x = 1$$ $$\sin x = \frac{1}{2}$$ $$x = \frac{\pi}{6} + 2\pi k \text{ or } x = \frac{5\pi}{6} + 2\pi k$$	Problem: Let $f(x) = 2\arccos(2x+3)$. Find all values of x in the domain of f such that $f(x) = \pi$. Answer: $x = \frac{\pi}{2} + 2\pi k, x = \frac{3\pi}{2} + 2\pi k$ Solution: $$2\arccos(2x+3) = \pi$$ $$\arccos(2x+3) = \frac{\pi}{2}$$ $$\cos(\arccos(2x+3)) = \cos\left(\frac{\pi}{2}\right)$$ $$2x + 3 = 0$$ $$x = -\frac{3}{2}$$

Reflection

Suggested Class Time. 60-70 minutes

Prerequisites. A lot of algebraic and arithmetic skill.

Instructional Strategies.
Some of my students finished, but most got to around 7 or 8 complete. I will say that my students seemed to enjoy the circuit, and I'll consider adding more of them next year if time permits.

It turns out that the biggest "trap" was on Cell 1: multiple students tried to convert $4\cos^2 x - 1$ into $2\cos(2x)$. I merely encouraged them to work carefully and to actually write down their substitutions. Additionally, I made a general note for the class that "you typically only want to introduce a double angle identity if you're *reducing the number of distinct trigonometric functions*".

Be warned that Cell 8 is *really* hard for students. It may be worthwhile to go over this one as a class, or to give some hints along the way. I considered changing this to Cell 10 or removing it altogether, but it just isn't that difficult. It's a system of equations created by working from the desired result back to the given expression. Students can do it!

Technology. None was required.

Topic 3.13 ~ Trigonometry and Polar Curves

Learning Objectives

1. (3.13.A) Determine the location of a point in the plane using both rectangular and polar coordinates.

Success Criteria

1. I can convert between polar and rectangular coordinates.
2. I can graph a complex number in the complex plane using rectangular and polar coordinates.

Suppose you were asked to graph the point $P(-1.2, 1.6)$ in the coordinate plane. You would, with your eyes or even possibly with your pencil, trace out a path: starting at the origin, you would go 1.2 to the left and then 1.6 up, where you would draw a small circle for P, as shown below.

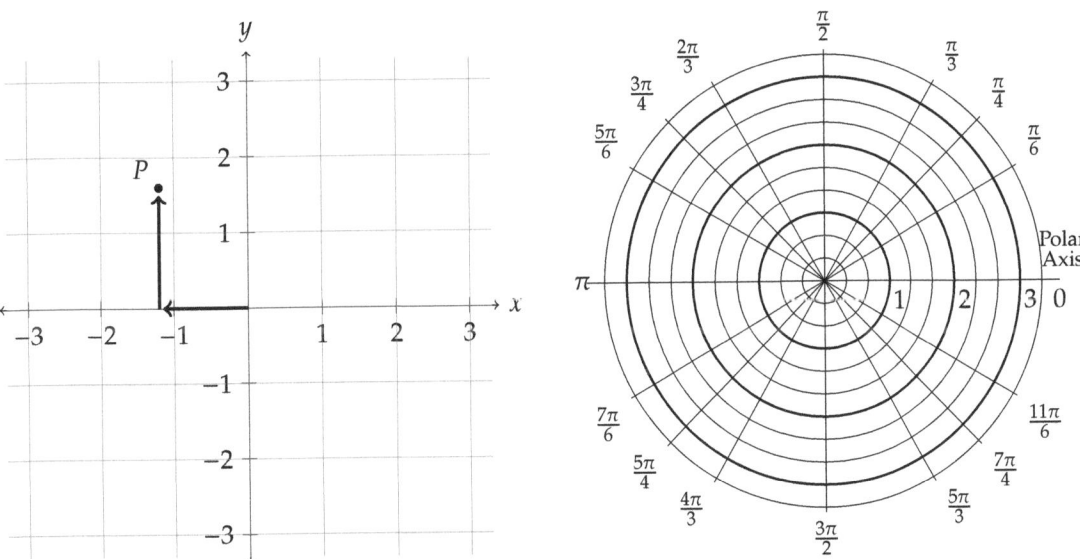

This way (x, y) of locating (and being able to draw) a point in the plane is called *rectangular coordinates*.

1. Can you see from the diagram one reason why (x, y) might be called "rectangular" coordinates?

However, based on all of what you've learned in Unit 3, we know that there is *another* way to identify the location of a point: using an angle θ and a distance from the origin r.

2. (▨) Draw the ray from the origin through the point P in the left diagram and label the angle θ formed by this ray and the positive x-axis. Find the measure of the angle θ of this angle in standard position.

3. (▨) Draw the circle centered at the origin that goes through P. Find the radius of this circle using the rectangular coordinates of P.

We now have a different way of specifying the location of P: using the radius r and the angle θ. These coordinates, written (r, θ), are called POLAR COORDINATES. The diagram on the right is of what is called the POLAR PLANE.

4. Draw P in the polar plane.

5. Let A, B, and C be the points with polar coordinates $A\left(3, \frac{\pi}{2}\right)$, $B\left(\frac{2}{3}, \frac{3\pi}{4}\right)$, and $C\left(-2, \frac{11\pi}{6}\right)$. Draw A, B, and C on the polar plane above. (What do you think you do for a *negative* radius?)

Converting between rectangular and polar coordinates makes use of the trigonometry we learned in Unit 3 as well as the Pythagorean Theorem.

Rectangular & Polar Coordinate Conversions

Let P be a point with rectangular coordinates (x, y) and polar coordinates (r, θ). Then

$$x = r \cos \theta \qquad\qquad r = \sqrt{x^2 + y^2}$$

$$y = r \sin \theta \qquad\qquad \theta = \arctan\left(\frac{y}{x}\right) \text{ if } x > 0$$

$$\qquad\qquad\qquad\qquad = \arctan\left(\frac{y}{x}\right) + \pi \text{ if } x < 0$$

---

6. (🖩) Convert the point $(2, 3)$ into polar coordinates.

7. Convert the point $\left(4, \frac{2\pi}{3}\right)$ into rectangular coordinates.

8. Write both the polar and rectangular coordinates of the point K shown in the polar plane below.

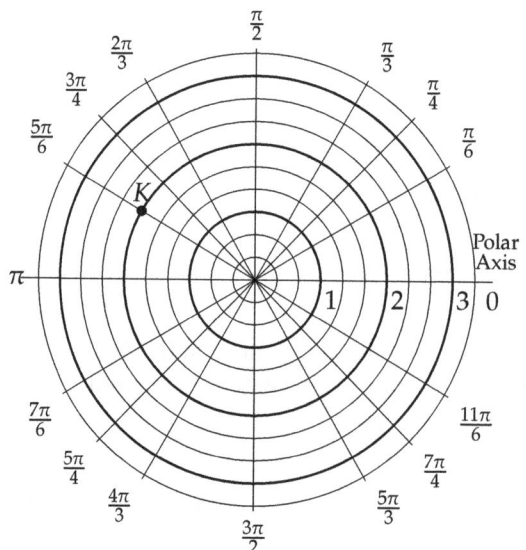

As you work with polar coordinates, here are a few facts that will commonly appear.

Useful Tips for Polar Coordinates

1. An isosceles triangle with legs both of length x will have a hypotenuse of $x\sqrt{2}$.
2. $\arctan\left(\pm\sqrt{3}\right) = \pm\dfrac{\pi}{3}$.
3. $\arctan\left(\pm\dfrac{1}{\sqrt{3}}\right) = \pm\dfrac{\pi}{6}$.
4. Knowledge of the unit circle may render the arctan formulas superfluous.

---

9. Find the polar coordinates of a point P with rectangular coordinates $(-4, -4\sqrt{3})$.

10. Write the polar coordinates of the point A with rectangular coordinates $(-4, 4)$.

We end with one interesting application of rectangular and polar coordinates: complex numbers.

Every complex number is of the form $z = a + bi$. It turns out, if we look at the xy-plane slightly differently, we can *graph* complex numbers. The resulting plane is called the COMPLEX PLANE. For an example, the complex number $z = -2 + 3i$ is graphed below.

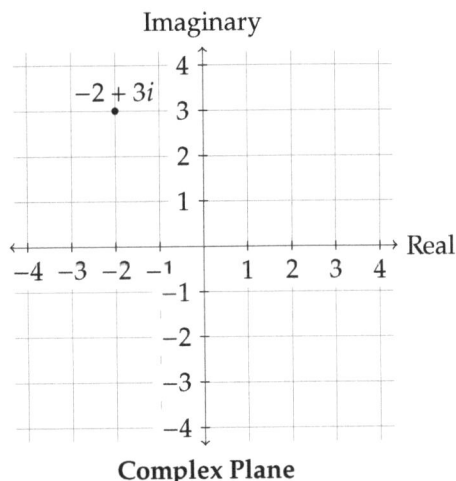

Complex Plane

11. How does the placement of $z = -2 + 3i$ in the plane above make sense?

12. Graph the complex numbers $4 + i, 2 - 2i, 3,$ and $2i$ in the complex plane above.

Because we can graph/write complex numbers $z = x + yi$ using rectangular coordinates (x, y), it means we can also graph/write them using polar coordinates.

For example, the complex number $z = 2i$ is a distance of 2 away from the origin at an angle of $\dfrac{\pi}{2}$. Therefore, it can be written as
$$z = \left(2\cos\dfrac{\pi}{2}\right) + i\left(2\sin\dfrac{\pi}{2}\right)$$

13. Write the polar coordinates of the complex number $z = 3 - 3i$.

14. Write the polar coordinates of the complex number $z = -2 + 2\sqrt{3}i$.

Topic 3.13 Homework

1. Four points - $A, B, C,$ and D - are graphed in the polar plane below.

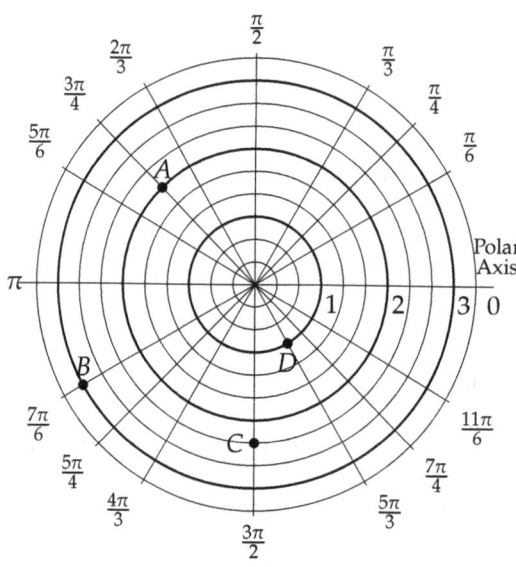

   Find both the polar and rectangular coordinates for each point.

2. Convert each point from rectangular coordinates to polar coordinates with $-2\pi < \theta < 2\pi$.

   (a) $(2, -2)$

   (b) $\left(\frac{1}{4}, \frac{\sqrt{3}}{4}\right)$

   (c) $(-6\sqrt{3}, 6)$

   (d) $(-4\sqrt{2}, 4\sqrt{2})$

   (e) $(-1, -\sqrt{3})$

   (f) $\left(5\cos\frac{\pi}{7}, 5\sin\frac{\pi}{7}\right)$

3. In the formulae for converting from rectangular to polar coordinates, we wrote
$$\theta = \arctan\frac{y}{x} \text{ if } x > 0 \quad \text{and} \quad \theta = \arctan\frac{y}{x} + \pi \text{ if } x < 0.$$
   Explain why we need to add π for $x < 0$.

4. Kisha is answering an online homework question that asks for the polar coordinates of the point with rectangular coordinates $(3, -3\sqrt{3})$. She will enter an r in simplified form and a θ with $0 \leq \theta < 2\pi$. Kisha enters $r = 6$ and $\theta = -\frac{\pi}{3}$, and the online program tells Kisha she is incorrect. She triple checks her work and has confirmed that $r = 6$ and $\theta = -\frac{\pi}{3}$ will generate a point with rectangular coordinates $(3, -\sqrt{3})$. What did Kisha do "wrong," and what can she type in to get the problem correct?

5. The points P and Q have polar coordinates $\left(4, \frac{5\pi}{12}\right)$ and $\left(-4, \frac{5\pi}{12}\right)$, respectively. Find the distance between P and Q.

6. Suppose a point X has polar coordinates (r, θ). If X is to be rotated counterclockwise about the origin by α radians to create a point X', write an expression for the polar coordinates of X'.

Topic 3.13 Solutions/Notes

Activity

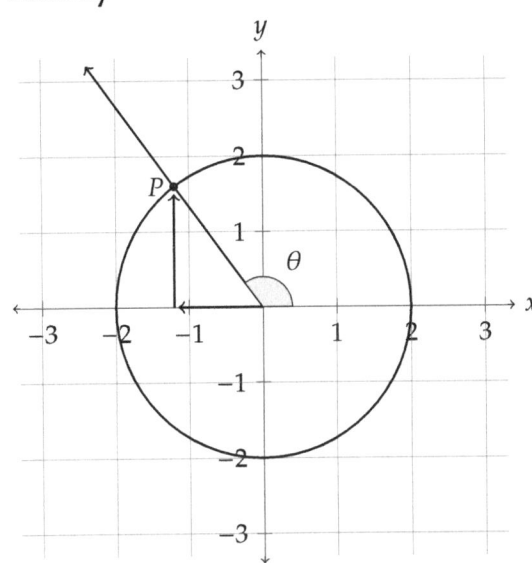 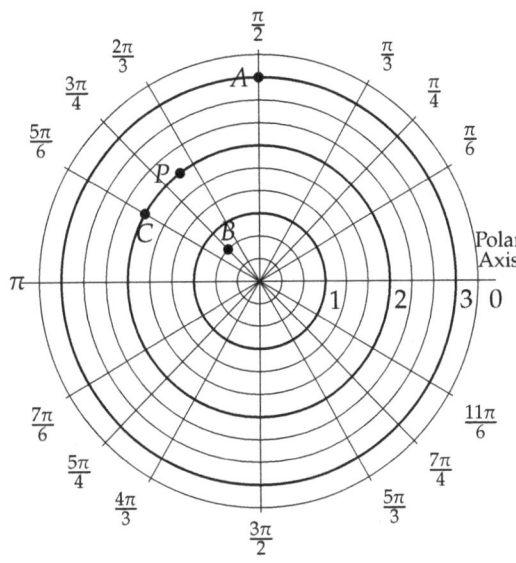

1. The triangle forms half of a rectangle.

2. $\arctan\left(\frac{1.6}{-1.2}\right) = -0.927$ radians, but this would give us an angle in Quadrant IV. To find the angle in Quadrant II, we can add π, giving $\theta = -0.927 + \pi \approx 2.214$ radians.

3. We use the Pythagorean Theorem: $r = \sqrt{(-1.2)^2 + 1.6^2} = 2$.

5. Shown above. For a negative radius, we draw the terminal ray in the opposite direction of the angle.

6. $r = \sqrt{2^2 + 3^2} = \sqrt{13}$ and $\theta = \arctan \frac{3}{2} \approx 0.983$, so the polar coordinates are $(\sqrt{13}, 0.983)$.

7. $x = 4\cos\frac{2\pi}{3} = 4\left(-\frac{1}{2}\right) = -2$, $y = 4\sin\frac{2\pi}{3} = 4 \cdot \frac{\sqrt{3}}{2} = 2\sqrt{3}$, so the rectangular coordinates are $(-2, 2\sqrt{3})$.

8. The polar coordinates are $\left(2, \frac{5\pi}{6}\right)$. The rectangular coordinates can be found with

$$x = 2\cos\frac{5\pi}{6} = 2\left(-\frac{\sqrt{3}}{2}\right) = -\sqrt{3} \quad \text{and} \quad y = 2\sin\frac{5\pi}{6} = 2 \cdot \frac{1}{2} = 1,$$

giving $(-\sqrt{3}, 1)$.

9. If we rely on the unit circle, we can see that

$$(-4, -4\sqrt{3}) = \left(8\left(-\frac{1}{2}\right), 8\left(-\frac{\sqrt{3}}{2}\right)\right) = \left(8\cos\frac{4\pi}{3}, 8\sin\frac{4\pi}{3}\right).$$

This gives the polar coordinates as $(8, \frac{4\pi}{3})$. On the other hand, we can compute $r = \sqrt{(-4)^2 + (-4\sqrt{3})^2} = \sqrt{16 + 16 \cdot 3} = \sqrt{64} = 8$ and $\theta = \arctan\frac{-4\sqrt{3}}{-4} + \pi = \arctan\sqrt{3} + \pi = \frac{\pi}{3} + \pi = \frac{4\pi}{3}$.

10. As $|x| = |y|$, we'll forgo the formulae in favor of isosceles triangle geometry: the angle is $\theta = \frac{3\pi}{4}$ and the radius is $4\sqrt{2}$ giving $\left(4\sqrt{2}, \frac{3\pi}{4}\right)$.

11. The x-coordinate is the real part -2, and the y-coordinate is the coefficient 3 of the imaginary part.

12.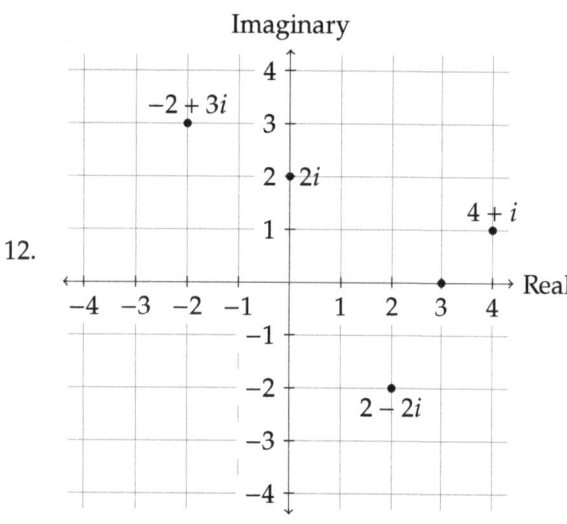

13. $\theta = -\frac{\pi}{4}$ (or $\theta = \frac{7\pi}{4}$) and $r = 3\sqrt{2}$, so $z = \left(3\sqrt{2}\cos\left(-\frac{\pi}{4}\right)\right) + i\left(3\sqrt{2}\sin\left(-\frac{\pi}{4}\right)\right)$.

14. Again, we can rely on unit circle knowledge or the formulae. We get $z = \left(4\cos\frac{2\pi}{3}\right) + i\left(4\sin\frac{2\pi}{3}\right)$.

Homework

1. $A: \left(2, \frac{3\pi}{4}\right) \to \left(2\cos\frac{3\pi}{4}, 2\sin\frac{3\pi}{4}\right) \to \left(2\left(-\frac{\sqrt{2}}{2}\right), 2\cdot\frac{\sqrt{2}}{2}\right) = (-\sqrt{2}, \sqrt{2})$

    $B: \left(3, \frac{7\pi}{6}\right) \to \left(3\cos\frac{7\pi}{6}, 3\sin\frac{7\pi}{6}\right) \to \left(3\left(-\frac{\sqrt{3}}{2}\right), 3\left(-\frac{1}{2}\right)\right) \to \left(-\frac{3\sqrt{3}}{2}, -\frac{3}{2}\right)$

    $C: \left(3, \frac{3\pi}{2}\right) \to (0, -3)$

    $D: \left(1, \frac{5\pi}{3}\right) \to \left(\cos\frac{5\pi}{3}, \sin\frac{5\pi}{3}\right) \to \left(\frac{1}{2}, -\frac{\sqrt{3}}{2}\right)$

2. (a) $\theta = \arctan\frac{-2}{2} = \arctan(-1) = -\frac{\pi}{4}, r = 2\sqrt{2}$, so $\left(2\sqrt{2}, -\frac{\pi}{4}\right)$

    (b) $\theta = \arctan\frac{\sqrt{3}/4}{1/4} = \arctan\sqrt{3} = \frac{\pi}{3}, r = \sqrt{\left(\frac{1}{4}\right)^2 + \left(\frac{\sqrt{3}}{4}\right)^2} = \sqrt{\frac{4}{16}} = \frac{1}{2}$, so $\left(\frac{1}{2}, \frac{\pi}{3}\right)$

    (c) $\theta = \arctan\frac{6}{-6\sqrt{3}} + \pi = \arctan\left(-\frac{1}{\sqrt{3}}\right) + \pi = -\frac{\pi}{6} + \pi = \frac{5\pi}{6}, r = \sqrt{(-6\sqrt{3})^2 + 6^2} = \sqrt{108 + 36} = \sqrt{144} = 12$, so $\left(12, \frac{5\pi}{6}\right)$

    (d) $\theta = \arctan\frac{4\sqrt{2}}{-4\sqrt{2}} + \pi = \arctan(-1) + \pi = -\frac{\pi}{4} + \pi = \frac{3\pi}{4}, r = 4\sqrt{2}\cdot\sqrt{2} = 8$, so $\left(8, \frac{3\pi}{4}\right)$

    (e) $\theta = \arctan\frac{-\sqrt{3}}{-1} + \pi = \arctan\sqrt{3} + \pi = \frac{\pi}{3} + \pi = \frac{4\pi}{3}, r = \sqrt{(-1)^2 + (\sqrt{3})^2} = \sqrt{4} = 2$, so $\left(2, \frac{4\pi}{3}\right)$

    (f) $\left(5, \frac{\pi}{7}\right)$

3. The domain restriction of $-\frac{\pi}{2} \leq \theta \leq \frac{\pi}{2}$ for $\tan x$ to have an inverse means the range of $\arctan x$ is restricted to only angles in Quadrant I and Quadrant IV. The period of $\tan\theta$ is π, though, so adding π to $\arctan x$ will yield an angle θ in either Quadrant II or Quadrant III with the same output of $\tan\theta$. These are precisely the quadrants where $x < 0$.

4. Kisha did not pay attention to the restriction $0 \leq \theta < 2\pi$. She simply needs to add 2π to her angle to get $\theta = -\frac{\pi}{3} + 2\pi = \frac{5\pi}{3}$.

5. These points are on opposite ends of a line through the origin making an angle of $\frac{5\pi}{12}$ with the positive x-axis. Therefore, the distance between them is simply the sum of the radii, or $4 + 4 = 8$.

6. $(r, \theta + \alpha)$

Reflection

Suggested Class Time.

Prerequisites. Comfort with unit circle and very preferably special right triangles

Instructional Strategies. By and large, students were fine with the lesson. I introduced it to students at the beginning, talking about the fact that rectangular coordinates are actually somewhat awkward: who would ever describe going from point A to point B in terms of horizontal and vertical distances? What about straight-line paths? I also went ahead and prefaced the complex plane. It was important for me to tell students that imaginary numbers aren't some set of "other" numbers; adding in the imaginaries to the reals creates the complex numbers, which include every number the students ever learned. We defined the real and imaginary parts and discussed how having two components allows for graphing.

During the activity, students were definitely under the impression they could use their calculator for more than they should have. I had to remind them, unfortunately, that $4 \cos \frac{2\pi}{3}$ is not, in fact, a calculator-active computation.

As for the conversions from rectangular to polar... I am conflicted. The College Board seems to want to de-emphasize special right triangles and focus on periodic behavior, but rectangular-to-polar conversions are *so much* easier using pattern recognition and special right triangles. I made a point of showing my students multiple examples after the activity was completed of using this kind of recognition.

For instance, I had them convert $(-\sqrt{5}, \sqrt{5})$ to polar, and we quickly (a) drew the point in the plane and (b) finished our conversion with

$$\left(\sqrt{5} \cdot \sqrt{2}, \frac{3\pi}{4}\right) \rightarrow \left(\sqrt{10}, \frac{3\pi}{4}\right)$$

For another example, we did $\left(-6\sqrt{3}, -6\right)$, to which we (a) observed the $\sqrt{3}$ was part of the horizontal component, thereby making a reference angle of $\frac{\pi}{6}$ appropriate, (b) drew the point in the plane, and (c) deduced that $\left(-\frac{\sqrt{3}}{2}, \frac{1}{2}\right)$ would need to be scaled by a factor of 12 to get the given point. We finished with $\left(12, \frac{5\pi}{6}\right)$.

I also emphasized how negative radii work a bit more than the lesson might indicate, as I'm already foreseeing how unnatural (or easy to forget) it will be for students in the following Topics.

To aid students, I created a supplemental Delta Math assignment. While some Delta Math assignments are very limited (degrees only, no general solutions, etc.), the conversions for polar/rectangular are outstanding.

Technology. Students needed their calculators for certain questions.

Problems.

1. Values of the polar function $r = f(\theta)$ are provided in the table for certain angles θ.

θ (rad.)	1	3	4	6
$r = f(\theta)$	-2	4	-1	-3

   For which of the angles θ will the point (r, θ) on the graph of $r = f(\theta)$ lie in Quadrant III?
   (A) $\theta = 1$ **(B)** $\theta = 3$ **(C)** $\theta = 4$ **(D)** $\theta = 6$

2. The complex number z can be expressed as $z = \sqrt{3} - i$. Which of the following is an expression for z using polar coordinates?

   (A) $z = \cos\left(-\frac{\pi}{6}\right) + i\left(\sin\left(-\frac{\pi}{6}\right)\right)$

   (B) $z = \cos\left(-\frac{\pi}{3}\right) + i\left(\sin\left(-\frac{\pi}{3}\right)\right)$

   (C) $z = 2\cos\left(-\frac{\pi}{6}\right) + i\left(2\sin\left(-\frac{\pi}{6}\right)\right)$

   (D) $z = 2\cos\left(-\frac{\pi}{3}\right) + i\left(2\sin\left(-\frac{\pi}{3}\right)\right)$

3. Point P lies on the line $y = -x$ in the rectangular coordinate plane. The x-coordinate of point P is $x = -\sqrt{2}$. Which of the following is the representation of point P in the polar plane?

   (A) $\left(\sqrt{2}, \frac{\pi}{4}\right)$ **(B)** $\left(\sqrt{2}, \frac{3\pi}{4}\right)$ **(C)** $\left(2, \frac{\pi}{4}\right)$ **(D)** $\left(2, \frac{3\pi}{4}\right)$

4. A point P is graphed in the polar plane, as shown below.

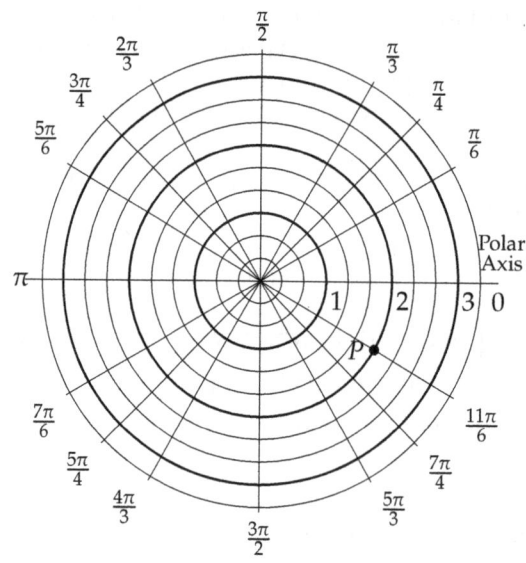

   Which of the following is the representation of P in the rectangular coordinate plane?
   (A) $\left(1, \sqrt{3}\right)$ **(B)** $\left(1, -\sqrt{3}\right)$ **(C)** $\left(\sqrt{3}, 1\right)$ **(D)** $\left(\sqrt{3}, -1\right)$

5. (■) The complex number z can be expressed as $z = -4 + 3i$. Which of the following is the value of θ in the polar representation of z?

   (A) -3.7851 **(B)** -0.6435 **(C)** 2.4981 **(D)** 5

Solutions.

1. An angle of 1 radian lies in Quadrant I, but the associated radius is -2. This places the point (r, θ) in the diametrically opposite Quadrant III. The correct answer is therefore **(A)**.

2. We have $r = \sqrt{(\sqrt{3})^2 + (-1)^2} = \sqrt{3+1} = \sqrt{4} = 2$ and $\theta = \arctan(-1/\sqrt{3}) = -\pi/6$. Then $z = 2\cos(-\pi/6) + i(2\sin(-\pi/6))$. The correct answer is therefore **(C)**.

3. The y-coordinate of P is $y = \sqrt{2}$. This point lies in the second quadrant, so $\theta = \arctan(-\sqrt{2}/\sqrt{2}) = \arctan(-1) + \pi = -\pi/4 + \pi = 3\pi/4$. Also, $r = \sqrt{(-\sqrt{2})^2 + (-\sqrt{2})^2} = \sqrt{2+2} = \sqrt{4} = 2$. The polar representation of P is $(2, 3\pi/4)$. The correct answer is therefore **(D)**.

4. We have $x = r\cos\theta = 2\cos(11\pi/6) = 2 \cdot \sqrt{3}/2 = \sqrt{3}$ and $y = r\sin\theta = 2\sin(11\pi/6) = 2 \cdot (-1/2) = -1$. The coordinates are $(\sqrt{3}, -1)$. The correct answer is therefore **(D)**.

5. The angle is $\theta = \arctan(-3/4) + \pi = -0.6435 + 3.14159 = 2.4981$. The correct answer is therefore **(C)**.

Topic 3.14 ~ Polar Function Graphs

Learning Objectives

1. (3.14.A) Construct graphs of polar functions.

Success Criteria

1. I can plot points in the polar coordinate system given a polar function.
2. I can graph a function in the polar coordinate system.
3. I can analyze the behavior of a polar function to predict the behavior of its graph in the polar coordinate system.
4. I can use the graph of a polar function to predict its analytical form.

For who knows how many years now, you have worked with one kind of function, where inputs and outputs, no matter what they represent, end up being horizontal distances x and vertical distances y in this grid we call the xy-, or Cartesian, plane. Now, however, we have a new type of coordinate system – the _____ system – which means we can graph an entirely new type of function.

Definition: A _____ $r = f(\theta)$ takes _____ as inputs and outputs _____

_____.

Let's look at an example of how polar functions differ from rectangular functions. Let $f(x) = x$.

Example 1: On the left, we'll graph $f(x) = x$ in the xy-plane for $0 \leq x \leq \pi$. On the right, we'll graph $r = f(\theta)$, where $f(\theta) = \theta$ in the polar plane.

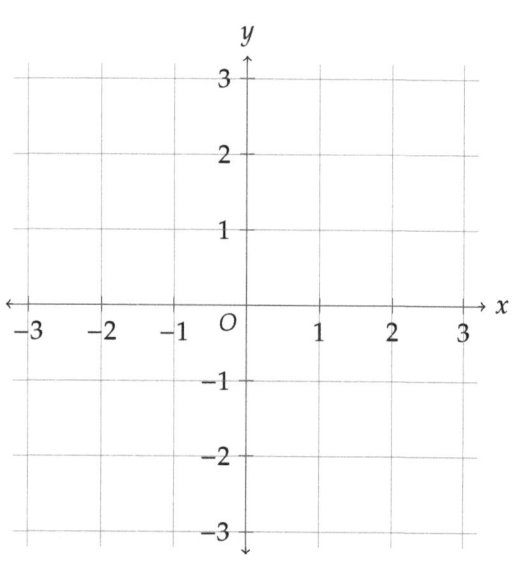 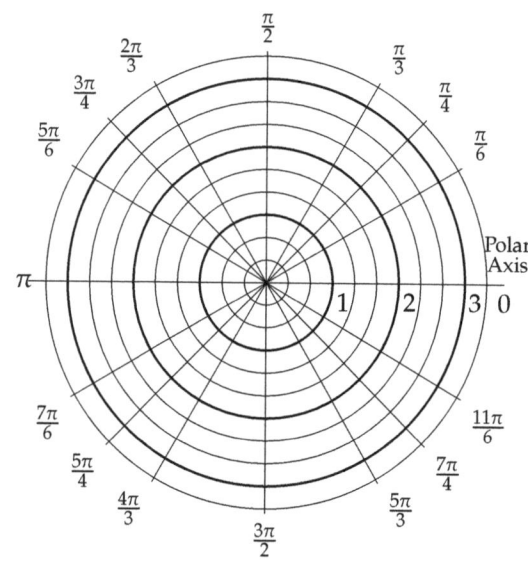

In the xy-plane, $f(x) = x$ means that the output is equal to the input: this makes the vertical displacement equal to the horizontal displacement. On the polar plane, though, $r = \theta$ means that the output – the _____ – will equal the input, the _____.

Practice

(a) What does the graph of $y = 2$ look like in the xy-plane?

(b) What do you think the graph of $r = 2$ looks like in the polar plane?

Let's graph another polar function.

Example 2: Let $f(\theta) = \cos\theta + 1$. Sketch a graph of the polar function $r = f(\theta)$.

Solution: It helps to make a table of values and then sketch our curve.

θ	$r(\theta) = \cos(\theta) + 1$
0	
$\frac{\pi}{3}$	
$\frac{\pi}{2}$	
$\frac{2\pi}{3}$	
π	
$\frac{4\pi}{3}$	
$\frac{3\pi}{2}$	
$\frac{5\pi}{3}$	
2π	

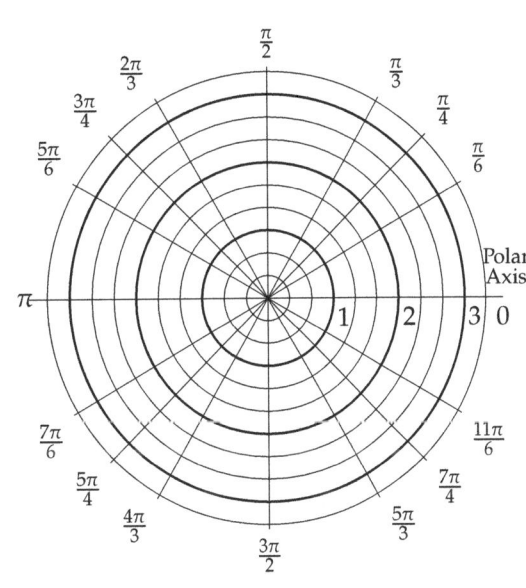

Drawing precise points can be a bit challenging when the outputs are irrational like $\frac{\sqrt{2}}{2}$ or $\frac{\sqrt{3}}{2}$, but you can typically find at least 5 angles that will have outputs of 0, ±1, or at worst, ±$\frac{1}{2}$. This is why we chose the angles $\frac{\pi}{3}$, $\frac{2\pi}{3}$, and so on in the example.

Example 3: Shown is a graph of the polar function $r = f(\theta)$ for $0 \leq \theta \leq 2\pi$. Write an expression for $f(\theta)$.

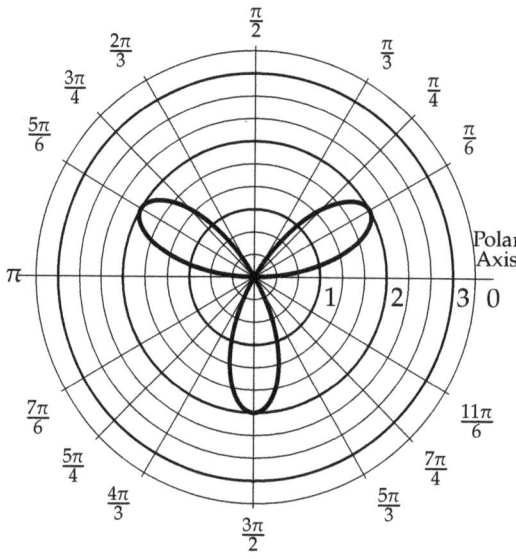

Solution:

Practice 1: Let $f(x) = \cos(2x)$. Sketch a graph of the polar function $r = f(\theta)$ on the polar coordinate system below.

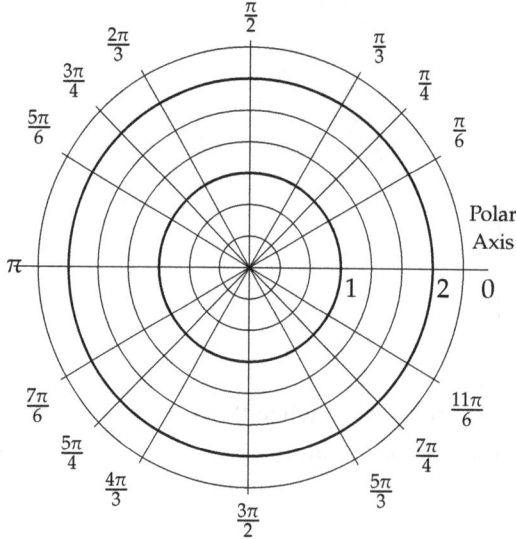

Practice 2: Shown below is the graph of a function $y = f(x)$ in the xy-plane with $0 \leq x \leq 2\pi$. Use the graph from f to create a graph of the polar function $r = f(\theta)$ for $0 \leq \theta \leq 2\pi$.

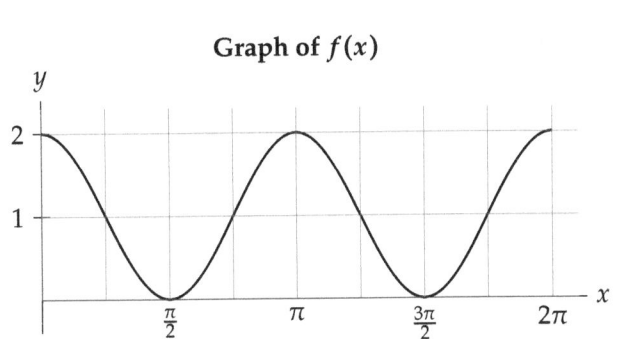

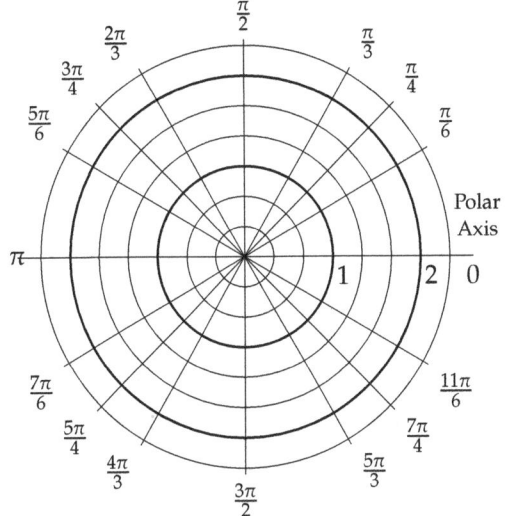

Topic 3.14 Homework

1. Sketch a graph of each of the polar functions below on the given domains.

   (a) $r = 2\sin(\theta), 0 \leq \theta \leq 2\pi$

   (b) $r = \frac{\theta}{2}, 0 \leq \theta \leq \pi$

   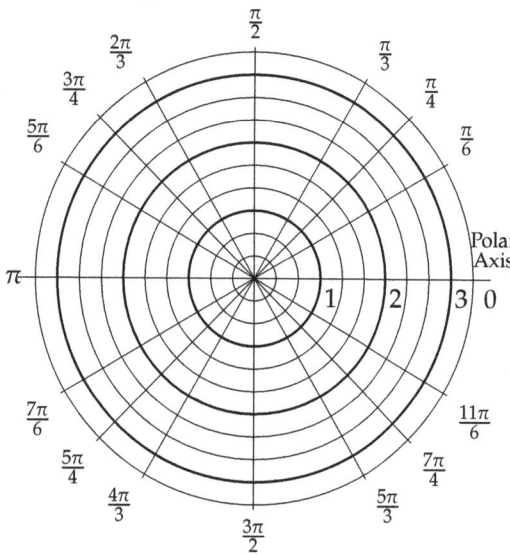
   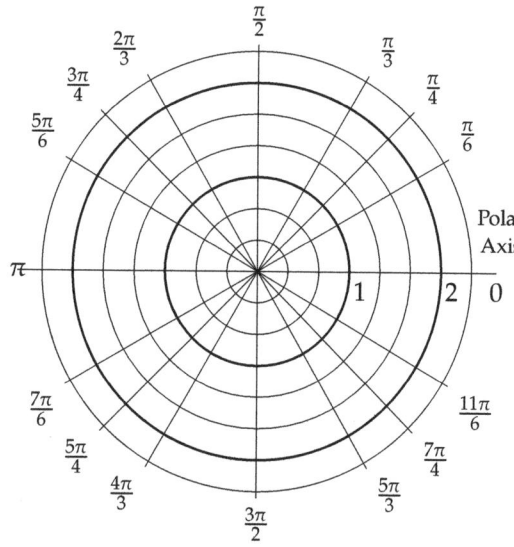

   (c) $r = 3\sin(2\theta), 0 \leq \theta \leq 2\pi$

   (d) $r = \frac{\pi}{\theta}, \pi \leq \theta \leq 2\pi$

   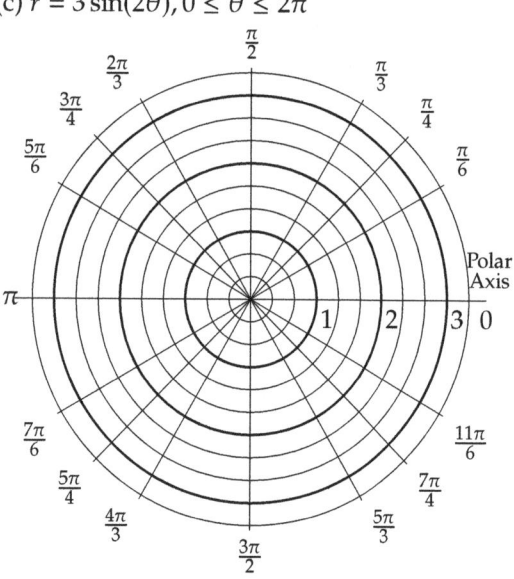
   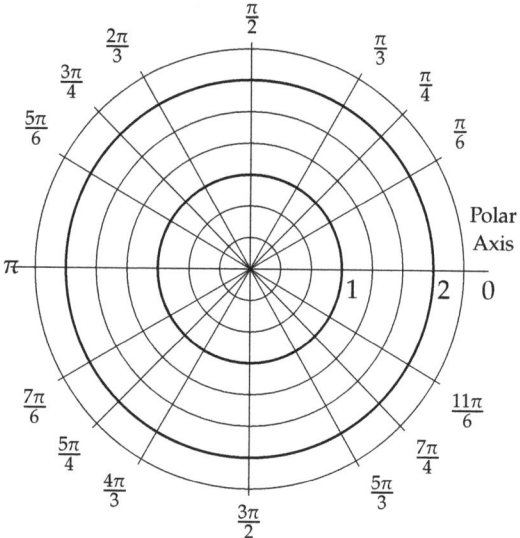

2. The graphs of the polar functions $r = \sin(\theta)$ and $r = \cos(\theta)$ make circles with a radius of $\frac{1}{2}$ centered at $\left(0, \frac{1}{2}\right)$ and $\left(\frac{1}{2}, 0\right)$, respectively.

   (a) Without any computations, predict what the graph of $r = \sin(-\theta)$ would look like in the polar coordinate system.

   (b) Without any computations, predict what the graph of $r = \cos(-\theta)$ would look like in the polar coordinate system.

3. Let f, g, h, and j be the functions

$$f(\theta) = \sin\left(\frac{\theta}{2}\right) \qquad g(\theta) = 1 - \cos(\theta) \qquad h(\theta) = 2\cos(3\theta) \qquad j(\theta) = 2 + \cos(2\theta)$$

Shown below are the graphs of $r = f(\theta), r = g(\theta), r = h(\theta)$, and $r = j(\theta)$ for $0 \leq \theta \leq 2\pi$. Determine which graph matches each function.

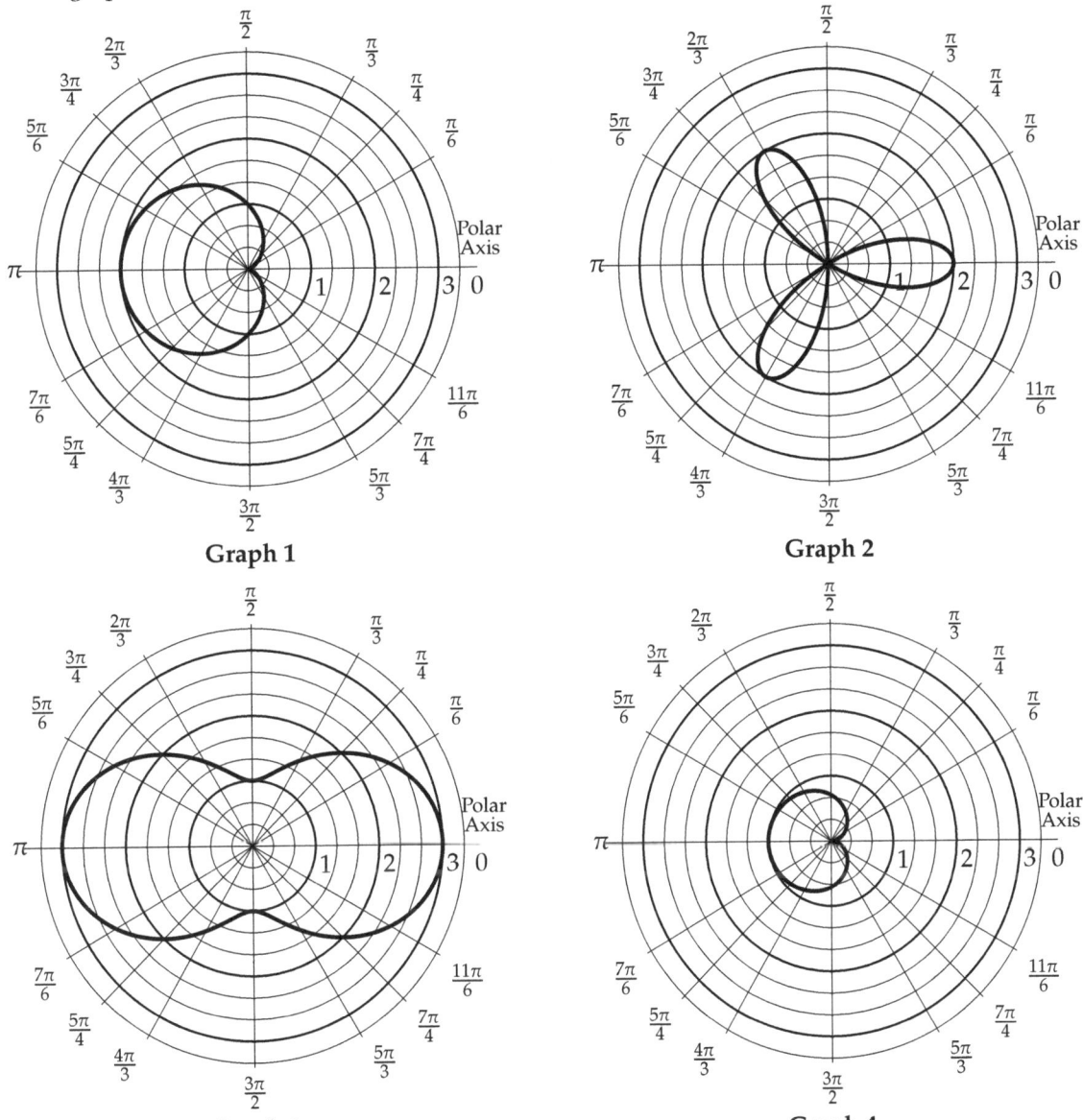

Graph 1

Graph 2

Graph 3

Graph 4

Topic 3.14 Solutions/Notes

Activity

Now, however, we have a new type of coordinate system – the polar system – which means we can graph an entirely new type of function.

Definition: A polar function $r = f(\theta)$ takes angles as inputs and outputs displacements from the origin.

Solution 1: On the left, we'll graph $f(x) = x$ in the xy-plane for $0 \leq x \leq \pi$. On the right, we'll graph $r = f(\theta)$, where $f(\theta) = \theta$ in the polar plane.

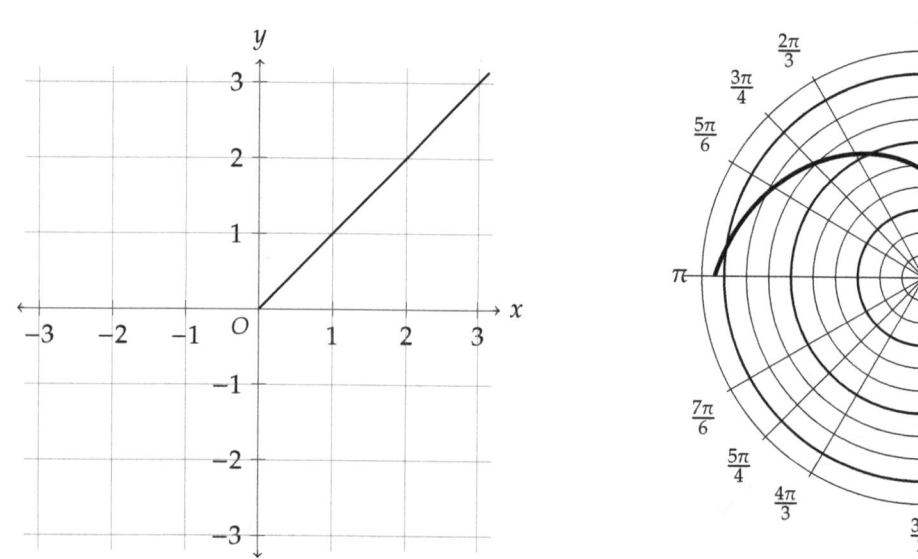

On the polar plane, though, $r = \theta$ means that the output – the displacement from the origin – will equal the input, the angle.

Practice Solutions

(a) The graph of $y = 2$ is a horizontal line, where the vertical displacement from the x-axis is always equal to 2.

(b) The graph of $r = 2$ will keep a constant displacement from the origin of 2, regardless of the angle. This will form a circle of radius 2.

Solution 2: It helps to make a table of values and then sketch our curve.

θ	$r(\theta) = \cos(\theta) + 1$
0	$1 + 1 = 2$
$\frac{\pi}{3}$	$\frac{1}{2} + 1 = \frac{3}{2}$
$\frac{\pi}{2}$	$0 + 1 = 1$
$\frac{2\pi}{3}$	$-\frac{1}{2} + 1 = \frac{1}{2}$
π	$-1 + 1 = 0$
$\frac{4\pi}{3}$	$-\frac{1}{2} + 1 = \frac{1}{2}$
$\frac{3\pi}{2}$	$0 + 1 = 1$
$\frac{5\pi}{3}$	$\frac{1}{2} + 1 = \frac{3}{2}$
2π	$1 + 1 = 2$

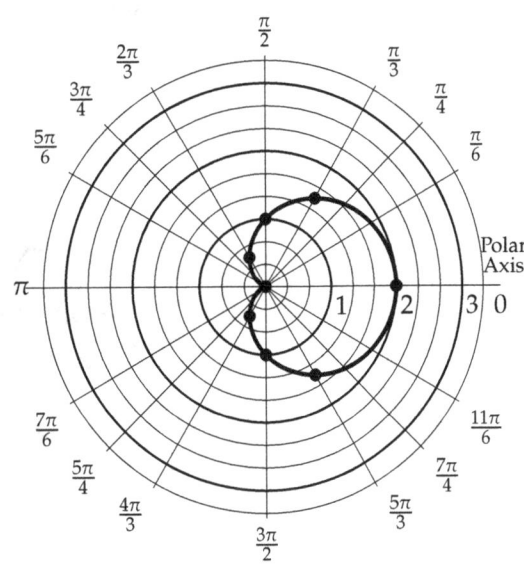

Drawing precise points can be a bit challenging when the outputs are irrational like $\frac{\sqrt{2}}{2}$ or $\frac{\sqrt{3}}{2}$, but you can typically find at least 5 angles that will have outputs of 0, ±1, or at worst, $\pm\frac{1}{2}$. This is why we chose the angles $\frac{\pi}{3}$, $\frac{2\pi}{3}$, and so on in the example.

Solution 3:

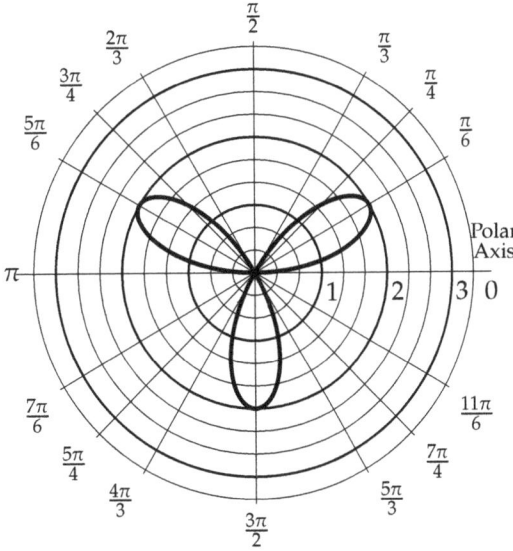

At $\theta = 0$, we have $r = 0$, which possibly indicates a sine function, since $\sin 0 = 0$. Now, the spiral repeats the same radial pattern 3 times, indicating a period of $\frac{2\pi}{3}$. This leads us to suspect $f(\theta) = \sin(3\theta)$. However, the maximum distance from the origin is 2, which would make $f(\theta) = 2\sin(3\theta)$. If we test out a value, like $\frac{\pi}{6}$, we get $f\left(\frac{\pi}{6}\right) = 2\sin\left(3 \cdot \frac{\pi}{6}\right) = 2\sin\left(\frac{\pi}{2}\right) = 2(1) = 2$. This agrees with our graph. We can test more points, but we will continue to get confirmation that $f(\theta) = 2\sin(3\theta)$.

Practice 1 Solution: We can make a table of values again. In order to get convenient values of r, we will use multiples of $\frac{\pi}{2}$ and $\frac{\pi}{6}$ (since $2 \cdot \frac{\pi}{6} = \frac{\pi}{3}$ and $\cos\left(\frac{\pi}{3}\right) = \frac{1}{2}$). Because the period of $\cos(2x)$ is $\frac{2\pi}{2}$, we can just use the outputs from the first period and replicate them for the second period (from π to 2π).

θ	$r = \cos(2\theta)$
0	1
$\frac{\pi}{6}$	$\frac{1}{2}$
$\frac{\pi}{2}$	-1
$\frac{5\pi}{6}$	$-\frac{1}{2}$
π	1
$\frac{7\pi}{6}$	$\frac{1}{2}$
$\frac{3\pi}{2}$	-1
$\frac{11\pi}{6}$	$-\frac{1}{2}$
2π	1

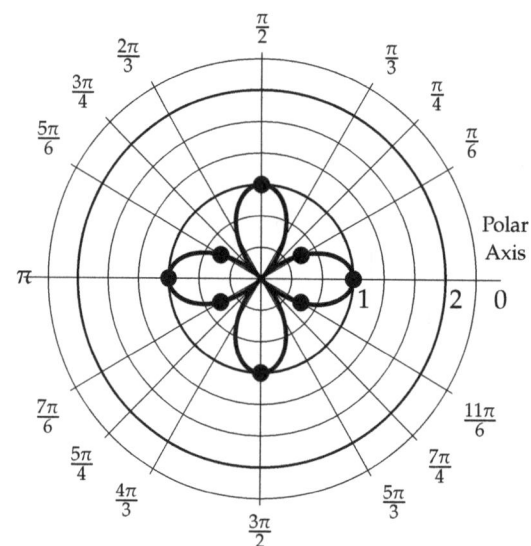

Practice 2 Solution: Shown below is the graph of a function $y = f(x)$ in the xy-plane with $0 \leq x \leq 2\pi$. Use the graph from f to create a graph of the polar function $r = f(\theta)$ for $0 \leq \theta \leq 2\pi$.

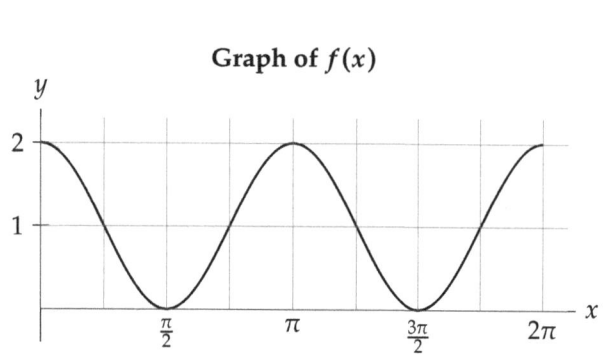
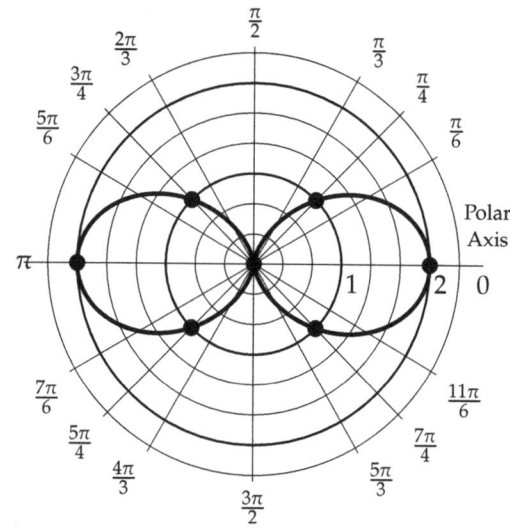

UNIT 3   TOPIC 3.14 ~ POLAR FUNCTION GRAPHS   **419**

Homework

1. Sketch a graph of each of the polar functions below on the given domains.

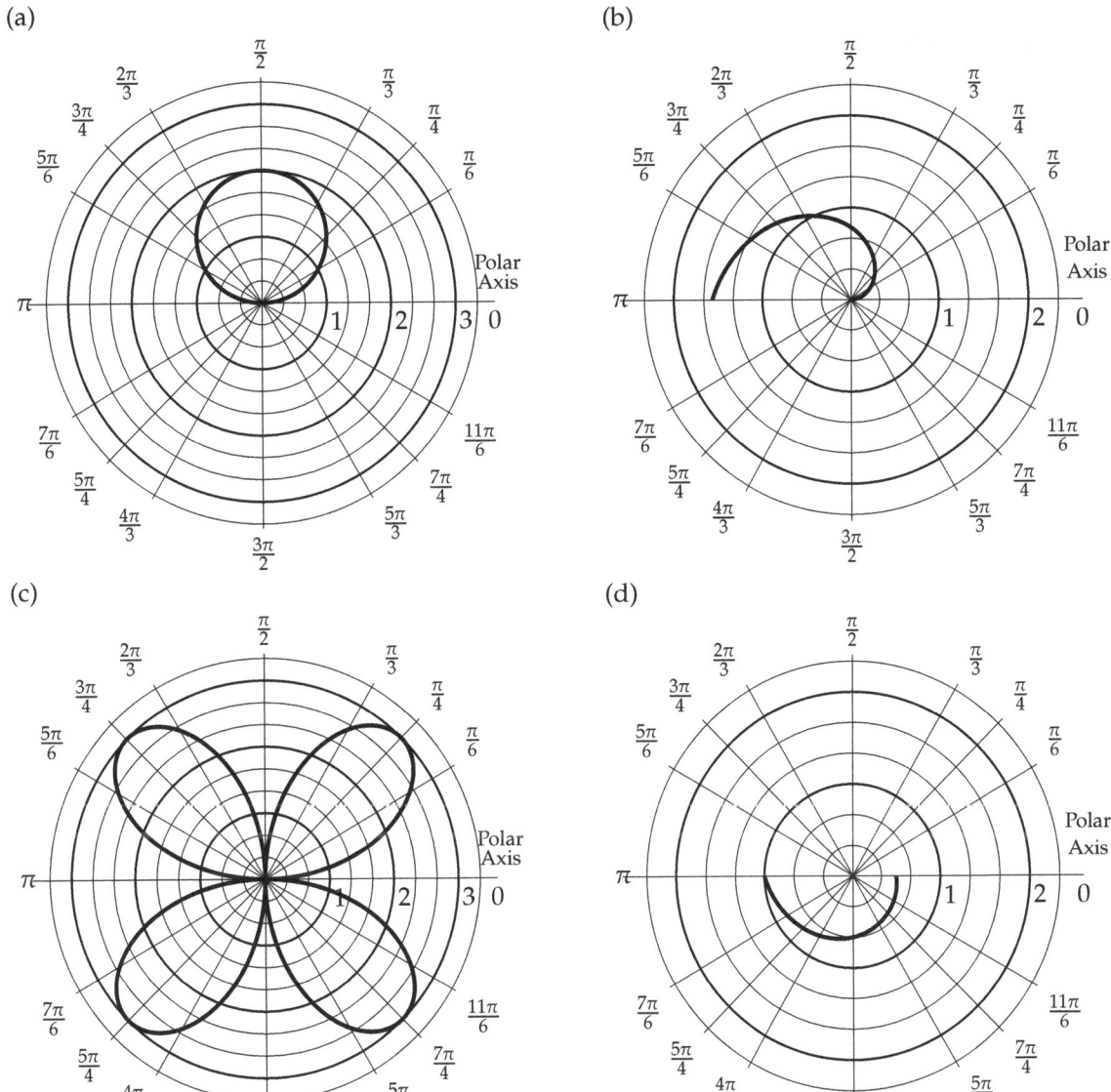

2. (a) Because $\sin\theta$ is odd, we have $\sin(-\theta) = -\sin(\theta)$. Therefore, for any angle θ, any point on the graph of $r = \sin(-\theta)$ will have displacement in the opposite direction of the corresponding point on the graph of $r = \sin(\theta)$. Since the original graph is a circle sitting atop the origin centered at $\left(0, \frac{1}{2}\right)$, the graph of $r = \sin(-\theta)$ will be a circle sitting directly *below* the origin, centered at $\left(0, -\frac{1}{2}\right)$.

   (b) Because $\cos x$ is even, we have $\cos(\theta) = \cos(-\theta)$ for any x. Therefore, the graph of $r = \cos(-\theta)$ is identical to the graph of $r = \cos(\theta)$, making a circle with radius $\frac{1}{2}$ centered at $\left(\frac{1}{2}, 0\right)$.

3. Graph 1 is $r = g(\theta)$; Graph 2 is $h(\theta)$; Graph 3 is $j(\theta)$; and Graph 4 is $f(\theta)$.

Reflection

Suggested Class Time. 60 minutes

Prerequisites. Students must be proficient with the unit circle (i.e. evaluating trigonometric functions).

Instructional Strategies. In class, my own explanations almost exactly mirrored those provided in the Solutions. That said, I did work very slowly through Example 1: I wanted to emphasize the connection between the function in the Cartesian plane and the polar plane. It feels like half the battle on most of the AP multiple choice questions is simply understanding what the inputs and outputs represent in a polar function.

Examples 2 and 3 went well enough – not much to say here.

Example 4, admittedly, took forever. Students were making mistakes left and right with what values to test out: a few were using facts like $\cos \frac{\pi}{3} = \frac{1}{2}$ and then, instead of recognizing that this would make $x = \frac{\pi}{6}$ an ideal input, would write something like $f\left(\frac{\pi}{3}\right) = 2 \cdot \frac{1}{2}$ - doubling the output. Another student used $\cos \frac{\pi}{3} = \frac{1}{2}$ and somehow decided that $\cos \frac{2\pi}{3} = \frac{1}{2}$ as well.

As I walked around the room, I realized that not all students will graph polar functions the same way. In one group, they had sketched the polar graph perfectly when most other students had at most 2 or 3 points. The former students clearly understood using my "method" of finding when the radius is 0, maximized/minimized, and using convenient values. The latter students, in some cases, did best with actually working their way around the circle, one angle at a time. They needed to see the function come to life, working slowly through angles of $0, \frac{\pi}{6}, \frac{\pi}{3}$, and so on.

In an ideal world, I'd spend an entire class day having students just graph polar functions - from an analytical representation, from a given graph in the Cartesian plane, and from a table. Unfortunately, I don't have time for this in my own class (or I find that time could better be spent elsewhere).

One comfort I found was that, given 3 Topic Questions from AP Classroom at the end of class, students did very well. I don't love teaching to the test, but for such complex material to be covered in such a short period of time, I could do worse than to teach to the test while also emphasizing the important concepts (the covariation of the inputs and outputs and what these represent).

Technology. None was required, though I did draw some graphs in Geogebra.

Problems.

1. The graph of a polar function $r = g(\theta)$ contains the point A, where the polar coordinates of A are given by $\left(-4, -\frac{2\pi}{3}\right)$. In which quadrant does point A lie, along with correct reasoning?

    (A) Because the terminal ray of θ is in Quadrant III and $r < 0$, the point A lies in Quadrant I.

    (B) Because the terminal ray of θ is in Quadrant III and $r < 0$, the point A lies in Quadrant III.

    (C) Because the terminal ray of θ is in Quadrant II and $r < 0$, the point A lies in Quadrant II.

    (D) Because the terminal ray of θ is in Quadrant II and $r < 0$, the point A lies in Quadrant IV.

2. The polar function $r = f(\theta)$ is graphed in the polar coordinate system in the figure.

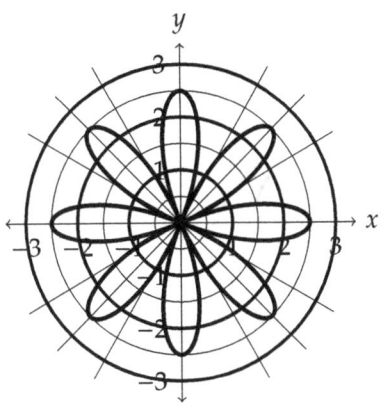

Which of the following could be an expression for $f(\theta)$?

(A) $f(\theta) = \dfrac{5}{2}\cos\left(\dfrac{\theta}{4}\right)$ **(B)** $f(\theta) = \dfrac{5}{2}\sin\left(\dfrac{\theta}{4}\right)$ **(C)** $f(\theta) = \dfrac{5}{2}\cos(4\theta)$ **(D)** $f(\theta) = \dfrac{5}{2}\sin(4\theta)$

3. The figure below shows the graph of the polar function $r = f(\theta)$, where $f(\theta) = \cos(2\theta)(2 - \sin\theta)$, in the polar coordinate system for $0 \le \theta \le 2\pi$. There are five points labeled A, B, C, D, and E.

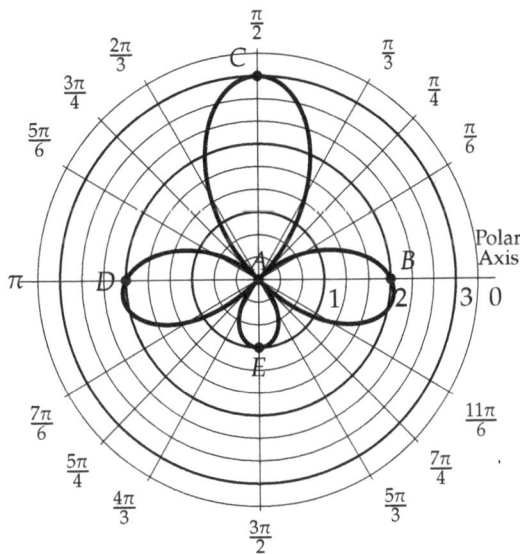

If the domain of f is restricted to $\pi \le \theta \le \dfrac{3\pi}{2}$, the portion of the given graph that remains consists of two pieces. One is the portion of the graph in Quadrant III from D to A. Which of the following describes the other remaining piece?

(A) The portion of the graph in Quadrant I from A to B

(B) The portion of the graph in Quadrant I from A to C

(C) The portion of the graph in Quadrant II from A to D

(D) The portion of the graph in Quadrant IV from A to E

4. Which of the following is the graph of the polar function $r = f(\theta)$, where $f(\theta) = 1 + 2\sin\theta$, in the polar coordinate system for $0 \leq \theta \leq 2\pi$?

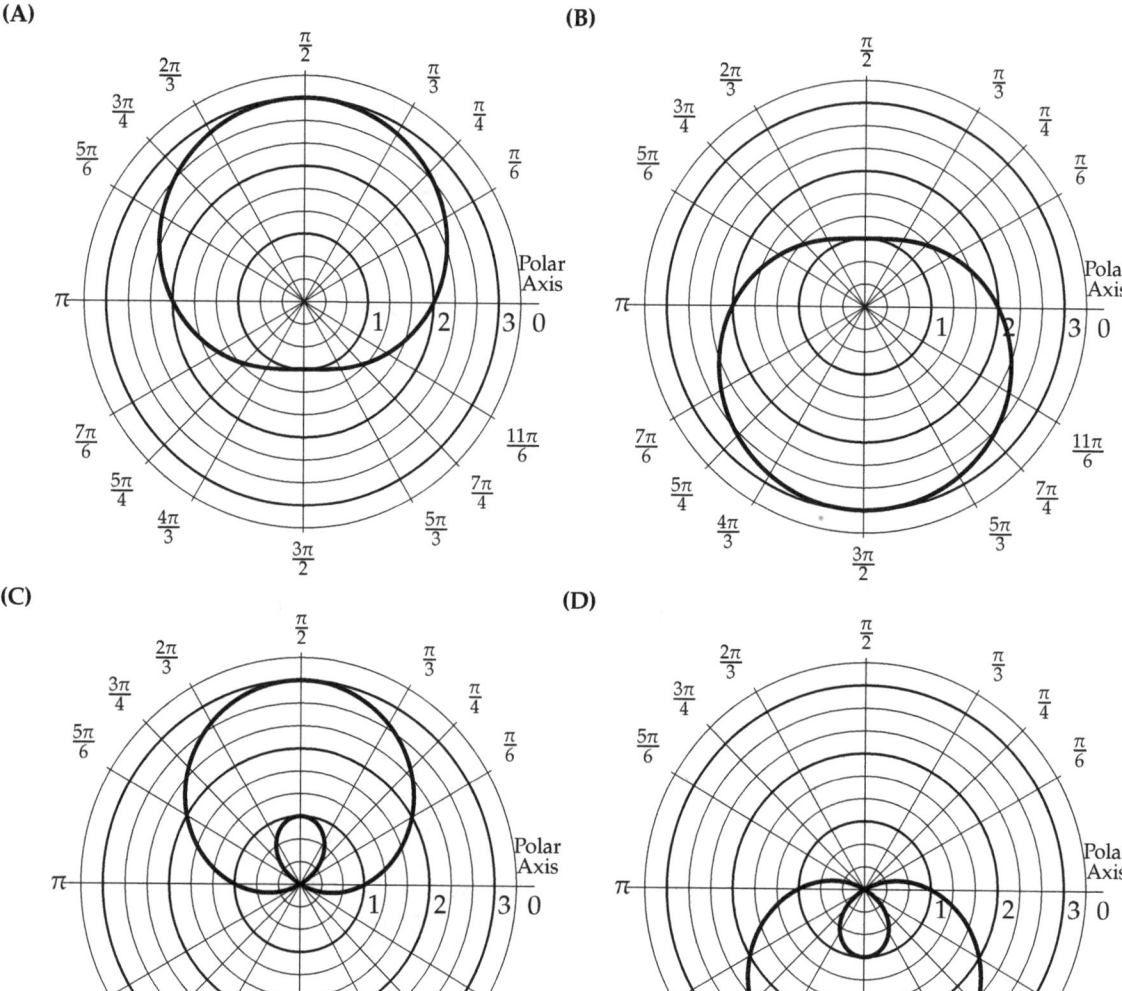

5. The graph of the polar function $r = f(\theta)$, where $f(\theta) = -2 + \cos\theta$, in the polar coordinate system for $0 \leq \theta \leq 2\pi$ is shown below.

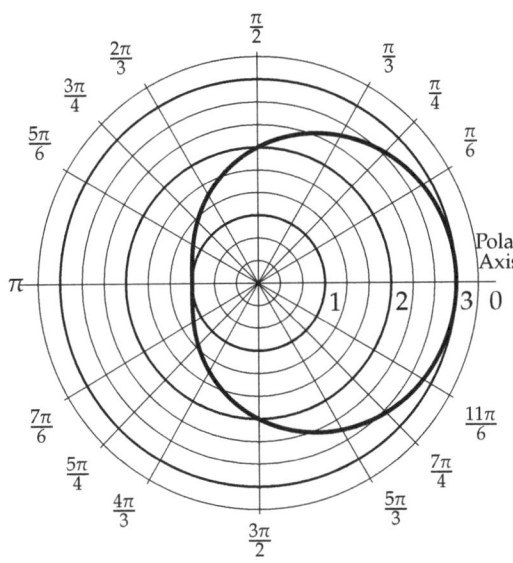

Which of the following polar functions $r = g(\theta)$, when graphed in the polar coordinate system for $0 \leq \theta \leq 2\pi$, produces the same graph?

(A) $g(\theta) = 2 - \cos\theta$ (B) $g(\theta) = 2 + \cos\theta$ (C) $g(\theta) = 1 - 2\cos\theta$ (D) $g(\theta) = 1 + 2\cos\theta$

Solutions.

1. The terminal ray of $\theta = -\frac{2\pi}{3}$ is in the third quadrant, and $r < 0$, so the the point is diametrically opposite from $(4, -2\pi/3)$ and lies in the first quadrant. The correct answer is therefore **(A)**.

2. The function chosen must pass through the polar points $(5/2, 0)$ and $(5/2, \pi)$. Only one of the choices does this. The correct answer is therefore **(C)**.

3. We calculate $r = f(3\pi/2) = \cos(3\pi)(2 - \sin 3\pi/2) = -1(2 - (-1)) = -3$. This is the polar point $(-3, 3\pi/2)$, which is labeled as C. The correct answer is therefore **(B)**.

4. We have $r = f(0) = 1$, $r = f(\pi/2) = 3$, and $r = f(\pi) = 1$ so the graph of f passes through the polar points $(1, 0), (3, \pi/2)$, and $(1, \pi)$. There is only one graph that passes through all three of these points. The correct answer is therefore **(C)**.

5. Let's consider two simple values of θ. We see that $r = f(0) = -1, r = f(\pi) = -3$, giving us the polar points $(-1, 0)$ and $(-3, \pi)$. The other representations of these points in the polar coordinate system are $(1, \pi)$ and $(3, 0)$, respectively. Hence, we want the graph of the function g to pass through these both of these points. This happens for $g(\theta) = 2 + \cos\theta$. The correct answer is therefore **(B)**.

Topic 3.15 ~ Rates of Change in Polar Functions

Learning Objectives
1. (3.15.A) Describe characteristics of the graph of a polar function.

Success Criteria
1. I can use rates of change of a function to describe how the graph of a polar function is changing and predict values within a given interval.

In Topic 3.14, we introduced polar functions. These functions take inputs of angles in radians and output displacements from the origin. Now, we're going to look at *rates of change* of polar functions.

1. On the xy-plane below, sketch a graph of $f(\theta) = \sin(2\theta)$.

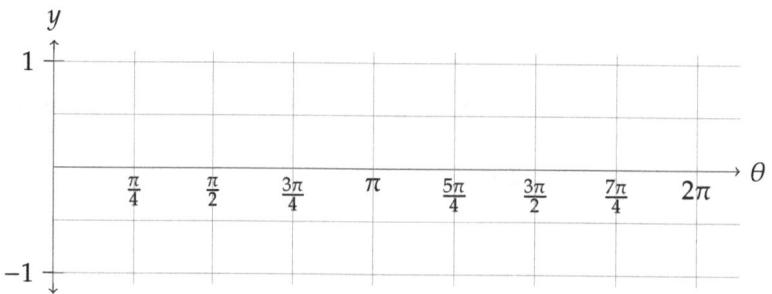

2. Fill in the table below to describe the outputs of $f(\theta) = \sin(2\theta)$ on each interval. **Fill out only the 2nd and 3rd columns.**

Interval	Sign of sin(2x) (+/-)?	Increasing/Decreasing?	Dist. from origin in $r = f(\theta)$ increasing/decreasing?
$0 < \theta < \frac{\pi}{4}$			
$\frac{\pi}{4} < \theta < \frac{\pi}{2}$			
$\frac{\pi}{2} < \theta < \frac{3\pi}{4}$			
$\frac{3\pi}{4} < \theta < \pi$			
$\pi < \theta < \frac{5\pi}{4}$			
$\frac{5\pi}{4} < \theta < \frac{3\pi}{2}$			
$\frac{3\pi}{2} < \theta < \frac{7\pi}{4}$			
$\frac{7\pi}{4} < \theta < 2\pi$			

3. Go to the Geogebra link at https://www.geogebra.org/m/hp6r3cqn. On the left will be the graph of $f(\theta) = \sin(2\theta)$. On the right will be the graph of $r = f(\theta)$.

4. Paying close attention to the x-axis (θ-axis), drag θ to the right until the graph of $f(\theta) = \sin(2\theta)$ is at its first peak - this will occur at $\theta = \frac{\pi}{4}$.

5. As you drag from $\theta = 0$ to $\theta = \frac{\pi}{4}$ (~ 0.79), what happens to the point on the polar function? Does its distance from the origin increase or decrease?

6. Fill in the 4th column of the table for the interval $0 < \theta < \frac{\pi}{4}$.

7. Now, from $\theta = \frac{\pi}{4}$, drag θ until the graph of $f(\theta) = \sin(2\theta)$ reaches the x-axis. This should occur at $\theta = \frac{\pi}{2}$ (~ 1.57).

8. As you drag from $\theta = \frac{\pi}{4}$ to $\theta = \frac{\pi}{2}$ (~ 1.57), what happens to the point on the polar function? Does its distance from the origin increase or decrease?

9. Fill in the 4th column of the table for the interval $\frac{\pi}{4} < \theta < \frac{\pi}{2}$.

10. Repeat the process of Problems 4-6 and Problems 7-9 for each of the remaining intervals in the table until you have completed the fourth column of the table.

11. You should notice a pattern in when the rate of change of the *polar function* is increasing or decreasing. Try to fill in the following.

Rates of Change of Polar Functions

Let $r = f(\theta)$ be a polar function with graph consisting of the points $(f(\theta), \theta)$. Then
- the distance from the origin will be <u>increasing</u> when $f(\theta)$ is
    ◊ _____ and _____, OR
    ◊ _____ and _____.
- The distance from the origin will be <u>decreasing</u> when $f(\theta)$ is
    ◊ _____ and _____, OR
    ◊ _____ and _____.

---

12. Try a couple of practice questions.

   (a) Consider the polar function $r = f(\theta)$, where $f(\theta) = (\theta + 1)(\theta - 3)$. For θ in the interval $1 < \theta < 3$, will the point $(f(\theta), \theta)$ on the graph of the polar function be getting closer to or farther from the origin in the polar coordinate system?

   (b) Consider the graph of $y = f(x)$ shown below.

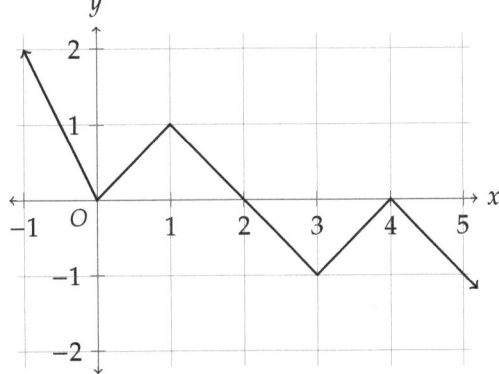

   The polar function r is given by $r = f(\theta)$. On what interval/s of $x = \theta$ will the point $(f(\theta), \theta)$ be getting farther from the origin?

We can also use the extrema of functions to predict when points will be the farthest from or closest to the origin.

13. Consider the function $f(\theta) = -\frac{1}{4}(\theta - 1)(\theta - 3)$ on the interval $0 \leq \theta \leq 5$.

    (a) For what value/s of θ will $f(\theta) = 0$?

    (b) For what value/s of θ will $f(\theta)$ reach a local or global maximum? How about a local or global minimum?

14. Go to the Geogebra sketch. Drag α back to 0. Then, input $f(\theta)$ into the box for $f(x) =$ (use x instead of θ).

15. Predict the angles at which the point $(f(\theta), \theta)$ will be the farthest from the origin and will be the closest to the origin.

16. Check your previous answer by dragging α accordingly (don't forget $0 \leq \alpha \leq 5$). Were you correct?

Notes

Topic 3.15 Homework

1. Each of the following is a polar function of the form $r = f(\theta)$. Determine for each function whether the point $(f(\theta), \theta)$ is getting closer to or further from the origin on the given interval.

    (a) $r = \sin(\theta), \pi < \theta < \frac{3\pi}{2}$

    (b) $r = -2\cos(\theta) + 2, \frac{3\pi}{2} < \theta < 2\pi$

    (c) $r = \theta(\theta + 2)^2, \theta < -2$

2. Consider the function $f(\theta)$ graphed below for $0 \leq x \leq 2\pi$.

    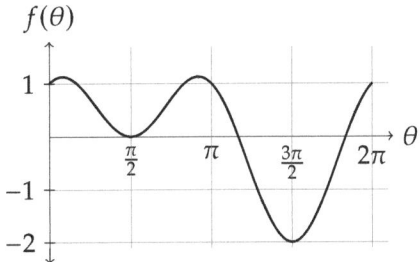

    Let r be the polar function given by $r = f(\theta)$. r is to be graphed in the polar coordinate system. For what θ in the interval $0 \leq \theta \leq 2\pi$ will the point $(f(\theta), \theta)$ be the farthest from the origin?

3. Let $r = f(\theta)$, where f is defined by
$$f(\theta) = \frac{2\theta^2}{(\theta + 1)(\theta + 2)}.$$
What will happen to the distance between the point $(f(\theta), \theta)$ and the origin as θ increases without bound?

4. It's worth knowing how to graph polar functions in your calculator. You can press mode and go to the row with FUNCTION PARAMETRIC POLAR SEQ. You'll now graph the function from Problem 3 as follows.

    - Select POLAR.
    - Press y=
    - Type in the function from Problem 3, using the X,T,θ,n button for θ.
    - Press window and change θmax to 72π. Also change Xmin=-4, Xmax=4, Ymin=-4, and YMax=4.
    - Press graph.

5. The function f given by
$$f(x) = \frac{2x^2}{(x + 1)(x + 2)}$$
has a horizontal asymptote at $y = 2$. How did this horizontal asymptote appear when we graphed f not in the xy-plane, but as the polar function in Problems 3 and 4? Why does this make sense?

6. Consider the polar function given by $r = f(\theta)$, where $f(\theta) = \sin(\theta) + \cos(2\theta)$.

    (a) (🖩) Compute the average rate of change of $r = f(\theta)$ over the interval $7 \leq \theta \leq 9$.

    (b) Interpret the value from (a) in the context of the graph of the polar function.

Topic 3.15 Solutions/Notes

Activity

1.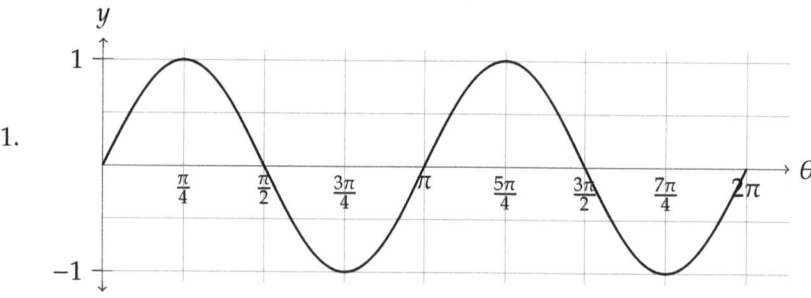

2.

Interval	Sign of sin(2x) (+/-)?	Increasing/Decreasing?	Dist. from origin in $r = f(\theta)$ increasing/decreasing?
$0 < \theta < \frac{\pi}{4}$	Positive	Increasing	
$\frac{\pi}{4} < \theta < \pi$	Positive	Decreasing	
$\pi < \theta < \frac{3\pi}{4}$	Negative	Decreasing	
$\frac{3\pi}{4} < \theta < \pi$	Negative	Increasing	
$\pi < \theta < \frac{5\pi}{4}$	Positive	Increasing	
$\frac{5\pi}{4} < \theta < \frac{3\pi}{2}$	Positive	Decreasing	
$\frac{3\pi}{2} < \theta < \frac{7\pi}{4}$	Negative	Decreasing	
$\frac{7\pi}{4} < \theta < 2\pi$	Negative	Increasing	

5. The distance increases.

8. The distance decreases.

10.

Interval	Sign of sin(2x) (+/-)?	Increasing/Decreasing?	Dist. from origin in $r = f(\theta)$ increasing/decreasing?
$0 < \theta < \frac{\pi}{4}$	Positive	Increasing	Increasing
$\frac{\pi}{4} < \theta < \pi$	Positive	Decreasing	Decreasing
$\pi < \theta < \frac{3\pi}{4}$	Negative	Decreasing	Increasing
$\frac{3\pi}{4} < \theta < \pi$	Negative	Increasing	Decreasing
$\pi < \theta < \frac{5\pi}{4}$	Positive	Increasing	Increasing
$\frac{5\pi}{4} < \theta < \frac{3\pi}{2}$	Positive	Decreasing	Decreasing
$\frac{3\pi}{2} < \theta < \frac{7\pi}{4}$	Negative	Decreasing	Increasing
$\frac{7\pi}{4} < \theta < 2\pi$	Negative	Increasing	Decreasing

11.
> **Rates of Change of Polar Functions**
> Let $r = f(\theta)$ be a polar function with graph consisting of the points $(f(\theta), \theta)$. Then
> - the distance from the origin will be increasing when $f(\theta)$ is
>   - ◇ positive and increasing, OR
>   - ◇ negative and decreasing.
> - The distance from the origin will be decreasing when $f(\theta)$ is
>   - ◇ positive and decreasing, OR
>   - ◇ negative and increasing.

12. (a) The graph of f in the xy-plane is a positively oriented quadratic with zeros at $\theta = -1$ and $\theta = 3$. Due to the symmetry of quadratics, there will be a global minimum halfway between the zeros at $\theta = 1$, somewhere less than 0. The function f will be increasing for $1 < \theta < 3$. Since f will be negative and increasing, the point $(f(\theta), \theta)$ on the graph of the polar function $r = f(\theta)$ will be getting closer to the origin.

    (b) The point $(f(\theta)$ will be getting farther from the origin when f is positive and increasing or negative and decreasing. This occurs on the intervals $0 < \theta < 1$, $2 < \theta < 3$, and $\theta > 4$.

13. (a) $f(\theta) = 0$ for $\theta = 1$ and $\theta = 3$.

    (b) f will reach a global maximum at $\theta = 2$, which is halfway between the zeros of $\theta = 1$ and $\theta = 3$. It will be a global maximum because f is a negatively oriented quadratic. There will be a local minimum at the endpoint $\theta = 0$ and a global minimum at $\theta = 5$ (global because we're only considering the interval $0 \leq \theta \leq 5$).

15. Student answers will vary.

16. The point $(f(\theta), \theta)$ will be farthest from the origin when $\theta = 5$, which corresponds to the extremum of f with the greatest absolute value, and closest to the origin (precisely *at* the origin) at $\theta = 1$ and $\theta = 3$, which is when $\theta = 0$.

> **Notes**
>
> Polar functions increase and decrease similarly to functions in the Cartesian plane. The distance from $(f(\theta), \theta)$ to the origin will be increasing when f is either positive and increasing or negative and decreasing; this distance will be decreasing when f is either positive and decreasing or negative and increasing.
>
> The extremum of a function $f(\theta)$ correspond to local or global *maxima* of a polar function $r = f(\theta)$. The farthest the point $(f(\theta), \theta)$ will be from the origin corresponds to the maximum or minimum with the greatest *absolute value*.

Homework

1. (a) $f(\theta) = \sin(\theta)$ is negative and decreasing on this interval, so $(f(\theta), \theta)$ is getting farther from the origin on this interval.

   (b) $f(\theta) = -2\cos(\theta) + 2$ is positive and decreasing on this interval, so $(f(\theta), \theta)$ is getting closer to the origin on this interval.

   (c) $f(\theta) = \theta(\theta+2)^2$ is a cubic with a positive leading coefficient and zeros at $\theta = -2$ (with multiplicity 2) and $x = 0$. A sketch of $f(x)$ in the xy-plane shows that $f(x)$ is negative and increasing for $x < -2$. Therefore, $(f(\theta), \theta)$ is getting closer to the origin for $\theta < -2$.

2. $(f(\theta), \theta)$ will be farthest from the origin at $\theta = \frac{3\pi}{2}$, which is when $f(\theta)$ reaches its extremum with the greatest absolute value ($|-2| = 2$).

3. $\lim_{\theta \to \infty} \dfrac{2\theta^2}{(\theta+1)(\theta+2)} = \lim_{\theta \to \infty} \dfrac{2\theta^2}{\theta^2 + 3\theta + 2} = \lim_{\theta \to \infty} \dfrac{2\theta^2}{\theta^2} = 2$. Therefore, as θ increases without bound, the distance between $(f(\theta), \theta)$ and the origin will approach 2.

4.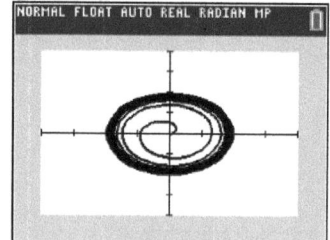

5. A horizontal asymptote of $y = 2$ means that, as inputs grow without bound, the outputs (y-values) approach a constant. In a polar function, the outputs are distance from the origin. Therefore, a horizontal asymptote of $y = 2$ would correspond to an asymptote of $r = 2$, which is a circle of radius 2.

6. (a) $\dfrac{f(9) - f(7)}{9 - 7} \approx 0.139$

   (b) As θ increases from 7 radians to 9 radians, the distance from $(f(\theta), \theta)$ to the origin increases by an average of 0.139 per radian.

Reflection

Suggested Class Time. 50 minutes for activity

Prerequisites. All of the function families in AP Precalculus

Instructional Strategies. This lesson went *really* well. Here were the only snags.

- The entire activity is predicated on students correctly graphing $f(\theta) = \sin(2\theta)$. Make sure they do so correctly early on.

- Students don't read carefully – three of the five groups in my class thought that "2nd and 3rd columns" meant the 2nd and 3rd *empty* columns. They just completely ignored the first column and tried to fill in the 4th column without looking at the polar graph.

I gave students 50 minutes, and everyone finished or got very close. As I walked around the room, I got a few questions about Problem 12a. What really helped was having students consider the function in rectangular coordinates. At first, I asked one student in particular, "Okay, so what kind of function is this?," and she was silent. When I replaced θ with an x and wrote $f(x) = (x+1)(x-3)$, she immediately said, "Oh, it's a quadratic." At first, I thought this "rely on x" was the typically disappointing inflexibility/reliance on x common in students, but with this very particular topic, it's actually quite smart and in the spirit of the Cartesian-to-polar connection in the topic.

After we did the notes, I wrote the following question for students:

"Let $r = f(\theta)$ be given by $f(\theta) = \dfrac{\theta^2 + 15}{2\theta^2 - 4}$.
Describe the behavior of the point $(f(\theta), \theta)$ as θ increases without bound."

I chose this specifically because the other two common end behaviors for polar functions – increasing without bound and decreasing to 0 – are pretty easy to visualize: points either "blow up" or collapse to the pole. Approaching a constant, though, is a bit trickier. The problem worked very well: I then graphed the function in Geogebra, hid it, graphed a single point on it, and then traced the point out as θ increased. The point rapidly approaches a circle, and it seemed to really click for students.

I found one aspect of this lesson very interesting – one that arguably contradicts the necessity of an applet. A student called me over and had filled in the 4th column of the table without having ever opened the applet - and she had done it correctly. I asked how, and she very calmly and quickly said, "Well, I looked at the graph [of sin(2x)] and saw it was getting closer to the x-axis here, then farther away from it here, and so on." What this student had internalized almost immediately was that the outputs of the function in rectangular coordinates corresponded to signed distances from the origin, and she was quickly able to use it to determine rates of change of polar functions. This kind of blew me away, and it led me to make this diagram for students when we took notes:

Rectangular function $y = f(x)$	Diagram	Polar function $r = f(\theta)$ (Point $(f(\theta), \theta)$)
$f(x) > 0$ and increasing or $f(x) < 0$ and decreasing		Increasing (getting farther from origin)
$f(x) > 0$ and decreasing or $f(x) < 0$ and increasing		Decreasing (getting closer to origin)

At the end of class, I gave 3 Topic Questions from AP Classroom for the topic, and the average percentage correct was 83.33%. Hopefully this will carry over to the exam!

Technology. Students used their laptops (and even some iPads) to access the Geogebra. It works on a phone, but it is far from ideal.

Problems.

1. A polar function is given by $r = f(\theta) = 2 - \cos\theta$. As θ increases on the interval $0 \leq \theta \leq \pi/2$, which of the following is true about the points on the graph of $r = f(\theta)$ in the xy-plane?

    (A) The points on the graph are above the x-axis and getting closer to the origin.

    (B) The points on the graph are above the x-axis and getting farther from the origin.

    (C) The points on the graph are below the x-axis and getting closer to the origin.

    (D) The points on the graph are below the x-axis and getting farther from the origin.

2. Consider the polar function $r = f(\theta)$. When the points $(f(\theta), \theta)$ are graphed in the polar coordinate system, the points approach a circle of radius 2 as θ increases without bound. Which of the following could be the function $f(\theta)$?

    (A) $f(\theta) = 2\theta$ (B) $f(\theta) = \theta + 2$ (C) $f(\theta) = \dfrac{2\theta^2}{\theta^2 + 1}$ (D) $f(\theta) = \dfrac{2\theta}{\theta^2 + 1}$

3. The polar function $r = f(\theta)$ is given by $f(\theta) = -1 + \cos(2\theta)$. As θ increases on the interval $\frac{\pi}{4} < \theta < \frac{\pi}{2}$, which of the following statements is true about the points on the graph of $r = f(\theta)$ in the xy-plane?

   (A) The points on the graph are above the x-axis and are getting closer to the origin.
   (B) The points on the graph are below the x-axis and are getting closer to the origin.
   (C) The points on the graph are above the x-axis and are getting farther from the origin.
   (D) The points on the graph are below the x-axis and are getting farther from the origin.

4. (▣) Selected values of the polar function $r = g(\theta)$ are given in the table below.

θ	0	$\frac{\pi}{4}$	$\frac{\pi}{2}$	$\frac{3\pi}{4}$	π	$\frac{5\pi}{4}$	$\frac{3\pi}{2}$	$\frac{7\pi}{4}$	2π
r	1	-1.293	2	8.121	3	-2.707	2	3.879	1

   The largest average decrease in the distance from $(g(\theta), \theta)$ to the origin between successive input values given in the table is over which of the following intervals?

   (A) $0 \le \theta \le \frac{\pi}{4}$ (B) $\frac{3\pi}{4} \le \theta \le \pi$ (C) $\pi \le \theta \le \frac{5\pi}{4}$ (D) $\frac{7\pi}{4} \le \theta \le 2\pi$

5. The function f is defined by $y = f(x) = \cos(2x)\sin(x)$. The graph of f is plotted in the xy-plane below.

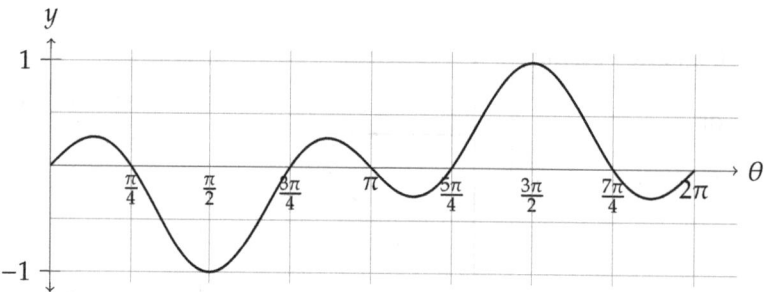

   The function $r = f(\theta) = \cos(2\theta)\sin(\theta)$ is plotted in the polar coordinate system. Which of the following is true concerning the distance between the point $(f(\theta), \theta)$ and the origin?

   (A) The distance is decreasing for $\frac{\pi}{4} \le \theta \le \frac{\pi}{2}$. (B) The distance is increasing for $\frac{3\pi}{4} \le \theta \le \pi$.
   (C) The distance is decreasing for $\pi \le \theta \le \frac{5\pi}{4}$. (D) The distance is increasing for $\frac{5\pi}{4} \le \theta \le \frac{3\pi}{2}$.

Solutions.

1. The points on the graph are above the x-axis since $2 - \cos x \ge 0$ on the interval $0 \le x \le \pi/2$. On this interval $\cos x$ decreases, so $2 - \cos x$ increases. Hence, the points for $f(\theta)$ are above the x-axis and get farther from the origin. The correct answer is therefore **(B)**.

2. The polar function must approach $f(\theta) = 2$ as θ increases without bound. Only $2\theta^2/(\theta^2 + 1)$ will resemble $2\theta^2/\theta^2 = 2$ as θ increases without bound. The correct answer is therefore **(C)**.

3. The sign of $f(x) = -1 + \cos(2x)$ is negative for $\frac{\pi}{4} < x < \frac{\pi}{2}$. Since $\cos(2x)$ increases on this interval, $-1 + \cos(2x)$ decreases on this interval. This implies that the points are getting farther away from the origin. Moreover, since $r = f(\theta)$ is negative on this interval which includes first quadrant angles, the points are in the third quadrant and therefore below the x-axis. The correct answer is therefore **(D)**.

4. We note that the change in θ between successive entries in the table is always $\pi/4$. So it remains to find the largest decrease in values of $|r|$. This happens for $8.121 - 3 = 5.121$. The correct answer is therefore **(B)**.

5. On the interval $5\pi/4 \leq \theta \leq 3\pi/2$, $f(x)$ is positive and increasing, so the distance between $(f(\theta), \theta)$ and the origin is increasing. The correct answer is therefore **(D)**.

Unit 3 Test ~ Trigonometric and Polar Functions

> **Part A: 6 multiple choice**
> A calculator may be required for some questions on this section.

1. (Topic 3.5) The function f is given by $f(x) = 3\sin(x - 1.3) + 2$ for $0 \le x \le 6$. On which of the following intervals is f increasing at a decreasing rate?

   (A) $0 < x < 1.3$ **(B)** $0 < x < 2.871$ **(C)** $1.3 < x < 2.871$ **(D)** $4.442 < x < 6$

2. (Topic 3.7) A student wanted to see how well a trigonometric function modeled the time of sunset each day in her city, so she recorded the time sunset occurred on the first day of the month for the first 6 months of the year. The table shows the sunset time S, in hours since 12:00 noon, for each day t since she began her research, where $t = 1$ is January 1 and $t = 32$ is February 1.

Day t	Sunset time S
1	5.633
32	6.1
61	6.55
92	7.95
122	8.317
153	8.683

   At this point, she ran a sinusoidal regression with a period of 365 to construct a function $S(t) = a\sin(bt + c) + d$ to try to predict future sunsets. On July 1 of that year, the 182nd day of the year, the sunset occurred at 8:50 p.m., or 8.833 hours after noon. By how many hours does the regression function S overestimate or underestimate the sunset on July 1?

   (A) $S(t)$ underestimates the sunset time by 1.994 hours.
   (B) $S(t)$ underestimates the sunset time by 0.557 hour.
   (C) $S(t)$ overestimates the sunset time by 0.557 hour.
   (D) $S(t)$ overestimates the sunset time by 1.994 hours.

3. (Topic 3.9) The function g is defined as $g(x) = a\sin(x)$, where $a \ne 0$. Which of the following gives the domain of the inverse function $g^{-1}(x)$?

   (A) $[-a, a]$ **(B)** $\left[-\dfrac{1}{a}, \dfrac{1}{a}\right]$ **(C)** $\left[-\dfrac{\pi}{2} \cdot a, \dfrac{\pi}{2} \cdot a\right]$ **(D)** $\left[-\dfrac{\pi}{2} \cdot \dfrac{1}{a}, \dfrac{\pi}{2} \cdot \dfrac{1}{a}\right]$

4. (Topic 3.10) Over the course of one year, a certain stock price increased and decreased periodically. The sinusoidal function $P(t) = 1.25\sin\left(\frac{2\pi}{182}(x - 1.1)\right) + 12.5$ models the price P, in dollars, of the stock on day t of the year. For approximately how many days in one year, for $0 \le t \le 365$, will the stock price be higher than \$13?

   (A) 102 **(B)** 134 **(C)** 160 **(D)** 262

5. (Topic 3.11) The function g is given by $g(x) = \sec(2x - k)$. If g is has a vertical asymptote at $x = 1.25$, which of the following could be the value of k?

   (A) $k = -3.783$ **(B)** $k = -0.642$ **(C)** $k = 0.929$ **(D)** $k = 2.5$

6. (Topic 3.12) The functions f and g are defined by $f(\theta) = \sin\theta$ and $g(\theta) = \cos\theta$. The values of f and g for an angle α are shown in the table.

θ	$f(\theta)$	$g(\theta)$
$\theta = \alpha$	0.28	0.96

Approximately what is the value of $f\left(\alpha - \frac{\pi}{3}\right)$?

(A) -0.691 **(B)** -0.237 **(C)** 0.722 **(D)** 0.971

Part B: 14 multiple choice and 2 free response
A calculator is prohibited on this section.

7. (Topic 3.1) The moon Io rotates around the planet Jupiter. Its distance from Jupiter can be modeled by a periodic function with a period of 42 hours. If an astronomer observes that Io is at its maximum distance from Jupiter at time $t = 3$ hours and continues observing Io, which of the following should be true?

   (A) Io will next be observed to be at its minimum distance from Jupiter at time $t = 42$ hours.
   (B) Io will next be observed to be at its minimum distance from Jupiter at time $t = 45$ hours.
   (C) Io will next be observed to be at its maximum distance from Jupiter at time $t = 42$ hours.
   (D) Io will next be observed to be at its maximum distance from Jupiter at time $t = 45$ hours.

8. (Topic 3.2) In the figure, point P lies at the intersection of a circle of radius 3 and the terminal ray of an angle measuring θ radians.

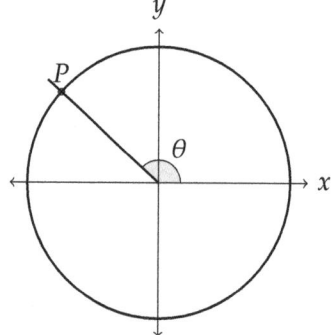

   If point P has coordinates $(-\sqrt{5}, 2)$, what is the value of $\cos(\theta)$?

   (A) $-\sqrt{5}$ (B) $-\dfrac{\sqrt{5}}{3}$ (C) $\dfrac{2}{3}$ (D) 2

9. (Topic 3.6) The function f is given by $f(x) = a\cos\left(\frac{\pi}{2}x\right) + d$. The graph of f has a midline of $y = 7$. If f obtains a maximum value at $(2, 9)$, which of the following is the value of a?

   (A) -9 (B) -2 (C) 2 (D) 9

10. (Topic 3.4) The point S lies at the intersection of the unit circle and the terminal ray of an angle θ. As θ increases, the function f outputs the horizontal displacement of S from the y-axis. Which of the following describes the behavior of f as θ increases from $\frac{\pi}{2}$ to $\frac{3\pi}{2}$?

   (A) f increases only.
   (B) f increases and then decreases.
   (C) f decreases and then increases.
   (D) f decreases only.

11. (Topic 3.7) Recall that Io, a moon of Jupiter, has a period of 42 hours. Io has a minimum distance of 420,000 kilometers (km) and a maximum distance of 423,400 km from Jupiter. The maximum distance occurs at time $t = 3$ hours after a certain observation begins. Which of the following functions $d(t)$ could model the distance d, in thousands of kilometers, Io is from Jupiter after t hours?

   (A) $d(t) = 1.7\cos\left(\dfrac{\pi}{21}(t+3)\right) + 421.7$
   (B) $d(t) = 1.7\cos\left(\dfrac{\pi}{21}(t+7.5)\right) + 421.7$
   (C) $d(t) = 1.7\sin\left(\dfrac{\pi}{21}(t+3)\right) + 421.7$
   (D) $d(t) = 1.7\sin\left(\dfrac{\pi}{21}(t+7.5)\right) + 421.7$

12. (Topic 3.8) In the coordinate plane, consider the terminal ray of an angle in standard position measuring $\frac{5\pi}{6}$ radians. What is the slope of this ray?

    (A) $-\sqrt{3}$ (B) $-\frac{\sqrt{3}}{3}$ (C) $\frac{\sqrt{3}}{3}$ (D) $\sqrt{3}$

13. (Topic 3.9) Let the function f be given by $f(x) = \arcsin x$. If g is the function defined by $g(x) = -2f(x+1) - 3$, what is the range of g?

    (A) $[-\pi - 3, \pi - 3]$ (B) $[-\pi - 1, \pi - 1]$ (C) $[-5, -1]$ (D) $[-2, 0]$

14. (Topic 3.10) Consider the function f defined by $f(x) = 3\sin x - 2$. One of the zeros of f on the domain $0 \le x \le 2\pi$ is $x = \arcsin \frac{2}{3}$. Which of the following correctly describes the existence of other zeros on this domain, along with correct reasoning?

    (A) $x = \arcsin \frac{2}{3}$ is the only zero on this domain because the range of $g(x) = \arcsin x$ is $[-1, 1]$.

    (B) $x = \arcsin \frac{2}{3}$ is the only zero on this domain because f completes only one period from $x = 0$ to $x = 2\pi$.

    (C) There is one other zero at $x = \pi - \arcsin \frac{2}{3}$ because this is the only other angle in the domain for which the output $f(x)$ equals $\frac{2}{3}$.

    (D) There is one other zero at $x = 2\pi - \arcsin \frac{2}{3}$ because this is the only other angle in the domain for which the output $f(x)$ equals $\frac{2}{3}$.

15. (Topic 3.12) The function f is given by $f(x) = (\sin x + \cos x)(\sin x - \cos x)$. A student is trying to solve $f(x) = 1$. Which of the following equations could be used to find the values of x that solve the equation $f(x) = 1$?

    (A) $\cos(2x) = -1$ (B) $\sin(2x) = -1$ (C) $\cos(2x) = 1$ (D) $\sin(2x) = 1$

16. (Topic 3.12) The functions f and g are given by $f(x) = \sin(2x)\tan x$ and $g(x) = \sin x$. Which of the following is an expression for f in terms of g for all x where $f(x)$ is defined?

    (A) $f(x) = (g(x))^2$ (B) $f(x) = 2(g(x))^2$ (C) $f(x) = (2g(x))^2$ (D) $f(x) = (g(2x))^2$

17. (Topic 3.13) The complex number z is graphed in the complex plane, as shown in the figure.

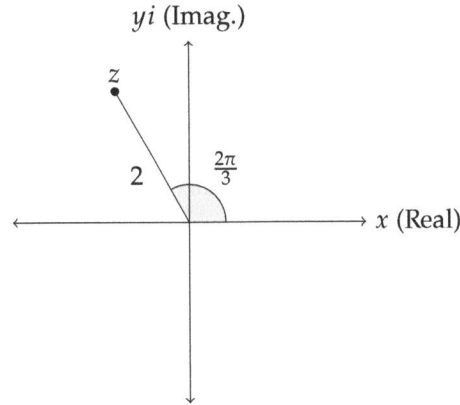

Which of the following is an expression for z?

(A) $z = -1 + \sqrt{3}i$ (B) $z = -\sqrt{3} + i$ (C) $z = -\frac{1}{2} + \frac{\sqrt{3}}{2}i$ (D) $z = -\frac{\sqrt{3}}{2} + \frac{1}{2}i$

18. (Topic 3.14) The graph of a polar function $r = g(\theta)$ contains the point A, where the polar coordinates of A are given by $\left(-4, -\frac{2\pi}{3}\right)$. In which quadrant does point A lie, along with correct reasoning?

   (A) Because the terminal ray of θ is in Quadrant III and $r < 0$, the point A lies in Quadrant I.
   (B) Because the terminal ray of θ is in Quadrant III and $r < 0$, the point A lies in Quadrant III.
   (C) Because the terminal ray of θ is in Quadrant II and $r < 0$, the point A lies in Quadrant II.
   (D) Because the terminal ray of θ is in Quadrant II and $r < 0$, the point A lies in Quadrant IV.

19. (Topic 3.14) The polar function $r = f(\theta)$ is graphed in the polar coordinate system in the figure.

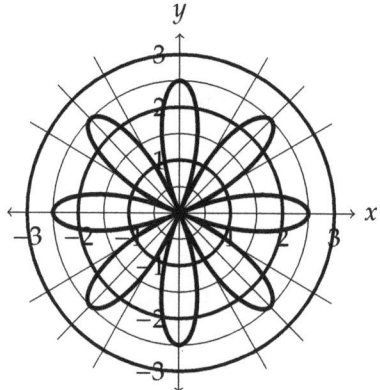

   Which of the following could be an expression for $f(\theta)$?

   (A) $f(\theta) = \frac{5}{2} \cos\left(\frac{\theta}{4}\right)$ (B) $f(\theta) = \frac{5}{2} \sin\left(\frac{\theta}{4}\right)$ (C) $f(\theta) = \frac{5}{2} \cos(4\theta)$ (D) $f(\theta) = \frac{5}{2} \sin(4\theta)$

20. (Topic 3.15) Consider the polar function $r = f(\theta)$. When the points $(f(\theta), \theta)$ are graphed in the polar coordinate system, the points approach a circle of radius 2 as θ increases without bound. Which of the following could be the function $f(\theta)$?

   (A) $f(\theta) = 2\theta$ (B) $f(\theta) = \theta + 2$ (C) $f(\theta) = \dfrac{2\theta^2}{\theta^2 + 1}$ (D) $f(\theta) = \dfrac{2\theta}{\theta^2 + 1}$

Free Response Part B - Instructions

- Show all of your work. Your work will be scored on the correctness and completeness of your responses as well as your answers. Answers without supporting work may not receive credit in cases where supporting work is requested.

- Unless otherwise specified, the domain of a function f is assumed to be the set of all real numbers x for which $f(x)$ is a real number.

UNIT 3 UNIT 3 TEST ~ TRIGONOMETRIC AND POLAR FUNCTIONS 439

Free Response #1: A string trimmer is an electric lawn tool that has a string that can be used to cut grass along a sidewalk. The figure shows a string trimmer rotating in a circular clockwise direction that completes one rotation every 2 milliseconds. Point S is on the edge of the string. As the string rotates at a constant speed, the height of S above the ground periodically increases and decreases. At time $t = 0$ milliseconds, S is at its highest position, 6 inches above the ground.

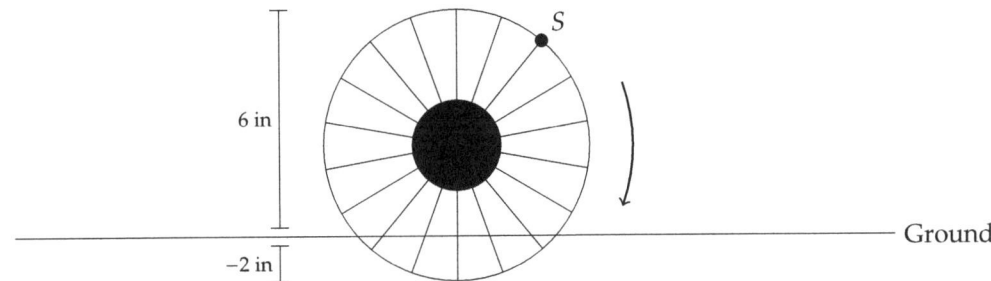

Note: Figure not drawn to scale.

The sinusoidal function h models the height of S above the ground, in inches, as a function of time t, in milliseconds. A positive value of $h(t)$ indicates S is above the ground; a negative value of $h(t)$ indicates S is below the ground.

(A) The graph of h and its dashed midline for two full cycles is shown. Five points, F, G, J, K, and P, are labeled on the graph. No scale is indicated, and no axes are presented. Determine possible coordinates $(t, h(t))$ for the five points: F, G, J, K, and P.

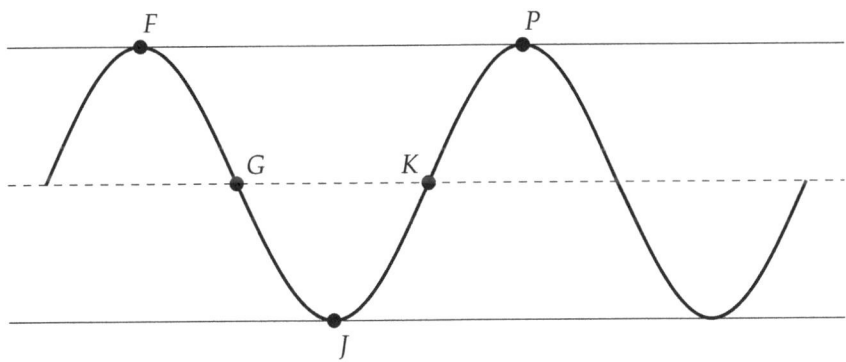

(B) The function h can be written in the form $h(t) = a\cos(b(t+c)) + d$. Find values of constants a, b, c, and d.

(C) Refer to the graph of h in part (A). The t-coordinate of F is t_1, and the t-coordinate of G is t_2.

  (i) On the interval (t_1, t_2), which of the following is true about h?

    a. h is positive and increasing.
    b. h is positive and decreasing.
    c. h is negative and increasing.
    d. h is negative and decreasing.

  (ii) Describe how the rate of change of h is changing on the interval (t_1, t_2).

Free Response #2:

(A) The functions g and h are defined by

$$g(x) = \sin x \cos x$$
$$h(x) = \cos(\arcsin(x))$$

(i) Solve $g(x) = \frac{1}{2}$ for values of x in the domain of g.

(ii) Solve $h(x) = \frac{1}{2}$ for values of x in the domain of h.

(B) The functions j and k are defined by

$$j(x) = \sin^2 x \csc x \cot x$$
$$k(x) = \frac{\cos x}{\sec x - \cos x}$$

(i) Rewrite $j(x)$ as a single trigonometric function.

(ii) Rewrite $k(x)$ as an expression involving a single trigonometric function raised to a power.

(C) The function m is given by

$$m(x) = 2\sin\left(x + \frac{\pi}{4}\right)$$

Find all values in the domain of m that yield an output value of $\sqrt{3}$.

Unit 3 Test Solutions and Scoring

1. The smallest positive period of the sine function increases at a decreasing rate between 0 and $\pi/2$. Thus, the function f increases at a decreasing rate for $0 < x - 1.3 < \pi/2$. This is $1.3 < x < 2.871$. Alternatively, we can graph f and the midline $y = 2$ and find the x-coordinates of the intersection and the maximum of f, which will result in the same interval. The correct answer is therefore **(C)**.

2. We input the data into lists, perform sinusoidal regression with a period of 365 and 3 iterations (default), and store the regression equation. Evaluating the regression function at $t = 182$ yields 8.276, which is $8.833 - 8.276 = 0.557$ hour below the actual 8.833 hours. The correct answer is therefore **(B)**.

3. The range of g is $[-a, a]$ so this is the domain of g^{-1}. The correct answer is therefore **(A)**.

4. We solve $P(t) \geq 13$ by graphing P and $y = 13$ in the calculator. The smallest positive intersection points are $x = 13.02$ and $x = 80.18$, which includes day 14 to day 80, for a total of 67 days. As we want the total over the interval $0 \leq t \leq 365$, there are two spans of 67 days during which the stock will be greater than \$13. Hence, $2 \cdot 67 = 134$ days. The correct answer is therefore **(B)**.

5. We know that g is undefined at $x = 1.25$, which means that $\cos(2(1.25) - k)$ must equal 0. We graph $y = \cos(2.5 - k)$ and find an intersection at $x = 0.929$. The correct answer is therefore **(C)**.

6. $f(\alpha - \frac{\pi}{3}) = \sin(\alpha - \frac{\pi}{3}) = \sin\alpha\cos\frac{\pi}{3} - \sin\frac{\pi}{3}\cos\alpha = 0.28 \cdot 0.5 - 0.866 \cdot 0.96 = -0.691$. The correct answer is therefore **(A)**.

7. The maxima of a periodic function are given at input values that differ by a multiple of the period. Since one maximum is at $t = 3$, the next maximum is at $t = 3 + 42 = 45$. The correct answer is therefore **(D)**.

8. We have $\cos(\theta) = -\dfrac{\sqrt{5}}{3}$. The correct answer is therefore **(B)**.

9. The period of f is $\frac{2\pi}{\pi/2} = 4$. Thus, the input value of half the period, 2, must give the minimum of a cosine function. However, this gives the maximum of f. Thus the value of a must be negative. Since $9 - 7 = 2$, we see that $a = -2$. The correct answer is therefore **(B)**.

10. Since f outputs the horizontal displacement of S from the y-axis, f must be the cosine function. From $\pi/2$ to $3\pi/2$, the cosine function decreases then increases. The correct answer is therefore **(C)**.

11. If we use a cosine function, we require $t = 3$ to make the argument of cosine 0. Neither of these cosine functions do so. If we use a sine function, we require $t = 3$ to make the argument of sine $\pi/2$. Only $\frac{\pi}{21}(t + 7.5)$ becomes $\pi/2$ when $t = 3$. The correct answer is therefore **(D)**.

12. The slope is $\tan\dfrac{5\pi}{6} = -\dfrac{\sqrt{3}}{3}$. The correct answer is therefore **(B)**.

13. The range of $\arcsin x$ is $[-\pi/2, \pi/2]$. Applying a horizontal shift does not affect the range. Applying a dilation by a factor of -2 makes the range $[-\pi, \pi]$. Finally, applying a vertical shift of -3 units makes the range $[-\pi - 3, \pi - 3]$. The correct answer is therefore **(A)**.

14. There is one other zero given by $\pi - \arcsin\frac{2}{3}$ because this is the only other angle in the domain for which $f(x) = \frac{2}{3}$. The correct answer is therefore **(C)**.

15. Carrying out the multiplication gives $\sin^2 x - \cos^2 x$, which is $-\cos(2x)$. Hence, we solve $-\cos(2x) = 1$, or $\cos(2x) = -1$. The correct answer is therefore **(A)**.

16. We have $f(x) = \sin(2x)\tan x = 2\sin x \cos x \cdot \frac{\sin x}{\cos x} = 2\sin^2 x = 2(g(x))^2$. The correct answer is therefore **(B)**.

17. $z = \left(2\cos\dfrac{2\pi}{3}\right) + i\left(2\sin\dfrac{2\pi}{3}\right) = \left(2\left(-\dfrac{1}{2}\right)\right) + i\left(2\dfrac{\sqrt{3}}{2}\right) = -1 + \sqrt{3}i$. The correct answer is therefore **(A)**.

18. The terminal ray of $\theta = -\dfrac{2\pi}{3}$ is in the third quadrant, and $r < 0$, so the the point is diametrically opposite from $(4, -2\pi/3)$ and lies in the first quadrant. The correct answer is therefore **(A)**.

19. The function chosen must pass through the polar points $(5/2, 0)$ and $(5/2, \pi)$. Only one of the choices does this. The correct answer is therefore **(C)**.

20. The polar function must approach $f(\theta) = 2$ as θ increases without bound. Only $2\theta^2/(\theta^2 + 1)$ will resemble $2\theta^2/\theta^2 = 2$ as θ increases without bound. The correct answer is therefore **(C)**.

Question 1 Scoring Guidelines

	Model Solution	Scoring	
(A)	F has coordinates $(0, 6)$. G has coordinates $(0.5, 2)$. J has coordinates $(1, -2)$. K has coordinates $(1.5, 2)$. P has coordinates $(2, 6)$.	t-coordinates $h(t)$-coordinates	1 point 1 point
		Total for part (A)	**2 points**
(B)	$h(t) = a\cos(b(t+c)) + d$ $a = 4$ $b = \dfrac{2\pi}{2} = \pi$ $c = 0$ $d = 2$ Note: Based on horizontal shifts and reflections, there are other correct forms for $h(t)$.	Vertical transformations (values of a and d) Horizontal transformations (values of b and c)	1 point 1 point
		Total for part (B)	**2 points**
(C) (i)	b	Function behavior	1 point
(ii)	Because the graph of h is concave down on the interval (t_1, t_2), the rate of change of h is decreasing on the interval (t_1, t_2).	Change in rate of change	1 point
		Total for part (C)	**2 points**
		Total for Question 1	**6 points**

We propose the following time limits for each part of the test, assuming similar timings as for the AP exam. That means each multiple-choice should be allowed 3 minutes and each free-response 15 minutes. This could vary slightly, depending on the difficulty of the multiple-choice: it is reasonable to expect 2 minutes per problem for the easier problems. This creates the range of timings as given in the table below.

Suggested Time Limits	
Part A	15-20 minutes
Part B	60-70 minutes

We choose 20 multiple-choice and 2 free-response to include on the test in order to equal the same proportions of those two types of problems as on the AP Exam. That means such a test can be given an "AP score" based on the number of points earned. Each multiple-choice problem is worth 1 point and each free-response is worth 6 points, for a total of 32 points available on the test. The table below shows the conversions to an "AP score."

Raw score	AP score
21-32	5
17-20	4
12-16	3
7-11	2
0-6	1

Question 2 Scoring Guidelines

	Model Solution	Scoring	
(A) (i)	We obtain: $$\sin x \cos x = \frac{1}{2}$$ $$2\sin x \cos x = 1$$ $$\sin(2x) = 1$$ $$2x = \frac{\pi}{2} + 2\pi k$$ $$x = \frac{\pi}{4} + \pi k, \text{ where } k \text{ is any integer.}$$	General solution	1 point
(ii)	We solve: $$\cos(\arcsin(x)) = \frac{1}{2}$$ $$\arcsin(x) = \frac{\pi}{3} \quad \text{or} \quad \arcsin(x) = -\frac{\pi}{3}$$ $$x = \frac{\sqrt{3}}{2} \quad \text{or} \quad x = -\frac{\sqrt{3}}{2}$$	Values	1 point
		Total for part (A)	**2 points**
(B) (i)	Rewriting, we get $$j(x) = \sin^2 x \csc x \cot x$$ $$= \sin^2 x \cdot \frac{1}{\sin x} \cdot \frac{\cos x}{\sin x} = \frac{\sin^2 x \cos x}{\sin^2 x}$$ $$= \cos x$$	Expression for $j(x)$	1 point
(ii)	Rewriting, we get $$k(x) = \frac{\cos x}{\sec x - \cos x} = \frac{\cos x}{\frac{1}{\cos x} - \cos x}$$ $$= \frac{\cos x}{\frac{1}{\cos x} - \frac{\cos^2 x}{\cos x}} = \frac{\cos x}{\frac{1-\cos^2 x}{\cos x}}$$ $$= \frac{\cos x}{\frac{\sin^2 x}{\cos x}} = \cos x \cdot \frac{\cos x}{\sin^2 x} = \frac{\cos^2 x}{\sin^2 x} = \cot^2 x$$	Expression for $k(x)$	1 point
		Total for part (B)	**2 points**
(C)	We solve: $$2\sin\left(x + \frac{\pi}{4}\right) = \sqrt{3}$$ $$\sin\left(x + \frac{\pi}{4}\right) = \frac{\sqrt{3}}{2}$$ $$x + \frac{\pi}{4} = \frac{\pi}{3} + 2\pi k \quad \text{or} \quad x + \frac{\pi}{4} = \frac{2\pi}{3} + 2\pi k$$ $$x = \frac{\pi}{12} + 2\pi k \quad \text{or} \quad x = \frac{5\pi}{12} + 2\pi k$$ where k is any integer.	One solution General solutions	1 point 1 point
		Total for part (C)	**2 points**
		Total for Question 4	**6 points**

AP Precalculus: The Teacher's Compendium ~ © 2024 David Hornbeck and Chuck Garner

Pacing and Scheduling

We teach in a school system located twenty-five miles east of Atlanta, Georgia, which starts the school year around August 1 and finishes the Friday before Memorial Day in May. Our school system uses a "modified block schedule" for high schools: students take eight courses each year, each day alternating between four courses one day (designated an "A" day) and four other courses the next day (a "B" day). Each class is approximately 90 minutes. So AP Precalculus meets every other day for 90 minutes, all year long. Obviously, not every teacher has this schedule! What follows are suggestions on how to pace your course using this book if you teach this course under the following schedules. In these various pacings, we consider the "end-of-the-year" to be the week before the AP Precalculus exam.

- 90-minute periods for 80 days. This corresponds to the author's modified block schedule, but would also correspond to a traditional block schedule where the class meets only first semester.

- 90-minute periods for 70 days. This corresponds to a modified block schedule for a school that starts in September, or for a traditional block where the class meets only second semester.

- 45-minute periods for 160 days. This corresponds to a daily, year-long schedule for a school that starts in August.

- 45-minute periods for 140 days. This corresponds to a daily, year-long schedule for a school that starts in September.

We offer suggested pacing for 70, 80, 140, and 160 days, instead of 90 and 180 days, because we all know there will be unforseen interruptions to the schedule. If you are lucky and have no interruptions, then you will have even more time to go in depth, do more practice, spend more time reviewing for the AP exam, or even present some of the Unit 4 content.

80 days, 90-minute periods

Topic/Assessment	Days
1.1	2
1.2-1.3	1
1.4	1
Review, Quiz	1
1.5-1.6	2
Calculator Skills	1
Review, Progress Check	1
Test 1A (1.1-1.6)	1
1.7	1
1.8-1.10	2
1.11	1
1.7-1.11 Review	1
Quiz	1
1.12	1
1.13	1
1.14	1
Review, Progress Check	1
Test 1B (1.1-1.14)	1
2.1	1
2.2-2.3	1
2.4	1
2.5-2.6	1
Quiz	1
2.7	1
2.8	1
Review, Progress Check	1
Test 2A (2.1-2.8)	1
2.9-2.10	1
2.11	1
2.12	1

Topic/Assessment	Days
2.13	1
2.14	1
2.15	1
Review, Progress Check	1
Test 2B (2.1-2.15)	1
Midterm Review	1
Midterm Exam (units 1 and 2)	1
3.1-3.2	1
3.2-3.3	2
Review, Quiz	1
3.4-3.5	1
Practice	1
Review, Quiz	1
3.6	1
3.7	1
Practice FRQs	1
Review, Progress Check	1
Test 3A (3.1-3.7)	1
3.8	1
3.9	1
3.10	2
3.11	1
3.12	2
3.10-3.12 Circuit	1
3.13	1
Quiz	1
3.14-3.15	2
Review, Progress Check	1
Test 3B (3.1-3.15)	1
AP Exam Review	14

Note: The Mid-Term exam is the 40th day on this schedule. This corresponds to the conclusion of first semester on a year-long modified block. Therefore, there are also 40 days second semester, 14 of them being AP Exam Review.

70 days, 90-minute periods

Topic/Assessment	Days	Topic/Assessment	Days
1.1	1	2.12	1
1.2-1.3	1	2.13	1
1.4	1	2.14	1
Review, Quiz	1	2.15	1
1.5-1.6	2	Review, Progress Check	1
Calculator Skills	1	Midterm Review	1
Review, Progress Check	1	Midterm Exam (units 1 and 2)	1
Test 1A (1.1-1.6)	1	3.1-3.2	1
1.7	1	3.2-3.3	2
1.8-1.10	2	Review, Quiz	1
1.11	1	3.4-3.5	1
1.7-1.11 Review	1	Practice	1
Quiz	1	Review, Quiz	1
1.12	1	3.6	1
1.13	1	3.7	1
1.14	1	Review, Progress Check	1
Review, Progress Check	1	Test 3A (3.1-3.7)	1
Test 1B (1.1-1.14)	1	3.8	1
2.1	1	3.9	1
2.2-2.3	1	3.10	2
2.4	1	3.11	1
2.5-2.6	1	3.12	2
Quiz	1	3.13	1
2.7	1	Quiz	1
2.8	1	3.14-3.15	2
Review, Progress Check	1	Review, Progress Check	1
Test 2A (2.1-2.8)	1	Test 3B (3.1-3.15)	1
2.9-2.10	1	AP Exam Review	8
2.11	1		

Note: The Mid-Term exam is the 38th day on this schedule. Therefore, there are 32 days after the Mid-Term, 8 of them being AP Exam Review.

160 days, 45-minute periods

Topic/Assessment	Days
1.1	3
1.2-1.3	3
1.4	2
Review, Quiz	2
1.5-1.6	4
Calculator Skills	2
Review, Progress Check	2
Test 1A (1.1-1.6)	2
1.7	2
1.8-1.10	4
1.11	2
1.7-1.11 Review	3
Review, Quiz	2
1.12	2
1.13	2
1.14	2
Review, Progress Check	2
Test 1B (1.1-1.14)	2
2.1	2
2.2-2.3	2
2.4	4
2.5-2.6	2
Review, Quiz	2
2.7	2
2.8	2
Review, Progress Check	2
Test 2A (2.1-2.8)	2
2.9-2.10	2
2.11	2
2.12	2

Topic/Assessment	Days
2.13	4
2.14	2
2.15	3
Review, Progress Check	2
Test 2B (2.1-2.15)	2
Midterm Review	2
Midterm Exam (units 1 and 2)	2
3.1-3.2	2
3.2-3.3	3
Review, Quiz	2
3.4-3.5	2
Practice	1
Review, Quiz	2
3.6	2
3.7	2
Practice FRQs	1
Review, Progress Check	2
Test 3A (3.1-3.7)	2
3.8	2
3.9	2
3.10	4
3.11	2
3.12	5
3.10-3.12 Circuit	1
3.13	2
Review, Quiz	2
3.14-3.15	4
Review, Progress Check	2
Test 3B (3.1-3.15)	2
AP Exam Review	25

Note: The Mid-Term exam is the 85th and 86th day on this schedule. Therefore, there are 74 days after the Mid-Term, 25 of them being AP Exam Review.

140 days, 45-minute periods

Topic/Assessment	Days
1.1	2
1.2-1.3	3
1.4	2
Quiz	1
1.5-1.6	4
Calculator Skills	2
Review, Progress Check	2
Test 1A (1.1-1.6)	2
1.7	2
1.8-1.10	4
1.11	2
1.7-1.11 Review	2
Quiz	1
1.12	2
1.13	2
1.14	2
Review, Progress Check	2
Test 1B (1.1-1.14)	2
2.1	2
2.2-2.3	2
2.4	3
2.5-2.6	2
Quiz	1
2.7	2
2.8	2
Review, Progress Check	2
Test 2A (2.1-2.8)	2
2.9-2.10	2
2.11	2
2.12	2

Topic/Assessment	Days
2.13	3
2.14	2
2.15	2
Review, Progress Check	2
Test 2B (2.1-2.15)	2
Midterm Review	2
Midterm Exam (units 1 and 2)	2
3.1-3.2	2
3.2-3.3	2
Quiz	1
3.4-3.5	2
Practice	1
Quiz	1
3.6	2
3.7	2
Practice FRQs	1
Review, Progress Check	2
Test 3A (3.1-3.7)	2
3.8	2
3.9	2
3.10	3
3.11	2
3.12	3
3.10-3.12 Circuit	1
3.13	2
Review, Quiz	2
3.14-3.15	3
Review, Progress Check	2
Test 3B (3.1-3.15)	2
AP Exam Review	20

Note: The Mid-Term exam is the 77th and 78th day on this schedule. Therefore, there are 62 days after the Mid-Term, 20 of them being AP Exam Review.

So You Want to Write a Test?

The problems in this book are original. We (David and Chuck) spent a long time on writing the problems so that they 1) are at an appropriate level, 2) assess the intended topic, and 3) are in the style of an AP exam question. What follows is a short guide to our process to enable you to write your own. We strongly encourage you to write your own problems for tests and quizzes because you know how well you have taught your students, you know how well your students have learned, and therefore you know what is a fair and appropriate problem to ask your students. (This can be particularly useful if you have time to include the fourth unit of AP Precalculus in your course: you may still want to write problems in the AP Exam style.)

It's getting near that time to give an assessment in AP Precalculus, and you want to put one together for your students. Where to begin?

You could, of course, mine AP Classroom, online resources, and textbooks for good problems. You could screenshot them, type them up yourself, or even cut them out and glue them onto paper for copying. Maybe someone else gives you access to their test and you just give that. These are all valid options, but there is another, even greater one: writing your own test!

Writing AP problems can be intimidating, though: how do you know that the items are rigorous enough, or that they are actually similar to those that might appear on the AP exam? How do you create the diagrams? Really, where do you even begin?

Before even beginning, it's worth first getting a feel for what may be assessed on the AP exam. While we will never have access to the actual exam, we have a tremendous number of released questions in AP Classroom and the (currently) 3 practice exams. If you want to write good items, it really helps to *see* good items. I cannot emphasize this enough: **work the problems in AP Classroom and the practice exams!**

Further, I have found that, when I wanted to begin writing questions for specific topics, it helped to look at the Topic Questions. These are generally more challenging than items that would appear on an exam, but that's alright: better to over-prepare students than under-prepare them. Looking at all of the Topic Questions would give me a feel for the style and rigor of AP exam style questions. Often, first attempts at writing items may even just be mimicking items you've seen elsewhere; to quote T.S. Eliot, "Immature poets imitate; mature poets steal; bad poets deface what they take, and good poets make it into something better, or at least something different." I am in no way saying to copy questions – simply that we all need inspiration at first, and that you will progress from an immature aper to a creative writer before you know it.

Now that you've worked through dozens or hundreds of AP sample problems, it's time to write your own assessment. There is no one perfect method, but here is my approach to writing assessments.

I. **Figure out the test structure**
This begins with figuring out how many multiple choice (MC) and how many free response questions (FRQ) you'd like to give. [3] A good rule of thumb is that students should get between

---

[3] All of my tests include only these types of items, as these are the only ones assessed on the AP exam. Other item types are great

2-and-a-half and 3-and-a-half minutes per MC and 15 minutes per FRQ. If you're testing in 45 minutes, this might be 15-20 MC; 10-12 MC and 1 FRQ; or 5-6 MC and 2 FRQs.

Once you've decided upon the number of items, it's now time to determine how much of the assessment will be calculator active. Depending on the topics you're assessing, you may decide the entire assessment should be non-calculator. For more cumulative assessments, I tend to stick with the exact breakdown of the AP exam: half calculator active for FRQs and 2/5 calculator active for MC. Occasionally, especially during Unit 1, I would write assessments that were nearly all non-calculator with one or two MC items having a calculator icon.

II. **Write out a skeleton detailing the topic (or LO/EK) and type (calculator active or non calculator) for each item.** This was a game changer for me. Before I started doing this (in other courses), I would just write tests in order: #1, then #2, and so on. By the end of the test, I would simply hope or trust that I had a good balance of concepts and skills from whatever I was teaching. More often than not, though, I would realize after giving said assessments that I had forgotten to assess one or more important concepts or skills.

For this course, though, I started creating an outline in advance. This was as simple as laying out all the item numbers and then putting "Topic #" on each one. My test would then begin looking something like this...

1. (Topic 2.1)
2. (Topic 2.2)
3. (Topic 2.3)
4. (Topic 2.4)
⋮

This made it to where, before I wrote a single item, I knew that I would have a good balance of material from whatever segment of the course I was assessing.

It's worth noting here: when you write an assessment, **have the CED on hand!** Furthermore, if you want to *really* model the AP exam, don't just write items that assess a topic, but even more specifically, a learning objective (LO) and essential knowledge statement (EK). Every item in AP Classroom is coded according to these, so it might be worthwhile to try to do the same on your assessment.

III. **Write a short description of each item, possibly including the difficulty and presentation type (graphical, tabular, analytical, verbal).**
Another game changer! Rather than simply having which topics I would assess prepared before writing, I would start writing a short description for each item. Even on some assessments I've written where the *material* was a good mix, I have found afterward that I had no questions with a graph, or maybe no problems in context, or some other lack of variety. Other times, I've sat down to begin an assessment and had some (in my mind) ingenious idea for a problem, all to forget it when I come back later to finish the assessment.

Writing ideas in advance eliminates both of these issues. Before I start writing, my assessments will then look something like this...

1. (Topic 2.1) write general term from table – geometric
2. (Topic 2.1) find function that would define arithmetic sequence, include domain!
3. (Topic 2.2) concavity of exponential with reasons
4. (Topic 2.4) rewrite – negative, split up sum
⋮

- short answer, fill-in-the-blank for vocabulary, even matching - but they are somewhat outside the purview of "good AP items" and hence this guide.

SO YOU WANT TO WRITE A TEST?

It's worth noting here that, should you want to get *really* detailed, you could actually write items for specific skills. The CED outlines skills on page 14, and this may further help you balance your tests.

For a beginning item writer, it may really help to go with an alternating structure G-T-A-V (graphical, tabular, analytical, verbal) or some permutation of this. This will help you gain proficiency and comfort in writing all different items in all different sorts of presentation styles.

IV. **Write MC items.**

You've got your assessment outline, including descriptions of each time, and now it's just a matter of writing the actual items. Here, I've got a list of pointers.

- AP Precalculus is all about functions (inputs and outputs), meaning items are often quite wordy. You will very rarely see something like this:

  1. Solve for x: $9x^3 - 6x^2 + 4 = 3$.

  Instead, you will probably see something involving a function and inputs/outputs, like this:

  1. The function f is defined by $f(x) = 9x^3 - 6x^2 + 4$. Find all input values x for which $f(x) = 3$.

- Speaking of vocabulary, there are a few common phrases that are ubiquitous throughout the course and sample items.
    ◦ increasing/decreasing without bound
    ◦ input-output pairs
    ◦ over consecutive/successive equal-length intervals of the inputs/outputs
    ◦ terminal ray of angle in standard position
    ◦ vertical/horizontal displacement (for sine and cosine)
    ◦ equivalent expression

- When it comes to transformations, pay close attention to the CED. You will not see "stretch," "shrink," or the directions left/right/up/down. Instead, you'll see everything expressed quantitatively. So, for instance:
    ◦ "a shift up 1" would be "a vertical translation of +1"
    ◦ "stretch by 2" would be "dilation by a factor of 2"
    ◦ "translate left 1 and horizontally shrink by 3" would be (depending on how you teach "shrink") "horizontal translation of −1 and a horizontal dilation by $\frac{1}{3}$"

- Angles on the AP exam will always be in radians – never, absolutely never, degrees.

- Incorrect answer choices should generally reflect students' incorrect attempts to solve the problem: in theory, you want every student to feel like they were able to answer every question without guessing. I think most teachers are comfortable with writing good distractors, but just in case, here are some common ways of generating good distractors.
    ◦ Changing signs (of the answer, or in the problem solving process)
    ◦ Giving an x-coordinate instead of a y-coordinate (or vice-versa)
    ◦ Reversing the roles of inputs and outputs (in limits or function model construction, for instance, as well as horizontal and vertical transformations)
    ◦ Giving a correct answer with an incorrect reason (or an incorrect answer with the correct reason)
    ◦ Changing operations (if the answer is $\frac{f(x)}{g(x)}$ for an exponent rules item, you could make distractors include $f(x) + g(x)$, $f(x) - g(x)$, and $f(x) \cdot g(x)$)
    ◦ Giving reciprocal values (for a transformation question, if the answer is $2\cos\theta$, you could make one distractor $\frac{1}{2}\cos\theta$)

For calculator active questions, I especially want to emphasize that you should work the problem as if you were the student. What incorrect button might they press or choice might they make, and what answer will they get as a result of these steps? It's my experience that calculator questions take much longer to write because of this immersive attempt to create good distractors.

- When you are writing real world contexts, try to make them realistic. Cutesy questions may be fun for your students (and all power to you!) but the AP exam won't have questions about Beyonce songs or your teacher's dog or a superhero. Good contexts can often teach you interesting things, too! Don't worry, you may have to find or invent semi-realistic data that might be cleaner or smoother than raw data, so there is certainly creation at play, but these creations can be rooted in reality.

- If you decide to write your test in LaTeX, you'll be blown away by the graphing capabilities of packages like TikZ, PGF Plots, and PSTricks. Each of these comes with a tremendous manual that are freely available online.

- If you end up writing your test in Word, I would heavily recommend Geogebra for creating all diagrams. Desmos is wonderful for graphing functions, but Geogebra is, too -= and it is also excellent for graphing *geometric* figures, which Desmos certainly is not. There are other graphing utilities out there, but for my money (Geogebra is free, by the way), there is nothing that matches the combination of power and simplicity that Geogebra offers.

V. Write FRQs

Writing FRQs in AP Precalculus is actually quite easy: it's just about *knowing* the FRQs that will appear on the exam (sorry, the *task models*). There are 4 FRQs on the exam, and each is described in fabulous detail in the CED. There are videos in AP Classroom going over each task model, and with the 3 released/secure practice exams, you can see 3 examples of each!

The only thing I'll add here is that, early in the year, I myself eschewed specific task models in favor for other types of questions within FRQs. Part of this was a lack of familiarity with the FRQs at the time, but it was more due to the fact that even task models 1 and 2 dip into Unit 2. Really, it's not the most important thing for your FRQs to perfectly model the task models for the entire year. Yes, your students need practice, but you can work your way up to them.

VI. Clean things up and adhere to some best practices.

At this point, you have an assessment written, and now you may be looking to make it look professional for your students (or yourself!). The level to which you want to take this depends on your meticulousness, but there are certain non-negotiables.

- Numerical answer choices should be in increasing or decreasing order. This may mean that you have too many correct answers of, say, A, on your test. So what? The SAT may make an effort to equally distribute the number correct answer choices, but you do not need to. (In fact, most students expect an equal distribution, so it really throws them off where there is not one!)

- All answer choices should have the same number of decimal places. In most cases, this will be 3 decimal places.

- Fonts should be consistent throughout the assessment.

- All variables should be italicized.

- Trigonometric and logarithmic functions should not be italicized. (In Word, this amounts to needing to hit the spacebar after typing sin, cos, tan, ln, etc.)

Now that you've dealt with those, there are some more precise stylistic improvements you may choose to make.

- On the AP exam, figures will be *above* question stems.

- Answer choices will be written capitalized inside of parentheses or a bubble, i.e. (A) or Ⓐ.

- Axes will have labels.

- Exponential expressions with multi-term exponents will be written in parentheses, e.g. $3^{(x+2)}$ as opposed to 3^{x+2}.

- The origin will typically be notated with an O and angles markers will typically be shaded.

In truth, this list can go on and on, and my own exams may have plenty of shortcomings. You can rest assured, though, that should you address all of the above, your test will look positively professional.

VII. **Write a key; proofread.**

Ideally, I like to write solutions for my assessments – and I also apparently really enjoy making mistakes. The best way I've found to tackle both of these is to work out my own tests. It is so easy to gloss over things when you're simply finding an answer to a MC item, for example, that you may miss a typo within the question stem or the fact that two incorrect answer choices were actually identical.

When I write solutions, though, it helps me to read much more carefully and pay more attention to those small details. I will nearly always write these solutions before worrying about less consequential typos, thereby taking care of the math first.

If you're scatterbrained like I am, you might then make yourself a quick list of common mistakes you make. Then, you can systematically peruse your test for each. This might look like...

1. No missing #s or repeats
2. No missing answer choice letters
3. All answer choices are different
4. Title of the test, calculator/non-calculator headers, instructions are all accurate
$\vdots$

I have never made a formal list like this myself, but I have found that systematically looking for specific common kinds of typos is helpful. (Also, if you use LaTeX, you won't have to worry about the first 2 of those issues!)

Once you've proofread a test, it's always a good idea to proofread again (I write this so I'll use this advice more myself!) or to have someone else look at it.

We'd like to make a note here about writing "AP-style" items. One of the criticisms of AP is that it is too easy for teachers to "teach to the test". With a high-stakes assessment at the end of the course, the perception is that teachers of AP courses will teach students only what is required to "pass" the AP exam. It is easy to fall into this trap, particularly if one wants to give classroom assessments that model the style and rigor of the AP exam. But this not what we advocate. By writing your own assessments as early as possible into teaching the content, you are creating instructional goalposts for yourself as you teach – benchmarks, if you will, in terms of preparing students for problem complexity, rigor, and content. Rather than "teaching to the test," you are instead "teaching *for* the test." You may write "AP-style" items for whatever content you may teach, whether it's on the AP exam or not – Unit 4, for instance, or extensions of concepts in the course you may have chosen to teach. While, yes, certain formatting is utilized to resemble the high-stakes assessment, we honestly use "AP-style" synonymously with "rigorous, thought-provoking, and presented in a variety of formats."

Hopefully, you now have more confidence to make a beginning at writing your own AP Precalculus assessments. Remember that no assessments are perfect, no matter how experienced the writer is! If you're capable of teaching the class, which you are, by definition, then you are most certainly qualified to write high-quality assessments.

Also by David Hornbeck
The AP Statistics Study Companion
Precalculus, 2nd edition
Acing the AP Precalculus Exam

www.davidhornbeck.com

Also by Chuck Garner
Discrete Mathematics: A Gateway to the Mathematical Garden
Calculus: Dynamic Mathematics, For AB, 4th edition
Calculus: Dynamic Mathematics, For BC, 4th edition
Calculus: Dynamic Mathematics, For Multivariable, 4th edition
Five Weeks to a Five: Preparation for the AP Calculus AB Exam
Five Weeks to a Five: Preparation for the AP Calculus BC Exam
The AP Calculus Problem Book, 5th edition
The Rockdale Mathematics Competition Problem Book, 2nd edition
A Young Person's Guide to Competition Mathematics (with Debbie Poss and Don Slater)

www.drchuckgarner.com